Mathematical Problems of Statistical Mechanics and Dynamics

Mathematics and Its Applications (*Soviet Series*)

Mathematical Problems of Statistical Mechanics and Dynamics

A Collection of Surveys

Edited by
R. L. Dobrushin
Institute of Information Transmission Problems, Academy of Sciences, Moscow, U.S.S.R.

D. Reidel Publishing Company
A MEMBER OF THE KLUWER ACADEMIC PUBLISHERS GROUP
Dordrecht / Boston / Lancaster / Tokyo

Library of Congress Cataloging in Publication Data

CIP

Mathematical problems of statistical mechanics and dynamics.

(Mathematics and its applications (Soviet series))
Includes bibliographies and indexes.
1. Statistical mechanics. 2. Dynamics. 3. Stochastic processes. 4. Mathematical physics. I. Dobrushin, R. L., 1929– . II. Series: Mathematics and its applications (D. Reidel Publishing Company). Soviet series.
QC174.8.M36 1986 530.1'3 86-20230
ISBN-13: 978-94-010-8540-3 e-ISBN-13: 978-94-009-4592-0
DOI: 10.1007/978-94-009-4592-0

Published by D. Reidel Publishing Company,
P.O. Box 17,3300 AA Dordrecht, Holland.

Sold and distributed in the U.S.A. and Canada
by Kluwer Academic Publishers,
101 Philip Drive, Assinippi Park, Norwell, MA02061, U.S.A.

In all other countries, sold and distributed
by Kluwer Academic Publishers Group,
P.O. Box 322, 3300 AH Dordrecht, Holland.

Softcover reprint of the hardcover 1st edition 1986

Contents

Editor's Preface

Approach your problems from the right end and begin with the answers. Then one day, perhaps you will find the final question.

'The Hermit Clad in Crane Feathers' in R. van Gulik's *The Chinese Maze Murders.*

It isn't that they can't see the solution. It is that they can't see the problem.

G. K. Chesterton. *The Scandal of Father Brown* 'The point of a Pin'.

Growing specialization and diversification have brought a host of monographs and textbooks on increasingly specialized topics. However, the 'tree' of knowledge of mathematics and related fields does not grow only by putting forth new branches. It also happens, quite often in fact, that branches which were thought to be completely disparate are suddenly seen to be related.

Further, the kind and level of sophistication of mathematics applied in various sciences has changed drastically in recent years: measure theory is used (non-trivially) in regional and theoretical economics; algebraic geometry interacts with physics; the Minkowsky lemma, coding theory and the structure of water meet one another in packing and covering theory; quantum fields, crystal defects and mathematical programming profit from homotopy theory; Lie algebras are relevant to filtering; and prediction and electrical engineering can use Stein spaces. And in addition to this there are such new emerging subdisciplines as 'experimental mathematics', 'CFD', 'completely integrable systems', 'chaos, synergetics and large-scale order', which are almost impossible to fit into the existing classification schemes. They draw upon widely different sections of mathematics. This programme, Mathematics and Its Applications, is devoted to new emerging (sub)disciplines and to such (new) interrelations as exempla gratia:

- a central concept which plays an important role in several different mathematical and/or scientific specialized areas;
- new applications of the results and ideas from one area of scientific endeavour into another;
- influences which the results, problems and concepts of one field of enquiry have and have had on the development of another.

The Mathematics and Its Applications programme tries to make available a careful selection of books which fit the philosophy outlined above. With such books, which are stimulating rather than definitive, intriguing rather than encyclopaedic, we hope to contribute something towards better communication among the practitioners in diversified fields.

Because of the wealth of scholarly research being undertaken in the Soviet Union, Easter Europe, and Japan, it was decided to devote special attention to the work emanating from these particular regions. Thus it was decided to start three regional series under the umbrella of the main MIA programme.

Probability is definitely one area of mathematics where spectacular advances have been made in the last decennia both in terms of concepts and in terms of technique (or the ability to obtain formulas). As a result probabilistic techniques and ideas are penetrating into other areas, e.g., functional analysis and partial differential equations.

It is then natural to expect that in statistical physics, a domain which is closer to probabilistic thinking, there are also substantial advances to report. That is indeed the case. In this area which, as the editor notes, is in methods and philosophy sort of halfway between theoretical physics and mathematics, people like Dobrushin, Malyshev, Sinai and Minlos are starting to get a very good mathematical hold on such things as infinite particle systems, random fields and phase diagrams. A phase transition, e.g., turns out to be essentially a mathematical phenomenon.

At its present state of development this area of mathematics seems to need very long papers (or books) to establish the principal results, partly because so much is happening and much has to be recast. This volume, quite possibly the first of several, should give the reader a good up-to-date survey of what is happening in this particular leading school in mathematics.

The unreasonable effectiveness of mathematics in science ...

Eugene Wigner

Well, if you know of a better 'ole, go to it.

Bruce Bairnsfather

What is now proved was once only imagined.

William Blake

As long as algebra and geometry proceeded along separate paths, their advance was slow and their applications limited.

But when these sciences joined company they drew from each other fresh vitality and thenceforward marched on at a rapid pace towards perfection.

Joseph Louis Lagrange.

Bussum, December 1985 MICHIEL HAZEWINKEL

Introduction

Until the beginning of this century mathematics and physics developed in close contact with one another, with a constant interchange of ideas. Great achievements in both mathematics and physics are connected with the names of many outstanding scientists of the past.

However, this unity was abruptly broken at the beginning of the twentieth century. The centre of interest for mathematicians shifted to a domain of fascinating wide generalizations linked to the discovery of general structures of algebra, topology and functional analysis. Physicists interpreted this passion as a preoccupation with abstract games, with no relevance for them. Their own interests were meanwhile shifting to the newly opening fields of quantum mechanics, relativity theory and statistical physics. Traditional mathematical analysis was all they required here, and they felt no need to apply modern mathematical methods. Mathematicians, in their turn, showed almost no interest in the new physics; contact between mathematics and physics was maintained mainly at the level of the theory of differential equations (which is equivalent to the equations of mathematical physics), where mutual understanding had been established during the previous century.

The split was aggravated by the fact that mathematicians and physicists had begun to speak different languages. On one hand, after the advent of the set-theory approach a new standardized 'level of mathematical rigour' was elaborated, and constructions on other logical levels came to be perceived as being outside mathematics. On the other hand, the 'level of physical rigour' came to be generally accepted in modern theoretical physics. At this level, it is not considered obligatory to outline clearly the boundaries of correctness of a mathematical fact, and to prove it; the main arguments for its truthfulness are the possibility of deriving it in different ways, and an absence of essential contradictions with previously accepted physical concepts. An orthodox mathematician perceives such considerations of the physical level of rigour with a repulsion as little better than black magic: a typical physicist treats the efforts of a mathematician to prove a mathematical fact already known in physics as merely idle hair-splitting.

Of course, the break between mathematics and physics has never been a total one; it is possible to list many great scientists (mainly mathematicians) who have tried to bridge the gap. But until the time was ripe their efforts did not evoke much general response. However, in recent decades there has been a wind of change. The inner impulse for the development of mathematics has begun to diminish, and mathematicians increasingly turn to physics as a source of new problems and constructions which can be naturally incorporated into the structure of modern mathematics. Some physicists have begun to understand the fruitfulness of modern

R. L. Dobrushin (ed.), Mathematical Problems of Statistical Mechanisms and Dyanimcs. xi–xiv mathematical methods and ideas in solving problems of interest to them, and even to recognize at least a limited use for investigations into the mathematical foundations of generally accepted physical theories. New points of contact between physics and mathematics are emerging: field theory and differential geometry, quantum mechanics and functional analysis, statistical physics and the theory of probability. The parallel existence of the two languages continues. It is impossible to expect that mathematicians will give up the observance of mathematical rigour, but it is equally impossible to expect the physicists to give up their arguments for the physical level of rigour: mathematical problems arising at the growth points of modern physics are too complex and too urgent, and scientists who voluntarily bind themselves by the tenets of mathematical rigour will always lag far behind. The problem is finding a solution in new way which is only natural in these days of narrow scientific specialization. A new clan of mediators and translators is appearing, describing themselves as specialists in mathematical physics, who have begun to consider themselves as representatives of a special science differing both from physics and from mathematics. These are scientists capable of understanding both mathematical and physical languages, who devote themselves to the solution of problems of modern physics at the logical level of modern mathematics.

In the 1920s two concepts which had come into probability theory from physics – diffusion process and ergodicity – were among the most important influences on the creation of the theory of stochastic processes. Subsequently, in spite of the obvious proximity of their subjects, the theory of probability (including the stochastic theory of dynamical systems) and statistical mechanics have developed almost independently, taking little notice of one another's existence. Similiar ideas have often been worked out under different names in these disciplines. The pioneer investigations of Bogolubov and Hinchin, which could have laid the foundations of mutual understanding between the two sciences, remained almost unnoticed in the middle of this century. Comparing the results of this period it is possible to say that the theory of probability achieved more than statistical physics in the understanding of the nature of random processes (i.e., random functions of one variable), but lagged far behind in its understanding of the properties of random fields (i.e., random functions of several variables). Thus, only now it is becoming clear that the notion of phase transition, which long ago became a commonplace for physicists, is not specifically connected with molecular systems, but is in fact a quite general property of a very wide class of systems with a large number of interacting random elements. The situation is now changing rapidly. Probabilistic methods have found their place in mathematical physics, and at probabilistic conferences the sections on statistical physics and the theory of random fields do not any longer seem exotic. Research on methods of probabilistic substantiation of the postulates and findings of statistical physics is developing on a wide front.

It is interesting to note one paradoxical feature. In statistical physics investigations in the traditional style are usually founded on the use of complex and often cumbersome analytical techniques, and an intuitive interpretation is possible only for the final conclusions. On the other hand, much research in the mathematical direction is founded on systematic implementation of an intuitively visualized physical idea masked by the disguise of the obligatory (for this genre) language of measure theory. The physicist's love for formulas cannot be explained only by the desire to obtain numerical answers which can later be compared with experimental data. Explicit solutions are possible (at least in statistical physics) only for highly idealized models remote from reality, or based on rough approximations in the process of solution; hence a comparision with experiment is possible not for the formulas themselves but only for rough qualitative implications of these formulas. It seems that the main point is the following. Intuitively transparent physical reasoning is not convincing enough at the physical level of rigour even for a physicist, and formulas are necessary in order to remove the remaining suspicions. As every probabilist knows, the solution of a probabilistic problem starts with guessing the answer through visual probabilistic intuition: only after this comes the time-consuming stage of rigorous mathematical detail. As far as the problems of statistical physics are concerned, the physical and probabilistic intuitions are identical, and so the first stage of the probabilist's reasoning is equivalent to qualitative physical reasoning. It seems that the more visual quality of probabilistic intuition, in comparison with the quantum-mechanical intuition, and its more sophisticated methods of mathematically rigorous implementation, explain the popularity and success of the methods of stochastic integration in quantum mechanics and the euclidean theory of quantum fields.

Research on probabilistic methods of mathematical physics is intensively carried out in Moscow within the framework of the seminars conducted by R. L. Dobrushin, V. A. Malyshev, R. A. Minlos and Ya. G. Sinai. The present book is devoted to the exposition of certain results obtained by the members of these seminars.

The first contribution is by R. L. Dobrushin and M. Zahradnik. The theme of this extensive paper is a generalization of the Pirogov–Sinai method of investigation of the phase diagram for low temperatures to the case of continuous-spin systems. In order to do this it is necessary to supplement the techniques of the well-known Pirogov–Sinai papers by methods which give an opportunity to control the properties of Gibbsian fields which are constructed by small perturbations of Gaussian fields. In this paper such methods are developed for the case of Gaussian fields with finite-range potential and exponential decay of correlations (i.e., positive mass). No additional restrictions are imposed on the rate of decrease of correlations (i.e., on the mass), necessitating the development of a new variant of cluster techniques which forms the content of §2. The only prerequisite is the foundations of the theory of random fields, and the authors envisage that their §2 in particular may be used as a methodologically original introduction to the cluster and diagram methods of the theory of random fields. The second paper, by N. I. Chernov, is denoted to a proof of the existence of the entropy of equilibrium dynamics of infinite systems of Hamilton particles where, in contrast to the standard approach, the averaging is extended to the space and time coordinates simultaneously. The paper by R. A. Minlos and A. I. Mogilner deals with the spectral analysis of cluster operators. Such operators appear naturally in the analysis of infinite-particle systems of statistical physics and field theory.

Finally, two long contributions are devoted to the stochastic theory of dynamical systems. The paper by M. L. Blank, discusses the problem of stability of piecewise differentiable dynamical systems with respect to deterministic and stochastic perturbations. A general approach is developed, which can be applied to a wide class of situations previously studied by special methods. The connections between these problems and the computer simulation of dynamical systems are discussed. The last paper, by N. N. Čencova, proves the existence of two limit probability distribution: an eigendistribution on the expanding set and a natural invariant distribution on the invariant hyperbolic set for a smooth Smale horseshoe. Also ergodic properties of a smooth Smale horseshoe are considered.

R. L. Dobrushin

R. L. DOBRUSHIN and M. ZAHRADNÍK

Phase Diagrams for Continuous-Spin Models: An Extension of the Pirogov–Sinai Theory

0. Introduction

In a recent paper of Dobrushin and Shlosman [1] the existence of a phase transition for low temperatures was proved for lattice models with a Hamiltonian of the type

$$H(x) = \sum_{(s,t)} |x_t - x_s|^2 + \sum_t U(x_t) \tag{0.1}$$

where $x_t, x_s \in \mathbb{R}^k$, $t, s \in \mathbb{Z}^\nu$ (where it is assumed that $\nu \geqslant 2$ and $k \geqslant 1$), the first sum being taken over all pairs of nearest-neighbour spins.

Such phase transitions arise in situations when the potential U satisfies the following assumptions (see Figure 1):

(a) There are two points of relative minimum of U (denoted by σ_+ and σ_- in the following), which are separated by a sufficiently 'massive' barrier.

(b) The two potential 'wells' corresponding to the point of absolute minimum of the potential U (say σ_+) and the point of the second (relative) minimum σ_- satisfy the property that the second (less 'deep') well is wider than the first one.

Then, roughly speaking, the greater entropy of the second well compensates for its greater energy.

This balance is achieved for a particular value of the temperature, the *phase transition temperature*. For this transition temperature T_{tr} there are two Gibbsian states: the realizations of the first one are concentrated near σ_+, while the realizations of the second one are concentrated near σ_-.

The method used in [1] is the method of reflection positivity (see [2]).

This method admits quite general assumptions on the potential U. On the other

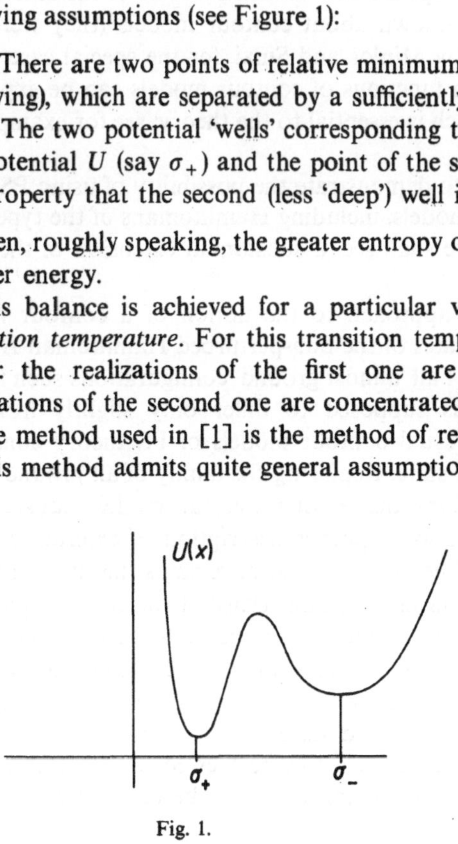

Fig. 1.

R. L. Dobrushin (ed.), Mathematical Problems of Statistical Mechanics and Dynamics, 1–123.

hand, the restriction to the nearest-neighbour quadratic interactions in the formula (0.1) is essential.

Further, this method does not give complete information about the behaviour of the phase diagram: for example, it gives no information about the continuity of T_{tr} in the case when the height of the relative minimum $U(\sigma_-)$ is continuously changed.

On the other hand, a complete picture of a phase diagram is given, for quite a general class of discrete-spin models at low temperatures, by the powerful theory of Pirogov and Sinai (PS theory) see [3, 4].

In this theory a family of Hamiltonians $\{H_\lambda\}$ depending on a vector parameter $\lambda \in \mathcal{U}(0) \subset \mathbb{R}^d$ is studied under the following assumptions:

(a) There are exactly $d + 1$ ground states of the Hamiltonian H_0.

(b) These ground states are separated by contours, which are supposed to satisfy a special condition called the Peierls condition in [3, 4] and the Gertzik – Pirogov – Sinai (GPS) condition in [5].

(c) The Hamiltonians H_λ are supposed to 'remove the degeneracy' of H_0.

Under these conditions. Pirogov and Sinai succeeded in constructing a phase diagram.

The main idea of their method is to express the external contour probabilities in the given model by choosing an appropriate contour model. One of the reasons for this approach is that quite a lot is known about contour models (they were first introduced and thoroughly studied by Minlos and Sinai, for the special case of the Ising model – see [6]). The partition functions of contour models can be expressed up to the 'surface tension' terms, which is essential to the theory; see for example the discussion in Chapter 2 of [4].

It is the aim of the present paper to demonstrate the possibility of using PS ideas for a wide class of continuous-spin models, including Hamiltonians of the type (0.1). In order to do this we require, as we shall see, a substantial extension of the tools used in [3, 4].

We will suggest a more general approach to the notion of a contour model: instead of the family of all ground states of the non-perturbed Hamiltonian H_0, our 'point of departure' will be a family of all 'almost ground' configurations such that in each site, the value of the spin is supposed to differ only slightly from the corresponding 'ground' value. The usual contour models of PS theory consist of contours 'floating' in a given ground state. Replacing the family of all ground states by our 'almost ground' configurations makes our contour models substantially richer objects, providing possibilities for a proper description of external contour probabilities of the original model. The crucial problem remains the same as in the original PS theory: to estimate the 'surface tension terms' of the contour partition functions. To solve this problem we use the following methods. First, we construct Gaussian approximations of the Gibbsian states, simply by replacing the potential by its second-order Taylor expansion near the ground states (each Gaussian approximation corresponding to some ground state).

Further, we show that our contour models can be viewed as perturbations of these Gaussian fields. Perturbations of Gaussian fields can be handled by a variety of methods, such as integration by parts or expansion into Wick polynomials and estimation of their semi-invariants. (A standard reference for the integration-by-parts methods is [7].) We will follow, however, the latter method based on estimates of semi-invariants (also called truncated expectations in the literature), as explained in [8]. Using this approach, the problem can be seen as the study of some abstract 'polymer model'. Such models are studied by cluster expansion methods, widely used in the literature (see e.g. [8, 9] among the more recent references). The cluster expansion is usually based on equations of the Kirkwood–Salsburg (or Minlos–Sinai) type.

Our §2 gives a self-contained and complete exposition of all the technical tools used in the paper. The technique developed in §2 also has other interesting applications, unrelated to PS theory. For example, we use it in the proof of the uniqueness of a Gibbs field which is a perturbation of a (positive-mass) Gaussian one. (See Theorem 2.3.15.)

Returning to the main theme of the paper, we now mention some limitations and some advantages of our methods:

We study here the case of two ground states of the unperturbed Hamiltonian, for simplicity. Thus, by the nature of the problem, we do not need to introduce contour models with a parameter (or other devices which might be needed for the construction of the full diagram in a general case). This simplifies some parts of the original PS reasoning. On the other hand, it enables us to concentrate on the novel features of our approach. The generalization of our arguments to the case of the maximal number of phases is straightforward, even in the general case of more than two ground states. The investigation of the whole phase diagram and the proof of completeness of the PS picture requires, however, some further constructions (e.g. that of [10]), not treated there.

It is hoped that the more general approach to the notion of a contour model will turn out to be useful in other situations also where the construction of the phase diagram is lacking: i.e. for various discrete-spin models where the original PS method does not work (for example in situations where there are several separate regions of 'almost ground states' but no finite choice of ground states) as well as for continuum-spin models which are not 'nearly Gaussian' (in this case one must use other technical tools for the study of the contour model). What seems to be necessary for applying our scheme is:

(a) the existence of several, 'sufficiently rich' regions of 'almost ground' configurations, with a strong correlation decay in each restricted Gibbs field;

(b) a sufficiently 'strong' barrier between those regions.

Problems similar to that considered here have been solved, independently, using different methods, by other workers. The recent papers of Imbrie [11] devoted to quantum fields contain, among other results, some results comparable to ours. The methods used in [11] are also based on a cluster expansion technique, but different expansions are used and the concept of a contour model is used in a different stage of the proof. In general terms it can be said that Imbrie first writes cluster expansions of the model and then studies these expansions using the usual form of PS theory, whereas in our approach more general contour models are defined at the very beginning of the proof and the cluster expansion technique is used to control these contour models.

Cluster expansion of perturbed Gaussian fields, the main technical theme of this paper, presented in §2.3, is also developed in the recent book by Malyshev and Minlos [12]. The same authors combined their methods with PS theory in [13].

The paper of Dinaburg and Sinai [14] also deals with similar problems, but it develops, another approach based on the notion of a contour model with interactions between contours. Note that in both [13] and [14] the case of sufficiently large mass of the unperturbed Gaussian field (i.e. of a fast enough decay of correlations) was considered, and some further restrictions (such as the nearest-neighbour interactions) which are not used in the present paper were also imposed.

Finally, the recent preprint [15] is also relevant to the subject of the present paper.

The investigations presented here were carried out independently of [11, 12, 13, 14, 15]. It would be interesting to compare all these approaches.

1. Formulation of the Main Result

1.1. CONFIGURATIONS

In this section, we consider a k-dimensional spin model where the values of 'spins' are taken from the Euclidean space $\mathbb{R}^k, k \geqslant 1$ on the ν-dimensional lattice $\mathbb{Z}^\nu, \nu \geqslant 2$. (Only technical difficulties arise if $\mathbb{R}^k$ is replaced by a manifold.) We will treat $\mathbb{Z}^\nu$ as a graph with edges (s, t) such that $|t - s| = 1$, the norm $|t|$ being given as

$$|t| = \sum_{i=1}^{\nu} |t_i|, \qquad t = (t_1, \ldots, t_\nu)$$

everywhere. We say that $A \subset \mathbb{R}^\nu$ is connected if it can be treated as a connected subgraph of $\mathbb{Z}^\nu$.

Given $\Lambda \subset \mathbb{Z}^\nu$, denote by $\partial_r \Lambda$ the set

$$\partial_r \Lambda = \{t \in \Lambda \colon \operatorname{dist}(t, \Lambda^c) \leqslant r\}, \qquad r \in \mathbb{N}. \tag{1.1}$$

We denote by $|\cdot|$ and $(\cdot, \cdot)$ the usual Euclidean norm and scalar product on $\mathbb{R}^k$. By $|\Lambda|$ we also denote the cardinality of a set $\Lambda \subset \mathbb{Z}^\nu$.

Our basic space of configurations is the space $(\mathbb{R}^k)^{\mathbb{Z}^\nu}$ denoted also by X in the following, the configurations being denoted by $x = \{x_t \in \mathbb{R}^k, t \in \mathbb{Z}^\nu\}$ etc. We also use the notation $X(A) = (\mathbb{R}^k)^A$, for any $A \subset \mathbb{Z}^\nu$. If $A \subset B \subset \mathbb{Z}^\nu$, $x \in X(B)$ we denote by x_A the restriction of x to A. For any $x_A \in X(A)$, $x_{\tilde{A}} \in X(\tilde{A})$, $A \cap \tilde{A} = \emptyset$ we define $x_A \cup x_{\tilde{A}} \in X(A \cup \tilde{A})$ such that $(x_A \cup x_{\tilde{A}})_A = x_A$ and $(x_A \cup x_{\tilde{A}})_{\tilde{A}} = x_{\tilde{A}}$.

1.2. STATES

By a state on X we mean, as usual, some probability measure P on $(X, \mathscr{B})$ where $\mathscr{B}$

denotes the smallest σ-algebra on X such that all the projections

$$\{x \rightsquigarrow x_t\}: X \to \mathbb{R}^k; \qquad t \in \mathbb{Z}^\nu$$

are measurable.

A sequence $\{P_n\}$ of states is said to be *vaguely convergent* to a state P if for any finite $A \subset \mathbb{Z}^\nu$ and any measurable bounded function φ of $x_A \in X(A)$ the relation

$$\lim_{n \to \infty} \int_X \varphi(x_A)\, \mathrm{d}\, P_n(x) = \int_X \varphi(x_A)\, \mathrm{d}\, P(x)$$

holds. For each $t \in \mathbb{Z}^\nu$ define a shift, denoted by S_t:

$$\{\{x_s, s \in A\} \rightsquigarrow \{\tilde{x}_s, s \in A + t\}\}: X(A) \to X(A + t), \qquad A \subset \mathbb{Z}^\nu$$

where $\tilde{x}$ denotes the configuration $\tilde{x}_s = x_{s-t}$, $s \in A$. A state P is said to be *translation-invariant* if $P(B) = P(S_t(B))$ for each $B \in \mathscr{B}$ $S_t(B)$ (being defined in an obvious sense) and each $t \in \mathbb{Z}^\nu$.

More generally, let $\tilde{\mathbb{Z}}^\nu$ be a subgroup of $\mathbb{Z}^\nu$ such that the factor group $\mathbb{Z}^\nu/\tilde{\mathbb{Z}}^\nu$ is finite. We say that $x \in X$ is $\tilde{\mathbb{Z}}^\nu$-periodic if $S_t x = x$ for each shift S_t, $t \in \tilde{\mathbb{Z}}^\nu$. The definition of a $\tilde{\mathbb{Z}}^\nu$-periodic state is analogous.

1.3. HAMILTONIANS

Suppose some family of interactions (called briefly an *interaction*) to be given on X, i.e. some family $\{\Phi_A, A \subset \mathbb{Z}^\nu\}$ where each Φ_A is a measurable function $\Phi_A: X(A) \to \{\mathbb{R} \cup +\infty\}$. We say that $\{\Phi_A\}$ is *translation-invariant* if

$$\Phi_A(x_A) = \Phi_{A+t}(S_t x_A)$$

for each $A \subset \mathbb{Z}^\nu$ and each $x_A \in X(A)$. The $\tilde{\mathbb{Z}}^\nu$-periodicity of $\{\Phi_A\}$ is defined analogously.

1.4. GIBBS STATES

Given any configuration $X_\Lambda \in X(\Lambda)$, Λ finite $\subset \mathbb{Z}^\nu$, and any boundary condition $x_{\Lambda^c} \in X(\Lambda^c)$ (where Λ^c denotes the complement of Λ) we define the (relative) Hamiltonian

$$H(x_\Lambda \mid x_{\Lambda^c}) = \sum_{A \not\subset \Lambda^c} \Phi_A(x_A). \tag{1.2}$$

The partition function in a volume Λ, corresponding to the Hamiltonian (1.2), temperature T and a boundary condition x_{Λ^c} is defined by the formula

$$Z^T(\Lambda, x_{\Lambda^c}) = \int_{X(\Lambda)} \exp(-T^{-1} H(x_\Lambda \mid x_{\Lambda^c}))\, \mathrm{d}\, x_\Lambda. \tag{1.3}$$

(The conditions guaranteeing the existence of $Z^T(\Lambda, x_{\Lambda^c})$ and its finiteness will be formulated later.)

Given a state P on X, denote by $P(\cdot \mid x_{\Lambda^c})$ the conditional probability given by

$x_{\Lambda^c} \in X(\Lambda^c)$. As usual, P will be called a Gibbs state corresponding to the Hamiltonian (1.2) and the temperature T if $Z^T(\Lambda, x_{\Lambda^c}) > 0$ holds for P-almost all configurations $x_{\Lambda^c} \in X(\Lambda^c)$ and if the density of $P(\cdot | x_{\Lambda^c})$ with respect to Lebesgue measure is expressed, for almost all x_{Λ^c}, by the formula $(Z^T(\Lambda . x_{\Lambda^c}))^{-1} \exp(-T^{-1} H(x_\Lambda | x_{\Lambda^c}))$.

1.5. ASSUMPTIONS OF THE MAIN THEOREM

We now introduce a series of assumptions on $\{\Phi_A\}$.

ASSUMPTION 1 (general conditions). The interactions Φ_A are translation-invariant continuous functions bounded from below (with values in $\mathbb{R} \cup +\infty$). They have a finite range, i.e. there is some $r \in \mathbb{N}$ such that $\Phi_A \equiv 0$ whenever diam $A > r$, where r is the range of interactions of the given model.

Under Assumption 1, we will often write (1.2) as $H(x_\Lambda \mid x_{\partial_r \Lambda^c})$.

If $\Lambda \subset \mathbb{Z}^\nu$ is finite and if $x_\Lambda \in X(\Lambda)$ we denote by

$$H(x_\Lambda) = \sum_{A \subset \Lambda} \Phi_A(x_A) \tag{1.4}$$

the (absolute) Hamiltonian of x_Λ. (This notion will be much less frequently used than (1.2).)

Further denote by $V(\Lambda, e)$ the Lebesgue measure of the set

$$\{X_\Lambda \in X(\Lambda) \colon H(x_\Lambda) \leqslant e |\Lambda|\}. \tag{1.5}$$

We now impose some bounds on the entropies of the 'energy shells' of our model: we will suppose that there are some finite constants $c \in \mathbb{R}$ and $\tilde{c} > 0$ such that the inequality

$$\ln V(\Lambda, e) \leqslant (c + \tilde{c}e) |\Lambda| \tag{1.6}$$

holds for each $e \in \mathbb{R}$ and each finite $\Lambda \subset \mathbb{Z}^\nu$.

NOTE. This is a very mild restriction on $\{\Phi_A\}$. In fact, it is possible to show that (1.6) holds in the situation when the single-spin interaction Φ_0 satisfies the following assumption, for some $\kappa > 0$:

$$\int_{\mathbb{R}^k} \exp(-\kappa \Phi_{\{0\}}(x)) \, \mathrm{d}\, x < \infty. \tag{1.7}$$

(We leave the proof of this fact to the reader.)

Now let x denote some periodic configuration. The mean energy of x is defined by the formula

$$e(x) = \lim_{\Lambda} |\Lambda|^{-1} H(x_\Lambda), \tag{1.8}$$

the limit being taken in the Van Hove sense, i.e. such that $|\partial_1 \Lambda| (|\Lambda|)^{-1} \to 0$.

We write

$$\bar{e} = \inf e(x), \tag{1.9}$$

the infimum being taken over all periodic configurations. We say that a periodic configuration $\bar{x}$ is a ground state of $\{\Phi_A\}$ if $e(\bar{x}) = \bar{e}$.

ASSUMPTION 2 (existence of ground states). There are exactly two ground states of $\{\Phi_A\}$, and they have a common period $\tilde{\mathbb{Z}}^\nu$.

NOTE. It is evident that any two periodic configurations have a common period. Also, for any subgroup $\tilde{\mathbb{Z}}^\nu \subset \mathbb{Z}^\nu$ which has a finite factor group $\mathbb{Z}^\nu/\tilde{\mathbb{Z}}^\nu$ there are some $d^1, \dots, d^\nu \in \mathbb{Z}^\nu$ such that

$$\tilde{\mathbb{Z}}^\nu = \left\{ \sum_{i=1}^{\nu} t_i d^i ; t_i \in \mathbb{Z}, i = 1, \dots, \nu \right\}. \tag{1.10}$$

We will not prove this simple algebraic fact.

The ground states given by Assumption 2 are denoted by $x^+ = \{x_t^+; t \in \mathbb{Z}^\nu\}$ and $x^- = \{x_t^-; t \in \mathbb{Z}^\nu\}$.

Say that a configuration $x = \{x_t, t \in \mathbb{Z}^\nu\}$ coincides with x^+ (x^-) almost everywhere if $x_t = x_t^+$ $(x_t = x_t^-)$ holds everywhere except for some finite set of $t \in \mathbb{Z}^\nu$.

DEFINITION. Fix some $\delta > 0$. Given any configuration $x \in X$ we say that $t \in \mathbb{Z}^\nu$ is a *+correct point* of x if $|x_s - x_s^+| < \delta$ holds for each $s \in \mathbb{Z}^\nu$ such that $|s - t| \leqslant r$. Similarly define the *−correct points* of x. The points $t \in \mathbb{Z}^\nu$ which are neither + nor −correct points of x are called *boundary points* of x. The union of all boundary points of x is denoted by $B_\delta(x)$.

NOTE. We also assume (without loss of generality) that

$$r \geqslant \max_{i=1,\dots,\nu} |d^i| \tag{1.11}$$

(see (1.10)).

This is reasonable because we require that the following property holds: if $x \in X$ is such that $B_\delta(x) = \emptyset$ for each $\delta > 0$ then either $x = x^+$ or $x = x^-$.

ASSUMPTION 3 (GPS-type condition). For each sufficiently small $\delta > 0$ there is a $\tau = \tau(\delta)$ such that

$$\sum_{A \subset \mathbb{Z}^\nu} (\Phi_A(x_A) - \Phi_A(X_A^+)) > \tau |B_\delta(x)| \tag{1.12}$$

holds, for each $x \in X$ which coincides with x^+ almost everywhere, and the same inequality holds if + is replaced by −.

NOTE. Even in the case of the discrete-spin models, such a condition does not follow from Assumption 2, as shown in [16].

We now state some assumptions about the behaviour of the interactions near x^+ and x^-.

ASSUMPTION 4 (Gaussian approximations for low temperatures). The functions Φ_A are finite in a neighbourhood of

$x_A^+ = \{x_t^+, t \in A\}$ and $x_A^- = \{x_t^-, t \in A\}$ and twice differentiable in x_A^+, x_A^- for each A.

It is a simple consequence of Assumption 2 that the equation

$$\sum_{A \subset \mathbb{Z}^\nu} \frac{\partial}{\partial x_t} \Phi_A(x_A) = 0 \tag{1.13}$$

holds, at $x_A = x_A^+$ and $x_A = x_A^-$, for each $t \in \mathbb{Z}^\nu$. Define, for each $s, t \in \mathbb{Z}^\nu$, the matrices

$$b_{s,t}^+ = \frac{1}{2} \sum_{A \subset \mathbb{Z}^\nu} \frac{\partial^2}{\partial x_t \partial x_s} \Phi_A(x_A^+). \tag{1.14}$$

The symbols $\partial/\partial x_t(\cdot)$ and $\partial^2/\partial x_t \partial x_s$ stand for

$$\frac{\partial}{\partial x_t}(\cdot) = \left(\frac{\partial}{\partial x_t^i}(\cdot),\ i = 1, \ldots, k\right). \tag{1.15}$$

and

$$\frac{\partial^2}{\partial x_t \partial x_s}(\cdot) = \left(\frac{\partial^2}{\partial x_t^i \partial x_s^j}(\cdot);\ i,j = 1, \ldots, k\right). \tag{1.16}$$

The quantities $b_{s,t}^-$ are defined analogously.

It is another consequence of Assumption 2 that for each finite Λ the matrix

$$b_\Lambda^+ = \{b_{s,t}^+ : s, t \in \Lambda\} \tag{1.17}$$

is non-negative definite, i.e. for any $x_\Lambda \in X(\Lambda)$,

$$(b_\Lambda^+ x_\Lambda, x_\Lambda) \geqslant 0 \tag{1.18}$$

and the same is true for the (analogously defined) matrix b_Λ^-. It is also clear that both $\{b_{s,t}^+\}$ and $\{b_{s,t}^-\}$ are $\tilde{\mathbb{Z}}^\nu$-periodic, i.e. $b_{s,t}^\pm = b_{s+\tau, t+\tau}$, $\tau \in \tilde{\mathbb{Z}}^\nu$.

Fix, for any class $\hat{t} \equiv t + \tilde{\mathbb{Z}}^\nu$, some $w_{\hat{t}} \in \mathbb{Z}^\nu$. Let W denote the set of all those $w_{\hat{t}}$. For example – see (1.10) – we can take

$$W = \{w \in \mathbb{Z}^\nu : w = \sum_{i=1}^{\nu} t_i d^i : 0 \leqslant t_i < 1\}. \tag{1.19}$$

Put, for any $t, s \in W$ and any $\xi \in [-\pi, \pi]^\nu$

$$f_{s,t}^+(\xi) = \sum_{u \in \tilde{\mathbb{Z}}^\nu} \exp(i(\xi, u)) b_{s,t+u}^+. \tag{1.20}$$

Similarly define the matrices $f_{s,t}^-(\xi)$.

Finally, let $F^+(\xi)$ denote the matrix (of the type $k|W| \times k|W|$) consisting of the blocks $f_{s,t}^+$: $s, t \in W$. Analogously define the matrix $F^-(\xi)$.

It follows from (1.18) that both matrices $F^+(\xi)$ and $F^-(\xi)$ are non-negative definite for all $\xi \in [-\pi, \pi]^\nu$, but we now make further assumptions.

ASSUMPTION 5 (positive mass of the Gaussian approximations). Both $F^+(\xi)$ and $F^-(\xi)$ are positive for all values of $\xi \in [-\pi, \pi]^\nu$, i.e. there is some $c > 0$ such that for each $\xi \in [-\pi, \pi]^\nu$ and each $x \in (\mathbb{R}^k)^W$,

$$(F^\pm(\xi)x, x) > c|x|^2. \tag{1.21}$$

NOTE. $(F^\pm(\xi))^{-1}$ are the spectral density matrices of periodical Gaussian fields defined as Gibbsian ones, with interactions

$$\Phi_A(x_A) = (b^\pm_{s,t}x_s, x_t) \quad \text{if } A = \{s,t\}, s \neq t,$$
$$\Phi_A(x_A) = \tfrac{1}{2}(b^\pm_{t,t}x_t, x_t) \quad \text{if } A = \{t\},$$
$$\Phi_A(x_A) = 0 \text{ otherwise.}$$

The existence of such Gaussian fields and their uniqueness (in the class of periodic fields) is an immediate consequence of Assumption 5 (see e.g. [17]). Moreover, one also finds from Assumption 5 the exponential decay of correlations in these Gaussian fields (again see [17], and also §2.2 below). This fact (the positive mass of the Gaussian field, if we use the language of quantum field theory) will be significant in the following considerations.

So far, the family $\{\Phi_A\}$ has been fixed. Suppose further, from now on, that there is an additional family of interactions $\{\Psi^\lambda_A;\ A \subset \mathbb{Z}^\nu\}$ (a perturbation of the original Hamiltonian), depending on a real parameter $\lambda \in \mathcal{U}(0)$, $\mathcal{U}(0) \subset \mathbb{R}$ being some neighbourhood of zero. Suppose that for each $A \subset \mathbb{Z}^\nu$, $\Psi^0{}_A \equiv 0$ and set

$$\Phi^\lambda_A = \Phi_A + \Psi^\lambda_A. \tag{1.22}$$

Suppose that the interactions $\{\Psi^\lambda_A\}$ satisfy the following properties.

NOTATION FOR CONSTANTS. We denote by $c_{(1.21)}$ the constant c from (1.21), etc. This convention will be used throughout.

ASSUMPTION 6. (analyticity of perturbations). Suppose that all the properties of Assumption 1 hold uniformly also for all the interactions Φ^λ_A, $\lambda \in \mathcal{U}(0)$, with all the constants r, $c_{(1.6)}$, $\tilde{c}_{(1.6)}$, not depending on λ, and with a uniform boundedness from below of all Φ^λ_A. Denote by $X^\infty(A)$ the domain of $X(A)$ where $\Phi^\lambda_A(x_A) = \infty$. Suppose that for any $A \subset \mathbb{Z}^\nu$, $X^\infty(A)$ does not depend on λ, and outside this domain Φ^λ_A depends analytically on λ in the sense that there is some open set $\mathcal{V}(0) \subset \mathbb{C}$, $\mathcal{V}(0) \supset \mathcal{U}(0)$ such that for each $x_A \in X(A) \setminus X^\infty(A)$ the function $\Phi^\lambda_A(x_A)$ can be extended to an analytical function on $\mathcal{V}$. Suppose further that there is a constant c such that

$$\left| \frac{\partial \Phi^\lambda_A}{\partial \lambda}(x_A) \right| < c(|\Phi^\lambda_A(x_A)| + 1) \tag{1.23}$$

holds for any $A \subset \mathbb{Z}^\nu$, $x_A \in X(A) \setminus X^\infty(A)$ and $\lambda \in \mathcal{V}(0)$. Assume also the joint continuity of $\partial/\partial\lambda\, \Phi_A(x_A)$ at $x_A \in X(A) \setminus X^\infty(A)$, $\lambda \in \mathcal{V}(0)$. (Again, (1.23) is a very mild assumption which can be further weakened, as will become clear later.) Write

$$u^+_A = \frac{\partial}{\partial \lambda} \Psi^\lambda_A(x^+_A)|_{\lambda=0}$$

and

$$u^+ = \lim_{\Lambda} |\Lambda|^{-1} \sum_{A \subset \Lambda} u_A^+. \tag{1.24}$$

(The limit is taken in the Van Hove sense.)
Similarly define the quantities u_A^- and u^-.

ASSUMPTION 7 (degeneracy-removing condition).

$$u^+ \neq u^-.$$

EXAMPLE. Consider the situation (0.1) with U replaced by $\{U_\lambda\}$, the potentials U_λ satisfying the following conditions:

(a) U_0 grows sufficiently fast near infinity (e.g. by (1.7)).

(b) U_0 has exactly two points (σ_+ and σ_-) of absolute minima. Then the constant configurations $x^+ = \{x_t \equiv \sigma_+\}$ and $x^- = \{x_t \equiv \sigma_-\}$ are the only ground states of the Hamiltonian, corresponding to the quadratic pair interaction and a single spin interaction U_0.

(c) If Ω is any open set containing both σ_+ and σ_- then $\inf_{\sigma \notin \Omega} U_0(\sigma) > U_0(\sigma_\pm)$.

(d) U_0 is continuous for all x and twice differentiable at σ_+ and σ_-.

(e) $d^2/d\sigma^2 U_0 |_{\sigma=\sigma_+}$ and $d^2/d\sigma^2 U_0 |_{\sigma=\sigma_-}$ are positive definite.

In order to remove the degeneracy of U_0 suppose for example that the perturbation of U_0 is caused by an external field of the type

$$U_\lambda(\sigma) = U_0(\sigma) + \lambda(\sigma, \mu)$$

for some $\mu \in \mathbb{R}^k$, $\mu \neq 0$. Then (cf. Assumption 7)

$$u^+ - u^- = (\mu, (\sigma_+ - \sigma_-));$$

if this quantity does not vanish, Assumption 7 and all the previous assumptions are fulfilled, as it can easily be checked.

1.6. THE MAIN THEOREM

Suppose that Assumptions 1–7 are all satisfied. Then, for a sufficiently small $T_0 > 0$ there exists a function $\{T \rightsquigarrow \lambda(T)\}$ defined on $[0, T_0]$ such that $\lambda(0) = 0$, $\lambda(\cdot)$ is analytical on $(0, T_0]$ and for each $T \in (0, T_0]$ there are two different $\tilde{\mathbb{Z}}^\nu$-periodical Gibbs states P_T^+, P_T^- corresponding to the interactions $\{\Phi_A^{\lambda(T)}\}$ and the temperature T. The renormalized states $\hat{P}_T^+, \hat{P}_T^-$ defined by

$$\hat{P}_T^\pm(A) = P_T^\pm((A - x^\pm)\sqrt{T}), \qquad A \in \mathscr{B} \tag{1.25}$$

converge, for $T \to 0$, in the sense of vague convergence, to the $\tilde{\mathbb{Z}}^\nu$-periodical Gaussian fields with spectral densities F^+, F^- and zero mean values. The function $\{T \rightsquigarrow \lambda(T)\}$ is differentiable at 0_+ and the formula

$$\frac{d\lambda(T)}{dT} = \tfrac{1}{2}((u^- - u^+)(2\pi)^\nu |W|)^{-1} \int_{[-\pi,\pi]^\nu} \ln(\det F^+(\xi) \det^{-1}(F^-(\xi)) d\xi \tag{1.26}$$

holds, with u^+ and u^- defined by (1.24).

1.7. STRATEGY OF THE PROOF

In the next section we present, in a self-contained form, the main technical tools needed for the proof. These tools are the cluster expansion method, correlation decay estimates for Gaussian fields with a positive mass and cluster expansion of perturbed Gaussian fields (using the technique of expansion into the products of semi-invariants).

An important technical result of this paper is our Main Lemma, which is formulated in §3. The difference in the formulation of the Main Lemma and the Main Theorem is, roughly speaking, only in the change of variables in T and x, but the Main Lemma is a slightly more general result, not necessarily a 'low-temperature' one. The Main Lemma is proved in §4. A new type of contour model is introduced in this section, and is thoroughly studied using the tools introduced in §2.

2. Preliminaries

2.1. ABSTRACT POLYMER MODELS AND CLUSTER EXPANSIONS

In this section, an abstract version of the cluster expansion method is studied. We consider a general 'polymer' model, according to the generally accepted terminology. The method has been widely used by many workers (see for example the papers cited in [7, 8, 9]). The estimates proved in this section are suited to the needs of the subsequent §2.3 and §4.4. They are, in some respects, more general than usual.

2.1.1. *Notation and General Formulas of Cluster Expansion*

Let $\mathbb{W}$ be a countable set (typically, $\mathbb{W} \subseteq \mathbb{Z}^\nu$). Denote by $\mathrm{Fin}(\mathbb{W})$ the family of all finite non-empty subsets of $\mathbb{W}$. Let k_T be a function of $T \in \mathrm{Fin}(\mathbb{W})$, with complex number values. (We fix the family $\{k_T, T \in \mathrm{Fin}(\mathbb{W})\}$ throughout §2.1.)

Given $\Lambda \in \mathrm{Fin}\,(\mathbb{W})$ denote by $g(\Lambda)$ the family of all collections $\gamma = \{T_1, \ldots, T_n\}$ of mutually non-intersecting non-empty sets $T_i \subset \Lambda$. For any $\gamma = \{T_1, \ldots, T_n\}$ put

$$k_\gamma = \prod_{i=1}^{n} k_{T_i}; \quad k_\phi = 1. \tag{2.1.1}$$

The abstract *partition function* in a volume $\Lambda \in \mathrm{Fin}(\mathbb{W})$ is defined by the formula

$$Z_\Lambda = \sum_{\gamma \in g(\Lambda)} k_\gamma, \quad Z_\emptyset = 1. \tag{2.1.2}$$

Suppose that $Z_\Lambda \neq 0$ and define the *correlation function*

$$\rho_A^\Lambda = (Z_\Lambda)^{-1} Z_{\Lambda \setminus A}. \tag{2.1.3}$$

We will construct the cluster expansion of (2.1.3), representing ρ_A^Λ as a sum of a series of which the terms can be explicitly evaluated.

It is evident that for each $B \in \mathrm{Fin}(\mathbb{W})$ and any $t \in \mathbb{W} \setminus B$ we have the relation

$$Z_B = Z_{B \cup \{t\}} - \sum_{T \subset B \cup \{t\} : T \ni t} k_T Z_{B \setminus T}. \tag{2.1.4}$$

This gives the *Kirkwood–Salsburg equation:* for any $\Lambda \in \mathrm{Fin}(\mathbb{W})$ and any $A \subset \Lambda$, $t \in A$

$$Z_{\Lambda \setminus A} = Z_{(\Lambda \setminus A) \cup \{t\}} - \sum_{T \subset \Lambda : T \cap A = \{t\}} k_T Z_{\Lambda \setminus (A \cup T)}. \tag{2.1.5}$$

We introduce a partial ordering $<$ on Fin $(\mathbb{W})$ as follows: for any $A \in \mathrm{Fin}(\mathbb{W})$ fix a point $t_A \in A$ (in an arbitrary way). Say that $B < A$ iff there is a sequence $B_1 = A, B_2, \ldots, B_n = B$ such that $B_i = B_{i-1} \setminus \{t_{B_{i-1}}\}$ for $i = 2, \ldots, n$.

Using (2.1.5) repeatedly for $Z_{(\Lambda \setminus A) \cup \{t_A\}}$, etc. $|A|$ times we get the equation

$$Z_{\Lambda \setminus A} = Z_\Lambda - \sum_{B \leqslant A} \left(\sum_{T \subset \Lambda : T \cap B = \{t_B\}} k_T Z_{\Lambda \setminus (B \cup T)} \right). \tag{2.1.6}$$

This equation will be iterated: let $\Lambda \in \mathrm{Fin}(\mathbb{W})$. A sequence $\beta = \{B_1, T_1, \ldots, B_n, T_n\}$ where $B_i, T_i \subset \Lambda$; $T_i \cap B_i = t_{B_i}$ for $i = 1, \ldots, n$; $B_1 \leqslant A$, $B_i \leqslant B_{i-1} \cup T_{i-1}$ for $i = 2, \ldots, n$ will be called a *chain in* Λ beginning in A of length $n (n = 1, \ldots)$. The set of all such chains is denoted by $\mathscr{B}^\Lambda_{A,n}$. We put

$$\mathscr{B}^\Lambda_A = \bigcup_{n=1}^{\infty} \mathscr{B}^\Lambda_{A,n}. \tag{2.1.7}$$

Iterating (2.1.6) N times ($N \in \mathbb{N}$), we obtain the equation

$$Z_{\Lambda \setminus A} = Z_\Lambda \left(1 + \sum_{n=1}^{N-1} (-1)^n \sum_{\beta \in \mathscr{B}^\Lambda_{A,n}} k_{T_1} \cdots k_{T_n} \right) + (-1)^N \sum_{\beta \in \mathscr{B}^\Lambda_{A,N}} k_{T_1} \cdots k_{T_N} Z_{\Lambda \setminus (B_N \cup T_N)}. \tag{2.1.8}$$

For $Z_\Lambda \neq 0$ this equation is usually written as

$$\rho^\Lambda_A = 1 + \sum_{n=1}^{N-1} (-1)^n \sum_{\beta \in \mathscr{B}^\Lambda_{A,n}} k_{T_1} \cdots k_{T_n} + (-1)^N \sum_{\beta \in \mathscr{B}^\Lambda_{A,N}} k_{T_1} \cdots k_{T_N} \rho^\Lambda_{B_N \cup T_N}. \tag{2.1.9}$$

This suggests writing

$$\rho^\Lambda_A = 1 + \sum_{\beta \in \mathscr{B}^\Lambda_A} (-1)^n k_{T_1} \cdots k_{T_n}. \tag{2.1.10}$$

This series is called the *cluster expansion* for ρ^Λ_A, assuming that it is absolutely convergent.

NOTE. If the series (2.1.10) converges, then clearly

$$\lim_{N \to \infty} \sum_{\beta \in \mathscr{B}^\Lambda_{A,N}} |k_{T_1} \cdots k_{T_N} \rho^\Lambda_{B_N \cup T_N}| = 0$$

and so (2.1.9) implies (2.1.10).

2.1.2. *Investigation of the Convergence of* (2.1.10).

LEMMA. *Suppose that for some* $\varepsilon < \frac{1}{4}$ *the following condition is satisfied, for each* $t \in \mathbb{W}$ *and* $m \in \mathbb{N}$:

$$\sum_{T\in \mathrm{Fin}(\mathbb{W}):T\ni t,|T|=m} |k_T| \leqslant \varepsilon^m. \tag{2.1.11}$$

Then the series in (2.1.10) *is absolutely convergent,* $Z_\Lambda \neq 0$ *and the relation* (2.1.10) *holds.*

Further, let $d(T)$ *be some function of* $T\in \mathrm{Fin}(\mathbb{W})$ *with positive integer values. Suppose that for all* $t\in\mathbb{W}, m, d\in\mathbb{N}$ *the condition*

$$\sum_{T\in \mathrm{Fin}(\mathbb{W}):T\ni t,|T|=m,d(T)=d} |k_T| \leqslant \varepsilon^m q^d \tag{2.1.12}$$

is satisfied, with some $\varepsilon > 0$ *and* $0 < q < 1$. *Then the following estimate holds for each* $A\subset\Lambda\in \mathrm{Fin}(\mathbb{W})$: *if* $\bar{q}$ *is chosen such that* $\bar{q} > q + 4\varepsilon$ *then*

$$\sum_{\beta\in\mathscr{B}_A^\Lambda:d(T_1)+\cdots+d(T_n)=d} |k_{T_1}\cdots k_{T_n}| \leqslant \frac{8}{\bar{q}-q}\varepsilon\bar{q}^d 2^{|A|}. \tag{2.1.13}$$

NOTE 1. Consider the case $\mathbb{W}\subseteq\mathbb{Z}^\nu$. The function $d(T)$ can be chosen as diam T, for example. The usual estimates of the convergence of (2.1.10), known for models with connected sets T (i.e. models where $k_T = 0$ for non-connected T) are obtained for a trivial choice of d (e.g. $d(T) = |T|$).

Our more general models (with non-connected T) turn out to be important in applications of §2.1 later in the paper. The following choice of d will be used in §2.3 and §4.4:

$$d(A) = \min\ \{d(G)\} \tag{2.1.14}$$

where $d(G) = \Sigma_{\{s,t\}\in G}|s-t|$, the minimum being taken over all connected graphs G with the vertex set A.

The estimate (2.1.13) is crucial for the rest of §2.1.

Proof of Lemma. We will show in detail the implication (2.1.12) ⇒ (2.1.13).

The first part of the lemma is proven by similar, but simpler, arguments. Note that (2.1.12) implies (2.1.11) (with a new value of ε).

First notice that (2.1.12) implies the estimate

$$\sum_{\beta\in\mathscr{B}_{A,n}^\Lambda:|T_i|=m_i d(T_i)=d_i,|B_i|=b_i:i=1,\ldots,n} |k_{T_1}\cdots k_{T_n}| \leqslant \varepsilon^{\sum_{i=1}^n m_i} q^{\sum_{i=1}^n d_i} \tag{2.1.15}$$

for any fixed set of integers $n; d_1,\ldots,d_m; m_1,\ldots,m_n; b_1,\ldots,b_n$.

In fact, if we fix $\{B_1,T_1,\ldots,B_{n-1},T_{n-1}\}$ in any way, and if $|B_n| = b_n$ is given, then there is only one way to choose B_n and also for any possible T_n we must have $t_{B_n}\in T_n$. Using (2.1.12) we get, by the obvious induction argument, the estimate (2.1.15).

Summing (2.1.15) over all possible choices of $d_1,\ldots,d_n$ we obtain

$$\sum_{\beta\in\mathscr{B}_{A,n}^\Lambda:d(T_1)+\cdots+d(T_n)=d;|T_i|=m_i,|B_i|=b_i:i=1,\ldots,n} |k_{T_1}\cdots k_{T_n}|$$

$$\leqslant \varepsilon^{\sum_{i=1}^n m_i}\left(\sum_{(d_1,\ldots,d_n):d_1+\cdots+d_n=d} q^d\right) = \varepsilon^{\Sigma_{i=1}^n m_i} q^d \binom{d}{n-1}. \tag{2.1.16}$$

Now fix $|T_i| = m_i$; $i = 1,\ldots,n$. Denote by $r_i = |B_{i-1} \cup T_{i-1}| - |B_i|$, $i = 2,\ldots,n$; $r_1 = |A| - |B_1|$, $r_{n+1} = |B_n \cup T_n|$.

The numbers r_i define the numbers $|B_i|$ in a unique way. Given $m = m_1 + \cdots + m_n$, the number of all possible choices of $B_1,\ldots,B_n$ is no greater than $\binom{m+|A|}{n}$ because $r_i \geqslant 0$; $i = 1,\ldots,n+1$ and $r_1 + \cdots + r_{n+1} = |A| + m_1 + \cdots + m_n - n = m + |A| - n$. Thus,

$$\sum_{\beta\in\mathscr{B}^{\Lambda}_{A,n}:d(T_1)+\cdots+d(T_n)=d;|T_i|=m_i,i=1,\ldots,n} |k_{T_1}\cdots k_{T_n}| \leqslant \varepsilon^m q^d \binom{d}{n-1}\binom{m+|A|}{n}$$

and therefore

$$\sum_{\beta\in\mathscr{B}^{\Lambda}_{A,n}:d(T_1)+\cdots+d(T_n)=d,|T_1|+\cdots+|T_n|=m} |k_{T_1}\cdots k_{T_n}|$$

$$\leqslant \varepsilon^m q^d \binom{m}{n-1}\binom{d}{n-1}\binom{m+|A|}{n}. \tag{2.1.17}$$

The combinatorial factors in (2.1.17) can be estimated as follows. Note the inequality

$$\binom{n}{k} \leqslant \alpha^{-k}(1+\alpha)^n;\ n,k\in\mathbb{N};\ \alpha > 0.$$

Choosing an arbitrary $\alpha_1 > 0$, $\alpha_2 > 0$, $\alpha_3 > 0$ we estimate the right-hand side of (2.1.17) as

$$\sum_{\beta\in\mathscr{B}^{\Lambda}_{A,n}:d(T_1)+\cdots+d(T_n)=d,|T_1|+\cdots+|T_n|=m} |k_{T_1}\cdots k_{T_n}|$$

$$\leqslant \varepsilon^m q^d \frac{(1+\alpha_1)^m(1+\alpha_2)^d(1+\alpha_3)^{m+|A|}}{\alpha_3(\alpha_1\alpha_2\alpha_3)^{n-1}}. \tag{2.1.18}$$

The summation over all m and $n \leqslant m$ yields the bound

$$\sum_{\beta\in\mathscr{B}^{\Lambda}_{A,n}:d(T_1)+\cdots+d(T_n)=d} |k_{T_1}\cdots k_{T_n}| \leqslant (q(1+\alpha_2))^d(1+\alpha_3)^{|A|}$$

$$\sum_{n\geqslant 1}\sum_{m\geqslant n}\left(\frac{1}{\alpha_1\alpha_2\alpha_3}\right)^{n-1}\frac{(\varepsilon(1+\alpha_1)(1+\alpha_3))^m}{\alpha_3}. \tag{2.1.19}$$

Consider a particular choice of

$$(1+\alpha_2) = q^{-1}\bar{q}, \qquad \alpha_3 = 1, \qquad \alpha_1 = (\alpha_2)^{-1}.$$

For $\varepsilon < \frac{1}{4}(\bar{q} - q)$ we have $\varepsilon(1+\alpha_1)(1+\alpha_3) < \frac{1}{2}$ and the summation in (2.1.19) gives (2.1.13). □

NOTE. In later applications, we usually also assume that $\Sigma_i |T_i| \geqslant 2$. Then we can write $16\varepsilon^2(\bar{q} - q)^{-2}\bar{q}^d 2^{|A|}$ on the right-hand side of (2.1.13).

Similar arguments applied to the inequality (2.1.11) (in this case there is no summing over d_i) give the inequalities

$$\sum_{\beta\in\mathscr{B}^{\Lambda}_{A,n}:|T_1|+\cdots+|T_n|=m} |k_{T_1}\cdots k_{T_n}| \leqslant \varepsilon^m\binom{m}{n-1}\binom{m+|A|}{n}$$
$$< \varepsilon^m\frac{(1+\alpha_1)^m(1+\alpha_3)^{m+|A|}}{\alpha_3(\alpha_1\alpha_3)^{n-1}} \tag{2.1.20}$$

and therefore it suffices to choose α_1, α_3 such that $\alpha_1\alpha_3>1$ and $\varepsilon(1+\alpha_1)(1+\alpha_3)<1$. This is possible for $\varepsilon<\frac{1}{4}$, and summing over all m and $n\leqslant m$ in (2.1.20) gives the convergence of the sum (2.1.10).

At last we remove the condition $Z_\Lambda\neq 0$ included in some of the previous considerations: the relations (2.1.8), Note 2.1.1 and Lemma 2.1.2 give, for $\varepsilon<\frac{1}{4}$,

$$Z_{\Lambda\setminus A} = Z_\Lambda\left(1+\sum_{n=1}^{\infty}(-1)^n\sum_{\beta\in\mathscr{B}^{\Lambda}_{A,n}} k_{T_1}\cdots k_{T_n}\right). \tag{2.1.21}$$

Its application to the case $A=\Lambda$ shows that $Z_\Lambda\neq 0$ for all finite Λ, and (2.1.10) holds. □

NOTE 2. It can also be seen from (2.1.20) and (2.1.21) that there is some $\tilde{\varepsilon}>0$ such that for each $\varepsilon<\tilde{\varepsilon}$ and each $\Lambda,\Lambda'\in\mathrm{Fin}(\mathbb{W})$, the condition (2.1.11) implies that

$$|\ln Z_\Lambda - \ln Z_{\Lambda'}| \leqslant \tilde{c}\varepsilon|\Lambda\,\triangle\,\Lambda'| \tag{2.1.22}$$

where $\tilde{c}=\tilde{c}(\tilde{\varepsilon})$ is some constant. We will not explicitly estimate $\tilde{\varepsilon}$ and $\tilde{c}$.

NOTE 3. In (2.1.22) and all subsequent relations containing logarithms of complex numbers we assume that the relations are true for a suitable choice of the imaginary part of the logarithm.

2.1.3. *Dependence of the Correlation Functions ρ_A on Λ.*

Assume, in the rest of §2.1, that $\mathbb{W}\subseteq\mathbb{Z}^\nu$ and also that the function d (see Note 1 of §2.1.2) satisfies the condition

$$d(T)\geqslant \operatorname{diam} T. \tag{2.1.23}$$

LEMMA. Let $\Lambda\subset\tilde{\Lambda}\in\mathrm{Fin}\,(\mathbb{W})$, let $0<q<\bar{q}<1$ and let $\varepsilon<\frac{1}{4}(\bar{q}-q)$. Suppose that the conditions (2.1.12) and (2.1.23) are satisfied. Then for each $A\subset\Lambda$,

$$|\rho^{\tilde{\Lambda}}_A-\rho^{\Lambda}_A| \leqslant \frac{8\varepsilon}{(1-\bar{q})(\bar{q}-q)}(\bar{q})^{\operatorname{dist}(A,\tilde{\Lambda}\setminus\Lambda)}2^{|A|}. \tag{2.1.24}$$

Proof. The cluster expansion (2.1.10) implies that

$$\rho^{\tilde{\Lambda}}_A-\rho^{\Lambda}_A = \sum_{\beta\in\mathscr{B}^{\tilde{\Lambda}}_A\setminus\mathscr{B}^{\Lambda}_A}(-1)^n k_{T_1}\cdots k_{T_n}. \tag{2.1.25}$$

Let $\beta=\{(B_j,T_j);\,j=1,\ldots,n\}\in\mathscr{B}^{\tilde{\Lambda}}_A\setminus\mathscr{B}^{\Lambda}_A$. Then $T_j\cap(\tilde{\Lambda}\setminus\Lambda)\neq\emptyset$ for some j. On the other hand, $T_1\cap A\neq\emptyset$. Therefore,

$$d(T_1) + \cdots + d(T_m) \geqslant \operatorname{diam} T_1 + \cdots + \operatorname{diam} T_n \geqslant \operatorname{dist}(A, \tilde{\Lambda} \setminus \Lambda).$$

Using this inequality in (2.1.25), (2.1.24) follows from the relation (2.1.13). □

2.1.4. *Expression of* $\ln Z_\Lambda$.

THEOREM. *Denote by* $\leqslant$ *the lexicographic ordering on* $\mathbb{Z}^\nu$. *Given any* $\Lambda \subseteq \mathbb{Z}^\nu$ *and* $t \in \mathbb{Z}^\nu$ *denote by* Λ_t *the set*

$$\Lambda_t = \{s \in \Lambda : s \leqslant t\}. \tag{2.1.26}$$

Suppose that (2.1.11) *is satisfied, with* $\varepsilon < \frac{1}{4}$. *Then all the quantities*

$$h_t^\Lambda = -\ln\Bigl(1 + \sum_{\beta \in \mathscr{B}_{\{t\}}^{\Lambda_t}} (-1)^n k_{T_1} \cdots k_{T_n}\Bigr) \tag{2.1.27}$$

(*see* (2.1.10) *and for a finite* Λ *also* (2.1.33)) *are well defined and if* $\Lambda_1 \subset \Lambda_2 \subset \cdots$ *are such that* $\cup_{n=1}^\infty \Lambda_n = \Lambda$ *then*

$$h_t^\Lambda = \lim_{n\to\infty} h_t^{\Lambda_n}. \tag{2.1.28}$$

For each finite $\Lambda \subset \mathbb{Z}^\nu$,

$$\ln Z_\Lambda = \sum_{t \in \Lambda} h_t^\Lambda. \tag{2.1.29}$$

Write h_t *instead of* $h_t^{\mathbb{Z}^\nu}$. *If* k_T *are translation-invariant (i.e.* $k_T = k_{T+t}$ *for any* t, T) *and if the relation* $<$ *(based on the choice of* $t_A \in A$) *was also chosen translation-invariant, then*

$$h_t \equiv h = \lim_{\Lambda \uparrow \mathbb{Z}^\nu} |\Lambda|^{-1} \ln Z_\Lambda,\ t \in \mathbb{Z}^\nu. \tag{2.1.30}$$

The quantities h_t^Λ *depend analytically on each* k_T.

Moreover let (2.1.12) *be satisfied, with a sufficiently small value of* $\varepsilon(1-q)^{-1}$. *Then for each* $\bar{q} > q + 4\varepsilon$ *there is a constant* $c = c(q, \bar{q})$ *such that for each* $t \in \mathbb{Z}^\nu$ *and each* $\Lambda \subset \mathbb{Z}^\nu$,

$$|h_t - h_t^\Lambda| \leqslant c\varepsilon \bar{q}^{\operatorname{dist}(t, \Lambda^c)}. \tag{2.1.31}$$

In particular there is another constant c' *such that*

$$\Bigl|\ln Z_\Lambda - \sum_{t\in\Lambda} h_t\Bigr| \leqslant c'\varepsilon |\partial_1 \Lambda| \tag{2.1.32}$$

for each finite $\Lambda \subset \mathbb{Z}^\nu$ (*see* (1.1)).

NOTE. It is reasonable to call the quantities h_t^Λ the 'densities of the free energy' of the given polymer model.

There are also other methods of proving (2.1.31) and (2.1.32), not using the lexicographic order, yielding an expression of h_t^Λ other than (2.1.27). Most of these methods are also based on an inequality of the type (2.1.24).

Proof of Theorem. We will demonstrate in detail the statements (2.1.31) and (2.1.32). The proofs of the remaining statements of the Theorem are simple, and will only be sketched.

By (2.1.10) we obtain for each finite $\Lambda \subset \mathbb{Z}^\nu$ that

$$h_t^\Lambda = -\ln \rho_{\{t\}}^\Lambda. \tag{2.1.33}$$

Therefore (2.1.28) follows from the obvious relation

$$\ln Z_\Lambda = -\sum_{t \in \Lambda} \ln \rho_{\{t\}}^\Lambda. \tag{2.1.34}$$

Notice that (2.1.22) implies (assuming that ε is sufficiently small) the uniform boundedness of all $\rho_{\{t\}}^\Lambda$ and $(\rho_{\{t\}}^\Lambda)^{-1}$. Write

$$\rho_{\{t\}}^{\mathbb{Z}^\nu} = \lim_{\Lambda \uparrow \mathbb{Z}^\nu} \rho_{\{t\}}^\Lambda.$$

Lemma 2.1.3 implies that for any $\Lambda \subset \mathbb{Z}^\nu$ and any $t \in \Lambda$,

$$|\rho_{\{t\}}^{\mathbb{Z}^\nu} - \rho_{\{t\}}^\Lambda| < \frac{16\varepsilon}{(1-\bar{q})(\bar{q}-q)} \bar{q}^{\operatorname{dist}(t, \Lambda^c)}. \tag{2.1.35}$$

For $\alpha, \beta \in \mathbb{C}$ such that $|\alpha| > \omega$, $|\beta| > \omega$, $\omega > 0$ we have an elementary inequality

$$|\ln \alpha - \ln \beta| \leqslant K|\alpha - \beta|, \tag{2.1.36}$$

where $K = K(\omega)$. Applied to (2.1.33) this gives the estimate

$$|h_t - h_t^\Lambda| \leqslant \frac{16K\varepsilon}{(1-\bar{q})(\bar{q}-q)} \bar{q}^{\operatorname{dist}(t, \Lambda^c)} \tag{2.1.37}$$

which is (2.1.31). If we sum this inequality over all $t \in \Lambda$ we get

$$\left| \ln Z_\Lambda - \sum_{t \in \Lambda} h_t \right| \leqslant \frac{16K\varepsilon}{(1-\bar{q})(\bar{q}-q)} \sum_{t \in \Lambda} \bar{q}^{\operatorname{dist}(t, \Lambda^c)} \tag{2.1.38}$$

which gives, for

$$c' = \frac{16K}{(1-\bar{q})(\bar{q}-q)} \sum_{t \in \mathbb{Z}^\nu} \bar{q}^{|t|}$$

(and a suitable $\bar{q}$ such that $1 > \bar{q} > q + 4\varepsilon$), the desired estimate (2.1.32).

The analyticity of h_t^Λ is clear from (2.1.27) and Lemma 2.1.2. □

2.2. GAUSSIAN GIBBSIAN FIELDS; THE CORRELATION DECAY

2.2.1. *Basic Notations and Assumptions*

We now need some facts about the correlation decay of Gaussian fields with a positive mass, treated from the Gibbsian point of view. Some of these facts are studied in [17] (only the case of scalar fields was considered there, but the generalization to vector fields is not difficult). The following discussion is independent of [17].

Suppose that we have matrix function Φ_t of $t \in \mathbb{Z}^\nu$, the values of which are $k \times k$ real symmetric matrices. Suppose further that

$$\Phi_t = \Phi_{-t}, \qquad t \in \mathbb{Z}^\nu$$

and also that there is some $r \in \mathbb{N}$ (range of interactions) such that

$$\Phi_t \equiv 0 \qquad \text{whenever} \qquad |t| > r. \tag{2.2.1}$$

(We write simply $\partial\Lambda$ instead of $\partial_r\Lambda$ everywhere in the following.) Given any finite $\Lambda \subset \mathbb{Z}^\nu$ and any $x_{\Lambda^c} = \{x_t, t \in \Lambda^c\} \in X(\Lambda^c) \equiv (\mathbb{R}^k)^{\Lambda^c}$ consider the Hamiltonian

$$H(x_\Lambda \mid x_{\Lambda^c}) \equiv H(x_\Lambda \mid x_{\partial\Lambda^c}) = \sum_{s,t \in \mathbb{Z}^\nu : \{s,t\} \not\subset \Lambda^c} (\Phi_{s-t} x_t, x_s). \tag{2.2.2}$$

Consider also the Hamiltonian

$$H(x_\Lambda) = H(x_\Lambda \mid 0) = \sum_{s,t \in \Lambda} (\Phi_{s-t} x_t, x_s). \tag{2.2.3}$$

Denote by

$$|x_\Lambda| = \left(\sum_{t \in \Lambda} |x_t|^2 \right)^{1/2} \tag{2.2.4}$$

the usual Euclidean norm on $X(\Lambda) \equiv (\mathbb{R}^k)^\Lambda$. Suppose that there is some $c > 0$ such that the quadratic form $\{x_\Lambda \rightsquigarrow H(x_\Lambda)\}$ satisfies the following condition of positive definiteness: for any finite $\Lambda \subset \mathbb{Z}^\nu$ and any $x_\Lambda \in X(\Lambda)$,

$$c^{-1}|x_\Lambda|^2 \geqslant H(x_\Lambda) \geqslant c|x_\Lambda|^2. \tag{2.2.5}$$

NOTE. Clearly, the left-hand side of (2.2.5) follows from the relation (2.2.1), for a suitable $c > 0$. So the crucial assumption is on the right-hand side of (2.2.5). This assumption (the positivity of the mass, if we use 'physical' language) will play an important role in many parts of this paper.

For any finite $\Lambda \subset \mathbb{Z}^\nu$ and any boundary condition $x_{\Lambda^c} \in X(\Lambda^c)$ consider the Gibbsian probability density

$$P_{x_{\Lambda^c}}(x_\Lambda) = Z^{-1} \exp(-H(x_\Lambda \mid x_{\Lambda^c})) \tag{2.2.6}$$

where

$$Z = Z(\Lambda, x_{\Lambda^c}) = \int_{X(\Lambda)} \exp(-H(x_\Lambda | x_{\Lambda^c}))\, dx_\Lambda. \tag{2.2.6'}$$

Notice that $Z(\Lambda, x_{\Lambda^c})$ and $P_{x_{\Lambda^c}}$ depend only on $x_{\partial\Lambda^c}$. Denote by

$$d\mu_{x_{\Lambda^c}}(\cdot) = P_{x_{\Lambda^c}}(\cdot)\, d(\cdot) \tag{2.2.7}$$

the measure on $X(\Lambda)$ given by the density $P_{x_{\Lambda^c}}$. Write also $Z(\Lambda, x_{\partial\Lambda^c})$, $\mu_{x_{\partial\Lambda^c}}$ instead of $Z(\Lambda, x_{\Lambda^c})$, $\mu_{x_{\Lambda^c}}$, etc.

Notice that the density $P_{x_{\Lambda^c}}(\cdot)$ can be written as

$$P_{x_{\Lambda^c}}(x_\Lambda) = (\pi)^{-k|\Lambda|/2} (\det \Phi_\Lambda)^{1/2} \exp\left(- \sum_{s,t \in \Lambda} (\Phi_{s-t}(x_t - \bar{x}_t), (x_s - \bar{x}_s)) \right) \tag{2.2.8}$$

where Φ_Λ is a positive Toeplitz matrix (of the type $k|\Lambda| \times k|\Lambda|$) with matrix elements $\Phi_\Lambda(s,t) = \Phi_{s-t}$; $s, t \in \Lambda$ and where $\bar{x}_\Lambda$ is defined as the solution of the system of linear equations

$$\Phi_0 \bar{x}_s + \sum_{t \in \Lambda \setminus \{s\}} \Phi_{s-t} \bar{x}_t + \sum_{u \in \partial\Lambda^c} \Phi_{s-u} x_u = 0, \quad s \in \Lambda. \tag{2.2.9}$$

(Because of the positive definiteness of Φ_Λ this solution is unique.)

Alternatively, $\bar{x}_\Lambda$ can be defined as the configuration minimizing the Hamiltonian $H(x_\Lambda \mid x_{\Lambda^c})$ at the condition x_{Λ^c}. (2.2.8) says that $P_{x_{\Lambda^c}}$ is a Gaussian probability density on $X(\Lambda)$ with mean values $\bar{x}_t$, $t \in \Lambda$ and a covariance matrix

$$B_\Lambda = \{b^\Lambda_{s,t};\ s, t \in \Lambda\} = \tfrac{1}{2}\Phi_\Lambda^{-1}, \tag{2.2.10}$$

with matrix elements $b^\Lambda_{s,t}$.

The operator $x_{\Lambda^c} \rightsquigarrow \bar{x}_\Lambda$ is linear. Thus, it is possible to write

$$\bar{x}_t = \sum_{u \in \partial\Lambda^c} a^\Lambda_{t,u} x_u, \ t \in \Lambda \tag{2.2.11}$$

with some real matrices $a^\Lambda_{t,u}$, $t \in \Lambda$, $u \in \partial\Lambda^c$.

For any $\Lambda' \subset \Lambda$ denote by

$$B^{\Lambda'}_\Lambda = \{b^\Lambda_{s,t};\ s, t \in \Lambda'\} \tag{2.2.12}$$

the submatrix of B_Λ (of the type $k|\Lambda'| \times k|\Lambda'|$) corresponding to the volume Λ'. Finally, denote

$$C^{\Lambda'}_\Lambda = (B^{\Lambda'}_\Lambda)^{-1} \tag{2.2.13}$$

and denote by $c^{\Lambda,\Lambda'}_{s,t}$; $s, t \in \Lambda'$ the $k \times k$ matrix elements of $C^{\Lambda'}_\Lambda$. Clearly, $C^\Lambda_\Lambda = 2\Phi_\Lambda$ (see also (2.2.38) and (2.2.36)). Denote by $\mu_{\Lambda', x_{\Lambda^c}}$ the projection on $X(\Lambda')$, $\Lambda' \subset \Lambda$ of the probability $\mu_{x_{\Lambda^c}}$. Denote by $P^{\Lambda'}_{x_{\Lambda^c}}$ the corresponding probability density. Clearly, the following formula holds:

$$P^{\Lambda'}_{x_{\Lambda^c}}(x_{\Lambda'}) = (2\pi)^{-k|\Lambda'|/2} (\det C^{\Lambda'}_\Lambda)^{1/2} \exp\left(-\frac{1}{2} \sum_{s,t \in \Lambda'} (c^{\Lambda,\Lambda'}_{s,t}(x_t - \bar{x}_t), (x_s - \bar{x}_s)) \right). \tag{2.2.14}$$

2.2.2. *Theorem*

Consider the situation described in Section 2.2.1. *There are some constants* $K > 0$ *and* $0 < q < 1$ *which depend only on the constant c from* (2.2.5) (*also on* ν, *k*, *r*, *but we fix these values throughout the whole section*) *such that the following estimates hold*:

(a) *For any finite* $\Lambda \subset \mathbb{Z}^\nu$, *any* $s \in \Lambda$ *and* $u \in \partial\Lambda^c$,

$$|a^\Lambda_{s,u}| \leqslant K q^{|s-u|}. \tag{2.2.15}$$

(b) *For any finite* $\Lambda \subset \mathbb{Z}^\nu$, *any* $s, t \in \Lambda$,

$$|b^\Lambda_{s,t}| \leqslant K q^{|s-t|}. \tag{2.2.16}$$

(c) *For any finite* $\Lambda' \subset \Lambda \subset \mathbb{Z}^\nu$, *any* $s, t \in \Lambda'$,

$$|c^{\Lambda,\Lambda'}_{s,t}| \leqslant K q^{|s-t|}. \tag{2.2.17}$$

(d) *For any finite* Λ, $\tilde{\Lambda} \subset \mathbb{Z}^\nu$, *any* $s \in \Lambda \cap \tilde{\Lambda}$ *and* $u \in \partial\Lambda^c \cap \partial\tilde{\Lambda}^c$,

$$|a^{\Lambda}_{s,u} - a^{\tilde{\Lambda}}_{s,u}| \leqslant K q^{\operatorname{dist}(s,\Lambda\triangle\tilde{\Lambda}) + \operatorname{dist}(u,\Lambda\triangle\tilde{\Lambda})}. \tag{2.2.18}$$

(e) *For any finite* Λ, $\tilde{\Lambda} \subset \mathbb{Z}^\nu$ *and any* $s, t \in \Lambda \cap \tilde{\Lambda}$,

$$|b^{\Lambda}_{s,t} - b^{\tilde{\Lambda}}_{s,t}| \leqslant K q^{\operatorname{dist}(t,\Lambda\triangle\tilde{\Lambda}) + \operatorname{dist}(s,\Lambda\triangle\tilde{\Lambda})}. \tag{2.2.19}$$

(f) *For any finite* Λ, $\tilde{\Lambda}$, $\Lambda' \subset \mathbb{Z}^\nu$ *such that* $\Lambda' \subset \Lambda \cap \tilde{\Lambda}$, *and for any* s, $t \in \Lambda'$,

$$|c^{\Lambda,\Lambda'}_{s,t} - c^{\tilde{\Lambda},\Lambda'}_{s,t}| \leqslant K q^{\operatorname{dist}(s,\Lambda\triangle\tilde{\Lambda}) + \operatorname{dist}(t,\Lambda\triangle\tilde{\Lambda})}. \tag{2.2.20}$$

NOTE. Given a $k \times k$ matrix c we define $|c| = \sup_{x \in \mathbb{R}^k} |cx|/|x|$.

Proof. The basic estimate is (2.2.15). All the remaining estimates will be proved using it.

(a) The proof of (2.2.15) is based on the following elementary considerations. Let us start with some finite Λ, $M \subset \mathbb{Z}^\nu$, $\Lambda \cap M = \emptyset$. Denote by $\bar{x}_\Lambda = \bar{x}_\Lambda(x_M)$ the configuration on Λ minimizing the Hamiltonian $H(x_\Lambda \cup x_M)$ at the fixed condition x_M. (Such a configuration on Λ is unique because of (2.2.5).) The minimizing property of $\bar{x}_\Lambda(x_M)$ implies that the function

$$\varphi(\lambda) = H(\lambda\bar{x}_\Lambda \cup x_M) = \lambda^2 H(\bar{x}_\Lambda) + H(x_M) + \lambda H'(\bar{x}_\Lambda \mid x_M)$$

where $\lambda \in \mathbb{R}$ and $H'(\bar{x}_\Lambda|x_M) = 2\Sigma_{t\in\Lambda, u\in M}\,(\Phi_{t-u}\bar{x}_t, x_u)$ has $\lambda = 1$ as its minimizing value. Therefore,

$$H'(\bar{x}_\Lambda \mid x_M) = -2H(\bar{x}_\Lambda).$$

Using this relation and the positiveness of $H(\bar{x}_\Lambda \cup x_M)$ we find that

$$H(x_M) - H(\bar{x}_\Lambda) = H(x_M) + H'(\bar{x}_\Lambda \mid x_M) + H(\bar{x}_\Lambda) = H(\bar{x}_\Lambda \cup x_M) \geqslant 0. \tag{2.2.21}$$

Fix Λ, M and write $\bar{x}$ instead of $\bar{x}_\Lambda(x_M) \cup x_M$ for simplicity. Further let $\Lambda' \subset \Lambda$ and $M' \subset (\Lambda \setminus \Lambda') \cup M$ be such that $M' \supset \partial(\Lambda')^c \cap (\Lambda \cup M)$. It is clear from our definition of $\bar{x}$ that the following relation then holds:

$$(\bar{x})_{\Lambda'} = \bar{x}_{\Lambda'}(\bar{x}_{M'}). \tag{2.2.22}$$

The relations (2.2.21), (2.2.22) imply that

$$H(\bar{x}_{\Lambda'}) \leqslant H(\bar{x}_{M'}) \qquad \text{if} \qquad M' \supset \partial(\Lambda')^c \cap (\Lambda \cup M). \tag{2.2.23}$$

Now define the volumes $\Lambda^{(i)}$, $M^{(i)}$ inductively as follows (see Figure 2): $M^{(0)} = M$, $\Lambda^{(0)} = \Lambda$ and, for each $i = 1, \ldots$:

$$M^{(i)} = \partial\left(\bigcup_{j=0}^{i-1} M^{(j)}\right)^c \cap \Lambda, \qquad \Lambda^{(i)} = \Lambda \setminus \bigcup_{j=1}^{i} \mathbf{M}^{(j)}.$$

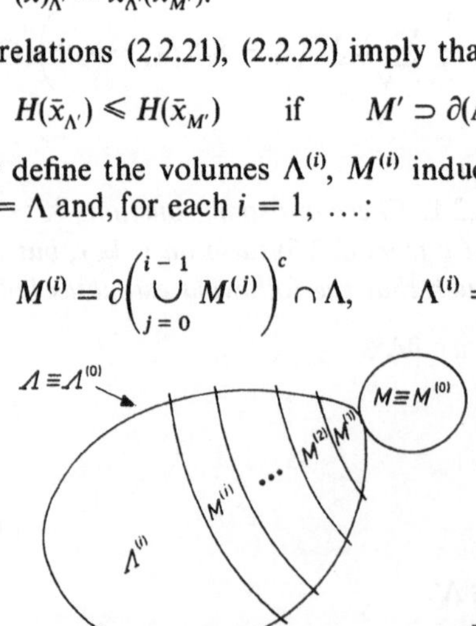

Fig. 2.

Using (2.2.23) with $M' = M^{(i)}, \Lambda' = \Lambda^{(i)}$ we obtain

$$H(\bar{x}_{M^{(i)}}) \geqslant H(\bar{x}_{\Lambda^{(i)}}),\ i = 1, 2, \ldots . \tag{2.2.24}$$

Combining (2.2.24) with the basic assumption (2.2.5) we get

$$\sum_{j \geqslant i+1} H(\bar{x}_{M^{(j)}}) \leqslant c^{-1} \sum_{j \geqslant i+1} |\bar{x}_{M^{(j)}}|^2$$
$$= c^{-1}|\bar{x}_{\Lambda^{(i)}}|^2 \leqslant c^{-2}H(\bar{x}_{\Lambda^{(i)}}) \leqslant c^{-2} H(\bar{x}_{M^{(i)}}) \tag{2.2.25}$$

with $c = c_{(2.2.5)}$

Using this inequality we can construct a monotone sequence $i_0 = 0, i_1, i_2, \ldots$ such that $i_{j+1} - i_j \leqslant 2c^{-2}, j \geqslant 0$ and

$$H(\bar{x}_{M^{(i_{j+1})}}) \leqslant \tfrac{1}{2}H(\bar{x}_{M^{(i_j)}}). \tag{2.2.26}$$

From (2.2.25) we obtain that the sequency $c^{2i}H(\bar{x}_{M^{(i)}})$ is monotone non-increasing. Thus, by (2.2.26) we get the existence of some $K > 0$ and $0 < q < 1$ such that

$$H(\bar{x}_{M^{(i)}}) \leqslant Kq^{\mathrm{dist}(M^{(i)},\, M)}H(x_M). \tag{2.2.27}$$

Now we apply (2.2.27) to the special case when $M = \{s\}$, $s \notin \Lambda$. By (2.2.11), (2.2.5) and (2.2.27) we obtain the desired relation (2.2.15).

(b) *Proof of* (2.2.16). Let $\Lambda' \subset \Lambda$. Choose the boundary condition $x_t \equiv 0$, $t \in \partial\Lambda^c$ and denote by $\{\xi_t, t \in \Lambda\}$ the Gaussian random variables (with values in $\mathbb{R}^k$), given by the density (2.2.8) with $x_t \equiv 0$, $t \in \partial\Lambda^c$. The usual consistency property of conditional Gibbs distributions implies that the conditional distribution of $\{\xi_t, t \in \Lambda'\}$ is for fixed $\{\xi_t, t \in \Lambda \setminus \Lambda'\}$ given by the density (2.2.8) with $x_t = 0$, $t \in \partial\Lambda^c$ and $x_t = \xi_t$, $t \in \Lambda \setminus \Lambda'$. Therefore,

$$\eta_s = \sum_{\mu \in \Lambda \cap \partial(\Lambda')^c} a^{\Lambda'}_{s,u}\xi_u, \qquad s \in \Lambda' \tag{2.2.28}$$

is the conditional mathematical expectation of ξ_s and $b^{\Lambda'}_{s,t}$ is the conditional covariance of ξ_s, ξ_t; $s, t \in \Lambda'$.

Note the following well-known fact (from prediction theory for Gaussian variables – see e.g. [18]). In the Hilbert space generated by the variables $\{\xi_t \in \Lambda\}$, η_s is the projection of ξ_s on the linear subspace generated by $\{\xi_t, t \in \Lambda \setminus \Lambda'\}$ and $b^{\Lambda'}_{s,t}$ is a matrix whose elements are scalar products of the components of corresponding perpendiculars $(\xi_s - \eta_s)$. Thus, for each $s, t \in \Lambda'$ and $i, j = 1, \ldots, k$

$$\langle \xi_s^{(i)}\xi_t^{(j)} \rangle = \langle (\xi_s - \eta_s)^{(i)}(\xi_t - \eta_t)^{(j)} \rangle + \langle \eta_s^{(i)}\eta_t^{(j)} \rangle$$

$$\langle \eta_s^{(i)}\eta_t^{(j)} \rangle = \left\langle \left(\sum_{u \in \Lambda \cap \partial(\Lambda')^c} a^{\Lambda'}_{s,u}\xi_u \right)^{(i)} \left(\sum_{v \in \Lambda \cap \partial(\Lambda')^c} a^{\Lambda'}_{t,v}\xi_v \right)^{(j)} \right\rangle$$

which implies the equation

$$b^{\Lambda}_{s,t} = b^{\Lambda'}_{s,t} + \sum_{u,v \in \Lambda \cap \partial(\Lambda')^c} a^{\Lambda'}_{s,u}\, b^{\Lambda}{}_{u,v}\,(a^{\Lambda'}_{t,v})^T; \quad s,t \in \Lambda' \tag{2.2.29}$$

where a^{T} denotes the transposed matrix a. Similarly we obtain the equation

$$b^{\Lambda}_{s,t} = \sum_{u \in \Lambda \cap \partial(\Lambda')^c} a^{\Lambda'}_{s,u} b^{\Lambda}_{u,t},\ s \in \Lambda', \quad t \in \Lambda \setminus \Lambda'. \tag{2.2.30}$$

For the special case $\Lambda \setminus \Lambda' = \{t\}$ this states that

$$b^{\Lambda}_{s,t} = a^{\Lambda'}_{s,t} b^{\Lambda}_{t,t}. \tag{2.2.31}$$

Because (2.2.15) holds it suffices to show that there is some $K' > 0$ such that for each finite $\Lambda \subset \mathbb{Z}^{\nu}$ and each $s, t \in \Lambda$,

$$b^{\Lambda}_{s,t} \leqslant K'. \tag{2.2.32}$$

But this is again a consequence of (2.2.5). In fact, (2.2.5) means that $\Phi_{\Lambda} \geqslant cE$, where E is the identity matrix, in the sense of inequalities between positive definite matrices. It implies that

$$B_{\Lambda} = \tfrac{1}{2}(\Phi_{\Lambda})^{-1} \leqslant \tfrac{1}{2}c^{-1}E$$

(see for example [19], 4, §12) and (2.2.32) follows. Clearly, (2.2.15), (2.2.31) and (2.2.32) imply the relation (2.2.16) (with a different K).

(c) *Proof of* (2.2.17). We represent the symmetric matrices B_{Λ} and $\Phi_{\Lambda} = \frac{1}{2}C^{\Lambda}_{\Lambda}$ in block form as

$$B_{\Lambda} = \begin{pmatrix} B^{\Lambda'}_{\Lambda} & L \\ L^{\mathrm{T}} & B^{\Lambda\setminus\Lambda'}_{\Lambda} \end{pmatrix}, \qquad \Phi_{\Lambda} = \begin{pmatrix} \Phi_{\Lambda'} & M \\ M^{\mathrm{T}} & \Phi_{\Lambda\setminus\Lambda'} \end{pmatrix} \tag{2.2.33}$$

where L, M are some $k|\Lambda'| \times k|\Lambda \setminus \Lambda'|$ matrices. The relation

$$B_{\Lambda}\Phi_{\Lambda} = \tfrac{1}{2}E$$

implies that

$$B^{\Lambda'}_{\Lambda}\Phi_{\Lambda'} + LM^{\mathrm{T}} = \tfrac{1}{2}E \qquad \text{and} \qquad B^{\Lambda'}_{\Lambda}M + L\Phi_{\Lambda\setminus\Lambda'} = 0. \tag{2.2.34}$$

Therefore,

$$L = -B^{\Lambda'}_{\Lambda}M(\Phi_{\Lambda\setminus\Lambda'})^{-1} = -\tfrac{1}{2}B^{\Lambda'}_{\Lambda}M\,B_{\Lambda\setminus\Lambda'}$$

(because $B_{\Lambda\setminus\Lambda'} = \frac{1}{2}(\Phi_{\Lambda\setminus\Lambda'})^{-1}$) and

$$(B^{\Lambda'}_{\Lambda})^{-1} = 2\Phi_{\Lambda'} - MB_{\Lambda\setminus\Lambda'}M^{\mathrm{T}}. \tag{2.2.35}$$

But this means that

$$c^{\Lambda,\Lambda'}_{s,t} = 2\Phi_{s-t} - \sum_{u,v\in\Delta\setminus\Lambda'} \Phi_{s-u} b^{\Lambda\setminus\Lambda'}_{u,v} \Phi_{v-t} \tag{2.2.36}$$

for each $s, t \in \Lambda'$.

Now we use the fact that Φ has a finite range r. Let $K' > 0$ be choosen such that $|\Phi_t| \leqslant K'$, $t \in \mathbb{Z}^{\nu}$. Let $|t - s| > r$. Then we obtain, by (2.2.16) and (2.2.36),

$$|c^{\Lambda,\Lambda'}_{s,t}| \leqslant (K')^2 \sum_{u\in\Lambda\setminus\Lambda':|u-s|\leqslant r} \quad \sum_{v\in\Lambda\setminus\Lambda':|v-t|\leqslant r} K\,q^{|u-v|} \leqslant K''q^{|s-t|} \tag{2.2.37}$$

for some $K'' = K''(r, q, K, K')$. If we drop the assumption $|t - s| > r$ we similarly obtain the inequality $|c^{\Lambda,\Lambda'}_{s,t}| \leqslant K' + K''q^{|t-s|}$. But this is the required estimate (2.2.17) (with some new value of K). The idea of using (2.3.34) in a similar situation is due to Guerra, Rosen and Simon (see [20], Th.VIII.1).

NOTE. It follows from (2.2.36) that if either dist $(s,(\Lambda')^c) > r$ or dist $(t,(\Lambda')^c) > r$

then for any finite $\Lambda \supset \Lambda'$ we have

$$c_{s,t}^{\Lambda,\Lambda'} = 2\Phi_{s-t}. \tag{2.2.38}$$

(d) *Proof of* (2.2.18). Define $a_{s,u}^{\Lambda} \equiv 0$ whenever $s \in \Lambda, u \in \Lambda^c \setminus \partial\Lambda^c$. It suffices to prove (2.2.18) in the special case $\bar{\Lambda} \subset \Lambda$; we can then compare $a_{s,u}^{\Lambda}$ with $a_{s,u}^{\bar{\Lambda}}$ through $a_{s,u}^{\Lambda\cap\bar{\Lambda}}$. Let $\bar{\Lambda} \subset \Lambda$. Denote by $\{\xi_s, s \in \Lambda\}$ the Gaussian random variables given by the density (2.2.6) with $x_t = 0$, $t \in \partial\Lambda^c \setminus \{u\}$. Then

$$\langle \xi_s \rangle = a_{s,u}^{\Lambda} x_u, \qquad s \in \Lambda$$

and the conditional mathematical expectation of ξ_s is, for a fixed $\{\xi_t, t \in \Lambda \setminus \bar{\Lambda}\}$, $\eta_s = \sum_{v \in \Lambda \cap \partial\bar{\Lambda}^c} a_{s,v}^{\bar{\Lambda}} \xi_v + a_{s,u}^{\bar{\Lambda}} x_u$ (compare (2.2.28)). Therefore, by averaging the variables ξ_v we find that

$$a_{s,u}^{\Lambda} = \sum_{v \in \Lambda \cap \partial\bar{\Lambda}^c} a_{s,v}^{\bar{\Lambda}} a_{v,u}^{\Lambda} + a_{s,u}^{\bar{\Lambda}}. \tag{2.2.39}$$

Using (2.2.39) and (2.2.15) we obtain

$$\begin{aligned} |a_{s,u}^{\Lambda} - a_{s,u}^{\bar{\Lambda}}| &\leqslant \sum_{v \in \partial\bar{\Lambda}^c \cap \Lambda} |a_{s,v}^{\bar{\Lambda}} a_{v,u}^{\Lambda}| \leqslant K^2 \sum_{v \in \bar{\Lambda}^c \cap \Lambda} q^{|s-v|+|v-u|} \\ &\leqslant K^2 \sum_{v \in \bar{\Lambda}^c \cap \Lambda} q^{|s-v|} \sum_{v \in \bar{\Lambda}^c \cap \Lambda} q^{|v-u|}. \end{aligned} \tag{2.2.40}$$

Write $q = q'q''$ with $0 < q' < 1$ and $0 < q'' < 1$. Then $\sum_{\sigma \in \bar{\Lambda}^c \cap \Lambda} q^{|v-u|} \leqslant (q')^{\operatorname{dist}(u, \bar{\Lambda}^c \cap \Lambda)} \sum_{v \in \mathbb{Z}^\nu} (q'')^{|u-v|} \leqslant (q')^{\operatorname{dist}(u, \bar{\Lambda}^c \cap \Lambda)} \sum_{t \in \mathbb{Z}^\nu} (q'')^{|t|}$. The sum $\sum_{v \in \bar{\Lambda}^c \cap \Lambda} q^{|s-v|}$ can be estimated in a similar way. So (2.2.40) proves (2.2.18), with some new constants q, K.

(e) *Proof of* (2.2.19). This is proven similarly, using the relation (2.2.29) and then (2.2.15), (2.2.16). For $\Lambda \supset \bar{\Lambda}$ we obtain

$$|b_{s,t}^{\Lambda} - b_{s,t}^{\bar{\Lambda}}| \leqslant K^3 \sum_{u,v \in \Lambda \cap \partial\bar{\Lambda}^c} q^{|s-u|+|u-v|+|v-t|} \leqslant K^3 \sum_{u \in \Lambda \setminus \bar{\Lambda}} q^{|s-u|} \sum_{v \in \Lambda \setminus \bar{\Lambda}} q^{|t-v|}. \tag{2.2.41}$$

The last two sums in (2.2.41) are estimated by exactly the same method as in (d), and we obtain (2.2.19), with new constants q, K.

(f) *Proof of* (2.2.20). For the estimate of $|c_{s,u}^{\Lambda,\Lambda'} - c_{s,u}^{\bar{\Lambda},\Lambda'}|$ we use the relation (2.2.36). We obtain

$$|c_{s,t}^{\Lambda,\Lambda'} - c_{s,t}^{\bar{\Lambda},\Lambda'}| \leqslant \sum_{u \in \Lambda'^c : |u-s| \leqslant r} |\Phi_{s-u}| \sum_{v \in \Lambda'^c : |v-t| \leqslant r} |\Phi_{t-v}| |b_{u,v}^{\Lambda\setminus\Lambda'} - b_{u,v}^{\bar{\Lambda}\setminus\Lambda'}|. \tag{2.2.42}$$

Applying (2.2.19) we obtain the relation (2.2.20), with a new constant K_0, □

2.2.3. *The Case of Infinite* Λ

Let $\Lambda \subset \mathbb{Z}^\nu$. Theorem 2.2.2 implies the existence of limits

$$\begin{aligned} a_{s,u}^{\Lambda} &= \lim_n a_{s,u}^{\Lambda\cap\Lambda_n}, && s \in \Lambda, u \in \partial\Lambda^c \\ b_{s,t}^{\Lambda} &= \lim_n b_{s,t}^{\Lambda\cap\Lambda_n}, && s,t \in \Lambda \\ c_{s,t}^{\Lambda,\Lambda'} &= \lim_n c_{s,t}^{\Lambda\cap\Lambda_n,\Lambda'}, && s,t \in \Lambda, \Lambda' \subset \Lambda, |\Lambda'| < \infty \end{aligned} \tag{2.2.43}$$

for any monotone sequence $\Lambda_1 \subset \Lambda_2 \subset \ldots$ such that $\cup_{n=1}^{\infty} \Lambda_n = \mathbb{Z}^\nu$, Λ_n is finite for each n. This limit is clearly independent of the choice of $\{\Lambda_n\}$.

PROPOSITION. *All the estimates of Proposition 2.2.2 remain true even for infinite* Λ.

Proof. Obvious, from (2.2.43) and Proposition 2.2.2. □

NOTE 1. In the special case $\Lambda = \mathbb{Z}^{\nu}$ we write $b_{s,t}, c_{s,t}^{(\Lambda)}$ instead of $b_{s,t}^{\mathbb{Z}^\nu}, c_{s,t}^{\mathbb{Z}^\nu,\Lambda}$. Clearly, the values $b_{s,t}$ depend on $|s-t|$ only. Denote them also by b_{s-t}.

The Gaussian field on $\mathbb{Z}^{\nu}$ defined by zero means and covariances $b_{s,t}$ is also a Gibbsian field corresponding to the Hamiltonian (2.2.2). Its finite-dimensional densities are

$$(P^{(\Lambda)}(x_\Lambda) = (2\pi)^{-k|\Lambda|/2}\,(\det C^{(\Lambda)})^{1/2}\exp\left(-\frac{1}{2}\sum_{s,t\in\Lambda}(c_{s,t}^{(\Lambda)}x_s,x_t)\right) \tag{2.2.44}$$

where $\Lambda \subset \mathbb{Z}^{\nu}$ is any finite set and $C^{(\Lambda)}$ is (compare (2.2.20)) a Toeplitz matrix

$$\begin{aligned} C^{(\Lambda)} &= \{c_{s,t}^{(\Lambda)};\, s,t\in\Lambda\} = C^{\Lambda}_{\mathbb{Z}^\nu} \\ &= \{b_{s-t};\, s,t\in\Lambda\}^{-1}. \end{aligned} \tag{2.2.45}$$

COROLLARY. *There are some* $K > 0$ *and* $0 < q < 1$ *depending only on* $c_{(2.2.5)}$ *such that*

(*a*) $|b_s| \leqslant Kq^{|s|}$ *for any* $s\in\mathbb{Z}^{\nu}$. (2.2.46)

(b) *for any finite* $\Lambda \subset \mathbb{Z}^{\nu}$,

$$|c_{s,t}^{(\Lambda)}| \leqslant Kq^{|s-t|} \text{ for any } s,t\in\Lambda. \tag{2.2.47}$$

Proof. Immediate. □

NOTE 2. One can also generalize the notion of measure $\mu_{x_{\Lambda^c}}$ (see (2.2.7)) for any (not necessarily finite) $\Lambda \subset \mathbb{Z}^{\nu}$ and any bounded configuration $x_{\Lambda^c}\in X(\Lambda^c)$. It suffices to take the measures $\mu_{\tilde{x}_{(\Lambda\cap\bar{\Lambda}_n)^c}}$, $\tilde{x}_{(\Lambda\cap\bar{\Lambda}_n)^c}\in X((\Lambda\cap\bar{\Lambda}_n)^c)$ being chosen uniformly bounded, arbitrary but such that $\tilde{x}_t = x_t$ for $t\in\Lambda^c\cap(\Lambda\cap\bar{\Lambda}_n)^c$. The limit $\lim_n \mu_{\tilde{x}(\Lambda\cap\bar{\Lambda}_n)^c}$ (taken in the sense of vague convergence – see §1.2) clearly does not depend on the choice of $\tilde{x}_{(\Lambda\cap\bar{\Lambda}_n)^c}$. The measure $\mu_{x_{\Lambda^c}}$ will be used in later parts of the paper, even for infinite Λ, $\Lambda \neq \mathbb{Z}^{\nu}$. In the special case $\Lambda = \mathbb{Z}^{\nu}$ we will use the notation $\mu_{\emptyset} = \mu$.

2.2.4. *Expressions of Partition Functions*

Given any finite $\Lambda \subset z^{\nu}$ and any $a = a_{\partial\Lambda^c}\in X(\partial\Lambda^c)$, define the partition function $Z(\Lambda,a)$ by (2.2.6′), with $x_{\partial\Lambda^c} = a$.

PROPOSITION. *The following expressions of* $Z(\Lambda,a)$ *hold:*

(a) $Z(\Lambda,a) = \exp(-H(\bar{a}_\Lambda|a))Z(\Lambda,0)$, (2.2.48)

where $\bar{a}_\Lambda$ *(the mean value of the field conditioned by* $x_{\partial\Lambda^c} = a$*) is given as the solution* $\bar{x}_\Lambda = \bar{a}_\Lambda$ *of the system (2.2.9), with* $x_{\partial\Lambda^c} = a$.

(b) $Z(\Lambda,0) = (\pi)^{k|\Lambda|/2}(\det\Phi_\Lambda)^{-1}$. (2.2.49)

(c) *The limit*

$$h = \lim_{\Lambda \uparrow \mathbb{Z}^\nu} |\Lambda|^{-1} \ln Z(\Lambda,0) \tag{2.2.50}$$

exists, in the Van Hove sense.

(d) *There is a finite constant* $K = K(c_{(2.2.5)})$ *such that*

$$|\ln Z(\Lambda,0) - h|\Lambda|| \leqslant K|\partial_1 \Lambda^c| \tag{2,2,51}$$

for all finite $\Delta \subset \mathbb{Z}^\nu$.

Proof. (a) and (b) are simple and well known. (c) is an obvious consequence of (d). We sketch two methods of proving (d), and also of how to find explicit formulas for h. Both methods are based on the estimates of Proposition 2.2.2, especially on (2.2.16) and (2.2.19).

(i) *The method of Künsch* (see [21]). We consider the family of Hamiltonians $H^{(\lambda)}$ depending on $\lambda \in [0,1]$, given by the formula (2.2.3) with

$$\Phi_t^{(\lambda)} = \lambda \Phi_t, \quad t \neq 0$$

$$\Phi_0^{(\lambda)} = \Phi_0$$

instead of Φ. It is evident that all the assumptions of §2.2.1 are satisfied for $\Phi^{(\lambda)}$, $\lambda \in [0,1]$ with c not depending on λ. In fact, $\Phi^{(\lambda)} = (1-\lambda)\Phi^{(0)} + \lambda\Phi$, both Φ and $\Phi^{(0)}$ being positive definite Hamiltonians. Denote by an additional index (λ) the analogous objects constructed for $\Phi^{(\lambda)}$ instead of Φ. Obviously,

$$\frac{\mathrm{d}}{\mathrm{d}\lambda} \ln Z^{(\lambda)}(\Lambda,0) = \sum_{s,t \in \Lambda; s \neq t} \mathrm{Tr}(b_{s,t}^{\Lambda(\lambda)} \Phi_{s-t}) \tag{2.2.52}$$

and

$$\ln Z^{(0)}(\Lambda,0) = |\Lambda| h^{(0)} = |\Lambda| \ln (\pi^{k/2} (\det \Phi_0)^{-1/2}).$$

By (2.2.52) and (2.2.19) there is some constant $L > 0$ not depending on λ such that for each finite $\Lambda \subset \mathbb{Z}^\nu$,

$$\left| \frac{\mathrm{d}}{\mathrm{d}\lambda} \ln Z^{(\lambda)}(\Lambda,0) - \sum_{s,t \in \Lambda; s \neq t} \mathrm{Tr}(b_{s-t}^{(\lambda)} \Phi_{s-t}) \right| \leqslant L|\partial \Lambda^c|. \tag{2.2.53}$$

Thus, we find that (2.2.51) holds, with

$$h = h^{(0)} + \int_0^1 \sum_{t \in \mathbb{Z}^\nu : |t| \leqslant r, t \neq 0} (\mathrm{Tr}(b_t^{(\lambda)} \Phi_t))\, \mathrm{d}\lambda. \tag{2.2.54}$$

(ii) *The method used in Theorem* 2.1.4. Consider the sets Λ_t, $\mathbb{Z}_t^\nu$ (see (2.1.26)). Notice that (2.2.19) implies that the limit $b_{u,v}^{\mathbb{Z}_t^\nu} = \lim b_{u,v}^{\Lambda_t^n}$ exists for any $u,v \in \mathbb{Z}^\nu$ and any sequence $\Lambda^1 \subset \Lambda^2 \subset \ldots$ such that $\cup_{n=1}^\infty \Lambda^n \supset \mathbb{Z}_t^\nu, \Lambda^n$ finite. Moreover,

$$|b_{u,v}^{\Lambda_t} - b_{u,v}^{\mathbb{Z}_t^\nu}| \leqslant K q^{\mathrm{dist}(u,\Lambda^c) + \mathrm{dist}(\sigma,\Lambda^c)}. \tag{2.2.55}$$

Denote by

$$Z_{\mathrm{rel}}(\Lambda_t, 0) = Z(\Lambda_t, 0)(Z(\Lambda_{t*}, 0))^{-1} \tag{2.2.56}$$

where t^* is the maximal element (in $\leqslant$) of the set $\Lambda_t \setminus \{t\}$. Clearly,

$$\ln Z(\Lambda,0) = \sum_{t\in\Lambda} \ln Z_{\mathrm{rel}}(\Lambda_t,0). \tag{2.2.57}$$

Express

$$Z_{\mathrm{rel}}(\Lambda_t,0) = \int \exp\left(- \sum_{s\in\Lambda_t} (\Phi_{s-t}x_s,x_t) \right) \mathrm{d}\mu_{\Lambda_{t^*}}(x_{\Lambda_{t^*}})\, \mathrm{d}x_t$$

where $\mu_{\Lambda_{t^*}}$ is the Gaussian Gibbsian measure given by (2.2.7) with $\Lambda = \Lambda_{t^*}, x_{\Lambda^c} \equiv 0$. Computing these integrals we obtain

$$Z_{\mathrm{rel}}(\Lambda_t,0) = (\pi)^{k/2} \det(\Phi_0 - \tfrac{1}{2}B^{\Lambda}_{t^*})^{-1/2} \tag{2.2.58}$$

where $B^{\Lambda}_{t^*}$ is the correlation matrix of the variable $y = \Sigma_{s\in\Lambda_{t^*}} \Phi_{t-s}x_s$ with respect to the measure $\mu_{\Lambda_{t^*}}$. Noting that $|B^{\Lambda}_{t^*} - B^{\mathbf{Z}^\nu}_{t^*}| \leqslant Kq^{\mathrm{dist}(t^*,\Lambda^c)}$ where $B^{\mathbf{Z}^\nu}_{t^*} = \lim_{\Lambda\uparrow\mathbf{Z}^\nu} B^{\Lambda}_{t^*}$ we obtain by (2.2.57), the desired estimate (2.2.51) (with a new value of K). The free energy is expressed as

$$h = \ln(\pi)^{k/2} - \tfrac{1}{2}\ln\det(\Phi_0 - \tfrac{1}{2}B^{\mathbf{Z}^\nu}_{t^*}). \tag{2.2.59}$$

2.2.5. *Free Energy Expression*

The expressions of h obtained in (2.2.54) and (2.2.59) are quite complicated. In this section we give another formula for the free energy h, more suitable in computations. This formula goes back to Grenander and Szegö [22]: see also Künsch [21].

PROPOSITION. *The free energy* (2.2.50) *can be expressed as*

$$h = \tfrac{1}{2}(k \ln \pi) - (2\pi)^{-\nu} \int_{[-\pi,\pi]^\nu} \ln\det\tilde{\Phi}_\xi\, \mathrm{d}\xi \tag{2.2.60}$$

where

$$\tilde{\Phi}_\xi = \sum_{t\in\mathbf{Z}^\nu} \Phi_t \exp(i(\xi,t)). \tag{2.2.61}$$

Proof. See [21] (for $k = 1$, but the generalization is obvious). The method of [21] is founded on the relation (2.2.54). □

To give some insight into the formula (2.2.60) we will, however, also formulate the simplest variant of such formula which arises if $\mathbf{Z}^\nu$ is replaced by a finite group. This includes the case of periodical boundary conditions.

LEMMA. *Let* Λ *be a finite Abelian group. Denote by* Λ^* *the dual group of all characters on* Λ. *Suppose that some family* $\{\Phi_t, t\in\Lambda\}$ of $k \times k$ *matrices* (*'interactions'*) *is given such that* $\Phi_t = \Phi_{-t}, t\in\Lambda$. *Define*

$$\Phi_\kappa = \sum_{t\in\Lambda} \kappa(t)\Phi_t;\ \kappa\in\Lambda^*.$$

For any $x_\Lambda \in X(\Lambda)$ *put* $x_K = \Sigma_{t\in\Lambda} \kappa(t)x_t$, $\kappa\in\Lambda^*$. *Then*

$$\sum_{s,t\in\Lambda} (\Phi_{t-s}x_s,x_t) = |\Lambda| \sum_{\kappa\in\Lambda^*} (\Phi_\kappa x_\kappa, x_\kappa)$$

(≡*additivity of the Hamiltonian on characters*) *and if we write*

$$Z = \int_{X(\Lambda)} \exp\left(- \sum_{s,t \in \Lambda} (\Phi_{s-t} x_s, x_t) \right) \mathrm{d}x_\Lambda$$

then the following formula holds:

$$\ln Z = \tfrac{1}{2}(k|\Lambda|\ln \pi) + \sum_{\kappa \in \Lambda^*} \ln \det(\Phi_\kappa)^{-1/2}. \tag{2.2.62}$$

Proof. By standard calculations, using the mutual orthogonality of characters. □

2.2.6. *General Linear-Quadratic Hamiltonians*

DEFINITION 1. We say that a Hamiltonian H given by a family of interactions $\{\Phi_A\}$ *is linear-quadratic* if each Φ_A can be expressed as

$$\Phi_A(x_A) = \sum_{t,s \in A} (\varphi_{t,s,A} x_t, x_s) + \sum_{t \in A} (\varphi_{t,A}, x_t) + \varphi_A \tag{2.2.63}$$

with some symmetric real $k \times k$ matrices $\varphi_{t,s,A}$ and some $\varphi_{t,A} \in \mathbb{R}^k$ and $\varphi_A \in \mathbb{R}$. We further assume that $\varphi_{t,s,A} \equiv \varphi_{s,t,A}$.

NOTE. In the rest of this paper we will deal with general Hamiltonians of the type (2.2.63) rather than with the Hamiltonians considered in §2.2.1. The reason is that our linear-quadratic Hamiltonians will arise as second-order Taylor expansions of general finite-range Hamiltonians.

DEFINITION 2. We say that a linear-quadratic Hamiltonian has a *standard form* if $\varphi_A \equiv 0$; $\varphi_{t,A} \equiv 0$ whenever $A \neq \{t\}$ and $\varphi_{s,t,A} \equiv 0$ whenever $A \neq \{t,s\}$.

Given a general linear-quadratic Hamiltonian H we define the *standardized Hamiltonian* H_{st} associated with H as the Hamiltonian defined by the family of interactions

$$\begin{aligned} \Phi^*_A(x_A) &= 2(\varphi_{t,s} x_t, x_s) && \text{if } A = \{t,s\} \\ \Phi^*_A(x_A) &= (\varphi_t, x_t) + (\varphi_{t,t} x_t, x_t) && \text{if } A = \{t\} \\ \Phi^*_A(x_A) &= 0 \text{ for other } A, \end{aligned} \tag{2.2.64}$$

where

$$\varphi_{t,s} = \sum_A \varphi_{t,s,A}$$

and

$$\varphi_t = \sum_A \varphi_{t,A}. \tag{2.2.65}$$

The generalization of the results obtained so far in §2.2 to the general linear-quadratic Hamiltonians is based on the following simple observation.

LEMMA 1. *Let H be a linear-quadratic Hamiltonian and H_{st} be its standardized form. Let $\Lambda \subset \mathbb{Z}^\nu$ be finite, let $x_\Lambda, y_\Lambda \in X(\Lambda)$ and $x_{\Lambda^c} \in X(\Lambda^c)$ be arbitrary configurations. Then, in the notations of Definition* 2, *the following holds:*

$$H(x_\Lambda | x_{\Lambda^c}) - H(y_\Lambda | x_{\Lambda^c}) = H_{st}(x_\Lambda | x_{\Lambda^c}) - H_{st}(y_\Lambda | x_{\Lambda^c}). \tag{2.2.66}$$

In particular, the Gibbsian probability densities in any finite volume with any fixed boundary conditions are the same for H as for H_{st}.

Proof. This is an obvious consequence of (2.2.65). □

DEFINITION 3. Consider the interactions

$$\begin{aligned} \tilde{\Phi}_A(x_A) &= 2(\varphi_{s,t}x_t, x_s) \quad \text{if } A = \{s,t\} \\ \tilde{\Phi}_A(x_A) &= (\varphi_{t,t}x_t, x_t) \quad \text{if } A = \{t\}. \end{aligned} \tag{2.2.67}$$

The resulting Hamiltonian is called the *quadratic standardized Hamiltonian* associated with H, and denoted by H_{qst}.

LEMMA 2. *Let H be a linear-quadratic Hamiltonian and* H_{qst} *be its quadratic standardized form. Suppose that there is a solution* $\{\xi_t \in \mathbb{R}^k, t \in \mathbb{Z}^\nu\}$ *of the system of linear equations*

$$\varphi_t + \sum_{s \in \mathbb{Z}^\nu} \varphi_{t,s}\xi_s = 0, \quad t \in \mathbb{Z}^\nu. \tag{2.2.68}$$

Then for any $x_\Lambda \in X(\Lambda)$ *and* $x_{\Lambda^c} \in X(\Lambda^c)$,

$$\begin{aligned} H(x_\Lambda | x_{\Lambda^c}) = H_{qst}(x_\Lambda - \xi_\Lambda | x_{\Lambda^c} - \xi_{\Lambda^c}) + H(\xi_\Lambda | \xi_{\Lambda^c}) + \\ + \sum_{t \in \Lambda} \sum_{s \in \Lambda^c} (\varphi_{t,s}\xi_t, x_s - \xi_s). \end{aligned} \tag{2.2.69}$$

In particular $H(x_\Lambda | x_{\Lambda^c}) - H_{qst}(x_\Lambda - \xi_\Lambda | x_{\Lambda^c} - \xi_{\Lambda^c})$ *does not depend on the concrete choice of* x_Λ.

Proof. Obvious, e.g. by computing all partial derivatives $\partial/\partial x_t$ and values at $\xi_t \equiv x_t$ of both parts of (2.2.69). □

NOTE. Usually, we will consider translation-invariant Hamiltonians. Then $\varphi_{t,s}$ and φ_t are also translation-invariant.

If, moreover, the Hamiltonian H_{qst} satisfies the condition (2.2.5) then the solution of (2.2.68) clearly exists and is given by the formula

$$\xi_t = -\left(\sum_{s \in \mathbb{Z}^\nu} \varphi_{t,s}\right)^{-1} \varphi_t. \tag{2.2.70}$$

We summarize as follows.

COROLLARY. *Let H be a translation-invariant linear-quadratic Hamiltonian such that* H_{qst} *satisfies (2.2.5). Given any finite* $\Lambda \subset \mathbb{Z}^\nu$ *and any* $\bar{x}_{\Lambda^c} \in X(\Lambda^c)$ *write* $\bar{y}_{\Lambda^c} = \bar{x}_{\Lambda^c} - \xi_{\Lambda^c}$ *and denote by* $\bar{y}_\Lambda$ *the configuration on* Λ *minimizing the value* $H_{qst}(y_\Lambda | \bar{y}_{\Lambda^c})$. *Put* $\bar{x}_\Lambda = \xi_\Lambda + \bar{y}_\Lambda$. *Then for any* $x_\Lambda \in X(\Lambda)$ *the following relation holds:*

$$H(x_\Lambda | \bar{x}_{\Lambda^c}) = H(\bar{x}_\Lambda | \bar{x}_{\Lambda^c}) + H_{qst}(x_\Lambda - \bar{x}_\Lambda). \tag{2.2.71}$$

Proof. By Lemma 2, if we notice moreover the relation

$$H_{qst}(y_\Lambda | \bar{y}_{\Lambda^c}) = H_{qst}(\bar{y}_\Lambda | \bar{y}_{\Lambda^c}) + H_{qst}(y_\Lambda - \bar{y}_\Lambda).$$ □

2.2.7. *Configurations Minimizing the Hamiltonian*

The aim of this section is to formulate some estimates which will be of use in §3 and §4.3 in the investigation of conditions of the Peierls (GPS) type. Essentially, these estimates are more precise variants of the estimates (2.2.15) and (2.2.27).

NOTE. We consider quadratic standardized Hamiltonians throughout this section. The generalization to arbitrary linear-quadratic Hamiltonians is straightforward.

DEFINITION 1. Let $\Lambda \subset \mathbb{Z}^\nu$ be finite, and let $\bar{x}_\Lambda \in X(\Lambda)$. Let $W \subset \mathbb{Z}^\nu$ be such that $W \supset \Lambda$. Define

$$\mathring{H}^W(\bar{x}_\Lambda) = \inf\left\{\sum_{s,t \in W} (\Phi_{s-t}x_t, x_s)\right\} \tag{2.2.72}$$

the infimum being taken over all configurations $x \in X(W)$ which vanish almost everywhere and which satisfy the condition $x_\Lambda = \bar{x}_\Lambda$.

PROPOSITION 1. *For any finite* $\Lambda \subset \mathbb{Z}^\nu$, *any* $\bar{x}_\Lambda \in X(\Lambda)$ *and* $W \supset \Lambda$,

$$\mathring{H}^W(\bar{x}_\Lambda) = \sum_{s,t \in W} (\Phi_{s-t}\bar{x}_t, \bar{x}_s) \tag{2.2.73}$$

where

$$\bar{x}_t = (\bar{x}_\Lambda)_t, \qquad t \in \Lambda$$

$$\bar{x}_t = \sum_{u \in \Lambda} a_{t,u}^{W\setminus\Lambda} \bar{x}_u, \qquad t \in W\setminus\Lambda. \tag{2.2.74}$$

NOTE. The series in (2.2.73) is absolutely convergent by (2.2.15).

Proof. Let $W_1 \subset W_2 \subset \cdots$ be finite, such that $W_1 \supset \Lambda$ and $W = \cup_{n=1}^\infty W_n$. It is clear that the sequence $\mathring{H}^{W_n}(\bar{x}_\Lambda)$ is monotone non-increasing and

$$\mathring{H}^W(\bar{x}_\Lambda) = \lim_{n\to\infty} \mathring{H}^{W_n}(\bar{x}_\Lambda).$$

The considerations of §2.2.1 show that $\mathring{H}^{W_n}(\bar{x}_\Lambda) = H(\bar{x}_{W_n}^{(n)})$ where

$$\bar{x}_t^{(n)} = \bar{x}_t, \qquad t \in \Lambda,$$

$$\bar{x}_t^{(n)} = \sum_{u \in \Lambda} a_{t,u}^{W_n\setminus\Lambda} \bar{x}_u, \qquad t \in W_n\setminus\Lambda.$$

Now it suffices to apply the estimates (2.2.15) and (2.2.18), and (2.2.73) follows.

PROPOSITION 2. *Let* Λ, $\Lambda' \subset \mathbb{Z}^\nu$ *be finite*, $\Lambda \cap \Lambda' = \emptyset$, *let* $x_\Lambda \in (\mathbb{R}^k)^\Lambda$ *and* $x_{\Lambda'} \in (\mathbb{R}^k)^{\Lambda'}$. *Let* $\Lambda \cup \Lambda' \subset W \subset \mathbb{Z}^\nu$. *Then the following estimates hold*:

(a) *There is some constant* $K > 0$ *depending only on* $c_{(2.2.5)}$ *such that*

$$K^{-1}|x_\Lambda|^2 \leqslant \mathring{H}^W(x_\Lambda) \leqslant K|x_\Lambda|^2. \tag{2.2.75}$$

(b) *There are some* $K' > 0$ *and* $0 < q < 1$ *depending only on* $c_{(2.2.5)}$ *such that*

$$|\mathring{H}^W(x_\Lambda \cup x_{\Lambda'}) - \mathring{H}^W(x_\Lambda) - \mathring{H}^W(x_{\Lambda'})| \leqslant K' q^{\mathrm{dist}(\Lambda,\Lambda')}(|x_\Lambda|^2 + |x_{\Lambda'}|^2). \tag{2.2.76}$$

Proof. The statement (a) follows immediately from (2.2.73) and (2.2.5) if we notice the trivial relation $\mathring{H}^W(x_\Lambda) \leqslant H(x_\Lambda)$.

Proof of (b): First notice that by (2.2.15) there are some $K'' > 0$ and $0 < q < 1$ depending only on $c_{(2.2.5)}$ such that for any finite $\Lambda \subset \mathbb{Z}^\nu$ and any $\Lambda \subset M \subset W$,

$$|\mathring{H}^W(\bar{x}_\Lambda) - H(\bar{x}_M)| \leqslant K'' q^{\operatorname{dist}(\Lambda, M^c)} |\bar{x}_{\partial\Lambda}|^2 \tag{2.2.77}$$

(where $\bar{x}$ is defined in (2.2.74)). Choose now the greatest possible d such that $2d + r \leqslant \operatorname{dist}(\Lambda, \Lambda')$. Choose some sets $M \supset \Lambda$, $M' \supset \Lambda'$ such that $M \cap M' = \emptyset$ and $\operatorname{dist}(\Lambda, M^c) = \operatorname{dist}(\Lambda', M'^c) = d$. Using the definition of $\bar{x}$ (see (2.2.74)) denote by $\bar{x}^{(\Lambda)}$ $(\bar{x}^{(\Lambda')})$, $(\bar{x}^{(\Lambda \cup \Lambda')})$ the minimizing configurations on W corresponding to the given $\bar{x}_\Lambda (\bar{x}_{\Lambda'}), (\bar{x}_\Lambda \cup \bar{x}_{\Lambda'})$. Obviously,

$$\mathring{H}^W(\bar{x}_\Lambda) + \mathring{H}^W(\bar{x}_{\Lambda'}) \leqslant H(\bar{x}^{(\Lambda \cup \Lambda')}_{M \cup M'}). \tag{2.2.78}$$

On the other hand,

$$H(\bar{x}^{(\Lambda)}_M) + H(\bar{x}^{(\Lambda')}_{M'}) \geqslant \mathring{H}^W(\bar{x}_\Lambda \cup \bar{x}_{\Lambda'}). \tag{2.2.79}$$

Using (2.2.77), we obtain from (2.2.78) that

$$\mathring{H}^W(\bar{x}_\Lambda) + \mathring{H}^W(\bar{x}_{\Lambda'}) \leqslant \mathring{H}^W(\bar{x}_\Lambda \cup \bar{x}_{\Lambda'}) + K'' q^d |\bar{x}_{\partial\Lambda} \cup \bar{x}_{\partial\Lambda'}|^2 \tag{2.2.80}$$

and from (2.2.79) we obtain that

$$\mathring{H}^W(\bar{x}_\Lambda) + \mathring{H}^W(\bar{x}_{\Lambda'}) + 2k'' q^d (|\bar{x}_{\partial\Lambda}|^2 + |\bar{x}_{\partial\Lambda'}|^2) \geqslant \mathring{H}^W(\bar{x}_\Lambda \cup x_{\Lambda'}). \tag{2.2.81}$$

Clearly, the last two relations prove the desired relation (2.2.76) (with a different q). □

2.3. ESTIMATES OF SEMI-INVARIANTS: CLUSTER EXPANSIONS OF PERTURBED GAUSSIAN FIELDS

2.3.1. *Semi-Invariants*

Let $f = (f_1, \ldots, f_n)$ be a system of random variables having all mutual moments $\langle f_1^{m_1} \cdots f_n^{m_n} \rangle$, $m_i \in \mathbb{N}$, $i = 1, \ldots, n$.

NOTE. We use the abbreviations

$$(m_1, \ldots, m_n) = m, \qquad \prod_{i=1}^n m_i! = m! (0! = 1),$$

$$\lambda_1^{m_1} \cdots \lambda_n^{m_n} = \lambda^m, \qquad f_1^{m_1} \cdots f_n^{m_n} = f^m. \tag{2.3.1}$$

The symbol $\langle f \rangle$ denotes, here and everywhere in what follows, the mathematical expectation of the variable f. By a random variable we mean an $\mathbb{R}$-valued variable, throughout §2.3.

The moments $\langle f_1^{m_1} \cdots f_n^{m_n} \rangle$ can be defined as Taylor coefficients of the generating function of variables $f_1, \ldots, f_n$:

$$\left\langle \exp\left(\sum_{i=1}^n \lambda_i f_i \right) \right\rangle = \sum_{m \in \mathbb{N}^n} (m!)^{-1} \langle f^m \rangle \lambda^m \tag{2.3.2}$$

if the generating function exists.

The semi-invariants (otherwise known as cumulants, or truncated correlation functions) can be defined as Taylor coefficients of logarithms of the generating function: using (2.3.1) write

$$\ln \left\langle \exp\left(\sum_{i=1}^{n} \lambda_i f_i \right) \right\rangle = \sum_{m \in \mathbb{N}^n} (m!)^{-1} \langle f^m_{,} \rangle \lambda^m \tag{2.3.3}$$

$\langle f^m, \rangle$ being the notation for semi-invariants. In the special case $m = (1, \ldots, 1)$ we use the notation

$$\langle f_, \rangle = \langle f_1, \ldots, f_n \rangle. \tag{2.3.4}$$

The relations (2.3.2) and (2.3.3) imply the well-known formula

$$\langle f_1 \cdots f_n \rangle = \sum_{\mathscr{D} \in \mathfrak{D}} \prod_{D \in \mathscr{D}} \langle f_{D}, \rangle \tag{2.3.5}$$

where $\mathfrak{D}$ is the set of all partitions $\mathscr{D}$ of the set of indices $\{1, 2, \ldots, \mathrm{n}\}$ on subsets D and f_D is a family of variables $\{f_i, i \in D\}$. If we rewrite (2.3.5) as

$$\langle f_1, \ldots, f_n \rangle = \langle f_1 \cdots f_n \rangle - \sum_{\mathscr{D} \in \mathfrak{D} : \mathscr{D} \neq \mathscr{D}_0} \prod_{D \in \mathscr{D}} \langle f_{D}, \rangle \tag{2.3.6}$$

where $\mathscr{D}_0$ is the trivial one-element partition, we can treat this relation as a recurrent (with respect to n) definition of semi-invariants $\langle f_1, \ldots, f_n \rangle$. In fact, (2.3.6) can and will be used as a definition of semi-invariants in the case when only all moments $\langle f_1^{\alpha_1} \cdots f_n^{\alpha_n} \rangle$ where $\alpha_i = 1$ or 0 exist.

It is evident that $\langle f_1, \ldots, f_n \rangle$ is symmetric with respect to permutations of its arguments. Also, semi-invariants are multilinear forms:

$$\langle \lambda f_1 + \mu g_1, f_2, \ldots, f_n \rangle = \lambda \langle f_1, f_2, \ldots, f_n \rangle + \mu \langle g_1, f_2, \ldots, f_n \rangle.$$

This follows immediately from the inductive definition (2.3.6) and the multilinearity of moments.

2.3.2. *Wick Polynomials*

We recall some notions connected with Wiener–Ito–Wick representations of functions of Gaussian variables (Wiener–Ito integrals in the probabilistic tradition, Wick polynomials in the tradition of mathematical physics). See for example [7, 8, 20, 23].

Let ξ be a Gaussian variable on some probability space and let

$$H_n(x) = (-1)^n \exp\left(\frac{x^2}{2} \right) \frac{\mathrm{d}^n}{\mathrm{d}x^n} \exp\left(-\frac{x^2}{2} \right)$$

be a Hermite polynomial. As is usual in the literature of mathematical physics, we denote by $:\xi^n:$ the normalized variable $\langle \xi^2 \rangle^{-n/2} H_n(\langle \xi^2 \rangle^{-1/2} \xi)$. Such functions are called *Wick polynomials*.

For several Gaussian variables, this notion is generalized in the following way. Let $\{\xi_1, \ldots, \xi_n\}$ be a Gaussian family of random variables on some probability space (Ω, μ). Consider the orthogonal decomposition

$$\xi_1 \cdots \xi_n = \xi + \eta \tag{2.3.7}$$

where ξ is the projection of $\xi_1 \cdots \xi_n$ on the subspace of $L^2(\Omega, \mu)$ generated by all the functions $\xi_1^{m_1} \cdots \xi_n^{m_n}$ with $m_i \geqslant 0$ and $0 \leqslant \Sigma_{i=1}^n m_i < n$, and η is the corresponding perpendicular in the space $L^2(\Omega, \mu)$. The function η is called the *Wick polynomial* of the Gaussian variables $\xi_1, \ldots, \xi_n$ and denoted by $:\xi_1 \cdots \xi_n:$.

In the Hilbert space $L^2(\xi_1, \ldots, \xi_n)$ of all quadratic integrable functions measurable with respect to $\xi_1, \ldots, \xi_n$ the set of all Wick polynomials $:\Pi_1^n \xi_i^{k_i}:$, $k_i \in \mathbb{N}$, $i = 1, \ldots, n$ is complete. We omit the proof of this well-known fact (see [23], for example). Further, any two Wick polynomials $:\Pi_1^n \xi_i^{k_i}:, :\Pi_1^n \xi_i^{l_i}:$ are mutually orthogonal if $\Sigma_1^n k_i \neq \Sigma_1^n l_i$ (see Lemma 2 of §2.3.3 below).

2.3.3. *Diagrams: Expression of Semi-Invariants of Wick Polynomials*

The material of this section is again well known (see [7, 20]); we present it only for the convenience of the reader.

Let F be a finite set. A *diagram* on F is a partition of F consisting of sets of cardinality at most 2.

A *vacuum* diagram is a partition which does not contain sets of cardinality 1.

Let $\mathscr{F}$ be a diagram on F, let $\tilde{F} \subset F$. We say that $\mathscr{F}$ has *no loops* in $\tilde{F}$ if $A \in \mathscr{F}$ for no $A \subset \tilde{F}$. (This notion, and most of the others, will be used for vacuum diagrams only.)

Given any two partitions $\mathscr{F}$, $\mathscr{G}$ of a set F define the partition $\mathscr{F} \vee \mathscr{G}$ as the finest partition coarser than both $\mathscr{F}$ and $\mathscr{G}$. We say that a partion $\mathscr{F}$ (specifically a diagram) is *connected with respect to* $\mathscr{G}$ if $\mathscr{F} \vee \mathscr{G}$ is a trivial partition, i.e. if $\mathscr{F} \vee \mathscr{G}$ consists of only one set F.

It is convenient to represent a diagram as a graph having the set of vertices F, each vertex of F belonging to at most one bond of the graph (exactly one in the vacuum case).

We will consider mainly the case when F is a set of pairs $\omega = (i, k)$; $i = 1, \ldots, n$; $k = 1, \ldots, j_i$. Then we can look on F as a table with n columns and j_i vertices in each column. (See Figures 3(a)–6(a) where $n = 4$, $j_1 = j_2 = 3$, $j_3 = j_4 = 2$. Figure 3(a) represents a trivial non-vacuum diagram, while Figures 4–6 are examples of vacuum diagrams.) It is traditional in mathematical physics to depict F as a set of n points with j_i 'legs' at the ith point (see Figure 3(b)). The legs corresponding to vertices of a bond are united (see Figures 3(b)–6(b) where the same diagrams as in Figures 3(a)–6(a) are depicted). Figure 4(a) shows a diagram with loops in the first and second columns, while Figures 5(a) and 6(a) show a diagrams without loops in the columns. Figure 5(a) shows a connected diagram (connected with respect to the partition of F on columns) while the diagrams in Figures 3(a), 4(a) and 6(a) are non-connected.

Suppose now that F is an index set of some Gaussian family $\{\xi_j, j \in F\}$ of random variables. Assign, to each diagram $\mathscr{F}$ on F, the value

$$I(\mathscr{F}) = \prod_{\{j\} \in \mathscr{F}} \langle \xi_j \rangle \prod_{\{j,k\} \in \mathscr{F}} \langle \xi_j \xi_k \rangle. \tag{2.3.8}$$

Fig. 3(a).

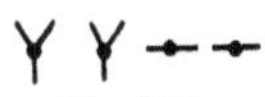
Fig. 3(b).

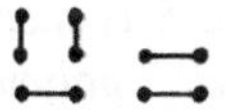
Fig. 4(a).

Fig. 4(b).

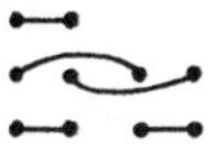
Fig. 5(a).

Fig. 5(b).

Fig. 6(a).

Fig. 6(b).

LEMMA 1. *Let $\{\xi_j, j \in F\}$ be a Gaussian family of random variables. Denote by $\mathscr{V}(F)$ the set of all diagrams on F. Then*

$$\left\langle \prod_{j\in F} \xi_j \right\rangle = \sum_{\mathscr{F}\in\mathscr{V}(F)} I(\mathscr{F}). \tag{2.3.9}$$

Proof. This is a consequence of the general formula (2.3.5). In fact, in the special case of Gaussian variables, (2.3.3) is a quadratic form and only the first two semi-invariants – the mean value and the covariance – have a non-zero value. □

To find an analogous formula to (2.3.9) for the products of Wick polynomials also, we will employ the generating-function method.

DEFINITION. Let F be a finite set and $\eta = \{\eta_j\}$ a family of random variables indexed by $j \in \mathbb{N}^F$. (We take $\eta_0 \equiv 1$ everywhere.)

Suppose that the series

$$f_\eta(\lambda) = \sum_{j\in\mathbb{N}^F} (j!)^{-1}\lambda^j \eta_j \tag{2.3.10}$$

(see the abbreviations (2.3.1)) converges, in the mean, in some neighbourhood of zero in $\mathbb{R}^F$. The function f_η will be called the *generating function* of the family η.

Given two systems η, $\tilde{\eta}$ of random variables with the generating function we obtain, in some neighbourhood of zero, the obvious formula

$$\langle f_\eta(\lambda) f_{\tilde{\eta}}(\mu)\rangle = \sum_{j,k\in\mathbb{N}F} (\gamma!k!)^{-1}\langle \eta_j\tilde{\eta}_k\rangle \lambda^j\mu^k \tag{2.3.11}$$

(assuming that for example the right-hand side is absolutely convergent in some neighbourhood of zero).

LEMMA 2. *Let $\{\xi_i, i \in F\}$ be a finite Gaussian family of random variables. Suppose that*

$$\langle \xi_i \rangle = 0 \textit{ for each } i \in F. \tag{2.3.12}$$

Put

$$\eta_j = \prod_{i\in F} \xi_i^{j_i}, \eta = \{\eta_j, j \in \mathbb{N}^F\}. \tag{2.3.13}$$

Given any $j \in \mathbb{N}^F$ denote by $\mathcal{T}$ the family of all indices $\omega = (i, k)$ where $i \in F$ is such that $j_i \neq 0$ and $k = 1, \ldots, j_i$. Put $\xi_\omega = \xi_i$ and consider the Wick polynomial $:\eta_j: = :\Pi_{\omega\in\mathcal{T}}\xi_\omega:$. Denote by

$$:\eta: = \{:\eta_j:, j \in \mathbb{N}^F\}. \tag{2.3.14}$$

Then the following relations hold:

(a) $f_\eta(\lambda) = \exp \sum_{i\in F} \lambda_i \xi_i$. (2.3.15)

(b) $f_{:\eta:}(\lambda) = \exp \sum_{i\in F} \lambda_i \xi_i \left\langle \exp \sum_{i\in F} \lambda_i \xi_i \right\rangle^{-1}$. (2.3.16)

(c) *Given $j, j^* \in \mathbb{N}^F$ take some mutually disjoint copies $\tilde{\mathcal{T}}, \tilde{\mathcal{T}}^*$ of the corresponding index sets $\mathcal{T}$, $\mathcal{T}^*$.*

Denote by $\mathcal{V}_{\mathcal{T}}(\mathcal{T} \,\&\, \mathcal{T}^*)$ the set of all vacuum diagrams on $\tilde{\mathcal{T}} \cup \tilde{\mathcal{T}}^*$ without loops in $\tilde{\mathcal{T}}$. Then

$$\langle :\eta_j: \eta_{j^*} \rangle = \sum_{\mathcal{F} \in \mathcal{V}_{\mathcal{T}}(\mathcal{T} \& \mathcal{T}^*)} I(\mathcal{F}). \tag{2.3.17}$$

(d) *Let $j, j^*, \mathcal{T}, \tilde{\mathcal{T}}, \mathcal{T}^*, \tilde{\mathcal{T}}^*$, be as before. Denote by $\mathcal{V}_{\mathcal{T},\mathcal{T}^*}(\mathcal{T} \,\&\, \mathcal{T}^*)$ the set of all vacuum diagrams on $\tilde{\mathcal{T}} \cup \tilde{\mathcal{T}}^*$ without loops either in $\tilde{\mathcal{T}}$ or in $\tilde{\mathcal{T}}^*$. Then*

$$\langle :\eta_j: :\eta_{j^*}: \rangle = \sum_{\mathcal{F} \in \mathcal{V}_{\mathcal{T}\mathcal{T}^*}(\mathcal{T} \& \mathcal{T}^*)} I(\mathcal{F}). \tag{2.3.18}$$

(e) *For $\emptyset \neq \mathcal{T}' \subset \mathcal{T}$, denote by $\xi_{\mathcal{T}'}$ the random variable $\Pi_{\omega\in\mathcal{T}'}\xi_\omega$. Analogously define $:\xi_{\mathcal{T}'}:$. Then*

$$\xi_{\mathcal{T}} = \sum_{\mathcal{T}: \emptyset \neq \mathcal{T}' \subset \mathcal{T}} \langle \xi_{\mathcal{T} \setminus \mathcal{T}'} \rangle :\xi_{\mathcal{T}'}:. \tag{2.3.19}$$

Proof. Obviously, (2.3.15) holds almost everywhere. The existence of a generating function for η follows from Lemma 1. Using (2.3.11) we get

$$\sum_{j, j' \in \mathbb{N}^F} (j!j'!)^{-1} \lambda^j \mu^{j'} \langle \eta_j \eta_{j'} \rangle = \langle f_\eta(\lambda) f_\eta(\mu) \rangle$$

$$= \langle \exp \sum_{i\in F} (\lambda_i + \mu_i)\xi_i \rangle = \exp(\tfrac{1}{2}(B\lambda, \lambda) + (B\lambda, \mu) + \tfrac{1}{2}(B\mu, \mu)) \tag{2.3.20}$$

where B denotes a covariance matrix with elements

$$b_{ii'} = \langle \xi_i \xi_{i'} \rangle.$$

Let $\tilde{\eta} = \{\tilde{\eta}_j, j \in \mathbb{N}^F\}$ be a system of random variables having the generating function $(\exp \Sigma_{i\in F} \lambda_i\xi_i)\langle \exp \Sigma_{i\in F} \lambda_i\xi_i\rangle^{-1}$. We want to prove that $\tilde{\eta}$ and $:\eta:$ coincide in distribution. As in (2.3.20) we find that

$$\sum_{j,j'\in\mathbb{N}^F} (j!j'!)^{-1}\lambda^j\mu^{j'}\langle \eta_j\tilde{\eta}_{j'}\rangle = \left\langle \exp \sum_{i\in F} (\lambda_i + \mu_i)\xi_i \right\rangle \left\langle \exp \sum_{i\in F} \lambda_i\xi_i \right\rangle^{-1}$$

$$= \exp(\tfrac{1}{2}(B\lambda, \lambda) + (B\lambda, \mu)). \tag{2.3.21}$$

Expanding the right-hand side of this equation into the power series we get

$$\langle \eta_j\tilde{\eta}_{j'}\rangle = \sum_{\mathscr{F}\in\mathscr{V}_{\mathscr{T}'}(\mathscr{T}\&\mathscr{T}')} I(\mathscr{F}). \tag{2.3.22}$$

In fact, the absence of $(B\mu, \mu)$ in (2.3.21) results in the vanishing of all the multipliers corresponding to the loops in $\mathscr{T}'$. Similarly, by using the relation

$$\sum_{j,j'\in\mathbb{N}^F} (j!j'!)^{-1}\lambda^j\mu^{j'}\langle \tilde{\eta}_j\tilde{\eta}_{j'}\rangle = \exp\,(B\lambda, \mu) \tag{2.3.23}$$

we prove the relation

$$\langle \tilde{\eta}_j\tilde{\eta}_{j'}\rangle = \sum_{\mathscr{F}\in\mathscr{V}_{\mathscr{T},\mathscr{T}'}(\mathscr{T}\&\mathscr{T}')} \mathrm{I}(\mathscr{F}). \tag{2.3.24}$$

The relations (2.3.22), (2.3.24) (cf. (2.3.17), (2.3.18)!) imply that $\langle \eta_j\tilde{\eta}_{j'}\rangle = 0$ whenever $\Sigma j_i < \Sigma j'_i$ (cf. (2.3.7)) and also $\langle \eta_j\tilde{\eta}_{j'}\rangle = \langle \tilde{\eta}_j\tilde{\eta}_{j'}\rangle$ whenever $\Sigma j_i = \Sigma j'_i$.

But this proves that $\tilde{\eta}_j = :\eta_j:$ and also the relations (2.3.17), (2.3.18).

Finally, (2.3.19) follows if we consider all the scalar products $\langle \xi_{\mathscr{T}}\xi_{\mathscr{T}'}\rangle$ and compare the right-hand and left-hand sides of the equation

$$\langle \xi_{\mathscr{T}}\xi_{\mathscr{T}'}\rangle = \sum_{\emptyset\neq\mathscr{T}''\subset\mathscr{T}} \langle \xi_{\mathscr{T}\setminus\mathscr{T}''}\rangle\langle :\xi_{\mathscr{T}''}:\xi_{\mathscr{T}'}\rangle \tag{2.3.25}$$

using the formulas (2.3.9) and (2.3.17). □

The following lemma (which will be frequently referred to later) is an example of the diagrammatic method in expressions of semi-invariants of (products of) Wick polynomials. The method is standard, but we discuss it in some detail here because of lack of suitable references.

LEMMA 3. *Let $\{\xi_{i,j,k}\}$ be a Gaussian family of random variables indexed by $i = 1,\ldots,n$; $j = 1,\ldots, n(i)$; $k = 1,\ldots,n(i,j)$. Put*

$$\xi_{i,j} = :\prod_{k=1}^{n(i,j)} \xi_{i,j,k}:$$

$$\xi_i = \prod_{j=1}^{n(i)} \xi_{i,j} \tag{2.3.26}$$

Denote by $\mathscr{V}(F)$ the set of all vacuum diagrams $\mathscr{F}$ on the set F of all triples (i,j,k).

Denote by

$$F_{i,j} = \{(i,j,k): k = 1,\ldots,n(i,j)\}$$
$$F_i = \bigcup_{j=1}^{n(i)} F_{i,j}. \tag{2.3.27}$$

Denote by $\tilde{\mathscr{V}}(F)$ *the subset of* $\mathscr{V}(F)$ *consisting of all* $\mathscr{F}$ *satisfying the following properties:* (i) $\mathscr{F}$ *has no loops in any set* $F_{i,j}$ *and* (ii) $\mathscr{F}$ *is connected with respect to the partition* $\{F_i\}$ *of* F. *Then the following formula holds:*

$$\langle \xi_1,\ldots,\xi_n\rangle = \sum_{\mathscr{F}\in\tilde{\mathscr{V}}(F)} I(\mathscr{F}). \tag{2.3.28}$$

Proof. The case $n = 1$ (as well as the case $n(i) = 1$ for each i) is proved in [8]. (It is possible, in the case $n = 1$, to use the same method as we used for the proof of (2.3.18) – which is the special case $n = 1$, $n(1) = 2$.). Thus, we can suppose that (2.3.28) is true for $n = 1$ and start with the formula

$$\langle \prod_{i=1}^{n} \xi_i\rangle = \sum_{\mathscr{F}\in\mathscr{V}'(F)} I(\mathscr{F}) \tag{2.3.29}$$

where $\mathscr{V}'(F)$ denotes the set of all $\mathscr{F}\in\mathscr{V}(F)$ which have no loops in any F_{ij}, $i = 1,\ldots,n$; $j = 1,\ldots,n(i)$.

Proceed now by induction on n, i.e. suppose that (2.3.28) holds for all $\langle\xi_1,\ldots,\xi_m\rangle$ such that $m < n$. Denote by $\mathscr{G}$ the partition $\{F_i: i = 1,\ldots,n\}$ of F. Let $\mathscr{H} = \{F_l^*: l = 1,\ldots,n^*\}$ be another partition of F, coarser than $\mathscr{G}$. Denote by $\mathscr{G}_l = \{F_i: F_i \subset F_l^*\}$ the induced partition of F_l^*.

With $\mathscr{H}$ fixed, denote by $\mathscr{V}_{\mathscr{H}}(F)$ the set of all diagrams $\mathscr{F}\in\mathscr{V}'(F)$ which have the property that (see §2.3.3 for the definition of $\mathscr{F} \vee \mathscr{G}$) $\mathscr{F} \vee \mathscr{G} = \mathscr{H}$. Using this notation we can write (2.3.29) as

$$\sum_{\mathscr{F}\in\mathscr{V}'(F)} I(\mathscr{F}) = \sum_{\mathscr{F}\in\tilde{\mathscr{V}}(F)} I(\mathscr{F}) + \sum_{\mathscr{H}\text{ non-trivial}} \sum_{\mathscr{F}\in\mathscr{V}_{\mathscr{H}}(F)} I(\mathscr{F}). \tag{2.3.30}$$

Denote by $\tilde{\mathscr{V}}(F_l^*)$ the set of all diagrams on F_l^* which have no loops in any $F_{ij} \subset F_l^*$ and are connected with respect to the partition $\mathscr{G}_l$. For any non-trivial $\mathscr{H}$ we have the decomposition

$$\sum_{\mathscr{F}\in\mathscr{V}_{\mathscr{H}}(F)} I(\mathscr{F}) = \prod_{l=1}^{m} \left(\sum_{\mathscr{F}_l\in\tilde{\mathscr{V}}(F_l^*)} I(\mathscr{F}_l)\right). \tag{2.3.31}$$

Now we use the induction argument

$$\sum_{\mathscr{F}_l\in\tilde{\mathscr{V}}(F_l^*)} I(\mathscr{F}_l) = \langle \xi_l^*,\rangle \tag{2.3.32}$$

where ξ_l^* is the collection of all ξ_i such that $F_i \subset F_l^*$. By substituting (2.3.31), (2.3.32) into (2.3.30) and by using the general formula (2.3.5) we obtain (2.3.28). □

2.3.4. *Estimation of Semi-Invariants. Formulation of the Basic Result*

The theorem below is an essential result of this section. Its consequences, formulated in later parts of §2.3, will be used, as one of the key technical tools, in §4.4.

THEOREM. *Let $\{\sigma_t, t\in S\}$ be a Gaussian family of random variables indexed by some finite set S. Suppose that*

$$\langle\sigma_t\rangle = 0, \qquad \langle\sigma_t^2\rangle = 1, \qquad t\in S. \tag{2.3.33}$$

Suppose that some positive function φ is given such that

$$|\langle\sigma_t\sigma_s\rangle| \leqslant \varphi(t,s), \qquad s,t\in S. \tag{2.3.34}$$

Suppose that the function φ satisfies the following condition: there is some $0<\alpha<1$ such that for each $t\in S$,

$$\sum_{s\in S: s\neq t} \varphi(t,s) \leqslant \alpha. \tag{2.3.35}$$

Let

$$\mathbb{S} = \{S_i, i=1,\ldots,m\} \tag{2.3.36}$$

be some partition of S. Suppose that some square-integrable functions f^i, $i=1,\ldots,m$ of random variables $\sigma_{S_i} = \{\sigma_t,\ t\in S_i\}$ are given. Suppose that we have the Wick series

$$f^i(\sigma_{S_i}) = \sum_{n_{S_i}\in\mathbb{N}^{S_i}} c_{n_{S_i}}(n_{S_i}!)^{-1/2} :\sigma_{S_i}^{n_{S_i}}: \tag{2.3.37}$$

where

$$:\sigma_{S_i}^{n_{S_i}}: = \prod_{t\in S_i} :\sigma_t^{n_t}:, \qquad n_{S_i} = \{n_t, t\in S_i\} \tag{2.3.38}$$

(this is not in general an orthogonal series!) *converging in the L^2 sense and moreover such that all the quantities*

$$M_i = \sup_{n_{S_i}}\{|c_{n_{S_i}}|\}, \qquad i=1,\ldots,m \tag{2.3.39}$$

are finite.

Denote by $\tilde{\mathscr{C}}(\mathbb{S})$ the set of all non-oriented graphs Γ on S which have the following property: there is a tree T on the set of indices $\{1,2,\ldots,m\}$ such that there is a one-to-one correspondence between bonds $\{i,j\}\in T$ and bonds $\{s,t\}\in\Gamma$, such that $s\in S_i$ and $t\in S_j$. Denote by

$$n_t(\Gamma) = |\{S:\{s,t\}\in\Gamma\}| \tag{2.3.40}$$

(the number of bonds of Γ containing the vertex t). Then all the moments $\langle\Pi_{i=1}^m (f^i)^{\alpha_i}\rangle$ where $\alpha_i = 0$ or 1 do exist, and the semi-invariant $\langle f^1,\ldots,f^m\rangle$ satisfies the estimate

$$|\langle f^1,\ldots,f^m\rangle| \leqslant c^{|S|} \prod_{i=1}^m M_i \sum_{\Gamma\in\tilde{\mathscr{C}}(\mathbb{S})} \varphi(\Gamma) \prod_{t\in S} (n_t(\Gamma)!)^{1/2} \tag{2.3.41}$$

where c is some constant depending only on α and

$$\varphi(\Gamma) = \prod_{\{s,t\}\in\Gamma} \varphi(s,t). \tag{2.3.42}$$

COROLLARY. *In the same situation as before,*

$$|\langle f^1,\ldots,f^m\rangle| \leqslant c^{|S|}\prod_{i=1}^{m} M_i \tag{2.3.43}$$

with another constant c.

Proof of Corollary. Use the following observation: given $\Gamma\in\mathscr{C}(\mathbb{S})$ it is possible to find $\tilde{S}=\tilde{S}(\Gamma)\subset S$ such that each bond of Γ contains exactly one vertex of $\tilde{S}$ and such that $\Pi_{t\in S\setminus\tilde{S}}n_t(\Gamma)! \leqslant \Pi_{t\in\tilde{S}}n_t(\Gamma)!$. (It is sufficient to find $\tilde{S}$ satisfying the first assumption and then change $\tilde{S}$ by $S\setminus\tilde{S}$ if necessary. Such a construction is possible because all the connected components of Γ are trees.) Now,

$$\sum_{\Gamma\in\tilde{\mathscr{C}}(\mathbb{S})}\varphi(\Gamma)\prod_{t\in S}(n_t(\Gamma)!)^{1/2} \leqslant \sum_{\Gamma\in\tilde{\mathscr{C}}(\mathbb{S})}\varphi(\Gamma)\prod_{t\in\tilde{S}}n_t(\Gamma)!$$

$$\leqslant \sum_{\{n_t\in\mathbb{N},t\in S\}}\sum_{\tilde{S}\subset S}\prod_{t\in\tilde{S}}\Big(\sum_{s\in S:s\neq t}\varphi(t,s)\Big)^{n_t} \leqslant 2^{|S|}(1-\alpha)^{-|S|}. \tag{2.3.44}$$

□

NOTE. The idea of using sums of products over trees in the estimates of semi-invariants is due to Dunea, Souillard and Iagolnitzer who introduced it (see [24]) for spin fields.

Equation (2.3.35) is Malyshev's assumption (5.1) in [8]. It is possible to deduce Lemma 5.3 of [8] (which is a stronger variant of the previous corollary) from our theorem.

2.3.5. *Proof of Theorem* 2.3.4

We start with the case when $c_{n_{S_i}}\equiv 0$ for enough large n_{S_i}. The generalization to the case of arbitrary $\{c_{n_{S_i}}\}$ requires some additional considerations, and we shall return to it later.

For any $n_S\in\mathbb{N}^S$ consider the sets

$$\Omega^t=\Omega^t(n_S)=\{(t,k),k=1,\ldots,n_t\}, t\in S$$

and also the sets

$$\Omega^i=\Omega^i(n_S)=\bigcup_{t\in S_i}\Omega^t$$

and

$$\Omega=\Omega(n_S)=\bigcup_{i=1}^{m}\Omega^i=\bigcup_{t\in S}\Omega^t. \tag{2.3.45}$$

Denote by $\mathscr{V}(n_S)$ the set of all vacuum diagrams on $\Omega(n_S)$ which have no loops inside any $\Omega^t, t\in S$. For each $\omega=(t,k)$ put $t(\omega)=t$ and define, for any diagram $\mathscr{F}\in\mathscr{V}(n_S)$, the value

$$\varphi(\mathscr{F})=\prod_{\{\omega,\tilde{\omega}\}\in\mathscr{F}}\varphi(t(\omega),t(\tilde{\omega})). \tag{2.3.46}$$

Given any graph $\Gamma\in\tilde{\mathscr{C}}(\mathbb{S})$ introduce the quantity

$$R_\Gamma = \sum_{n_S} \sum_{\mathscr{F} \in \mathscr{V}(n_S)} \varphi(\mathscr{F}) \prod_{t \in S} (n_t!)^{-1/2} ((n_t + 1) \cdots (n_t + n_t(\Gamma)))^{1/2}. \tag{2.3.47}$$

We will prove the following two lemmas.

LEMMA 1. *Consider the situation of Theorem 2.3.4 with a finite number of non-vanishing $c_{n_{S_i}}$. Then*

$$|\langle f_1, \ldots, f^m \rangle| \leqslant \prod_{i=1}^{m} M_i \sum_{\Gamma \in \tilde{\mathscr{C}}(\mathbb{S})} \varphi(\Gamma) R_\Gamma. \tag{2.3.48}$$

LEMMA 2. *In the same situation, there is some $c = c(\alpha)$ such that*

$$R_\Gamma \leqslant c^{|S|} \prod_{t \in S} (n_t(\Gamma)!)^{1/2}. \tag{2.3.49}$$

NOTE. It is clear that these two lemmas imply the desired estimate (2.3.41). Lemma 2 is one of the key estimates needed in the proof.

Proof of Lemma 1. Using (2.3.37), (2.3.39) and the multilinearity of semi-invariants we obtain, in the notation of (2.3.38), the estimate

$$|\langle f^1, \ldots, f^m \rangle| \leqslant \prod_1^m M_i \sum_{n_S} \left(\prod_{t \in S} n_t! \right)^{-1/2} |\langle :\sigma_{S_1}^{n_{S_1}}:, \ldots, :\sigma_{S_m}^{n_{S_m}}: \rangle| \tag{2.3.50}$$

i.e. we aim to find a reasonable bound for the quantity

$$\sum_{n_S} \prod_{t \in S} (n_t!)^{-1/2} |\langle :\sigma_{S_1}^{n_{S_1}}:, \ldots, :\sigma_{S_m}^{n_{S_m}}: \rangle|. \tag{2.3.51}$$

The semi-invariants $\langle :\sigma_{S_1}^{n_{S_1}}:, \ldots, :\sigma_{S_m}^{n_{S_m}}: \rangle$ will be estimated using Lemma 3 of §2.3.3. Denote by $\tilde{\mathscr{V}}(n_S)$ the subset of $\mathscr{V}(n_S)$ consisting of all diagrams $\mathscr{F}$ which are connected with respect to the partition $\{\Omega^i\}$ of Ω.

Using the formula (2.3.28) and the inequality (2.3.34) we get

$$|\langle :\sigma_{S_1}^{n_{S_1}}:, \ldots, :\sigma_{S_m}^{n_{S_m}}: \rangle| \leqslant \sum_{\mathscr{F} \in \tilde{\mathscr{V}}(n_S)} \varphi(\mathscr{F}). \tag{2.3.52}$$

Let $\Gamma \in \tilde{\mathscr{C}}(\mathbb{S})$. Denote by $\tilde{\mathscr{V}}_\Gamma(n_S)$ the class of all diagrams $\mathscr{F} \in \tilde{\mathscr{V}}(n_S)$ such that for any $\{s,t\} \in \Gamma$ there is some bond $\{(s,k),(t,l)\}$ in $\mathscr{F}$. Given such $\mathscr{F}$, assign one particular choice of $\{(s,k),(t,l)\}$ to each $\{s,t\} \in \Gamma$. The definition of $\tilde{\mathscr{V}}_\Gamma(n_S)$ shows that

$$\tilde{\mathscr{V}}(n_S) \subset \bigcup_{\Gamma \in \tilde{\mathscr{C}}(\mathbb{S})} \tilde{\mathscr{V}}_\Gamma(n_S)$$

and therefore

$$|\langle :\sigma_{S_1}^{n_{S_1}}:, \ldots, :\sigma_{S_m}^{n_{S_m}}: \rangle| \leqslant \sum_{\Gamma \in \tilde{\mathscr{C}}(\mathbb{S})} \sum_{\mathscr{F} \in \tilde{\mathscr{V}}_\Gamma(n_S)} \varphi(\mathscr{F}). \tag{2.3.53}$$

Remove from $\mathscr{F}$ all the bonds $\{(s, k_{s,t}), (t, l_{s,t})\}$ assigned to the bonds $\{s,t\}$ of Γ and denote the resulting diagram by $\mathscr{F}'$. Clearly $\mathscr{F}' \in \mathscr{V}(n'_S)$ where $n'_t = n_t - n_t(\Gamma), t \in S$

For a fixed Γ and fixed $\{(s, k_{s,t}), (t, l_{s,t})\}$ the correspondence $\mathscr{F} \in \tilde{\mathscr{V}}_\Gamma(n_S)$

$\Leftrightarrow \mathscr{F}' \in \mathscr{V}(n'_S)$ is one-to-one and there are

$$\prod_{t \in S : n_t(\Gamma) > 0} (n_t(n_t - 1) \cdots (n_t - n_t(\Gamma) + 1)$$

$$= \prod_{t \in S : n_t(\Gamma) > 0} ((n'_t + 1)(n'_t + 2) \cdots (n'_t + n_t(\Gamma))) \tag{2.3.54}$$

ways of choosing the values $\{\{k_{t,s}, l_{t,s}\}\}$. Obviously,

$$\varphi(\mathscr{F}) = \varphi(\Gamma)\varphi(\mathscr{F}'). \tag{2.3.55}$$

We substitute this relation into (2.3.53) and then into (2.3.50). Taking account of (2.3.54) and changing the notation $n'_t \to n_t$ we obtain the desired inequality (2.3.48). □

Proof of Lemma 2. We will use a method due to Malyshev [8], p. 18, in a slightly elaborated form. Introduce the following definitions.

DEFINITION 1. Let S be a finite set. Any sequence

$$P = ((s_i, t_i) \in S \times S;\ i = 1, \ldots, n) \tag{2.3.56}$$

will be called a *path* (with steps (s_i, t_i)) if $s_i \neq t_i$ and $t_{i-1} = s_i$ for each $i = 2, \ldots, n$. Denote by $\operatorname{supp} P = \cup_{i=1}^{n} \{s_i\} \cup \{t_n\}$ the support of a given path P, by $n(P) = n$ its length.

Further denote by $s(P) = s_1$ the starting point of a path, by $t(P) = t_n$ its terminal point.

Identify the set S with the set of integers $\{1, \ldots, |S|\}$ as follows.

DEFINITION 2. A path system on S is a sequence $\mathbb{P} = (P_1, \ldots, P_l)$ of paths satisfying the following conditions:

(a) There is no $i = 1, \ldots, l$ such that

$$t(P_i) \in \bigcup_{j > i} \operatorname{supp} P_j. \tag{2.3.57}$$

(b) for each $i = 1, \ldots, l$; the smallest element of $\bigcup_{j=1}^{l} \operatorname{supp} P_j$ is $s(P_i)$.

Fix some $n_S = \{n_t, t \in S\}$. Denote by $n = \frac{1}{2}\Sigma_{t \in S} n_t$ the number of bonds in $\mathscr{F} \in \mathscr{V}(n_S)$. For any diagram $\mathscr{F} \in \mathscr{V}(n_S)$ arrange the bonds of $\mathscr{F}$ into a sequence
arrange the bonds of $\mathscr{F}$ into a sequence

$$((s_1, m_1), (t_1, n_1)), \qquad ((s_2, m_2), (t_2, n_2)), \ldots \tag{2.3.58}$$

of ordered bonds $((s_i, m_i), (t_i, n_i))$ defined successively as follows:

(a) We choose $(s_1, m_1) = (s, 1)$ with the smallest possible s, and look for $a(t_1, n_1)$ which is in the same bond as $(s, 1)$.

(b) If some (t_{i-1}, k) has not yet been used in the steps previous to the ith step we put $(s_i, m_i) = (t_{i-1}, k)$ with the smallest possible free k. If all (t_{i-1}, k) have already been used we switch to the smallest possible s with free elements (s,k) and take $(s_i,m_i)=(s,k)$ with the smallest possible free k. Again, we look for a (t_i,n_i) which is in the same bond (s_i,m_i).

It is clear that the sequence $(s_1,t_1),(s_2,t_2),\ldots$ induced by the sequence (2.3.58) defines some path system $\mathbb{P}=\mathbb{P}(\mathscr{F})$. It is also clear that the number of possible diagrams $\mathscr{F}$ with the same $\mathbb{P}(\mathscr{F})$ does not exceed the value

$$\prod_{t\in S} n_t!! \tag{2.3.59}$$

where $n!!$ is defined as

$$n!!=n(n-2)\cdots 2,\quad n \text{ even } (0!!=1)$$
$$n!!=n(n-2)\cdots 1,\quad n \text{ odd.}$$

We will now estimate R_Γ using sums over path systems. Given $n_S=\{n_t,t\in S]$ denote by $\mathscr{P}(n_S)$ the family of all path systems $\mathbb{P}=(P_1,\ldots,P_l)$ such that for each $t\in S$,

$$n_t=n_t(\mathbb{P})=|\{(s_j^i,t_j^i)\in P_i,\, i=1,\ldots,l: t\in\{s_j^i,t_j^i\}\}|. \tag{2.3.60}$$

Notice that if n_i are the lengths of paths P_i then

$$\sum_{t\in S} n_t = 2\sum_{i=1}^{l} n_i \tag{2.3.61}$$

for any $\mathbb{P}=(P_1,\ldots,P_l)\in\mathscr{P}(n_S)$.

Further, it is clear from our construction of $\mathbb{P}(\mathscr{F})$ that $\mathbb{P}\in\mathscr{P}(n_S)$ if there is $\mathscr{F}\in\mathscr{V}(n_S)$ such that

$$\mathbb{P}=\mathbb{P}(\mathscr{F}).$$

For any path system $\mathbb{P}=(P_1,P_2,\ldots,P_l)$ denote by

$$\varphi(\mathbb{P})=\prod_{i=1}^{l}\prod_{(s_j^i,t_j^i)\in P_i}\varphi(s_j^i,t_j^i). \tag{2.3.62}$$

By (2.3.59),

$$\sum_{\mathscr{F}\in\mathscr{V}(n_S)}\varphi(\mathscr{F})\leqslant\sum_{\mathbb{P}\in\mathscr{P}(n_S)}\varphi(\mathbb{P})\prod_{t\in S} n_t!!. \tag{2.3.63}$$

We will use the following elementary inequality (easily proved by induction):

$$n!!\leqslant(2n)^{1/4}(n!)^{1/2},\qquad n>0. \tag{2.3.64}$$

Substituting (2.3.63) and (2.3.64) into (2.3.47) we obtain the following intermediate estimate:

$$R_\Gamma\leqslant\sum_{n_S}\sum_{\mathbb{P}\in\mathscr{P}(n_S)}\varphi(\mathbb{P})\prod_{t\in S:n_t(\Gamma)>0}(2n_t)^{1/4}$$
$$((n_t+1)\cdots(n_t+n_t(\Gamma)))^{1/2}. \tag{2.3.65}$$

The combinatorial factors in this relation can be estimated as follows.

LEMMA 3. *For each $\beta > 1$, each $k \in \mathbb{N}$ and $n \in \mathbb{N}$ the following inequality holds:*

$$(n+1)(n+2)\cdots(n+k) < k!\,\beta^n\left(\frac{\beta}{\beta-1}\right)^k. \tag{2.3.66}$$

Proof. Write $\beta = 1+\gamma, \gamma > 0$. Obviously, from the binomial formula for $(1+\gamma)^{n+k}$ we obtain

$$\binom{n+k}{k} < \gamma^{-k}(1+\gamma)^{n+k}$$

which is (2.3.66). □

Using (2.3.66) and the additional elementary estimate $n^{1/4} \leqslant c\beta^n, c = c(\beta), \beta > 1$ we can now estimate the factors in (2.3.65) as follows:

$$(2n_t)^{1/4}((n_t+1)\cdots(n_t+n_t(\Gamma)))^{1/2} < (n_t(\Gamma)!)^{1/2}\,\beta^{n_t}c^{n_t(\Gamma)} \tag{2.3.67}$$

with some $c = c(\beta), \beta > 1$. We obtain the inequality

$$\mathrm{R}_\Gamma \leqslant \sum_{n_S}\sum_{\mathbb{P}\in\mathscr{P}(n_S)} \varphi(\mathbb{P})\,\beta^{\sum_{t\in S} n_t} c^{|S|-1}\prod_{t\in S}(n_t(\Gamma)!)^{1/2}. \tag{2.3.68}$$

To prove (2.3.49) it remains to show that, for a suitable $\beta > 1$,

$$\sum_{\mathbb{P}\in\mathscr{P}(S)} \varphi(\mathbb{P})\beta^{2n(\mathbb{P})} < c'^{|S|} \tag{2.3.69}$$

where $\mathscr{P}(S)$ is the family of all path systems on $S, n(\mathbb{P}) = \frac{1}{2}\Sigma_{t\in S} n_t(\mathbb{P})$ and $c' > 0$ is a suitable constant.

Notice that given the sequence of all $t_j - s_j$ corresponding to the successive steps (s_j, t_j) of $\mathbb{P}$ (we recall that we identify S with $\{1, 2, \ldots, |s|\}$), we can reconstruct the path system $\mathbb{P}$ assuming that all the numbers $n_t(\mathbb{P})$ are also known.

In fact, it suffices to determine the value of s_{j+1} in the next step (s_{j+1}, t_{j+1}) at each stage of the construction. Denote by $n_t^j(\mathbb{P})$ the number of the steps $(s_{j'}, t_{j'})$, $j' \leqslant j$ such that $t \in \{s_{j'}, t_{j'}\}$. If $n_{t_j}^j(\mathbb{P}) < n_{t_j}(\mathbb{P})$ we choose $s_{j+1} = t_j$. In the case $n_{t_j}^j(\mathbb{P}) = n_{t_j}(\mathbb{P})$ we take $s_{j+1} = s$ with a smallest possible s such that $n_s^j(\mathbb{P}) < n_s(\mathbb{P})$ (thus starting a new path in $\mathbb{P}$). Using this observation and the condition (2.3.40) we find that

$$\sum_{\mathbb{P}\in\mathscr{P}(n_S)} \varphi(\mathbb{P})\beta^{2n(\mathbb{P})} = \sum_{\mathbb{P}\in\mathscr{P}(n_S)}\prod_{t\in S}\left(\prod_{j:S_j=t}\varphi(s_j,t_j)\beta^2\right)$$
$$\leqslant (\alpha\beta^2)^{2n}, \qquad n = \tfrac{1}{2}\sum_{t\in S} n_t. \tag{2.3.70}$$

Now sum over all n_S. We obtain

$$\sum_{\mathbb{P}\in\mathscr{P}(S)} \varphi(\mathbb{P})\beta^{2n(\mathbb{P})} \leqslant \sum_{n_S}(\alpha\beta^2)^{2n} = \left(\frac{1}{1-\alpha\beta^2}\right)^{|S|} \tag{2.3.71}$$

which is (2.3.69) if $\alpha\beta^2 < 1$. We obtain, therefore, also the statement (2.3.49) of Lemma 2. □

It remains to investigate the general case of arbitrary $\{c_{n_{S_i}}\}$, not necessarily vanishing for large n_{S_i}. We need the following result.

LEMMA 4. *Under the assumptions of the Theorem, there is a constant $K > 0$ such that*

$$\langle f^1 \cdots f^m \rangle \leqslant K^{|S|} \prod_{i=1}^{m} \langle (f^i)^2 \rangle^{1/2}. \tag{2.3.72}$$

Proof. In addition to the matrix $B = \{b_{s,t};\ s,t \in S\}$ where $b_{s,t} = \langle \sigma_s \sigma_t \rangle$ consider also the matrix $\hat{B} = \{\hat{b}_{s,t}; s,t \in S\}$ such that

$$\hat{b}_{s,t} = b_{s,t}, \qquad s,t \in S_i,$$

$$\hat{b}_{s,t} = 0, \qquad s \in S_i, \qquad t \in S_j, \qquad i \neq j.$$

The crucial fact here is that

$$(1 + \alpha)\hat{B} > B \tag{2.3.73}$$

(in the sense of inequalities between two positive-definite matrices). Assume that (2.3.73) is true, for a moment. Using this we get the estimates

$$\begin{aligned}
\langle f^1 \cdots f^m \rangle &= ((2\pi)^{|S|} \det B)^{-1/2} \int_{\mathbb{R}^S} \prod_{i=1}^{m} f^i(x_{S_i}) \exp(-\tfrac{1}{2}(B^{-1}x, x))\, dx \\
&\leqslant ((2\pi)^{|S|} \det B)^{-1/2} \int_{\mathbb{R}^S} \prod_{i=1}^{m} f^i(x_{S_i}) \exp\left(- \frac{1}{2(1+\alpha)} (\hat{B}^{-1}x, x)\right) dx \\
&\leqslant ((2\pi)^{|S|} \det B)^{-1/2} \left(\int_{\mathbb{R}^S} \prod_{1}^{m} (f^i(x_{S_i}))^2 \exp(-\tfrac{1}{2}(\hat{B}^{-1}(x,x))\, dx) \right)^{1/2} \times \\
&\times \left(\int_{\mathbb{R}^S} \exp\left(- \frac{1-\alpha}{2(1+\alpha)} (\hat{B}^{-1}x, x) \right) dx \right)^{1/2} = K^{|S|} \prod_{i=1}^{m} \langle (f^i)^2 \rangle^{1/2}
\end{aligned}$$

which proves this lemma.

Proof of (2.3.73). Remember the following Levy–Desplangues theorem:
If $D = \{d_{i,j}; i,j = 1, \ldots, n\}$ is a matrix such that

$$|d_{i,i}| > \sum_{j:j \neq i} |d_{i,j}|, \qquad i = 1, \ldots, n$$

then $\det D \neq 0$.

Suppose, on the contrary, that there is some non-zero vector $x = (x_1, \ldots, x_n)$ such that $Dx = 0$. Choose k such that $|x_k| = \max_j |x_j|$. Then

$$|(Dx)_k| = |d_{k,k} x_k + \sum_j \mathrm{d}_{k,j} x_j| > 0$$

which is a contradiction. □

Consider now the matrix $B_\gamma = \gamma\hat{B} - B$. Because $\hat{B}$ is positive-definite we see that also B_γ is positive-definite for a large γ. Further, it is easy to verify (from (2.3.35)) the assumptions of the Levy– Desplangues theorem for any $\gamma \in [(1 + \alpha), \infty)$. Thus $\det\ B_\gamma \neq 0$ for any $\gamma \in [(1 + \alpha), \infty)$ and therefore we obtain that each matrix $B_\gamma, \gamma \in [(1 + \alpha), \infty)$ is necessarily positive-definite. This proves (2.3.73). □

For any $n \in \mathbb{N}$ now define (compare (2.3.37))

$$f^i_{(n)}(\sigma_{S_i}) = \sum_{n_{S_i}\in \mathbb{N}^{S_i}:\Sigma n_t \leqslant n} c_{n_{S_i}}(n_{S_i}!)^{-1/2}:\sigma_{S_i}^{n_{S_i}}:, \quad i = 1,\ldots,m.$$

It is easy to see from Lemma 4 that

$$\langle f^1\cdots f^m\rangle = \lim_{n\to\infty}\langle f^1_{(n)}\cdots f^m_{(n)}\rangle.$$

A similar relation is obtained for any moment $\langle f^{i_1}\cdots f^{i_k}\rangle$, $1 < i_1 < \cdots < i_k \leqslant m$ and therefore for any semi-invariant. This completes the proof of theorem.

NOTE 1. We cannot apply the relation (2.3.50) directly in the general case because the series (2.3.37) need not to converge in L^p sense, $p > 2$. The corresponding statement in [8], relation (5.14) is erroneous. Lemma 4 provides a way of filling this gap.

NOTE 2. Notice that the condition $\langle \sigma_t^2\rangle = 1$ was used only in the proof of Lemma 4. So the estimates (2.3.41), (2.3.43) are true also for $\langle\sigma_t^2\rangle \neq 1$ if the series (2.3.37) has a finite number of non-zero terms. Recently Malyshev and Minlos [12] have proposed an alternative way of extending the estimates of semi-invariants to the case of infinite series (2.3.37), not using the condition $\langle\sigma_t^2\rangle = 1$ but using another set of *a priori* conditions $\langle (f^1\cdots f^j)^{2+\varepsilon}\rangle < \infty$ for some $\varepsilon > 0$ and all $j = 1,\ldots,m$.

2.3.6. *Other Formulations and Generalizations of Theorem* 2.3.4

The condition (2.3.35) is rather restrictive, being invalid in many interesting situations. We will show in this section that it is possible, by changing the variables and suitably modifying the notions of a correlation decay estimate φ and a Wick series (2.3.37), to replace (2.3.35) by a condition (2.3.76). This latter condition can be verified, for example, for the Gaussian fields considered in §2.2, if the distances between S_i are sufficiently large (see §2.3.7).

THEOREM 1. *Let* $\{\sigma_t,\ t\in S\}$, *S finite, be a Gaussian family of* ($\mathbb{R}$-valued) *random variables. For any* $t\in S$ *and* $\Lambda\subset S$ *denote by* $\sigma_t^{(\Lambda)}$ *the Gaussian variable*

$$\sigma_t^{(\Lambda)} = \sigma_t - E(\sigma_t|\{\sigma_s, s\in\Lambda\}) \tag{2.3.74}$$

where $E\,((\cdot)|(\cdot))$ *denotes the conditional expectation of* σ_t, *given* $\{\sigma_s, s\in\Lambda\}$. *Assume that the partition* $\mathbb{S}$ *and the functions* f^i; $i = 1,\ldots,m$ *are the same as in Theorem* 2.3.4. *Assume that the points of* S_i *are enumerated in some way:* $S_i = \{t_{i,1},\ldots,t_{i,j_i}\}$, $j_i = |S_i|$. *Denote by* $S_{i,j} = \{t_{i,1},\ldots,t_{i,j}\}$; $j = 1,\ldots,j_i$. *Suppose that some positive function* φ *is given such that the estimate*

$$|\langle\sigma_t^{(\Lambda)}\sigma_s^{(\Lambda)}\rangle| \leqslant \varphi(t,s)(\langle(\sigma_t^{(\Lambda)})^2\rangle\langle(\sigma_s^{(\bar\Lambda)})^2\rangle)^{1/2} \tag{2.3.75}$$

holds for any $\Lambda = S_{i,j}, t = t_{i,j+1}$, $i = 1,\ldots,m$, $j = 1,\ldots,j_i - 1$ *and any* $\bar\Lambda = S_{\bar\imath,\bar\jmath}$, $s = t_{\bar\imath,\bar\jmath+1}$, $\bar\imath = 1,\ldots,m$, $\bar\jmath = 1,\ldots,\bar\jmath_{\bar\imath} - 1$. *Suppose further that for each* $i = 1,\ldots,m$ *and each* $t\in S_i$,

$$\sum_{s\in S\setminus S_i}\varphi(t,s) \leqslant \alpha < 1. \tag{2.3.76}$$

Then, using the notations (2.3.40), (2.3.42) *and with the same constant as in* (2.3.41), *the following estimate holds*:

$$|\langle f^1,\dots,f^m\rangle| \leqslant \prod_1^m \langle (f^i)^2\rangle^{1/2} c^{|S|} \sum_{\Gamma\in\mathscr{G}(\mathbf{S})} \varphi(\Gamma) \prod_{t\in S} (n_t(\Gamma)!)^{1/2} \tag{2.3.77}$$

In particular,

$$|\langle f^1,\dots,f^m\rangle| \leqslant \prod_1^m \langle (f^i)^2\rangle^{1/2} c^{|S|} \tag{2.3.78}$$

with the same constant as in (2.3.43).

Proof. For each S_i, apply the usual orthonormalization procedure to the ordered Gaussian variables $\{\sigma_t, t\in S_i\}$. Denote the resulting variables as

$$\xi_t = N_t^{-1}(\sigma_t - E(\sigma_t|\{\sigma_s, s\in\Lambda\})) \quad \text{where} \quad \Lambda = S_{i,j}, t = t_{i,j+1} \tag{2.3.79}$$

where N_t is chosen such that $\langle \xi_t^2\rangle = 1$. By (2.3.75) we get

$$|\langle \xi_t \xi_s\rangle| \leqslant \hat{\varphi}(t,s), \tag{2.3.80}$$

where

$$\begin{aligned}\hat{\varphi}(t,s) &= \varphi(t,s), \qquad t\in S_i, \qquad s\in S_j, \qquad i\neq j\\ &= 0 \qquad s,t\in S_i\end{aligned}$$

Clearly, by using (2.3.76) we obtain the condition (2.3.35) for the correlation decay estimate $\hat{\varphi}$.

Noting that the functions f^i are measurable with respect to the new variables $\{\zeta_t, t\in S_i\}$, we can now apply Theorem 2.3.4, with $M_i = \langle (f^i)^2\rangle^{1/2}$.

In fact, $(n_{S_i}!)^{-1/2}{:}\zeta_{S_i}^{n_{S_i}}{:}$ is an orthonormal basis of the corresponding L^2 space of quadratic integrable functions which are measurable with respect to $\{\xi_t, t\in S_i\}$. Therefore,

$$c_{n_{S_i}} = \langle f^i (n_{S_i}!)^{-1/2}{:}\xi_{S_i}^{n_{S_i}}{:}\rangle.$$

So (2.3.41) gives (2.3.77) and (2.3.43) gives (2.3.78). □

A slight change in the function φ permits us to free ourselves of the multipliers $(\Pi n_t(\Gamma)!)^{1/2}$ in the main estimate (2.3.77).

THEOREM 2. *Consider the situation described in Theorem* 1. *Suppose that the functions* φ *can be written as*

$$\varphi(t,s) = \tilde{\varphi}(t,s)\,\tilde{\tilde{\varphi}}(t,s) \tag{2.3.81}$$

where $\tilde{\varphi}$ *satisfies the estimate* (2.3.76) *and the function* $\tilde{\tilde{\varphi}}$ *is such that the following holds : there is a constant* $K > 0$ *such that for any* $F \subset \{1,2,\dots,m\}$ *and any choice of* $t_i\in S_i$, $i\in F$ *and* $s_j\in S_j$, $j\notin F$,

$$\prod_{i\in F} \tilde{\tilde{\varphi}}(s_j, t_i) \leqslant K^n (n!)^{-1}, \; n = |F|. \tag{2.3.82}$$

Let $f^1,\dots,f^m$ *be as in Theorem* 1. *Then*

$$|\langle f^1,\dots,f^m\rangle| \leqslant \sum_1^m \langle (f^i)^2\rangle^{1/2} c^{|S|} K^{m-1} \sum_{\Gamma\in\mathscr{F}(\mathbf{S})} \tilde{\varphi}(\Gamma) \tag{2.3.83}$$

where $\tilde{\varphi}(\Gamma)$ *is defined by* (2.3.42), for $\varphi = \tilde{\varphi}$. □

NOTE. As an example, when (2.3.82) is valid assume that the condition

$$\sum_{i\in F} \tilde{\tilde{\varphi}}(s_j, t_i) \leqslant K, \qquad j \notin F, \qquad s_j \in S_j, \qquad t_i \in S_i \tag{2.3.84}$$

is satisfied. Then, obviously,

$$\prod_{i\in F} \tilde{\tilde{\varphi}}(s_j, t_i) n! \leqslant \left(\sum_{i\in F} \tilde{\tilde{\varphi}}(s_j, t_i) \right)^n \leqslant K^n$$

which is (2.3.82).

More concrete examples of functions φ, $\tilde{\varphi}$, $\tilde{\tilde{\varphi}}$ will be shown below.

Proof of Theorem 2. It is clear from (2.3.82) that

$$\varphi(\Gamma) \prod_{t\in S} (n_t(\Gamma)!)^{1/2} = \tilde{\varphi}(\Gamma) \left(\prod_{t\in S} (n_t(\Gamma)!) \prod_{s\in S:\{s,t\}\in\Gamma} \tilde{\tilde{\varphi}}(s,t) \right)$$

$$\leqslant \tilde{\varphi}(\Gamma) \left(\prod_{t\in S} K^{n_t(\Gamma)} \right)^{1/2} = \tilde{\varphi}(\Gamma) K^{m-1}. \qquad \square$$

The following result gives more concrete information about the choice of the function φ.

Theorem 3. *Let* $\{\sigma_t, t \in W\}$, $W \subset \mathbb{Z}^\nu$ *be a Gaussian family of random variables such that*

$$\langle \sigma_t \rangle = 0, \qquad \langle \sigma_t^2 \rangle = 1, \qquad t \in W.$$

For each finite $\Lambda \subset W$ *and each* $t \in W \setminus \Lambda$ *write*

$$E(\sigma_t \mid \{\sigma_s, s \in \Lambda\}) = \sum_{s\in\Lambda} a_{t,s}^{\Lambda} \sigma_s \tag{2.3.85}$$

where $a_{t,s}^{\Lambda} \in \mathbb{R}$. *Denote by*

$$B_\Lambda = \{b_{s,t} = \langle \sigma_s \sigma_t \rangle; s, t \in \Lambda\}$$

the covariance matrix. Suppose that for each finite $\Lambda \subset W$,

$$(B_\Lambda x_\Lambda, x_\Lambda) \geqslant B |x_\Lambda|^2, \qquad x_\Lambda \in \mathbb{R}^\Lambda \tag{2.3.86}$$

where $\beta > 0$ *is some constant.*

Let $\mathbb{S}$, f^i, $i = 1, \ldots, m$ *be as in Theorem* 1. *Enumerate again the points of each* S_i *as* $S_i = \{t_{i,1} \ldots, t_{i,j_i}\}$. *Let* α *be a positive function such that* $\alpha(0) = 1$ *and the inequality*

$$|a_{t,s}^{\Lambda}| \leqslant \alpha(|s - t|), s \in \Lambda \tag{2.3.87}$$

holds for the following choices of t, Λ:

(a) *for any set* $\Lambda = \{t_{i,1}, \ldots, t_{i,j}\}, j < j_i$ *and* $t = t_{i,j+1}$

(b) *for any* t, $s \in S$ *and* $\Lambda = \{s\}$ *(there we require, in fact, that* $\langle \sigma_t \sigma_s \rangle \leqslant \alpha(|t - s|)$ *for any* t, $s \in S$). *Suppose that the series* $\Sigma_{t\in\mathbb{Z}^\nu} \alpha(|t|)$ *converges. Define the function*

$$\varphi(t,s) = \sum_{t'\in \mathbf{Z}^\nu} \sum_{s'\in \mathbf{Z}^\nu} \alpha(|t-t'|)\alpha(|t'-s'|)\alpha(|s-s'|). \tag{2.3.88}$$

Suppose that there is a sufficiently large integer d (with respect to α) such that

$$\operatorname{dist}(S_i, S_{i'}) \geqslant d \qquad \textit{whenever } i \neq i'. \tag{2.3.89}$$

Then

$$|\langle f^1,\ldots,f^m\rangle| \leqslant \prod_1^m \langle |f^i)^2\rangle^{1/2} c^{|S|} \sum_{\Gamma\in\tilde{\mathscr{G}}(S)} \varphi(\Gamma) \prod_{t\in S} (n_t(\Gamma)!)^{1/2} \tag{2.3.90}$$

where $c > 0$ *depends on* β, α, d *only. Suppose moreover that the series*

$$\sum_{t\in \mathbf{Z}^\nu} (\alpha(|t|))^{1/2}$$

converges. Then, if (2.3.89) holds with a sufficiently large d,

$$|\langle f^1,\ldots,f^m\rangle| \leqslant \prod_1^m \langle (f^i)^2\rangle^{1/2} c^{|S|} \sum_{\Gamma\in\tilde{\mathscr{G}}(S)} (\varphi(\Gamma))^{1/2} \tag{2.3.91}$$

with another constant $c = c(\beta, \alpha, d)$.

Proof. The relations (2.3.85), (2.3.87) imply that (for the above choices of Λ, t)

$$|\langle \sigma_t^{(\Lambda)}\sigma_s^{(\bar\Lambda)}\rangle| \leqslant \sum_{t'\in\Lambda\cup\{t\}} \sum_{s'\in\bar\Lambda\cup\{s\}} \alpha(|t-t'|)|\langle \sigma_{t'},\sigma_{s'}\rangle|\alpha(|s-s'|) \leqslant$$

$$\leqslant \sum_{t'\in\Lambda\cup\{t\}} \sum_{s'\in\bar\Lambda\cup\{s\}} \alpha(|t-t'|)\alpha(|t'-s'|)\alpha(|s'-s|). \tag{2.3.92}$$

From (2.3.86) we get, for any Λ, t, the inequality

$$\langle \sigma_t^{(\Lambda)}\sigma_t^{(\Lambda)}\rangle = \left\langle \left(\sigma_t - \sum_{t'\in\Lambda} a^\Lambda_{t,t'}\sigma_{t'}\right)\left(\sigma_t - \sum_{t'\in\Lambda} a^\Lambda_{t,t'}\sigma_{t'}\right)\right\rangle$$

$$\geqslant \beta\left(1 + \sum_{t'\in\Lambda} (a_{t,t'})^2\right) \geqslant \beta. \tag{2.3.93}$$

So (2.3.75) holds, with φ altered to $\beta^{-1}\varphi$, φ being defined by (2.3.88).

We require (2.3.76) to hold. Choose a sufficiently large $d\in\mathbb{N}$. Note that if $|s-t| \geqslant 3d$ then either $|s-s'| \geqslant d$ or $|s'-t'| \geqslant d$ or $|t'-t| \geqslant d$, i.e.

$$\sum_{t:|t-s|\geqslant 3d} \varphi(t,s) \leqslant 3\left(\sum_{t\in\mathbf{Z}^\nu} \alpha(|t|)\right)^2 \sum_{t:|t|\geqslant d} \alpha(|t|). \tag{2.3.94}$$

But this implies (2.3.76) for d sufficiently large. By Theorem 2, we obtain the estimate (2.3.90).

To prove (2.3.91) also put

$$\tilde\varphi(s,t) = \tilde{\tilde\varphi}(s,t) = \beta^{-1/2}\varphi(s,t)^{1/2}.$$

As in (2.3.94), using the obvious inequality (for $x_i \geqslant 0$) $(\Sigma x_i)^{1/2} \leqslant \Sigma x_i^{1/2}$ we obtain that

$$\sum_{t:|t-s| \geqslant 3d} \tilde{\varphi}(s,t) \leqslant 3\beta^{-3/2} \left(\sum_{t \in \mathbb{Z}^\nu} \alpha(|t|)^{1/2} \right)^2 \sum_{t:|t| \geqslant d} \alpha(|t|)^{1/2}.$$

So $\tilde{\varphi}$ satisfies (2.3.76) for d large enough and $\tilde{\tilde{\varphi}}$ satisfies (2.3.84). Thus Theorem 2 implies the estimate (2.3.91). □

2.3.7. *Applications to the Gaussian fields of* §2.2

THEOREM. *Let* $\{\sigma_t \in \mathbb{R}^k, t \in W\}$, $W \subseteq \mathbb{Z}^\nu$, $k \in \mathbb{N}$ *be a Gaussian vector field with probability distribution* μ_a, $a \in X(W^c)$ *bounded – see* §2.2.3, Note 2. *Let* S *be a finite subset of* W. *Let* $\mathbb{S} = \{S_i, i = 1, \ldots m\}$ *be a partition of* S *such that* $\operatorname{dist}(S_i, S_{i'}) \geqslant d$ *for any* $i \neq i'$ *where* d *is a sufficiently large integer (depending on* $c_{(2.2.5)}$ *only). Let some square-integrable functions* f^i of $\sigma_{S_i} = \{\sigma_t, t \in S_i\}$ *be given. Then the following estimate holds*:

$$|\langle f^1, \ldots, f^m \rangle| \leqslant c^{|S|} \prod_1^m \langle (f^i)^2 \rangle^{1/2} \sum_{\Gamma \in \tilde{\mathscr{C}}(\mathbb{S})} q^{d(\Gamma)} \tag{2.3.95}$$

where $\tilde{\mathscr{C}}(\mathbb{S})$ *is the same as in Theorem* 2.3.4, *and where* $c > 0$ *and* $0 < q < 1$ *are some constants depending only on the constant* $c_{(2.2.5)}$ *(and the interaction range* r). *The function* $d(\cdot)$ *is defined as follows*:

$$d(\Gamma) = \sum_{\{s,t\} \in \Gamma} |s - t|. \tag{2.3.96}$$

Proof. Note that only scalar fields were considered earlier in §2.3. We therefore first identify the given vector field $\{\sigma_t \in \mathbb{R}^k, t \in W\}$, $W \subset \mathbb{Z}^\nu$ with an appropriate scalar field on some subset of $\mathbb{Z}^{\nu+1}$, as follows. Consider the sets

$$\hat{S}_i = \{(s,j), s \in S_i, j = 1, \ldots, k\}, \hat{S} = \bigcup_{i=1}^m \hat{S}_i. \tag{2.3.97}$$

Consider the Gaussian family of random variables $\{\sigma_{\hat{t}}, \hat{t} \in \hat{S}\}$ defined, for $\hat{t} = (t,j)$, as the jth coordinate of σ_t. Then the f^i can be interpreted as functions of $\{\sigma_{\hat{t}}, \hat{t} \in \hat{S}_i\}$. It is clear that

$$\operatorname{dist}(S_i, S_{i'}) = \operatorname{dist}(\hat{S}_i, \hat{S}_{i'}). \tag{2.3.98}$$

From (2.2.17) we obtain uniform boundedness of the maximal eigenvalues of all the matrices $C_\Lambda^{\tilde{\Lambda}}$, $W \supset \tilde{\Lambda} \supset \Lambda$, but this proves (2.3.86) because of the relation $C_\Lambda^{\tilde{\Lambda}} = (B_\Lambda^{\tilde{\Lambda}})^{-1}$. Enumerate the points $\hat{t}=(t,j)$ of $\hat{S}_i$ in a lexicographic way, using the given enumeration of S_i. We therefore need to consider, in case (a) of (2.3.87), the following choices of $\hat{t}$, $\hat{\Lambda}$ only:

$$\hat{\Lambda} = \{(t_{i,1}, 1), \ldots, (t_{i,1}, k), \ldots, (t_{i,j}, 1), \ldots, (t_{i,j}, l)\}\ j < j_i, l \leqslant k$$
$$\hat{t} = (t_{i,j}, l+1) \text{ (or } \hat{t} = (t_{i,j+1}, 1) \quad \text{if} \quad l = k).$$

Applying Theorem 2.2.2 and Proposition 2.2.3 we see that it is possible to take

$$\alpha(|\hat{t} - \hat{s}|) = cq^{|t-s|} \tag{2.3.99}$$

in (2.3.87), with $c > 0$ and $0 < q < 1$ depending on $c_{(2.2.5)}$ (and r) only. Then we define φ by (2.3.88). It is elementary to show that for any $q' > q_{(2.3.99)}$ there is some $c' > 0$ depending only on $q', q_{(2.3.99)}$, and $c_{(2.3.99)}$ such that

$$\varphi(\hat{t}, \hat{s}) \leqslant c' \, q'^{|t-s|}. \tag{2.3.100}$$

By (2.3.91), we obtain the desired estimate (2.3.95) (with new values of the constants q, c), but with graphs $\hat{\Gamma}$ on the set $\hat{S}$ rather than on S.

By summing up all the possible $\hat{\Gamma} \in \tilde{\mathscr{C}}(\hat{\mathbb{S}})$ with the same projection

$$\Gamma = \{\{s,t\}:\{\hat{s},\hat{t}\} \in \hat{\Gamma} \text{ for some } \hat{s} = (s,j), \hat{t} = (t,j')\}$$

we obtain an additional factor k^{m-1}, if we want to sum over $\Gamma \in \tilde{\mathscr{C}}(\mathbb{S})$ only. By changing the constant c we thus prove the relation (2.3.95) even in the general case $k \neq 1$. □

2.3.8. *Cluster Expansions of Perturbed Gaussian Fields*

This section is the culmination of the whole of §2. We derive the expressions of partition functions of perturbed models, enabling us to estimate the 'surface tension' terms of these partition functions. These estimates will play a crucial role in §4 below (in §4.4, where contour models are investigated).

Consider a vector Gaussian field μ on X, as constructed in §2.2 (see Note 2, §2.2.3), and also the conditioned fields μ_a, $a \in X(\partial\Lambda^c)$, $\Lambda \subset \mathbb{Z}^\nu$. ($\partial \equiv \partial_r$, $r = r_{(2.2.1)}$) We will construct, in the forthcoming Definitions 1 and 2, a perturbation of such Gaussian fields, of the following type.

DEFINITION 1. Let $\mathfrak{A}$ be an arbitrary system of objects denoted by symbols A, such that to each A, a finite set supp $A \subset \mathbb{Z}^\nu$ (the support of A) is assigned. Denote by $N(A)$ the number of $A' \in \mathfrak{A}$ such that supp A' = supp A.

NOTE. A very simple example is $\mathfrak{A} = \text{Fin}(\mathbb{Z}^\nu)$; the set of all finite subsets of $\mathbb{Z}^\nu$, with supp $A = A$. More complex systems $\mathfrak{A}$ (of 'contours') will be used in §4.4.

DEFINITION 2. Suppose that some symmetric relation)((compatibility) is given on $\mathfrak{A}$. Suppose that the relation)(is such that for some integer d,

$$\text{dist}(\text{supp}\, A, \text{supp}\, A') \geqslant d \Rightarrow A)(A'; A, A' \in \mathfrak{A}. \tag{2.3.101}$$

We say that a subsystem $\mathscr{A} \subset \mathfrak{A}$ is compatible if $A)(A'$ for each $A, A' \in \mathscr{A}$.

DEFINITION 3. Let a group $\{T_t, t \in \mathbb{Z}^\nu\}$ of transformations (shifts) be given on $\mathfrak{A}$, such that $\text{supp}\, T_t A = \text{supp}\, A + t$ for each $A \in \mathfrak{A}$, $t \in \mathbb{Z}^\nu$ and $A)(A' \;\langle=\rangle\; T_t A)(T_t A'; t \in \mathbb{Z}^\nu$, $A, A' \in \mathfrak{A}$. Then we will say that the system $\mathfrak{A}$ is translation-invariant (with respect to the given system $\{T_t\}$).

DEFINITION 4. Let $\Lambda \subset \tilde{\Lambda}$ be finite subsets of $\mathbb{Z}^\nu$, let $x_{\partial\tilde{\Lambda}} = \tilde{a} \in X(\partial\tilde{\Lambda}^c)$ be a boundary condition. Let $\mathfrak{A}(\Lambda) \subset \mathfrak{A}$ be such that supp $A \subset \tilde{\Lambda} \cup \partial\tilde{\Lambda}^c$ and supp $A \cap \Lambda \neq \emptyset$ for each $A \in \mathfrak{A}(\Lambda)$. Let $\mathfrak{F} = \{f_A, A \in \mathfrak{A}(\Lambda)\}$ be a system of measurable functions depending on $x_A \in X(\text{supp}\, A)$. Denote by $\hat{\mathfrak{A}}(\Lambda)$ the family of all compatible system $\mathscr{A} \subset \mathfrak{A}(\Lambda)$. Define the following partition function:

$$Z(\bar{\Lambda}, \bar{a}, \mathfrak{F}) = \sum_{\mathscr{A} \in \hat{\mathfrak{A}}(\Lambda)} \int_{X(\bar{\Lambda})} \left(\prod_{A \in \mathscr{A}} f_A(x_A) \right) \mathrm{d}\mu_{\bar{a}}(x_{\bar{\Lambda}}). \tag{2.3.102}$$

EXAMPLE. In the study of Gibbsian perturbations of Gaussian fields the following partition functions appear: for $x_{\partial\Lambda^c} \equiv a \in X(\partial\Lambda^c)$,

$$Z(\Lambda, a, \{\Phi_A\}) = \int_{X(\Lambda)} \exp\left(- \sum_{A : A \cap \Lambda \neq \emptyset, \operatorname{diam} A \leqslant r} \Phi_A(x_A) \right) \mathrm{d}\mu_a(x_\Lambda) \tag{2.3.103}$$

where $\{\Phi_A\}$ are perturbing potentials. Writing

$$f_A(x_A) = \exp(-\Phi_A(x_A)) - 1, \tag{2.3.104}$$

we can express $Z(\Lambda, a, \{\Phi_A\})$ in the form (2.3.102) as $Z(\Lambda, a, \mathfrak{F})$ where $\mathfrak{A} = \{A \subset \mathbb{Z}^\nu : \operatorname{diam} A \leqslant r\}$; $A)(A'$ for all $A, A' \in \mathfrak{A}$ and $\mathfrak{A}(\Lambda) = \{A \in \mathfrak{A} : A \cap \Lambda \neq \emptyset\}$.

THEOREM. *Let ε, q be such that $0 < q < 1$ and $\varepsilon > 0$ is sufficiently small (depending on q and $c_{(2.2.5)}$). Suppose that some boundary condition $\bar{a} \in X(\partial\bar{\Lambda}^c)$ is given, for some finite $\bar{\Lambda} \subset \mathbb{Z}^\nu$. Suppose that some systems $\mathfrak{A}(\Lambda)$, $\Lambda \subset \bar{\Lambda}$ and $\mathfrak{F} = \{f_A, A \in \mathfrak{A}(\Lambda)\}$ are given, satisfying for each $\mathscr{A} \in \hat{\mathfrak{A}}(\Lambda)$ the condition*

$$\left(\int_{X(\bar{\Lambda})} \prod_{A \in \mathscr{A}} (f_A(x_A))^2 \, \mathrm{d}\mu_{\bar{a}}(x_{\bar{\Lambda}}) \right)^{1/2} \leqslant \varepsilon^{\sum_{A \in \mathscr{A}} |\operatorname{supp} A|} q^{\sum_{A \in \mathscr{A}} \mathrm{d}(\operatorname{supp} A)(N(\mathscr{A}))^{-1}} \tag{2.3.105}$$

where d(supp A) *is given by* (2.1.14) *and*

$$N(\mathscr{A}) = \prod_{A \in \mathscr{A}} N(A).$$

Then

$$\ln Z(\bar{\Lambda}, \bar{a}, \mathfrak{F}) = \sum_{t \in \Lambda} v_t \tag{2.3.106}$$

where

$$v_t = v_t(\bar{\Lambda}, \bar{a}, \mathfrak{F}) \tag{2.3.107}$$

satisfy the following conditions:

(a) $|v_t| \leqslant \varepsilon'$ (2.3.108)

where $\varepsilon' = \varepsilon'(\varepsilon, c_{(2.2.5)}, q)$ is such that

$$\lim_{\varepsilon \to 0} \varepsilon' = 0 \ (\textit{for fixed } c_{(2.2.5)}, q)$$

(b) *Let $\mathfrak{F}^1 = \{f_A^1, A \in \mathfrak{A}_1(\Lambda_1)$ and $\mathfrak{F}^2 = \{f_A^2, A \in \mathfrak{A}_2(\Lambda_2)\}$ be two systems of functions for which the previous conditions are valid and let*

$$v_t^i = v_t(\bar{\Lambda}, \bar{a}, \mathfrak{F}^i), \ i = 1, 2.$$

Let $\Lambda_0 \subset \Lambda_1 \cap \Lambda_2$ be such that for any A, $\operatorname{supp} A \subset \Lambda_0$ the conditions $A \in \mathfrak{A}_1(\Lambda_1)$ and $A \in \mathfrak{A}_2(\Lambda_2)$ are equivalent and

$$f_A^1 \equiv f_A^2 \text{ if } \operatorname{supp} A \subset \Lambda_0.$$

Then for any $t \in \Lambda_0$,

$$|v_t^1 - v_t^2| \leqslant c\varepsilon' q'^{\operatorname{dist}(t, \Lambda_0^c)} \tag{2.3.109}$$

where $c > 0$ *and* $0 < q' < 1$ *are some constants depending on* q *and* $c_{(2.2.5)}$ *only.*

(c) *Let* $\Lambda^n \to \Lambda$ *in the sense that* $t \in \Lambda$ *iff* $t \in \Lambda^n$ *for all except a finite number of n. Let* $\tilde{\Lambda}^n \to \tilde{\Lambda}$ *and also* $\mathfrak{A}(\Lambda^n) \to \mathfrak{A}(\Lambda)$ *in an analogous sense,* $\tilde{\Lambda}^n \supset \Lambda^n$. *Let* $\tilde{a}_n \in \mathcal{U}^{\partial \tilde{\Lambda}^{nc}}$, $\mathcal{U}$ *bounded, be such that all the limits* $\lim_n \tilde{a}_n(t)$ $t \in \partial \tilde{\Lambda}^c$ *exist. Let* $\mathfrak{F}^n = \{f_A^n, A \in \mathfrak{A}(\Lambda^n)\}$ *and* $\mathfrak{F} = \{f_A, A \in \mathfrak{A}(\Lambda)\}$ *be systems of functions such that* $A \in \mathfrak{A}(\Lambda^n) \cap \mathfrak{A}(\Lambda) \Rightarrow f_A^n = f_A$. *Assume that* (2.3.105) *holds for all systems* $\mathfrak{F}^n$, *all* $\tilde{\Lambda}^n$ *and all* $\tilde{a}_n \in \mathcal{U}^{\partial \tilde{\Lambda}^{nc}}$, *with the same* ε *and* q. *Assume also the uniform integrability of any* $(f_A)^2$ *with respect to all* $\mu_{\tilde{a}_n}$ *(i.e. the uniform convergence of* $\min(K, |f_A|)$, $K \to \infty$ *with respect to all* $L^2(\mu_{\tilde{a}_n})$*). Then for each* $t \in \Lambda$ *the quantities*

$$v_t^{(n)} = v_t(\tilde{\Lambda}^n, \tilde{a}_n, \mathfrak{F}^n)$$

converge for $n \to \infty$.

(d) *Let an infinite system* $\mathfrak{F} = \{f_A, A \in \mathfrak{A}\}$ *of measurable functions of* $x_A \in X$ (supp *A*) *be given. Denote by* $\mathfrak{F}_{\tilde{\Lambda}} = \{f_A, A \in \mathfrak{A}(\tilde{\Lambda})\}$*'where* $\mathfrak{A}(\tilde{\Lambda})$ *is the set of all* $A \in \mathfrak{A}$ *such that* $\operatorname{supp} A \cap \tilde{\Lambda} \neq \emptyset$ *and* $\operatorname{supp} A \subset \tilde{\Lambda} \cup \partial \tilde{\Lambda}^c$. *Denote by*

$$u_t = v_t(\tilde{\Lambda}, \tilde{a}, \mathfrak{F}_{\tilde{\Lambda}}). \tag{2.3.110}$$

Let $\mathcal{U} \subset \mathbb{R}^k$ *be a bounded set such that* (2.3.105) *holds uniformly for each finite* $\tilde{\Lambda}$ *and each* $\tilde{a} \in \mathcal{U}^{\partial \tilde{\Lambda}^c}$, *for the system* $\mathfrak{F}_{\tilde{\Lambda}}$. *Suppose the uniform integrability of any* $(f_A)^2$ *with respect to all* $\mu_{\tilde{a}}$. *Then there exists a limit*

$$\bar{u}_t = \bar{u}_t(\mathfrak{F}) = \lim_{\tilde{\Lambda} \uparrow \mathbb{Z}^\nu} u_t \tag{2.3.111}$$

uniformly with respect to $\tilde{\Lambda}$ *and* $\tilde{a} \in \mathcal{U}^{\partial \tilde{\Lambda}^c}$ *as* $\operatorname{dist}(t, \tilde{\Lambda}^c) \to \infty$.

(e) *Let the system* $\mathfrak{A}$ *be translation-invariant. If moreover the system* $\mathfrak{F} = \{f_A, A \in \mathfrak{A}\}$ *is invariant with respect to shifts* T_t, $t \in \mathbb{Z}^\nu$ *and if the condition in* (d) *is also fulfilled, then all* $\bar{u}_t$ *are the same;*

$$\bar{u}_t \equiv \bar{u}, \qquad t \in \mathbb{Z}^\nu. \tag{2.3.111'}$$

The proof of the theorem will be given in §§2.3.9–2.3.12 below.

The following series of notes explains in more detail some additional features of the quantities v_t. Most of these features are contained in the proof of the theorem, but deserve further discussion.

NOTE 1. The quantities v_t are by no means unique. Nevertheless we will use quite a concrete choice of v_t – see (2.3.151). In fact the quantities v_t constructed in (2.3.151) have some additional important properties not formulated in the theorem. They can be expressed as follows:

$$v_t = \sum_{T \ni t} k_T \tag{2.3.112}$$

where k_T satisfies a condition of the type (2.3.125) and depends only on f_A, supp $A \subset T$.

We will not use these expressions in this paper, but it seems that in further problems (such as the study of non-translation-invariant Gibbs states corresponding to the Hamiltonians of the main lemma) they cannot be avoided.

NOTE 2. In our formulation of the theorem there is no explicit information about the dependence of v_t on $\tilde{a} \in X(\partial\tilde{\Lambda}^c)$. In fact, because of the non-unicity of v_t it is more reasonable to study the dependence of $Z(\tilde{\Lambda}, \tilde{a}, \mathfrak{F})$ on $\tilde{a}$ only. We will prove statements of this type in §4.4 below, in more concrete situations.

NOTE 3. The proof of the theorem (in particular, of its essential estimate (2.3.109)) combines most of the material of §2, namely Theorem 2.1.4 (and its proof), Theorem 2.2.2 and Theorem 2.3.7.

NOTE 4. In most (but not all) applications of the theorem, only those compatibile objects A appear whose support is connected or whose diameter does not exceed some fixed number r. In these cases the term $q^{\sum_{A \in \mathscr{A}} \mathrm{d}(\operatorname{supp} A)(\mathscr{N}(A))^{-1}}$ can be omitted in the assumption (2.3.105), if ε is lowered correspondingly. Then the constants c, q' and ε' from (a), (b) depend on $c_{(2.2.5)}$ and ε only.

NOTE 5. Concerning an application of the theorem to the problem of uniqueness of perturbed Gaussian fields, see §2.3.15. In this section, Theorem 2.3.8 is applied in such a way that it proves the unicity of a perturbed Gaussian field, of the type (2.3.103). (The technique developed in Theorem 2.3.8 enables us also to prove some more precise correlation decay estimates for these perturbed models, which are, however, omitted there.)

In the important special case when the functions f_A are translation-invariant we obtain the following 'limit version' of Theorem 2.3.8, obtained for $\tilde{\Lambda} \uparrow \mathbb{Z}^\nu$, $\tilde{a} \in \mathscr{U}^{\partial\tilde{\Lambda}^c}$, $\mathscr{U}$ bounded.

COROLLARY. *Let $\mathfrak{F}$ be an infinite translation-invariant system, as in* (e). *Let $\mathfrak{A}(\Lambda)$ and $\Lambda' \subset \Lambda$, be such that all $A \in \mathfrak{A}$ satisfying the condition* $\operatorname{supp} A \subset \Lambda'$ *are contained in $\mathfrak{A}(\Lambda)$. Let $\mathfrak{F}_\Lambda$ satisfy* (2.3.105) *with respect to μ. Denote by (compare* (2.3.102))

$$Z(\{f_A, A \in \mathfrak{A}(\Lambda)\}) = \sum_{\mathscr{A} \in \mathfrak{A}(\Lambda)} \int_X \left(\prod_{A \in \mathscr{A}} f_A(x_A) \right) \mathrm{d}\mu(x) \tag{2.3.113}$$

where μ is the limit Gaussian measure of §2.2.3. *Then*

$$|\ln Z(\{f_A, A \in \mathfrak{A}(\Lambda)\}) - \bar{u}|\Lambda|| \leqslant c\varepsilon'|\partial\Lambda' \cup (\Lambda \setminus \Lambda')| \tag{2.3.114}$$

where c is some new constant depending on $c_{(2.2.5)}$ and q, and $\bar{u} = \lim_{\Lambda' = \Lambda \uparrow \mathbb{Z}^\nu} |\Lambda|^{-1} \ln Z(\{f_A, A \in \mathfrak{A}(\Lambda)$ (is the Van Hove sense).

Proof of Corollary. We can assume that all f_A are uniformly bounded, with uniformly bounded supports (the generalization of (2.3.114) then being obvious). Take $\tilde{\tilde{\Lambda}}$ such that $\operatorname{dist}(\Lambda, \tilde{\tilde{\Lambda}}^c) \to \infty$ and $\operatorname{dist}(\tilde{\tilde{\Lambda}}, \tilde{\Lambda}^c) \to \infty$. Taking the basic estimate (2.3.109) for the pair $v_t^1 = v_t(\tilde{\Lambda}, \tilde{a}, \{f_A, A \in \mathfrak{A}(\tilde{\tilde{\Lambda}})\})$ and $v_t^2 = v_t(\tilde{\Lambda}, \tilde{a}, \{f_A, A \in \mathfrak{A}(\Lambda) \cap \mathfrak{A}(\tilde{\tilde{\Lambda}})\})$ where $\mathfrak{A}(\tilde{\tilde{\Lambda}}) = \{A \in \mathfrak{A} \colon \operatorname{supp} A \subset \tilde{\tilde{\Lambda}}\}$ and using also the statement (c) of the theorem we obtain, in the limiting case $\tilde{\Lambda} \uparrow \mathbb{Z}^\nu$, $\tilde{a} \in \mathscr{U}^{\partial\tilde{\Lambda}^c}$, $\mathscr{U}$ bounded, the following expression: $\ln Z(\{f_A, A \in \mathfrak{A}(\Lambda)\}) = \sum_{t \in \Lambda} \bar{v}_t$ where

$$|\bar{v}_t - \bar{u}| \leqslant c\varepsilon' q^{\operatorname{dist}(t, \Lambda'^c)} \tag{2.3.115}$$

(it is clear that $\bar{u}_{(2.3.114)} = u_{(2.3.111)}$ in this case). It suffices now to use an estimate of the type (2.1.38). □

It now remains is to prove the theorem, and this will be the subject of §§2.3.9–2.3.12. Section 2.3.13 will show when the condition (2.3.105) holds. Some analyticity features of the quantities v_t, u_t are briefly explained in §2.3.14.

2.3.9. *Expression of* $Z(\bar{\Lambda}, \tilde{a}, \mathfrak{F})$ *as a Polymer Partition Function*

Consider a symmetric relation $\sim$ on the set of all $A \in \mathfrak{A}$, defined as follows.

Fix some integer $\bar{d}$ no smaller than d in (2.3.101). If A, A' are two compatible elements of $\mathfrak{A}$ write

$$A \sim A' \text{ iff } \operatorname{dist}(\operatorname{supp} A, \operatorname{supp} A') \leqslant \bar{d}. \tag{2.3.116}$$

DEFINITION 1. We interpret $\mathfrak{A}$ as a graph with bonds $\{A, A'\}$ where $A \sim A'$. Any compatible family $\mathscr{G} = \{A_i \in \mathfrak{A}\}$ will be called a *gang* if the corresponding subgraph on $\mathscr{G}$ is connected. We call two gangs non-interacting if their function is a compatible family but not a gang. In other words, two gangs $\mathscr{G}$, $\mathscr{G}'$ are non-interacting iff $A \sim A'$ for no $A \in \mathscr{G}$, $A' \in \mathscr{G}'$.

For any gang $\mathscr{G}$ denote by $f_{\mathscr{G}}$ the function

$$f_{\mathscr{G}}(x_{\bigcup_{A \in \mathscr{G}} \operatorname{supp} A}) = \prod_{A \in \mathscr{G}} f_A(x_A), \; x_A \in X(\operatorname{supp} A). \tag{2.3.117}$$

Given a finite $\Lambda \subset \mathbb{Z}^{\nu}$ and some system $\mathfrak{A}$ denote by $\hat{\mathscr{G}}(\mathfrak{A})$ the class of all gangs $\mathscr{G}$ such that $A \in \mathscr{G} \Rightarrow A \in \mathfrak{A}$.

PROPOSITION 1. *Using these notions we have the formula*

$$Z(\bar{\Lambda}, \tilde{a}, \mathfrak{F}) = \sum_{\{\mathscr{G}_j\}} \int \prod_j f_{\mathscr{G}_j}(x_{\bar{\Lambda}} \cup \tilde{a}) \mathrm{d}\mu_{\tilde{a}}(x_{\bar{\Lambda}}), \tag{2.3.118}$$

the sum being taken over all families $\{\mathscr{G}_j\}$ *of mutually non-interacting gangs from* $\hat{\mathscr{G}}(\mathfrak{A}(\Lambda))$.

Proof. This is a consequence of the following two observations:

(a) Any compatible system $\mathscr{A} \in \hat{\mathfrak{A}}(\Lambda)$ can be represented in a unique way as a sum of gangs $\mathscr{G}_j \in \hat{\mathscr{G}}(\mathfrak{A}(\Lambda))$.

(b) Any union of mutually non-interacting gangs is a compatible set (because of the condition $\bar{d} \geqslant d$). □

Now we use the general formula (2.3.5).

DEFINITION 2. Let $\mathscr{H} = \{\mathscr{G}_j, j = 1, 2, \ldots, n\}$ be a family of gangs $\mathscr{G}_j \in \hat{\mathscr{G}}(\mathfrak{A}(\Lambda))$. Denote by

$$\langle f_{\mathscr{H}} \rangle_{\tilde{a}} = \langle f_{\mathscr{G}_1}, \ldots, f_{\mathscr{G}_n} \rangle_{\tilde{a}} \tag{2.3.119}$$

the semi-invariant of the variables $f_{\mathscr{G}_1}, \ldots, f_{\mathscr{G}_n}$ being taken with respect to the measure $\mu_{\tilde{a}}$.

PROPOSITION 2.

$$Z(\bar{\Lambda}, \tilde{a}, \mathfrak{F}) = \sum_{\{\mathscr{G}_j\}} \sum_{\mathscr{D}} \prod_{\mathscr{H} \in \mathscr{D}} \langle f_{\mathscr{H}} \rangle_{\tilde{a}} \tag{2.3.120}$$

the first sum being taken over the same families of gangs as in (2.3.118) *and the second sum being taken over all partitions* $\mathscr{D}$ *of the given family* $\{\mathscr{G}_j\}$.

Proof. Immediate, by (2.3.5) and (2.3.118).

DEFINITION 3. Assume that $\bar{d}_{(2.3.116)}$ is even, for simplicity. Define the support of a gang $\mathscr{G}$ (denoted by supp $\mathscr{G}$) as the $\bar{d}/2$-neighbourhood of the set $\cup_{A\in\mathscr{G}} \operatorname{supp} A$. Given $T \subset \mathbb{Z}^\nu$ denote by $\hat{\mathscr{H}}(\mathfrak{A}(\Lambda), T)$ the set of all families of mutually non-interacting gangs $\mathscr{H} = \{\mathscr{G}_j, j\in\mathscr{T}\}$ such that

$$T = \bigcup_j \operatorname{supp} \mathscr{G}_j, \qquad \mathscr{G}_j \in \mathscr{G}(\mathfrak{A}(\Lambda)). \tag{2.3.121}$$

Put

$$k_T^{\Lambda,\tilde{a}} = \sum_{\mathscr{H}\in\mathscr{H}(\mathfrak{A}(\Lambda), T)} \langle f_{\mathscr{H}} \rangle_{\tilde{a}}. \tag{2.3.122}$$

PROPOSITION 3.

$$Z(\bar{\Lambda}, \tilde{a}, \mathfrak{F}_\Lambda) = \sum_{\{T_i\}} \prod_i k_{T_i}^{\Lambda,\tilde{a}} \tag{2.3.123}$$

where the sum is taken over all families of mutually disjoint sets $T_i \subset \mathbb{Z}^\nu$.

Proof. Note that two gangs $\mathscr{G}, \mathscr{G}'$ are mutually non-interacting iff supp $\mathscr{G} \cap$ supp $\mathscr{G}' = \emptyset$. Then we use (2.3.120) and the definition (2.3.122) of $k_T^{\Lambda,\tilde{a}}$. □

NOTE. In fact, the sum in (2.3.123) can be taken over only those T only satisfying the condition

$$T \cap \Lambda \neq \emptyset \text{ and } \operatorname{dist}(t, \Lambda) \leqslant \tfrac{\bar{d}}{2} \text{ for all } t \in T. \tag{2.3.124}$$

2.3.10. *Estimates of the Polymer Weights* $k_T^{\bar{\Lambda},\tilde{a}}$

PROPOSITION. *Choose* $\bar{d}$ *so large that the estimates* (2.3.95) *hold, for any* $\{S_i\}$ *and* $\{f^i\}$ *such that* $\operatorname{dist}(S_i, S_{i'}) \geqslant \bar{d}$, $i \neq i'$. *Then there are constants* $c, c' > 0$ *and* $0 < q' < 1$ *such that*

$$\sum_{T: T\ni t, |T|=n, d(T)=m} |k_T^{\Lambda,\tilde{a}}| \leqslant (c\varepsilon)^{c'n}(q')^m, \tag{2.3.125}$$

with $d(T)$ *given by* (2.1.14), *for any* $t\in\Lambda$ *and any* $n, m \in \mathbb{N}$. *The constants* c, q' *depend only on* $q_{(2.3.105)}$, r *and* $c_{(2.2.5)}$ (*through* $c_{(2.3.95)}$) *and the constant* c' *depends on* $\bar{d}$. *The constant* ε *is from* (2.3.105).

Proof. Essentially, what is needed is to estimate the terms $\langle f_{\mathscr{H}} \rangle_{\tilde{a}}$ in (2.3.122). If $\mathscr{G}_1, \ldots, \mathscr{G}_n$ is a sequence of mutually non-interacting gangs then (2.3.95) gives the estimate

$$|\langle f_{\mathscr{G}_1}, \ldots, f_{\mathscr{G}_n} \rangle_{\tilde{a}}| \leqslant c^{|T|} \prod_{j=1}^{n} \langle f_{\mathscr{G}_j}^2 \rangle_{\tilde{a}}^{1/2} \sum_{\Gamma\in\mathscr{C}(\mathbb{S})} q^{d(\Gamma)} \tag{2.3.126}$$

with

$$T = S = \bigcup_{j=1}^{n} S_j, \qquad S_j = \operatorname{supp} \mathscr{G}_j, \qquad \mathbb{S} = \mathbb{S}(\{\mathscr{G}_j\}) = \{S_j, j = 1, \ldots, n\}.$$

Substituting this into (2.3.122) we get the estimate

$$|k_T^{\Lambda,\bar{a}}| \leqslant c^{|T|} \sum_{\mathscr{H} \in \mathscr{H}(\mathfrak{A}(\Lambda),T)} \prod_j \langle f_{\mathscr{G}_j}^2 \rangle_{\bar{a}}^{1/2} \sum_{\Gamma \in \tilde{\mathscr{C}}(\mathbb{S})} q^{d(\Gamma)}. \tag{2.3.127}$$

Estimate the right-hand side of this relation: put

$$q = \max(q_{(2.3.105)}, q_{(2.3.126)}).$$

Write the gangs $\mathscr{G}_j$ as $\mathscr{G}_j = \{A_i^j\}$. Write $N(\mathscr{H}) = N(\cup_{j=1}^n \mathscr{G}_j)$. By (2.3.105), the right-hand side of (2.3.127) is no greater than

$$c^{|T|} \sum_{\mathscr{H} \in \hat{\mathscr{H}}(\mathfrak{A}(\Lambda),T)} (N(\mathscr{H}))^{-1} \varepsilon^{\Sigma_{j,i}|\operatorname{supp} A_i^j|} q^{\Sigma_{j,i} d(\operatorname{supp} A_i^j)} \sum_{\Gamma \in \tilde{\mathscr{C}}(\mathbb{S})} q^{d(\Gamma)}. \tag{2.3.128}$$

NOTE. $d(\operatorname{supp} A)$ was defined in (2.1.14), whereas $d(\Gamma)$ was defined in (2.3.96). We have $d(\Lambda) = \min_\Gamma \{d(\Gamma)\}$, the minimum being taken over all trees Γ on $\Lambda, \Lambda \subset \mathbb{Z}^\nu$.

Now write

$$q = q'q'', \qquad \varepsilon = \varepsilon'\varepsilon'' \tag{2.3.129}$$

with some $0 < q' < 1$, $0 < q'' < 1$ and some $\varepsilon', \varepsilon'' > 0$ to be specified later.

Estimate (from below) the powers of ε, q in (2.3.128). Write $S_j = \operatorname{supp} \mathscr{G}_j$. Notice that, by the definition of a gang, there is some $c' > 0$ depending on $\bar{d}$ such that for any j,

$$\sum_i |\operatorname{supp} A_i^j| \geqslant c'|S_j|. \tag{2.3.130}$$

On the other hand,

$$\sum_i d(\operatorname{supp} A_i^j) \geqslant d(S_j) - \bar{d}|S_j| \tag{2.3.131}$$

because for any system of trees Γ_i^j on $\operatorname{supp} A_i^j$ the graph $\cup_i \Gamma_i^j \cup S_j$ is connected (if we interpret S_j as a graph $\{\{s,t\}: s,t \in S_j; |s-t| \leqslant \bar{d}\}$).

For any S_j fix a tree Γ_j on S_j such that $d(S_j) = d(\Gamma_j)$. Given any $\Gamma \in \tilde{\mathscr{C}}(\mathbb{S})$ consider the graph $\Gamma^* = \cup_j \Gamma_j \cup \Gamma$. Obviously, this is a tree and the mapping $\{\Gamma \rightsquigarrow \Gamma^*\}$ is clearly one-to-one, if the Γ_j are fixed. From (2.3.131) it is clear that

$$d(\Gamma^*) = d(\Gamma) + \sum_j d(\Gamma_j) \leqslant \bar{d}|T| + d(\Gamma) + \sum_{i,j} d(\operatorname{supp} A_i^j). \tag{2.3.132}$$

Write q, ε in (2.3.128) by (2.3.129) and use the preceding observation. We obtain

$$|k_T^{(\Lambda,\bar{a})}| \leqslant c^{|T|} \varepsilon'^{c'|T|} q'^{-\bar{d}|T|} \sum_{\Gamma^* \in \tilde{\mathscr{C}}(T)} q'^{d(\Gamma^*)} \sum_{\mathscr{H} \in \hat{\mathscr{H}}(\mathfrak{A}(\Lambda),T)} (N(\mathscr{H}))^{-1}$$
$$\varepsilon''^{\Sigma_{j,i}|\operatorname{supp} A_i^j|} q''^{\Sigma_{j,i} d(\operatorname{supp} A_i^j)} \tag{2.3.133}$$

the summation over all $\Gamma^* = \cup_j \Gamma_j \cup \Gamma$ being now replaced by summation over a broader class $\tilde{\mathscr{C}}(T)$ of all trees on T. Fix some $t \in \Lambda$ and sum (2.3.133) over all $T \ni t$ such that $|T| = n$ and $d(T) = m$. Denote by $\tilde{\mathscr{C}}_{0,n,m}$ the set of all trees $\Gamma \ni 0$ such that $m = d(\Gamma)$ and $n-1$ is the number of bonds in Γ. Denote by $\hat{\mathscr{A}}(T)$ the family of all compatible systems $\mathscr{A} = \{A_i\}$, $\operatorname{supp} A_i \subset T$. We obtain

$$\sum_{T:T\ni t,|T|=n,d(T)=m} |k_T^{\Lambda,\tilde{a}}| \leqslant (c''\varepsilon'^{c'})^{|T|} \sum_{\Gamma\in\mathscr{C}_{0,n,m}} q'^{d(\Gamma)}$$

$$\sum_{\mathscr{A}\in\hat{\mathscr{A}}(T)} (N(\mathscr{A}))^{-1}\varepsilon''^{\Sigma_i|\text{supp}\,A_i|} q''^{\Sigma_i d(\text{supp}\,A_i)} \tag{2.3.134}$$

where $c'' = c(q')^{-d}$. (We have used the fact that each $\mathscr{A}\in\hat{\mathscr{A}}(T)$ can be uniquely represented as a set of mutually non-interacting gangs.)

We will show in §2.3.11 below that for each $0 < q < 1$ and each $q^* > q$ there is some $c^* = c^*(q^*q^{-1})$ such that

$$\sum_{\Gamma\in\mathscr{C}_{0,n,m}} q^{d(\Gamma)} \leqslant c^{*n}q^{*m}. \tag{2.3.135}$$

The estimate (2.3.135) gives the following upper bound for the second sum in (2.3.134): if $q'' < q^* < 1$, $c^*\varepsilon'' < 1$ then

$$\sum_{\mathscr{A}\in\mathscr{A}(T)} (N(\mathscr{A}))^{-1}\varepsilon''^{\Sigma_i|\text{supp}\,A_i|} q''^{\Sigma_i d(\text{supp}\,A_i)}$$

$$\leqslant \prod_{t\in T}\left(\sum_{B:B\ni t}\varepsilon''^{|B|}q''^{d(B)}\right)$$

$$\leqslant \left(\sum_{\tilde{n},\tilde{m}}\sum_{\Gamma\in\mathscr{C}_{0,\tilde{n},\tilde{m}}}\varepsilon''^{\tilde{n}}q''^{\tilde{m}}\right)^{|T|}$$

$$\leqslant \left(\sum_{\tilde{n},\tilde{m}}(c^{*\prime\prime}\varepsilon'')^{\tilde{n}}(q^{*\prime\prime})^{\tilde{m}}\right)^{|T|} \leqslant ((1-c^{*\prime\prime}\varepsilon'')(1-q^{*\prime\prime}))^{-|T|}. \tag{2.3.136}$$

If we now collect the estimates (2.3.134), (2.3.135), (2.3.136) we obtain, for $c^{*\prime\prime}\varepsilon'' < 1$, the estimate

$$\sum_{T:T\ni t,|T|=n,d(T)=m} |k_T^{\Lambda,\tilde{a}}| \leqslant (c^{*\prime}c''\varepsilon'^{c'}(1-c^{*\prime\prime}\varepsilon'')^{-1}(1-q^{*\prime\prime})^{-1})^n(q^{*\prime})^m \tag{2.3.137}$$

where

$$q^{*\prime} > q', \qquad q^{*\prime\prime} > q'', \qquad c^{*\prime} = c^*(q^{*\prime}(q')^{-1}), \qquad c^{*\prime\prime} = c^*(q^{*\prime\prime}(q'')^{-1}).$$

Therefore, for $\varepsilon' = \varepsilon'' = \varepsilon^{1/2}$ we obtain (2.3.125). □

2.3.11. *Estimate of* $\Sigma_\Gamma q^{d(\Gamma)}$

Denote by $\mathscr{C}_0$ the set of all trees with vertices in $\mathbb{Z}^\nu$ having 0 among the set of vertices. Denote by $|\Gamma|$ the number of bonds of Γ.

PROPOSITION. *For any* $0 < \kappa < 1$ *and all sufficiently small* $\omega = \omega(\kappa)$,

$$\sum_{\Gamma\in\mathscr{C}_0} \omega^{|\Gamma|}\kappa^{d(\Gamma)} < 1. \tag{2.3.138}$$

NOTE. This obviously proves the required estimate (2.3.135), if we take $\mathscr{C}_{0,n,m}$ instead of $\mathscr{C}_0$ in (2.3.138) and choose the values $\kappa = q(q^*)^{-1}$, $\omega = (c^*)^{-1}$ there.

Proof of Proposition. Assign to each vertex of the tree the 'level' of the vertex defined as the number of vertices on the string connecting the given vertex with 0. Denote by $\mathscr{C}_0^{(l)}$ the set of all trees $\Gamma \in \mathscr{C}_0$ which contain no vertices of the level $l+1$. Denote by

$$Q_l = \sum_{\Gamma \in \mathscr{C}_0^{(l)}} \omega^{|\Gamma|} \kappa^{d(\Gamma)}.$$

Note that

$$\sum_{\Gamma \in \mathscr{C}_0^{(1)}: |\Gamma| = k} \omega^{|\Gamma|} \kappa^{d(\Gamma)} \leqslant \frac{Q^k}{k!} \tag{2.3.139}$$

where

$$Q = \omega \sum_{t \in Z^\nu \setminus \{0\}} \kappa^{|t|} < \infty. \tag{2.3.140}$$

It is clear that

$$Q_1 \leqslant \exp Q - 1. \tag{2.3.141}$$

For any $\Gamma \in \mathscr{C}_0^{(1)}$ denote by $\mathscr{C}_0^{(l)}(\Gamma)$ the class of all $\Gamma^* \in \mathscr{C}_0^{(l)}$ such that after removing from Γ^* all the vertices of the levels $l \geqslant 2$ we obtain the graph Γ. Note that each $\Gamma^* \in \mathscr{C}_0^{(l)}(\Gamma)$ can be split into Γ and some number m (not exceeding the value of $|\Gamma|$) of trees from $\mathscr{C}_0^{(l-1)}$. Thus

$$\sum_{\Gamma^* \in \mathscr{C}_0^{(l)}(\Gamma)} \omega^{|\Gamma^*|} \kappa^{d(\Gamma^*)} \leqslant \omega^{|\Gamma|} \kappa^{d(\Gamma)} (1 + Q_{l-1})^{|\Gamma|} \tag{2.3.142}$$

and using also (2.3.139) we find that

$$\begin{aligned} Q_l &= \sum_{\Gamma \in \mathscr{C}_0^{(1)}} \sum_{\Gamma^* \in \mathscr{C}_0^{(l)}(\Gamma)} \omega^{|\Gamma^*|} \kappa^{d(\Gamma^*)} \leqslant \sum_{k=1}^{\infty} \frac{1}{k'} (Q(1 + Q_{l-1}))^k \\ &= \exp(Q(1 + Q_{l-1})) - 1. \end{aligned} \tag{2.3.143}$$

Denote by λ_Q the smallest positive solution of the equation

$$\lambda = \exp(Q(1 + \lambda)) - 1. \tag{2.3.144}$$

Such a solution exists for $0 < Q < l^{-1}$, as elementary calculus shows. Note also that

$$\lambda_Q < 1 \quad \text{if} \quad Q < \frac{\ln 2}{2}$$

and

$$\exp(Q(1 + \lambda)) - 1 \leqslant \exp(Q(1 + \lambda_Q)) - 1 = \lambda_Q \qquad \text{if} \qquad \lambda < \lambda_Q. \tag{2.3.145}$$

Therefore $Q_1 < \lambda_Q$ if $Q < \ln 2/2$ and the recurrent relation (2.3.143) gives the inequality $Q_l < \lambda_Q$ for all l.

So we have proved that

$$\sum_{\Gamma \in \mathscr{C}_0} \omega^{|\Gamma|} \kappa^{d(\Gamma)} < \lambda_Q < 1$$

if

$$2\omega < \ln 2\left(\sum_{t\in \mathbf{Z}^{\nu}\setminus\{0\}} \kappa^{|t|}\right)^{-1}. \qquad \square$$

2.3.12. *Proof of Theorem* 2.3.8

Now we can conclude the proof of Theorem 2.3.8.

We will use in particular the results of §2.1, Theorem 2.1.4 (and the method of its proof).

Denote by Λ' the $\bar{d}/2$ neighbourhood of Λ. Notice that (2.3.123) can be written, using the notation (2.1.2), as $Z_{\Lambda'}$ with the polymer weights

$$k_T = k_T^{(\Lambda,\tilde{a})}.$$

Define, by (2.1.27),

$$v_t' = h_t^{\Lambda'}, \qquad t \in \Lambda'. \tag{2.3.146}$$

The estimate (2.3.125) implies that the condition (2.1.12) of Lemma 2.1.4 is true, if $\varepsilon_{(2.3.105)}$ is small enough.

By the very definition of $h_t^{\Lambda'}$ (see (2.1.29))

$$Z(\bar{\Lambda}, \tilde{a}, \mathfrak{F}) = \sum_{t\in\Lambda'} v_t'. \tag{2.3.147}$$

To obtain (2.3.106) from (2.3.147) choose, for any $s \in \Lambda'$, some $t = t(s) \in \Lambda$ which is the closest possible point to s. Put

$$v_t = \sum_{s\in\Lambda' : t(s) = t} v_s'. \tag{2.3.148}$$

Note that

$$v_t = v_t' \text{ if } \operatorname{dist}(t, \Lambda^c) > \frac{d}{2} \tag{2.3.149}$$

and also that

$$|v_t| \leqslant c \ \sup\{|v_s'|, s \in \Lambda'\} \tag{2.3.150}$$

with some $c = c(\bar{d})$. This proves, together with (2.1.27) and (2.1.13), the relation (2.3.108) (if $\varepsilon_{(2.3.105)}$ is sufficiently small).

To prove the essential estimate (2.3.109) analyse the formula (with $k_T = k_T^{\Lambda,\tilde{a}}$)

$$v_t = -\ln\left(1 + \sum_{\beta\in\mathscr{B}_t^{\Lambda t}} (-1)^n k_{T_1} \cdots k_{T_n}\right) \tag{2.3.151}$$

which follows from (2.3.146) and (2.1.27) (for $\operatorname{dist}(t, \Lambda^c) > d/2$). Denote by v_t^i and k_T^i the corresponding quantities for the system $\mathfrak{F}^i$. Looking at the definition (2.3.122) of k_T^i we see that the following relation holds:

$$\operatorname{dist}(t, \Lambda_0) \leqslant \frac{\bar{d}}{2} \text{ for all } t \in T \Rightarrow k_T^1 = k_T^2. \tag{2.3.152}$$

Now denote by $\mathscr{B}_t^{\Lambda_t}(\Lambda_0)$ the subset of $\mathscr{B}_t^{\Lambda_t}$(see §2.1.2) consisting of all chains $(B_1, T_1, \ldots, B_n, T_n)$ such that $k_{T_j}^1 \neq k_{T_j}^2$ for some j. Because $\operatorname{dist}(t_j, \Lambda_0) > \bar{d}/2$ for these j and some $t_j \in T_j$ we obtain the inequality

$$\sum_{j=1}^{n} d(T_j) \geqslant \operatorname{dist}(t, \Lambda_0^c) + \frac{\bar{d}}{2}$$

and therefore the following consequence of (2.1.13):

$$\sum_{\beta \in \mathscr{B}_t^{\Lambda_t}(\Lambda_0)} |k_{T_1}^i \cdots k_{T_n}^{\hat{v}}| \leqslant c\varepsilon \bar{q}^{(\operatorname{dist}(t, \Lambda_0^c) + \bar{d}/2)}, \tag{2.3.153}$$

with $c = 16(\bar{q} - q')^{-1}$, $q' = q'_{(2.3.125)}$ and with $\varepsilon = (c\varepsilon)^{c'}_{(2.3.125)}$. (We emphasize that the bounds for ε and q needed in (2.1.12) are guaranteed by (2.3.125) for both k_T^i, $i = 1, 2$.)

We substitute these estimates into (2.3.151). The estimate (2.3.109) now follows easily from the same considerations as we used in the proof of (2.1.37), with $c\varepsilon'_{(2.3.109)}$ depending also on $\bar{d}$. (Taking $\Lambda_0 = \emptyset$, $\mathfrak{F}^2 \equiv 0$ one obtains (2.3.108) from (2.3.109). We will not specify $c_{(2.3.109)}$). The appropriate choice of $\bar{d}$ needed in Proposition 2.3.10 can be made, however, with a knowledge of $c_{(2.2.5)}$ only (see §2.3.7). Thus $c\varepsilon'_{(2.3.109)}$ depends, in that way, only on the constant $c_{(2.2.5)}$. This proves (2.3.109).

It is quite easy to prove the remaining statements of Theorem 2.3.8, having established (2.3.151) and some uniform estimates of the type (2.3.153). Namely, in (c) it suffices to prove convergence of the quantities $k_T^{(\Lambda, \tilde{a}_n)}$, $\tilde{a}_n \in \mathscr{U}^{\partial \tilde{\Lambda}^{nc}}$, as $n \to \infty$. The latter property obviously holds for bounded functions f_A, because the measures $\mu_{\tilde{a}_n, \Lambda}$ (projections on $X(\Lambda)$ of $\mu_{\tilde{a}_n}$) converge in variation for $n \to \infty$ and therefore also all the semi-invariants $\langle f_{\mathscr{H}'} \rangle_{\tilde{a}_n}$, $\tilde{a}_n \in \mathscr{U}^{\partial \tilde{\Lambda}^{nc}}$ in (2.3.122) converge (to the value $\langle f_{\mathscr{H}'} \rangle$, the latter semi-invariant being taken with respect to μ).

For general square-integrable functions f_A one first approximates them by bounded ones and then uses their uniform integrability (in $L^2(\mu_{\tilde{a}_n})$) and Lemma 4 of §2.3.5. The existence of a limit in (2.3.111) is proved by similar arguments.

2.3.13. *Complement to Theorem* 2.3.8

In this section we will show an example when the condition (2.3.105) is valid. This example will be used in §4.4 below.

PROPOSITION. *Let* $\mu_{\tilde{a}}$, $\tilde{a} \in \mathscr{U}^{\partial \tilde{\Lambda}^c}$ *be as in* §2.3.8, *and let* $\bar{r}$ *be an integer. Let some family of functions*

$$\mathfrak{F} = \{f_A, A \in \mathscr{A}\} \tag{2.3.154}$$

where $\mathscr{A} \subset \{A \subset \mathbb{Z}^\nu : \operatorname{diam} A \leqslant \bar{r},\ A \cap \tilde{\Lambda} \neq \emptyset\}$ *be given, each* f_A *depending on* $x_A \in X(A)$. *Then*

$$\left\langle \left| \prod_{A \in \mathscr{A}} f_A \right| \right\rangle_{\tilde{a}} \leqslant c^{\sum_{A \in \mathscr{A}} |A|} \left(\prod_{A \in \mathscr{A}} \langle |f_A|^p \rangle_{\tilde{a}} \right)^{1/p} \tag{2.3.155}$$

where c *and* p *depend only on* $\bar{r}$ *and* $c_{(2.2.5)}$.

Proof. Divide $\mathscr{A}$ into some number $n = n(\bar{r}, d)$ of subfamilies $\mathscr{A}_1, \ldots, \mathscr{A}_n$ such that for each $j = 1, \ldots, n$ and each pair $A, A' \in \mathscr{A}_j$, $A \neq A'$ the relation $\operatorname{dist}(A, A') \geqslant d$ holds, where d is from Theorem 2.3.7. Write

$$\prod_{A \in \mathscr{A}} f_A = \prod_{j=1}^{n} f_{\mathscr{A}_j}$$

where

$$f_{\mathscr{A}_j}(x_{\bar{A}}) = \prod_{A \in \mathscr{A}_j} f_A(x_A). \tag{2.3.156}$$

Use the Hölder inequality

$$\left\langle \left| \prod_{j=1}^{n} f_{\mathscr{A}_j} \right| \right\rangle_{\bar{a}} \leqslant \prod_{j=1}^{n} (\langle |f_{\mathscr{A}_j}|^n \rangle_{\bar{a}})^{1/n}. \tag{2.3.157}$$

It is clear from this inequality that it suffices to prove (2.3.156) for the families $\mathscr{A}_j$ only (with p being replaced by np in the general case). Thus, it suffices to prove (2.3.156) by means of the assumption

$$A, A' \in \mathscr{A}, \qquad A \neq A' \Rightarrow \operatorname{dist}(A, A') \geqslant d.$$

In this case it is possible to take $p = 2$ in (2.3.155): also from Theorem 2.3.7 we obtain, using also the relation (2.3.5), the estimate

$$\left\langle \prod_{A \in \mathscr{A}} |f_A| \right\rangle_{\bar{a}} \leqslant c^{|\mathbb{A}|} \prod_{A \in \mathscr{A}} \langle f_A^2 \rangle_{\bar{a}}^{1/2} \sum_{\mathscr{D} \in \hat{\mathscr{D}}} \prod_i \sum_{\Gamma_i \in \tilde{\mathscr{C}}(\mathscr{A}_i)} q^{d(\Gamma_i)} \tag{2.3.159}$$

where $q < 1$, $\mathbb{A} = \cup_{A \in \mathscr{A}} A$, $\hat{\mathscr{D}}$ is the family of all partitions $\mathscr{D} = \{A_i\}$ of $\mathscr{A}$ and $\tilde{\mathscr{C}}(\mathscr{A}_i)$ is the set of all graphs Γ_i on $\mathbb{A}_i = \cup_{A \in \mathscr{A}_i} A$ such that there is a tree Γ_i^* on $\mathscr{A}_i$ with a one-to-one map

$$\{s, t\} \in \Gamma_i \Rightarrow \{A, A'\} \in \Gamma_i^*; \qquad s \in A, \quad t \in A'.$$

Note that we can identify each Γ_i with an oriented graph $\vec{\Gamma}_i$ having the following property: for each $t \in \mathbb{A}_i$ there is no more than one bond of $\vec{\Gamma}_i$ having t as its starting point. So

$$\sum_{\mathscr{D}} \prod_i \sum_{\Gamma_i \in \tilde{\mathscr{C}}(A_i)} q^{d(\Gamma_i)} \leqslant \left(1 + \sum_{t \in \mathbb{Z}^\nu} q^{|t|}\right)^{|\mathbb{A}|}$$

which gives, together with (2.3.159), the desired inequality (2.3.155).

2.3.14. *Generalization of the Case of Complex Functions f_A*

Clearly, all the material of §§2.3.1, 2.3.4, 2.3.6 and 2.3.7 can also be reformulated to the case of complex functions f_A, with obvious changes (such as replacement of f^2 by $|f|^2$ in (2.3.77), etc.). In formulas such as (2.3.106) an appropriate value of Im(ln) must be chosen (this is usually the value prescribed by (2.1.29), (2.1.27)).

The following appendix to Theorem 2.3.8 will be useful in §4.5.

PROPOSITION. *Let $\{f_A^\lambda\}$, $\lambda \in \mathscr{V} \subset \mathbb{C}$ be a family of complex functions which are jointly measurable on $X(A) \times \mathscr{V}$, $\mathscr{V}$ being an open subset of $\mathbb{C}$. Suppose that for each*

$\lambda \in \mathscr{V}$, the functions $|f_A^\lambda|$ satisfy the condition (2.3.105), and have a common integrable majorant on $X(A)$. Suppose that for each fixed A and $x_A \in X(A)$, the function $f_A^{(\cdot)}(x_A)$ is analytical on $\mathscr{V}$. Then the quantities v_t constructed in (2.3.151) are analytical functions of $\lambda \in \mathscr{V}$. In particular, in the translation-invariant case, the quantity $\bar{u}$ depends analytically on $\lambda \in \mathscr{V}$.

Proof. This is easy to see if one inspects the proof of Theorem 2.3.8 bearing in mind that a uniform limit of analytical functions on $\mathscr{V}$ is analytical. (That is, one can approximate v_t uniformly by taking only a finite number of $\beta \in \mathscr{B}_t^{\Lambda'_i}$, as an easy consideration of the type (2.3.153) shows.) We omit the details of this proof. □

2.3.15. *Application: Uniqueness of a Perturbed Gaussian Field*

To illustrate some possibilities of applications of Theorem 2.3.8 (another important application is §4.4 below) we now consider the case of a Gibbsian field which is a small perturbation of a Gaussian field (with a positive mass).

THEOREM. *Let $\{\Phi_A^\gamma\}$, $\gamma > 0$ be some finite-range translation-invariant interaction such that*

$$\Phi_A^\gamma = \Phi_A + \gamma \Psi_A$$

where Φ_A is a quadratic interaction

$$\Phi_A(x_A) = (\Phi_{s-t} x_s, x_t) \qquad \text{if} \qquad A = \{s, t\}$$
$$\Phi_A(x_A) = 0 \text{ otherwise}$$

satisfying the conditions of §2.2.1, and Ψ_A is finite, bounded from below. Consider the Gibbsian density

$$P_\gamma(x_\Lambda \mid x_{\partial\Lambda^c}) = Z_\gamma^{-1}(\Lambda \mid x_{\partial\Lambda^c}) \exp(-H_\gamma(x_\Lambda \mid x_{\partial\Lambda^c}))$$

where H_γ, Z_γ are as in (1.2), (1.3) (with $\Phi = \Phi^\gamma$, $T = 1$). Then for any bounded $\mathscr{U} \subset \mathbb{R}^k$ there is some $\gamma_0 = \gamma_0(\Psi, \mathscr{U}, c_{(2.2.5)})$ such that if $\gamma \leqslant \gamma_0$ then the states $P_\gamma(\cdot \mid x_{\partial\Lambda_c})$ converge vaguely, if $\Lambda \uparrow \mathbb{Z}^\nu$ and $X_{\Lambda^c} \in \mathscr{U}^{\Lambda^c}$, to a unique Gibbs state P_γ on X i.e.

$$\int_X \varphi(x_B) \mathrm{d}P_\gamma(x) = \lim_{\Lambda \uparrow \mathbb{Z}^\nu} \int_{X(\Lambda)} \varphi(x_B) P_\gamma(x_\Lambda \mid x_{\partial\Lambda^c}) \mathrm{d}x_\Lambda \tag{2.3.160}$$

for any finite $B \subset \mathbb{Z}^\nu$ and any bounded measurable function φ.

Proof. We can write

$$\int_{X(\Lambda)} \varphi(x_B) P_\gamma(x_\Lambda \mid x_{\Lambda^c}) \mathrm{d}x_\Lambda = Z(\Lambda, x_{\partial\Lambda^c}, \{\gamma\Psi_A\})^{-1} Z'(\Lambda, x_{\partial\Lambda^c}, \{\gamma\Psi_A, \varphi\}) \tag{2.3.161}$$

where $Z(\Lambda, x_{\partial\Lambda^c}, \{\gamma\Psi_A\})$ is defined in (2.3.103) and where

$$Z'(\Lambda, x_{\partial\Lambda^c}, \{\gamma\Psi_A, \varphi\})$$
$$= \int_{X(\Lambda)} \exp\left(- \sum_{A: A \cap \Lambda \neq \emptyset;\ \mathrm{diam}\, A \leqslant r} \gamma\Psi_A(x_A) \right) \varphi(x_B) \mathrm{d}\mu_{x_{\partial\Lambda^c}}(x_\Lambda). \tag{2.3.161'}$$

As we noted in Example 2.3.8, the first partition function can be represented as

(2.3.102) with functions f_A defined by (2.3.104). Similarly (we may suppose that diam $B \leqslant r$ without loss of generality) we see that

$$Z'(\Lambda, x_{\partial\Lambda^c}, \{\gamma\Psi_A, \varphi\}) = Z(\Lambda, x_{\partial\Lambda^c}, \{f'_A, A \in \mathfrak{A}(\Lambda)\})$$

where

$$f'_A(\cdot) = f_A(\cdot) \quad \text{if} \quad A \neq B$$
$$f'_A(\cdot) = \exp(-\gamma\Psi(\cdot))\varphi(\cdot) - 1 \quad \text{if} \quad A = B.$$

To check the condition (2.3.105) of the theorem we apply Proposition 2.3.13, which reduces the problem to the proof of inequalities

$$\langle f'^p_A \rangle_a \leqslant \varepsilon(\gamma) \tag{2.3.162}$$

where $\lim_{\gamma\to 0}\varepsilon(\gamma) = 0$ uniformly with respect to A and $a \in \mathcal{U}^{\partial\Lambda^c}$. Notice that we can assume that $\sup_{x_B}\{|\varphi(x_B) - 1|\}$ is very small (in fact any φ is the linear combination of functions φ for which the latter condition is valid). It therefore suffices to prove (2.3.162) for the functions f_A.

To see this, note that by Theorem 2.2.2 and (2.2.14) we have for any A and any $x_{\partial\Lambda^c} \in \mathcal{U}^{\partial\Lambda^c}$ the estimate

$$\frac{d\mu_{A, x_{\partial\Lambda^c}}(x_A)}{dx_A} \leqslant p(x_A)$$

where $\mu_{A, x_{\partial\Lambda^c}}$ is the projection on $X(A)$ of the Gaussian Gibbsian measure μ, corresponding to the potential Φ, conditioned by $x_{\partial\Lambda^c}$, and $p(\cdot)$ is some integrable function. Then we use the boundedness of Ψ_A.

Now we can use the representation (2.3.106) for both $\ln Z$ and $\ln Z'$. Applying the estimate (2.3.109) and (2.3.110) we obtain the desired relation (2.3.160). □

NOTE 1. As in Section 2.3.14, we obtain the analyticity of the left-hand side of (2.3.160) in γ. We omit the proof of this.

It is also possible to prove the exponential decay of correlations in P_γ. Again we have left it to the reader to formulate and prove statements of this type.

NOTE 2. We notice that the considerations of §2.3.15 prove (under some general conditions on the Hamiltonian, similar to the conditions of the Main Theorem) the uniqueness of the Gibbs field in the case when there is only one ground state of the Hamiltonian (at sufficiently low temperatures).

It is reasonable to study this problem in a setting analogous to the main lemma (§3), making a change of variables $x \rightsquigarrow T^{-1/2}x$, but this, too is omitted here.

3. The Main Lemma

3.1. THE GEOMETRY OF CONFIGURATIONS

In this section, some geometrical notions needed for the formulation of the Main Lemma are defined. For other geometrical definitions (contours, etc.) needed in the proof of the Main Lemma see §4.1.

Recall that we treat $\mathbb{Z}^\nu$ as a connected graph with bonds (s, t) where $|s - t| = 1$, the norm $|t|$ being given as

$$|t| = \sum_{i=1}^{\nu} |t_i|.$$

Fix some integer $r \in \mathbb{N}$. Later, this integer will have the meaning of a range of interactions of the given model. In this case we will usually omit the subscript r in (1.1) and write

$$\partial_r \Lambda \equiv \partial \Lambda \tag{3.1.1}$$

for simplicity.

NOTE. It can be shown that $\partial\Lambda \cup \partial\Lambda^c$ is connected for any simple connected (i.e. such that both Λ and Λ^c are connected) set Λ. This fact will be used in some geometrical considerations. We omit its proof.

3.1.1. *Basic Notation*

Fix some $\delta > 0$ and some $\sigma_+, \sigma_- \in \mathbb{R}^k$ such that $|\sigma_+ - \sigma_-| > 2\delta$. Denote by $\mathscr{U}^\pm$ the δ neighborhood of σ_+. Denote by $\mathscr{U}^0 = \mathbb{R}^k \setminus (\mathscr{U}^+ + \cup \mathscr{U}^-)$.

3.1.2. *Definition*

Say that a point $t \in \mathbb{Z}^\nu$ is a $+$*correct* point of the configuration $x \in X \equiv (\mathbb{R}^k)^{\mathbb{Z}^\nu}$ if $x_s \in \mathscr{U}^+$ for all $s \in \mathbb{Z}^\nu$ satisfying the inequality $|s - t| \leqslant r$. The definition. of a $-$*correct* point of $x \in X$ is analogous. The points of $\mathbb{Z}^\nu$ which are neither $+$ nor $-$ correct points of x will be called non-correct or boundary points of the configuration x.

The set of all non-correct points of a configuration $x \in X$ will be called a *boundary of* x and denoted by $B(x)$. Given $\Lambda \subset \mathbb{Z}^\nu$, denote by $B(\Lambda)$ the subset of X consisting of all configurations $x \in X$ such that $B(x) = \Lambda$.

3.1.3. *Definition*

For any $x \in X$ define a function $\operatorname{sign}_x(\cdot)$ of $t \in \mathbb{Z}^\nu$ with values $+$, $-$, 0 such that $x_t \in \mathscr{U}^{\operatorname{sign}_x(t)}$, $t \in \mathbb{Z}^\nu$. It is evident that this function depends only on the values of x on $B(x)$ and has a constant value on each connected component of $(B(x))^c$. Depending on this value, we will distinguish the $+$components and $-$components of $(B(x))^c$. We denote by $(B(x))^{c\pm}$ the union of all $\pm$components of $(B(x))^c$. If $\Lambda \subset \mathbb{Z}^\nu$ is such that $\Lambda \supset B(x)$ then we also write $\Lambda^{c\pm}$ instead of $\Lambda^c \cap (B(x))^{c\pm}$.

We will often use the following notion of a $\pm$correct point of a configuration in a finite volume.

3.1.4. *Definition*

Let $\Lambda \subset \mathbb{Z}^\nu$, let $x_\Lambda \in X(\Lambda)$. Say that x_Λ is *outside correct* if it can be extended to the whole of $\mathbb{Z}^\nu$ such that $B(x) \subset \Lambda$ where $B(x)$ is the boundary of the extended configuration. (It is clear that the extension is unique in the sense of uniqueness of the function $\operatorname{sign}_x(t)$, $t \in \mathbb{Z}^\nu$.) Alternatively, $x_\Lambda \in X(\Lambda)$ is outside correct if for any $t \in \Lambda^c$, either $x_s \in \mathcal{U}^+$ for all $s \in \Lambda$, $|t - s| \leqslant r$ or $x_s \in \mathcal{U}^-$ for those $s \in \Lambda$. (We omit the proofs of all these geometrical facts.)

NOTE. If Λ^c is connected (in particular if Λ is simply connected) and $x_\Lambda \in X(\Lambda)$ is outside correct then clearly either $\operatorname{sign}_x(t) = +$ for all $t \in \Lambda^c$ or $\operatorname{sign}_x(t) = -$ for all $t \in \Lambda^c$. In the general case, $\operatorname{dist}(\Lambda^{c+}, \Lambda^{c-}) > r$.

3.1.5. *Definition*

Let $\Lambda \subset \mathbb{Z}^\nu$, let $x_\Lambda \in X(\Lambda)$ be outside correct. Because of the uniqueness of the function $\operatorname{sign}_x(t)$ we will denote it also by $\operatorname{sign}_{x_\Lambda}(t)$. We will also write $B(x_\Lambda)$ instead of $B(x)$. Each $t \in (B(x_\Lambda))^c$ will be called a correct point of x_Λ. Denoting $B(x_\Lambda) = \bar{\Lambda}$ we also write $x_\Lambda \in B(\bar{\Lambda})$, with a slight abuse of notation.

3.2. REDUCTION TO THE MAIN LEMMA

The aim of this section is to show that the proof of the Main Theorem can be reduced to the proof of the following result.

3.2.1. *Main Lemma*

Consider again (see §1) a spin model with the configuration space $X = (\mathbb{R}^k)^{\mathbb{Z}^\nu}$. Suppose that some family of interactions $\Phi^\lambda = \{\Phi_A^\lambda\}$ depending on some parameter $\lambda \in [\lambda_1, \lambda_2]$, $0 \in [\lambda_1, \lambda_2] \subset \mathbb{R}$ is given.

Suppose that some constant configurations $x^+ = \{x_t^+ \equiv \sigma^+, t \in \mathbb{Z}^\nu\}$ and $x^- = \{x_t^- \equiv \sigma^-, t \in \mathbb{Z}^\nu\}$ (*where* $\sigma^+, \sigma^- \in \mathbb{R}^k$) are fixed, and also two additional families of interactions $\Phi^{+,\lambda} = \{\Phi_A^{+,\lambda}\}$ and $\Phi^{-,\lambda} = \{\Phi_A^{-,\lambda}\}$ approximating Φ^λ in a sense explained below.

Suppose that all the interactions are translation-invariant and have the same finite interaction range r. We will use the constructions of §3.1, with these choices of r, σ^+, σ^-.

We now introduce some assumptions about these interactions.

ASSUMPTION 1. Suppose that the interactions $\Phi_A^{\pm,\lambda}$ are linear quadratic (see §2.2.6), i.e. that they can be expressed as follows:

$$\Phi_A^{\pm,\lambda}(x_A) = \sum_{s,t \in A} (\varphi_{s,t,A}^{\pm}(x_s - \sigma^\pm), (x_t - \sigma^\pm)) + \sum_{t \in A} (\varphi_{t,A}^{\pm}, (x_t - \sigma^\pm)) + \\ + \varphi_A^{\pm} + \lambda u_A^{\pm} \tag{3.2.1}$$

where $\varphi_A^\pm$, $u_A^\pm \in \mathbb{R}$, $\varphi_{t,A}^\pm \in \mathbb{R}^k$ and $\varphi_{s,t,A}^\pm$ are symmetric real $k \times k$ matrices. Put

$$\varphi_{s,t}^{\pm} = \sum_{A \subset \mathbb{Z}^\nu : \{s,t\} \subset A} \varphi_{s,t,A}^{\pm}, \qquad s, t \in \mathbb{Z}^\nu. \tag{3.2.2}$$

Note that $\varphi_A^\pm$, $u_A^\pm$, $\varphi_{t,A}^\pm$, $\varphi_{s,t,A}^\pm$ and therefore also $\varphi_{s,t}^\pm$, are translation-invariant. Write, therefore,

$$\varphi^{\pm}_{s,t} \equiv \varphi^{\pm}_{s-t}, \qquad s, t \in \mathbb{Z}^{\nu}.$$

Suppose that the quadratic Hamiltonians

$$H^{\pm}_{q}(x_{\Lambda}) = \sum_{s\in\Lambda, t\in\Lambda} (\varphi^{\pm}_{s-t} x_s, x_t)$$

are *positive definite*, in the following sense (compare (2.2.5)): there is some $c > 0$ such that for each $x_{\Lambda} \in X(\Lambda)$, $|\Lambda| < \infty$ the following inequality is satisfied:

$$c|x_{\Lambda}|^2 \leqslant H^{\pm}_{q}(x_{\Lambda}) \leqslant c^{-1}|x_{\Lambda}|^2. \tag{3.2.3}$$

Concerning the constants $\varphi^{\pm}_{A}$, $\varphi^{\pm}_{t,A}$ suppose that

$$\sum_{A\subset \mathbb{Z}^{\nu}: t\in A} \varphi^{\pm}_{t,A} = 0, \qquad \sum_{A\subset \mathbb{Z}^{\nu}: t\in A} |A|^{-1}\varphi^{\pm}_{A} = 0, \qquad t \in \mathbb{Z}^{\nu}. \tag{3.2.4}$$

Denote

$$u^{\pm} = \sum_{A\subset \mathbb{Z}^{\nu}: t\in A} |A|^{-1} u^{\pm}_{A}. \tag{3.2.5}$$

ASSUMPTION 2. There is some $\delta > 0$ such that $\delta < (|\sigma^+ - \sigma^-|)/2$ and some $\varepsilon > 0$, $\varepsilon' > 0$ such that for any $x_A \in X(A)$, A finite, satisfying the condition $|x_t - \sigma^+| \leqslant \delta$, $t \in A$ the following inequalities hold: for any parameter $\lambda \in [\lambda_1, \lambda_2]$:

$$|\Phi^{\lambda}_{A}(x_A) - \Phi^{+,\lambda}_{A}(x_A)| \leqslant \varepsilon, \tag{3.2.6}$$

$$\left| \frac{\partial}{\partial\lambda} \Phi^{\lambda}_{A}(x_A) - u^{+}_{A} \right| \leqslant \varepsilon'. \tag{3.2.7}$$

This also holds if $+$ is replaced by $-$. (Later it will be assumed that δ is sufficiently large and ε, ε' are sufficiently small.) We fix the value of δ throughout the proof of the Main Lemma. If not otherwise specified, we use this value of δ whenever the concepts of section 3.1 are discussed.

ASSUMPTION 3. Denote by $h^{\pm}_{\lambda}$ the free energy (see §§2.2.4 and 2.2.6) of the potential $\Phi^{\pm,\lambda}$:

$$h^{\pm}_{\lambda} = \lim_{\Lambda\uparrow \mathbb{Z}^{\nu}} |\Lambda|^{-1} \ln \int_{X(\Lambda)} \exp\left(- \sum_{A\subset\Lambda} \Phi^{\pm,\lambda}_{A}(x_A) \right) dx_{\Lambda}$$

(the limit is taken in the Van Hove sense). Note that $h^{\pm}_{\lambda} = h^{\pm}_{0} - \lambda u^{\pm}$ (see (3.2.1) and (3.2.5)). Suppose that there is some $\Delta > 0$ such that

$$h^{-}_{\lambda_1} - h^{+}_{\lambda_1} \geqslant \Delta \text{ and also } h^{+}_{\lambda_2} - h^{-}_{\lambda_2} \geqslant \Delta. \tag{3.2.8}$$

ASSUMPTION 4. For each finite $A \subset \mathbb{Z}^{\nu}$, the domain $X^{\infty}(A) = \{x_A \in X(A): \Phi^{\lambda}_{A}(x_A) = +\infty\}$ is closed and does not depend on λ, and there is some open $\mathscr{V} \subset \mathbb{C}$ suth that $\mathscr{V} \supset [\lambda_1, \lambda_2]$ and such that $\Phi^{(\cdot)}_{A}(x_A)$ extends to some analytical function on $\mathscr{V}$, for each $x_A \in X(A) \setminus X^{\infty}(A)$. The function $d/d\lambda \Phi^{\lambda}_{A}(x_A)$ is moreover continuous as a function of two variables $\lambda \in \mathscr{V}$, $x_A \in X(A) \setminus X^{\infty}(A)$.

Now we will formulate a condition of the Peierls (Gertzik–Pirogov–Sinai) type. Its relations to the analogous condition of §1 will be explained later in this section. Let $\Lambda \subset \mathbb{Z}^\nu$ be finite, let $x \in X$ be a configuration such that $x_t = x_t^+$ almost everywhere. Let $\Lambda^c = M^+ \cup M^-$ be a decomposition of Λ^c such that $\mathrm{dist}(M^+, M^-) > r$ and M^- is finite (if $B(x) \subset \Lambda$ we will usually take $M^+ = \Lambda^{c+}$, $M^- = \Lambda^{c-}$, see Definition 3.1.3). Put

$$H_{\mathrm{rel}}^{\lambda,\Lambda,M^+,M^-}(x) = \sum_{A \subset \Lambda} (\Phi_A^\lambda(x_A) - \Phi_A^{+,0}(x_A^+)) + $$

$$+ \sum_{A: A \cap M^+ \neq \emptyset} (\Phi_A^{+,0}(x_A) - \Phi_A^{+,0}(x_A^+)) +$$

$$+ \sum_{A: A \cap M^- \neq \emptyset} (\Phi_A^{-,0}(x_A) - \Phi_A^{+,0}(x_A^+)). \tag{3.2.9}$$

Analogously, substituting $\pm$ by $\mp$ we define the value $H_{\mathrm{rel}}^{\lambda,\Lambda,M^-,M^+}(x)$ for configurations x coinciding with x^- almost everywhere (in this case we assume that M^+ is finite!). Now let $x_\Lambda \in B(\Lambda)$ (see Definition 3.1.5). Put

$$H_{\mathrm{rel}}^\lambda(x_\Lambda) = \inf_{y: y_\Lambda = x_\Lambda} \{H_{\mathrm{rel}}^{\lambda,\Lambda,\Lambda^{c+},\Lambda^{c-}}(y)\}. \tag{3.2.10}$$

NOTE 1. This seems to be a more complicated notion than the quantities in (1.12). The reason for introducing $H_{\mathrm{rel}}^\lambda(x_\Lambda)$ is that it is more suitable to work with (see the later part of §3 and §4.3).

NOTE 2. We do not assume that $y \in B(\Lambda)$ in (3.2.10). This is reasonable if we look at the behaviour of an 'optimal' configuration y in (3.2.10). Such optimal configurations are studied (simultaneously for the + and − region) in §2.2.7. The minimizing configuration $\bar{\bar{x}}$ in (2.2.74) need not satisfy the condition $|\bar{x}_t| \leqslant \max_{w \in \Lambda} |\bar{x}_w|$, $t \in W$!

ASSUMPTION 5. There are some $\tau, \tau' > 0$ such that for each finite $\Lambda \subset \mathbb{Z}^\nu$ and each $\lambda \in [\lambda_1, \lambda_2]$,

$$\int_{B(\Lambda)} \exp(-H_{\mathrm{rel}}^\lambda(x_\Lambda)) \mathrm{d}x_\Lambda \leqslant \exp(-\tau|\Lambda|) \tag{3.2.11}$$

and

$$\frac{\mathrm{d}}{\mathrm{d}\lambda} \int_{B(\Lambda)} \exp(-H_{\mathrm{rel}}^\lambda(x_\Lambda)) \mathrm{d}x_\Lambda \leqslant \exp(-\tau'|\Lambda|). \tag{3.2.12}$$

The same is true for any $\lambda \in \mathscr{V}$ if we replace $\Phi_A^\lambda, \Phi_A^{+,\lambda}$ by $\mathrm{Re}\Phi_A^\lambda, \Phi_A^{+,\mathrm{Re}\lambda}$. Assume moreover that there is an integrable majorant for all $\exp(-H_{\mathrm{rel}}^\lambda(\cdot))$ and $\mathrm{d}/\mathrm{d}\lambda$ $(\exp(-H_{\mathrm{rel}}^\lambda(\cdot))$. (Later we will assume that τ and τ' are sufficiently large.)

We denote by $\mathscr{F}(\Phi^+, \Phi^-, \delta, \varepsilon, \varepsilon', \tau, \tau')$ the class of interactions $\Phi = \{\Phi^\lambda, \lambda \in [\lambda_1, \lambda_2]\}$ for which Assumptions 1 to 5 are satisfied, with fixed interactions $\Phi^+ = \{\Phi^{+,\lambda}, \lambda \in [\lambda_1, \lambda_2]\}$ $\Phi^- = \{\Phi^{-,\lambda}, \lambda \in [\lambda_1, \lambda_2]\}$ and fixed constants

$\delta, \varepsilon, \varepsilon', \tau, \tau'$. With this notations and these assumptions, the statement of the Main Lemma is as follows.

There are some δ_0, ε_0, ε'_0, τ_0, τ'_0 depending only on c, Δ, u^+, u^- (also on $r, v, k, \lambda_1, \lambda_2$, but we usually omit the dependence on these constants) such that the following holds: if $\Phi \in \mathscr{F}(\Phi^+, \Phi^-, \delta, \varepsilon, \tau, \tau')$ where $\delta \geqslant \delta_0$, $\varepsilon \leqslant \varepsilon_0$, $\varepsilon' \leqslant \varepsilon'_0$, $\tau \geqslant \tau_0(\ln \delta)^v$, $\tau' \geqslant \tau_0(\ln \delta)^v$ then there is some $\lambda = \lambda(\Phi) \in [\lambda_1, \lambda_2]$ such that

(a) *There are two different translation-invariant Gibbs states P_Φ^+, P_Φ^- corresponding to the interaction $\Phi^{\lambda(\Phi)}$.*

(b) *The value of $\lambda(\Phi)$ does not depend on the concrete choice of $\Phi^\pm, \delta, \varepsilon, \varepsilon', \tau, \tau'$.*

(c) *If $\beta\Phi \in \mathscr{F}(\Phi^+, \Phi^-, \delta, \varepsilon, \varepsilon', \tau, \tau')$ for each $\beta \in [\beta_1, \beta_2]$ then $\{\beta \rightsquigarrow \lambda(\beta\Phi)\}$ is a real analytical function of $\beta \in [\beta_1, \beta_2]$.*

(d) *For $\delta \to \infty$, $\varepsilon \to 0$, $\tau(\ln \delta)^{-v} \to \infty$ we have the asymptotic relation $\lambda(\Phi) \to \lambda_0$ where λ_0 is the solution of the equation (see Assumption 3)*

$$h^{+,\lambda} = h^{-,\lambda}.$$

(e) *With the same conditions as in (d) the states $P_\Phi^\pm$ are vaguely convergent to translation-invariant Gaussian fields, defined as Gibbsian fields corresponding to the linear quadratic pair interactions*

$$\begin{aligned}\Phi_A(x_A) &= \tfrac{1}{2}(\varphi^\pm_{t,t}(x_t - \sigma^\pm), (x_t - \sigma^\pm)) \qquad \text{if} \qquad A = \{t\}\\ &= (\varphi^\pm_{s,t}(x_t - \sigma^\pm), (x_s - \sigma^\pm)) \qquad \text{if} \qquad A = \{s, t\}\\ &= 0 \qquad \text{otherwise}\end{aligned}$$

(*see* §2.2.6).

3.2.2. *Proof of the Main Theorem*

In this section, we show how the Main Lemma implies the Main Theorem. The rest of the paper will be devoted to the proof of the Main Lemma.

First we show that the ground states of §1 can be assumed to be constant configurations: if the period of $x^\pm$ is non-trivial (i.e. if W in (1.19) is not equal to $\{0\}$) we define a new model on $\mathbb{Z}^v$ with 'spins' in $\mathbb{R}^{k|W|}$, defined as the blocks of the original spins

$$\tilde{x}_t \equiv \{x_s, s \in t + W\}, \qquad t \in \tilde{\mathbb{Z}}^v. \tag{3.2.13}$$

It is clearly possible to rewrite the interactions $\{\Phi_A^\lambda\}$ as interactions of the blocks, each new interaction $\Phi_{\tilde{A}}^\lambda$, $\tilde{A} \subset \tilde{\mathbb{Z}}^v$ being a sum of interactions Φ_A^λ such that

$$A \cap (t + W) \neq \emptyset \qquad \text{iff} \qquad t \in \tilde{A}.$$

In this new model, the ground states are constant configurations. Also, all the assumptions of the Main Theorem are satisfied in this new setting (with changed constants c, δ, etc.). By using the 'block' Gibbs states thus constructed we get the existence of the required Gibbs states in the original model. Thus, in the rest of this paper we assume that the ground states x^+ and x^- are constant configurations $\{x_t^+ \equiv \sigma^+\}$ and $\{x_t^- \equiv \sigma^-\}$.

We now construct objects satisfying the assumptions of the Main Lemma, using the analogous objects of the main theorem.

Define (we use the same notation as in (1.15), (1.16))

$$\varphi^{\pm}_{t,s,A} = \frac{\partial^2}{\partial x_t \partial x_s} \Phi^0_A(x_A)|_{x_A = \sigma^{\pm}_A}$$

$$\varphi^{\pm}_{t,A} = \frac{\partial}{\partial x_t} \Phi^0_A(x_A)|_{x_A = \sigma^{\pm}_A}$$

$$\varphi^{\pm}_A = \Phi^0_A(\sigma^{\pm}_A), \qquad u^{\pm}_A = \frac{\partial}{\partial \lambda} \Phi^{\lambda}_A(\sigma^{\pm}_A)|_{\lambda = 0}, \tag{3.2.14}$$

where

$$\sigma^{\pm}_A = \{x_t \equiv \sigma^{\pm}, t \in A\}.$$

Define the interactions

$$\Phi^{\pm}_A(x_A) = \tfrac{1}{2} \sum_{s,t \in A} (\varphi^{\pm}_{s,t,A}(x_s - \sigma^{\pm}), (x_t - \sigma^{\pm})) + \sum_{t \in A} (\varphi^{\pm}_{t,A}, (x_t - \sigma^{\pm})) + \varphi^{\pm}_A \tag{3.2.15}$$

for any $x_A \in X(A)$, diam $A \leqslant r$.

Using Assumption 4 of the Theorem we can find, for small enough δ, some $\varepsilon = \varepsilon(\delta)$ such that $\lim_{\delta \to 0} \varepsilon = 0$ and

$$|\Phi^0_A(x_A) - \Phi^{\pm}_A(x_A)| < \varepsilon |x_A - \sigma^{\pm}_A|^2 \tag{3.2.16}$$

whenever $x_A \in X(A)$, A finite is such that $|x_t - \sigma^{\pm}| < \delta$, $t \in A$. Using the continuity of $\partial/\partial_\lambda \Phi^{\lambda}_A(x_A)$ we can find some $\varepsilon' = \varepsilon'(\delta, \lambda)$ such that $\lim_{\delta \to 0/\lambda \to 0} \varepsilon' = 0$ and

$$|\Phi^{\lambda}_A(x_A) - \Phi^0_A(x_A) - \lambda u^{\pm}_A| \leqslant \lambda \varepsilon'(\delta, \lambda_0), \quad |\lambda| \leqslant \lambda_0 \tag{3.2.17}$$

whenever $|x_t - \sigma^+| < \delta$, $t \in A$ (or $|x_t - \sigma^-| < \delta, t \in A$), and λ_0 and δ are sufficiently small (see (1.24)). In fact, with the same condition on x_A we obtain the bound

$$\left| \frac{\partial}{\partial \lambda} \Phi^{\lambda}_A(x_A) - u^{\pm}_A \right| \leqslant \varepsilon'(\delta, \lambda_0), \quad \lambda \leqslant \lambda_0 \tag{3.2.18}$$

where $\varepsilon' = \varepsilon'(\delta, \lambda)$ is such that $\lim_{\delta \to 0/\lambda \to 0} \varepsilon' = 0$. This implies (3.2.17).

NOTE. In order to distinguish the concepts of the Main Theorem from the analogous ones of the Main Lemma, whenever there would be an ambiguity in notation in the rest of this section, we distinguish the notions of the Theorem by a tilde ($\sim$).

Fix some enough small $\delta > 0$ and $T > 0$ (to be specified later). We now introduce the following notation. Denote by

$$x = \tilde{T}^{-1/2} \tilde{x}, \qquad \sigma^{\pm} = \tilde{T}^{-1/2} \tilde{\sigma}^{\pm}, \qquad \delta = \tilde{T}^{-1/2} \tilde{\delta},$$

$$\lambda = M^{-1} \tilde{T}^{-1/2} \tilde{\lambda}, \quad \lambda_1 = -1, \quad \lambda_2 = 1,$$

$$\Phi^{\lambda}_A(x_A) = \tilde{T}^{-1} \tilde{\Phi}^{\tilde{\lambda}}_A(\tilde{x}_A) \tag{3.2.19}$$

where M is large enough and will be specified later. Define

$$\Phi^{\pm,\lambda}_A(x_A) = \tilde{T}^{-1} \tilde{\Phi}^{\pm,\tilde{\lambda}}(\tilde{x}_A) \tag{3.2.20}$$

where (see (1.24), (1.9) and (3.2.15))

$$\tilde{\Phi}_A^{\pm\tilde{\lambda}}(\tilde{x}_A) = \tilde{\Phi}_A^{\pm}(\tilde{x}_A) + \tilde{\lambda}\tilde{u}_A^{\pm} \qquad \text{if} \qquad |A| > 1$$

$$\tilde{\Phi}_A^{\pm,\tilde{\lambda}}(\tilde{x}_A) = \tilde{\Phi}_A^{\pm}(\tilde{x}_A) + \tilde{\lambda}\tilde{u}_A^{\pm} - \tilde{\bar{e}} \qquad \text{if} \qquad |A| = 1.$$

Write

$$\varphi_{s,t,A}^{\pm} = \tilde{\varphi}_{s,t,A}^{\pm}, \qquad \varphi_{t,A}^{\pm} = \tilde{T}^{-1/2}\tilde{\varphi}_{t,A},$$

$$\varphi_A^{\pm} = \tilde{T}^{-1}\tilde{\varphi}_A^{\pm}, \qquad u_A^{\pm} = M\tilde{u}_A^{\pm}.$$

With these definitions of $\Phi^{\lambda}, \Phi^{t,\lambda}$ we will show that the assumptions of Main Theorem (Th 1–Th 7) imply, for enough small T, all the assumptions of the Main Lemma (L 1–L 5), with properly chosen constants $\delta, \varepsilon, \varepsilon', \tau, \tau', c, \Delta$. We will demonstrate this in the following order:

(a) Th 2, 4, 5 $\Rightarrow$ L 1.
(b) Th 4, 6 $\Rightarrow$ L 2.
(c) Th 4, 6, L 1 $\Rightarrow$ L 3.
(d) Th 6 $\Rightarrow$ L 4.
(e) L 1, L 2, Th 1, 3 $\Rightarrow$ L 5.

(a) The first relation in (3.2.4) follows from (1.13). The second one follows from the definition of $\Phi_A^{\pm,\lambda}$ and $\bar{e}$ (see (3.2.20) and (1.9)).

To prove the relation (3.2.3) we use the Fourier transform formula

$$\sum_{s,t\in\Lambda} (\varphi_{s,t}^{\pm} x_s, x_t) = (2\pi)^{-\nu} \int_{[-\pi,\pi]^{\nu}} (f^{\pm}(\xi)x_{\xi}, x_{\xi})d\xi \tag{3.2.21}$$

where

$$f^{\pm}(\xi) = \sum_{t\in Z^{\nu}} \varphi_{t,0}^{\pm} \exp(i(\xi,t))$$

and (compare (1.20))

$$x_{\xi} = \sum_{t\in\Lambda} \exp(i(\xi,t))x_t.$$

By Parseval's equality and relation (1.21) we get the left-hand side of the inequality (3.2.3). The right-hand side of (3.2.3) follows trivially from the finiteness of the interaction range r of $\Phi^{\pm,\lambda}$.

(b) By (3.2.16), (3.2.17) and by definitions (3.2.19), (3.2.20) we get, for any x_A such that $|x_t - \sigma^{\pm}| \leqslant \delta$, $t\in A$ and any $\lambda\in[-1,1]$,

$$|\Phi_A^{\lambda}(x_A) - \Phi_A^{\pm,\lambda}(x_A)| \leqslant \tilde{T}^{-1}\tilde{\varepsilon}(\tilde{\delta})|A|\tilde{\delta}^2 + \tilde{\varepsilon}'(\tilde{\delta}, M\tilde{T})M. \tag{3.2.22}$$

Similarly, by using (3.2.18) we get, for the same x_A and λ,

$$\left|\frac{\partial}{\partial\lambda}\Phi_A^{\lambda}(x_A) - u_A^{\pm}\right| \leqslant \tilde{\varepsilon}'(\tilde{\delta}, M\tilde{T})M. \tag{3.2.23}$$

This clearly proves the desired relations (3.2.6) and (3.2.7) (with enough small $\varepsilon, \varepsilon'$), assuming that

$$\tilde{T}^{-1}\tilde{\varepsilon}(\tilde{\delta})\tilde{\delta}^2, \qquad M\tilde{\varepsilon}'(\tilde{\delta}, M\tilde{T}), \qquad \tilde{\varepsilon}'(\tilde{\delta}, M\tilde{T}) \tag{3.2.24}$$

are sufficiently small.

(c) Using (3.2.20) we easily get the relation

$$h^{\pm,\lambda} = h^{\pm,0} - \lambda u^{\pm} \tag{3.2.25}$$

where $u^{\pm} = M\tilde{u}^{\pm}$. Note that $h^{\pm,0}$ does not depend on the choice of $\tilde{T}$, because $\varphi^{\pm}_{\tilde{t},0}$ does not depend on $\tilde{T}$ and also because the first relation of (3.2.4) implies that the standardized Hamiltonian corresponding to $\Phi^{\pm,0}$ (see §2.2.6) is a pure quadratic one.

From our choice of λ_1, λ_2 (see (3.2.19)) it is clear that (3.2.8) will hold if (remember the assumption Th 6)

$$M|\tilde{u}^+ - \tilde{u}^-| > \Delta|h^{+,0} - h^{-,0}|. \tag{3.2.26}$$

(d) This is obvious.

(e) This is the only non-trivial part of the proof. It will be carried out in several steps.

(i) As a first step we will show that for all sufficiently small $\delta > 0$ it is possible to assume that the constant τ from (1.12) satisfies the relation

$$\tau \geqslant c\delta^2 \tag{3.2.27}$$

where c does not depend on δ.

NOTE. Because we are in the setting of the Main Theorem again in this part of the proof, we omit the tilde $\sim$ in all the considerations here and in the forthcoming two steps.

Write the expression on the left-hand side of (1.12) as

$$H^*_{\text{rel}}(x) = \sum_{A \subset \mathbf{Z}^\nu} (\Phi_A(x_A) - \Phi_A(x_A^+)). \tag{3.2.28}$$

(An analogous expression holds for $+$ altered to $-$ for a configuration x coinciding with x^- almost everywhere.)

Fix some enough small $\delta_0 > 0$, satisfying (1.12) with some $\tau_0 > 0$. Let $\delta < \delta_0$. In order to emphasize the value of δ used in the definitions of §3.1 write $B_\delta(x)$ instead of $B(x)$, $B_\delta(\Lambda)$ instead of $B(\Lambda)$, $\text{sign}^\delta_x(\cdot)$ instead of $\text{sign}_x(\cdot)$ etc.

Choose some $x \in X$ coinciding with x^+ almost everywhere, almost surely. Denote by $\Lambda_0 = B_{\delta_0}(x)$, $\Lambda = B_\delta(x)$. Obviously, $\Lambda \supset \Lambda_0$. Consider the functions $\text{sign}^\delta_{x\Lambda}(\cdot)$, $\text{sign}^{\delta_0}_{x\Lambda_0}(\cdot)$ (defined in an obvious sense see §3.1.5) and note that they are identical outside Λ. Recall the decomposition (see §3.1.3, with δ changed to δ_0) $\Lambda_0^c = \Lambda_0^{c+} \cup \Lambda_0^{c-}$.

Define by induction the sets $\Lambda_1, \Lambda_2, \ldots$ as follows:

$$\Lambda_i = \Lambda_{i-1} \cup M_i, \qquad M_i = \partial\Lambda_{i-1}^c, \qquad i = 1, \ldots. \tag{3.2.29}$$

Consider also the sets

$$\Lambda_i^{c\pm} = \Lambda_i^c \cap \Lambda_0^{c\pm}, \qquad M_i^{\pm} = M_i \cap \Lambda_0^{c\pm}. \tag{3.2.30}$$

Notice that $M_i^\pm$ are pairwise non-intersecting and also

$$\Lambda_i^c = \bigcup_{j>i} (M_j^+ \cup M_j^-).$$

Define the following auxiliary configurations:

$$x_i = x_{(M_i)^c} \cup x^+_{M_i^+} \cup x^-_{M_i^-}$$

$$y_i = x_{\Lambda_{i-1}} \cup x^+_{\Lambda^{c+}_{i-1}} \cup x^-_{\Lambda^{c-}_{i-1}},$$

$$z_i^+ = x_{\Lambda_i^c+} \cup x^+_{(\Lambda_i^{c+})^c}, \qquad z_i^- = x^{-}_{\Lambda_i^{c-}}{}' \cup x^-_{(\Lambda_i^{c-})^c}. \tag{3.2.31}$$

Using this notation we can write (see Definition (3.2.28))

$$H^*_{\text{rel}}(x) = H^*_{\text{rel}}(x_i) + H(x_{M_i} \mid x_{M_i^c}) - H(x^+_{M_i^+} \cup x^-_{M_i^-} \mid x_{M_i^c}) \tag{3.2.32}$$

and

$$H^*_{\text{rel}}(x_i) = H^*_{\text{rel}}(y_i) + H^*_{\text{rel}}(z_i^+) + H^*_{\text{rel}}(z_i^-) \tag{3.2.33}$$

By (1.12) we obtain, for $\tau_0 = \tau(\delta_0)$,

$$H^*_{\text{rel}}(y_i) \geqslant \tau_0 |\Lambda_0|. \tag{3.2.34}$$

The estimates of $H^*_{\text{rel}}(z_i^\pm)$ are based on the approximations (3.2.16) and the positive definiteness condition (1.21) written in the form (3.2.3) (see part (a) of this proof). We get

$$H^*_{\text{rel}}(z_i^\pm) \geqslant c_1 |(z_i^\pm)_{\Lambda_i^{c\pm}} - (x^\pm)_{\Lambda_i^{c\pm}}|^2 \tag{3.2.35}$$

where c_1 depends only on $c_{(1.21)}$, $c_1 > 0$. Denote by

$$a_i^\pm = |x_{M_i^\pm} - x^\pm_{M_i^\pm}|^2, \qquad a_i = a_i^+ + a_i^-. \tag{3.2.36}$$

Taking the Taylor expansion of Φ_A at $\bar{x}_A$ we see from (1.12) that, for enough small δ_0, there is a constant $c_2 > 0$ such that, for $i \geqslant 2$,

$$|H(x_{M_i} \mid x_{M_i^c}) - H(x^+_{M_i^+} \cup x^-_{M_i^-} \mid x_{M_i^c})| \leqslant c_2(a_{i-1} + a_i + a_{i+1}). \tag{3.2.37}$$

Combining the observations (3.2.32)–(3.2.37) we obtain, for $i \geqslant 2$, the following intermediate result:

$$H^*_{\text{rel}}(x) \geqslant \tau_0 |\Lambda_0| - 2c_2(a_{i-1} + a_i + a_{i+1}) + c_1 \sum_{j>i} a_j. \tag{3.2.38}$$

Now choose the smallest possible $i = i_0 \geqslant 2$ such that either

$$4c_2(a_{i-1} + a_i + a_{i+1}) \leqslant c_1 \sum_{j>i} a_j \tag{3.2.39}$$

$$4c_2(a_{i-1} + a_i + a_{i+1}) \leqslant \tau_0 |\Lambda_0|. \tag{3.2.40}$$

Because, for $2 \leqslant i < i_0$,

$$a_{i-1} + a_i + a_{i+1} > \tfrac{1}{4} c_1 c_2^{-1} \sum_{j>i+1} a_j$$

i.e.

$$\sum_{j \geq i-2} a_j \geq (1 + \tfrac{1}{4}c_1 c_2^{-1}) \sum_{j > i+1} a_j$$

we see that there is some $c_3 = c_3(c_1, c_2)$ and $q = q(c_1, c_2)$, $0 < q < 1$ such that

$$a_i \leq c_3(a_1 + a_2 + a_3)q^i \text{ for each } 2 \leq i < i_0. \tag{3.2.41}$$

On the other hand,

$$a_{i_0-2} + a_{i_0-1} + a_{i_0} > \frac{\tau_0}{4c_2}|\Lambda_0| \tag{3.2.42}$$

and combining this with (3.2.41) we obtain the inequality

$$\tau_0|\Lambda_0| < c_4(a_1 + a_2 + a_3)q^{i_0}, \qquad c_4 = c_4(c_2, c_3, q) > 0.$$

Noticing that there is some $c_5 = c_5(r, v)$ such that

$$a_1 + a_2 + a_3 \leq c_5\delta_0^2|\Lambda_0|$$

we get, for another constant $c_6 = c_4 \cdot c_5$, the inequality

$$\tau_0 \leq c_6\delta_0^2 q^{i_0}. \tag{3.2.43}$$

Return again to (3.2.38), for $i = i_0$. From the definition of i_0 we obtain

$$H^*_{\text{rel}}(x) \geq \frac{1}{2}\left(\tau_0|\Lambda_0| + c_1 \sum_{j > i_0} a_j\right). \tag{3.2.44}$$

Until now, we have not mentioned the set Λ. By the definition of Λ we can find, for each $t \in \Lambda \setminus \Lambda_{i+1}$, some $t' \in \Lambda \setminus \Lambda_i$ such that $|t' - t| \leq r$ and $|x_{t'} - x_{t'}^{\pm}| \geq \delta$ (depending on whether $t \in \Lambda_0^{c+}$ or $t \in \Lambda_0^{c-}$). Therefore, there is some $c_7 = c_7(r, v)$ such that $\Sigma_{j>i}\, a_j \geq c_7\, \delta^2|\Lambda \setminus \Lambda_{i+1}|$, $i \geq 2$. Substituting this and also the inequality

$$|\Lambda_i| \leq c_8\, i^{\,v}|\Lambda_0|, \qquad c_8 = c_8(v, r)$$

into (3.2.44) we finally obtain

$$H^*_{\text{rel}}(x) \geq \frac{\tau_0}{2c_8(i_0+1)^v}|\Lambda_{i_0+1}| + \tfrac{1}{2}c_1c_7\delta^2|\Lambda \setminus \Lambda_{i_0+1}|$$

which is the desired relation (3.2.27), because there on the upper bound (3.2.43) for the value of i_0, which does not depend on the choice of δ.

We conclude that there is some $c > 0$ depending only on $c_{(1.21)}$, τ_0, δ_0 and r such that

$$H^*_{\text{rel}}(x) \geq c\delta^2|B_\delta(x)|. \tag{3.2.45}$$

(ii) While remaining in the setting of Main Theorem we will prove, in the second step of the proof of (e), that the assumptions Th 4 and (3.2.45) imply the inequality

$$H^0_{\text{rel}}(x_\Lambda) \geq c\delta^2|\Lambda|, \qquad x_\Lambda \in B_\delta(\Lambda), \tag{3.2.46}$$

with another constant c depending on the same parameters as in (3.2.45).

NOTE. Here we use Definition (3.2.10), of course with $\tilde{\Phi}^0$, $\tilde{\Phi}^{\pm,0}$ instead of Φ^0, $\Phi^{\pm,0}$, etc. (We shall omit the tilde throughout step (ii).)

We will use, in the proof of (3.2.46), some results of §2.2. Fix some $\bar{x}_\Lambda \in B_\delta(\Lambda)$ and define the configuration (for a given M^+, M^-)

$$\bar{x}_t = \sum_{u\in\partial(M^+)^c} a_{t,u}^{+M^+}(\bar{x}_u - x_u^+) + x_u^+; \qquad t\in M^+$$

$$\bar{x}_t = \sum_{u\in\partial(M^+)^c} a_{t,u}^{-M^-}(\bar{x}_u - x_u^-) + x_u^-; \qquad t\in M^- \tag{3.2.47}$$

where $a_{t,u}^{\pm M^\pm}$ are coefficients from (2.2.11) (see also §2.2.3) corresponding to the standardized Hamiltonian (see (2.2.64)) $H_{\mathrm{st}}^{\pm,0}$. Take $M^\pm = \Lambda^{c\pm}$.

Using Proposition 2.2.7.1 we get

$$H_{\mathrm{rel}}^0(\bar{x}_\Lambda) = H_{\mathrm{rel}}^{0,\Lambda,\Lambda^{c+},\Lambda^{c-}}(\bar{x}) \tag{3.2.47'}$$

and therefore

$$H_{\mathrm{rel}}^*(\bar{x}) - H_{\mathrm{rel}}^0(\bar{x}_\Lambda)$$

$$= \sum_{A:A\cap\Lambda^{c+}\neq\emptyset}(\Phi_A^0(\bar{x}_A) - \Phi_A^{+,0}(\bar{x}_A)) + \sum_{A:A\cap\Lambda^{c-}\neq\emptyset}(\Phi_A^0(\bar{x}_A) - \Phi_A^{-,0}(\bar{x}_A)). \tag{3.2.48}$$

Note that we have proved, in §2.2, the exponential decay of $|\bar{x}_t - x_t^\pm|$: from (2.2.15) and (3.2.3) (which is (2.2.5) for the Hamiltonians $H^\pm$) it follows that there are some $0 < q < 1$ and $c > 0$ depending only on $c_{(1.21)}$ such that

$$|\bar{x}_t - x_t^\pm| \leqslant c\delta\, q^{\mathrm{dist}(t,\Lambda)}, \qquad t\in\Lambda^c. \tag{3.2.49}$$

Further, by (3.2.16) we obtain the following estimate: for A such that $A\cap M^+ \neq \emptyset$,

$$|\Phi_A^0(\bar{x}_A) - \Phi_A^{+,0}(\bar{x}_A)| \leqslant \varepsilon|\bar{x}_A - x_A^\pm|^2 \tag{3.2.50}$$

with some $\varepsilon = \varepsilon(\delta, c_{(1.21)})$ such that $\lim_{\delta\to 0}\varepsilon = 0$.

NOTE. This is not the same ε as in (3.2.16) because the inequality $|\bar{x}_t - x_t^\pm| \leqslant \delta$ is not necessarily true for t near Λ. (See also Note 2 in §3.2.1.)

Combining (3.2.49) and (3.2.50) we get the estimate

$$\sum_{A:\,A\cap\Lambda^{c\pm}\neq\emptyset} |\Phi_A^0(\bar{x}_A) - \Phi_A^{0,\pm}(\bar{x}_A)| \leqslant \varepsilon|\bar{x}_{\partial\Lambda} - x_{\partial\Lambda}^\pm|^2 \tag{3.2.51}$$

where $\varepsilon = \varepsilon(\delta, c_{(1.21)})$ is such that $\lim_{\delta\to 0}\varepsilon = 0$. It is now clear that (3.2.45), (3.2.48) and (3.2.51) imply, for a small δ, the desired relation (3.2.46). (Note that $B_\delta(\bar{x}) \supset \Lambda$.)

(iii) Now we can finish the proof of (3.2.11). By analogy with (1.5), denote by $\mathring{V}(\Lambda, h)$ the Lebesgue measure of the set

$$\{x_\Lambda \in B_\delta(\Lambda)\colon H_{\mathrm{rel}}^0(x_\Lambda) \leqslant h|\Lambda|\}. \tag{3.2.52}$$

To compare $H_{\mathrm{rel}}^0(x_\Lambda)$ with $H(x_\Lambda)$ notice that because of (3.2.47′) and the relation $e^+ = e^- = \Sigma_{A\ni 0}|A|^{-1}\Phi_A^0(x_A^\pm)$ we can write

$$H_{\mathrm{rel}}^0(x_\Lambda) = H(x_\Lambda) - H(x_\Lambda^+) + R^+ + R^- + S \tag{3.2.53}$$

where

$$R^{\pm} = R^{\pm}(x_\Lambda) = \sum_{A \cap \Lambda^{c\pm} \neq \emptyset} (\Phi_A^{\pm,0}(\bar{x}_A) - \Phi_A^{\pm,0}(x_A^{\pm})) \tag{3.2.54}$$

and

$$S = S(\Lambda) = \sum_{A \cap \Lambda^{c-} \neq \emptyset} (\Phi_A^{-,0}(x_A^{-}) - \Phi_A^{+,0}(x_A^{+}))$$

$$= \sum_{A \cap \Lambda^{c-} \neq \emptyset} \frac{|A \cap \Lambda|}{|A|} (\Phi_A^{+,0}(x_A^{+}) - \Phi_A^{-,0}(x_A^{-})) \tag{3.2.55}$$

(assuming that for example Λ^{c-} is finite – if Λ^{c+} is finite then $\pm$ must be changed to $\mp$ in the definition of S).

Clearly,

$$|S| \leqslant c|\partial\Lambda| \tag{3.2.56}$$

where c depends on the interaction only. Using the estimates of §2.2.7 (namely (2.2.75)) generalized in an obvious way to the case of linear quadratic Hamiltonians H^+, H^-) we further see from (3.2.54) that

$$|R^{\pm}| \leqslant c(\delta + \delta^2)|\partial\Lambda| \tag{3.2.57}$$

for another constant c depending on the interaction only. Thus, if δ is small enough,

$$H_{rel}^0(x_\Lambda) \geqslant H(x_\Lambda) - c|\Lambda| \tag{3.2.58}$$

with a new constant c, depending on the interaction only. From this inequality and from (1.6) it follows that

$$\ln \mathring{V}(\Lambda, h) \leqslant (c_1 h + c_2)|\Lambda| \tag{3.2.59}$$

for some c_1, $c_2 > 0$ depending on the interaction only.

We now return to the setting of the Main Lemma and again use the tilde to distinguish the analogous notions of the Main Theorem.

Estimate

$$\int_{B_\delta(\Lambda)} \exp(-H_{\text{rel}}^0(x_\Lambda))\,\mathrm{d}x_\Lambda = (\tilde{T})^{-k|\Lambda|/2} \int_{B_{\tilde{\delta}}(\Lambda)} \exp\left(-\frac{1}{\tilde{T}} \tilde{H}_{\text{rel}}^0(\tilde{x}_\Lambda)\right) \mathrm{d}\tilde{x}_\Lambda \tag{3.2.60}$$.

Using the estimates (3.2.46) and (3.2.59) and the obvious properties of the Stieltjes integral we get the relations

$$\int_{B_{\tilde{\delta}}(\Lambda)} \left(-\frac{1}{\tilde{T}} \tilde{H}_{\text{rel}}^0(\tilde{x}_\Lambda)\right) \mathrm{d}\tilde{x}_\Lambda \leqslant \int_{-\infty}^{\infty} \exp\left(-\frac{1}{\tilde{T}} h|\Lambda|\right) \mathrm{d}\tilde{\mathring{V}}(\Lambda, h)$$

$$= \int_{c\delta^2}^{\infty} \exp\left(-\frac{1}{\tilde{T}} h|\Lambda|\right) \mathrm{d}\mathring{V}(\Lambda, h) \text{ (where } c = c_{(3.2.46)})$$

$$= |\Lambda|(\tilde{T})^{-1} \int_{c\delta^2}^{\infty} \tilde{V}(\Lambda, h) \exp\left(-\frac{1}{\tilde{T}} h|\Lambda|\right) \mathrm{d}h$$

$$\leqslant |\Lambda|(\tilde{T})^{-1} \int_{c\delta^2}^{\infty} \exp\left((c_1 h + c_2)|\Lambda| - \frac{h}{\tilde{T}}|\Lambda|\right) \mathrm{d}h. \tag{3.2.61}$$

Suppose that $\tilde{T}$ is so small that

$$\tilde{T}^{-1} > c_1. \tag{3.2.62}$$

By (3.2.60) and (3.2.61) we get the final bound

$$\int_{B_{\delta}(\Lambda)} \exp(-H^0_{\text{rel}}(x_\Lambda))\,\mathrm{d}x_\Lambda \leqslant \exp((c_2 - c(\tilde{T}^{-1} - c_1)\delta^2)|\Lambda|)\frac{\tilde{T}^{-k|\Lambda|/2-1}}{(\tilde{T}^{-1} - c_1)} \tag{3.2.63}$$

which proves (3.2.11) for $\lambda = 0$ and all sufficiently small $\tilde{T}$ (with $\tau \geqslant c/2\ \tilde{T}^{-1}\tilde{\delta}^2 = c/2\delta^2$, $c = c_{(3.2.46)}$).

In the case of arbitrary λ we proceed as follows. Using Assumption Th 4 we explain below that (3.2.46) implies the relation

$$\tilde{H}^{\tilde{\lambda}}_{\text{rel}}(x_\Lambda) \geqslant c\tilde{\delta}^2|\Lambda|,\ \tilde{\lambda} \in (-c'\tilde{\delta}^2, c'\tilde{\delta}^2) \tag{3.2.64}$$

with come $c > 0$ and some enough small $c' > 0$ depending, like all the constants appearing in these reasonings, on the interaction only. (We again suppose that $\tilde{\delta}$ is small enough.) In fact, using the boundedness from below of all $\tilde{\Phi}^{\tilde{\lambda}}_A$ and using also (1.23), (3.2.58) we see that

$$\begin{aligned}\left|\frac{\mathrm{d}}{\mathrm{d}\lambda}\tilde{H}^{\tilde{\lambda}}_{\text{rel}}(x_\Lambda)\right| &= \left|\sum_{A\subset\Lambda}\frac{\mathrm{d}}{\mathrm{d}\lambda}\tilde{\Phi}^{\tilde{\lambda}}_A(x_A)\right| \leqslant c''\left(|\Lambda| + \sum_{A\subset\Lambda}|\tilde{\Phi}^{\tilde{\lambda}}_A(x_A)|\right)\\ &= c''|\Lambda| + c''|\tilde{H}^{\tilde{\lambda}}(x_\Lambda)| \leqslant c'''|\Lambda| +\\ &\quad + c''|\tilde{H}^{\tilde{\lambda}}_{\text{rel}}(x_\Lambda)|\end{aligned} \tag{3.2.65}$$

(for enough small $\tilde{\delta}$). It is easy to deduce (3.2.64) from (3.2.65), (3.2.46).

Finally, by repeating the arguments (3.2.60)–(3.2.63) we obtain the condition (3.2.11) for any $\lambda \in [\lambda_1, \lambda_2]$ if

$$M\tilde{T} \leqslant c'\tilde{\delta}^2 \tag{3.2.66}$$

Analogously, the relation (3.2.12) can be proved, by using (3.2.65), (3.2.64) (and replacing the estimates (3.2.61) by analogous estimates for $|\Lambda|\tilde{T}^{-1}$ $\int h \exp(-\tilde{T}^{-1}h|\Lambda|)\,\mathrm{d}\tilde{V}(\Lambda, h)$. We omit the details.

Thus, all the assumptions of the Main Lemma are satisfied if the constants $\tilde{\delta}$, M in (3.2.19) are suitably choosen.

We now explain our choice of these constants.

(a) First we choose $M > 0$ large enough such that (3.2.26) holds.

(b) Then, having fixed M, we choose another sufficiently large constant M' and relate $\tilde{\delta}$, $\tilde{T}$ by $\tilde{\delta}^2 = M'\tilde{T}$. Note that our choice of M' is determined by the assumptions on δ in the Main Lemma (δ is required to be sufficiently large) and also by the condition $M \leqslant C'M'$ which implies (3.2.66).

(c) For a fixed M, M' we can now choose any sufficiently small $\tilde{T} > 0$, such that all the quantities in (3.2.24) would be sufficiently small and for τ (guaranteed by (3.2.63) and a similar relation for $\lambda \neq 0$), the ratio $\tau(\ln\delta)^{-\nu}$ would be sufficiently large, as well as the ratio $\tau'(\ln\delta)^{-\nu}$. Note that in our construction we obtain the relations $\delta \to \infty$. $\tau(\ln\delta)^{-\nu} \to \infty$, $\tau^-(\ln\delta)^{-\nu} \to \infty$, $\varepsilon \to 0$, $\varepsilon' \to 0$ if $\tilde{T} \to 0$. The constants Δ and $c_{(3.2.3)}$ do not depend on $\tilde{T}$.

It is now easy to show that Main Lemma implies the Main Theorem: denote by $P_T^+ P_T^-$ the Gibbs states guaranteed by the Main Lemma when applied at temperature $\tilde{T}$. Denote by $\lambda = \lambda(\tilde{T})$ the corresponding value of the parameter λ.

Transform these states 'back' to the setting of the Main Theorem using the inverse transformation $\{x \to \tilde{T}^{1/2} x\}$. We obtain states $\tilde{P}_{\tilde{T}}^{\pm}$ on X, which are clearly Gibbs states with potential $\{\Phi_A^{\tilde{\lambda}}\}$, $\tilde{\lambda} = M\tilde{T}\lambda(\tilde{T})$ corresponding to the temperature $\tilde{T}$.

The mapping $\{\tilde{T} \rightsquigarrow \lambda(\tilde{T})\}$ is analytical in the neighborhood of any sufficiently small $\tilde{T}$, therefore analytical in some $(0, \tilde{T})$. Here we use, in addition to statement (c) of the Main Lemma the fact that $\lambda(\tilde{T})$ does not change if we transform the potential Φ by a homothety

$$\Phi_A^{\lambda,\gamma}(x_A) \equiv \Phi_A^{\lambda}(\gamma x_A), \qquad \gamma > 0.$$

(This will become obvious in §§4.2, 4.5 below.)

The differentiability of $\{\tilde{T} \rightsquigarrow \lambda(\tilde{T})\}$ at 0_+ follows clearly from the statement (d) of the Main Lemma. The convergence of the normalized states $P_{\tilde{T}}^+$, $P_{\tilde{T}}^-$ to the corresponding Gaussian states follows from the statement (e). Thus, it remains to prove the formula (1.26).

This is, however, a simple consequence of the formula (2.2.60), if one adds the following observations.

Note first that under the transition to the 'block' model (defined on $\tilde{\mathbb{Z}}^\nu$ with values in $\mathbb{R}^{k|W|}$, see the beginning of §3.2.2) the quantities det $F^{\pm}(\xi)$ and $\lambda(T)$ do not change but $u^{\pm}$ is multiplied by a factor of $|W|^{-1}$. It is therefore enough to consider the case $W = \{0\}$ when the Main Lemma is directly applicable.

If we express the condition $h^{+,\lambda} = h^{-,\lambda}$ of the Main Lemma in the setting of the Main Theorem we get the condition

$$\tfrac{1}{2}(k \ln \pi - (2\pi)^{-\nu} \int_{[-\pi,\pi]^\nu} \ln \det F^+(\xi)\, \mathrm{d}\xi - \lambda u^+$$
$$= \tfrac{1}{2}(k \ln \pi - (2\pi)^{-\nu} \int_{[-\pi,\pi]^\nu} \ln \det F^-(\xi)\, \mathrm{d}\xi - \lambda u^- \qquad (3.2.67)$$

which is (1.26). □

4. Proof of The Main Lemma

4.1. CONTOURS

The terminology of this section is based on the corresponding notions of §3.1. To emphasize the values r, δ used in the definitions we will talk about (r, δ) boundaries, write $B_r(x)$ or even $B_r^\delta(x)$ instead of $B(x)$, etc. The points from $\mathbb{Z}^\nu \setminus B_r^\delta(x)$ are called (r, δ) *correct* points of x.

Naively, one could define contours as, for example, the connected components of the set $B_r^\delta(x)$. This choice is technically not very convenient, however, and we therefore choose the following, more intricate one (its advantages will emerge only later, in §§4.4 and 4.5).

First choose some additional integer $\tilde{r} > 2r$, to be specified in §§4.3 and 4.4 below. (Roughly speaking we will choose $\tilde{r}$ such that any boundary condition from $\mathcal{U}^+$ (or $\mathcal{U}^-$) will be only slightly felt at the distance $\tilde{r}$ in the Gaussian approximation.)

Second, in addition to $\mathscr{U}^+, \mathscr{U}^-$ consider also some sets $\tilde{\mathscr{U}}^+$, $\tilde{\mathscr{U}}^+$ defined as a $\tilde{\delta}$ neighbourhood of σ^+, σ^- where $\tilde{\delta} = K\delta$, with a suitable $K = K(c_{(3.2.3)}) < 1$, will be specified in §§4.4 and 4.5 below. (Roughly speaking we will choose $\tilde{\mathscr{U}}^\pm$ such that the mean values of the Gaussian approximation will stay in $\mathscr{U}^\pm$ everywhere in Λ, for any boundary condition $x_{\partial\Lambda^c} \in \tilde{\mathscr{U}}^{\pm \partial\Lambda^c}$. As we explain later this is not necessarily true for $\tilde{\mathscr{U}}^\pm = \mathscr{U}^\pm$).

Say that a point $t \in \mathbb{Z}^\nu$ is affiliated to $B_{\bar{r}}^{\delta}(x)$ if there is a string $t_0 = t, t_1, \ldots, t_n = s$ such that $|t_i - t_{i-1}| \leqslant \bar{r}$ for each $i = 1, \ldots, n$; $s \in B_{\bar{r}}^{\tilde{\delta}}(x)$ and either $x_{t_i} \in \mathscr{U}^+ \setminus \tilde{\mathscr{U}}^+$ for each $i = 0, \ldots, n-1$ or $x_{t_i} \in \mathscr{U}^- \setminus \tilde{\mathscr{U}}^-$ for all those i. (Thus, we 'enrich' the set $B_{\bar{r}}^{\tilde{\delta}}(x)$ by some points from $B_{\bar{r}}^{\delta}(x)$, i.e. by those that are affiliated to it.) Denote by $B_{\text{aff}}(x)$ the set of all $t \in \mathbb{Z}^\nu$ which are affiliated to $B_{\bar{r}}^{\tilde{\delta}}(x)$.

4.1.1. *Definition*

Let $x \in X$ be a configuration, and let B be a finite connected component of $B_{\text{aff}}(x)$. The restriction $\Gamma = (x)_B$ is called a *contour* of the configuration x. We will talk simply about a contour Γ if it is a contour of some (non-specified) x. The set B will be called the support of Γ and denoted by supp Γ.

NOTE. Having defined the concept of a contour, we will not use the value $\bar{r}$ and $\tilde{\delta}$ in most of the following notions and constructions. Only later in §§4.3–4.5 do we discuss in more detail the structure of contours and benefit from our definition. For the rest of §4.2 we apply the geometrical notions of §3.1 with the original values or r and δ, if not specified otherwise. See also Note 4.1.2.

4.1.2. *Definition*

Denote by $\partial\Gamma$ (the r-boundary of the contour) the restriction of Γ to the set (see (3.1.1))

$$\partial \operatorname{supp} \Gamma \equiv \partial_r \operatorname{supp} \Gamma.$$

Denote by int Γ the union of all finite (i.e. 'interior') connected components of the set $(\operatorname{supp} \Gamma)^c$. Denote by

$$V(\Gamma) = \operatorname{supp} \Gamma \cup \operatorname{int} \Gamma. \tag{4.1.1}$$

NOTE. We emphasize that $\partial\Gamma$ (even $\partial_{\bar{r}}\Gamma$) attains values from $\tilde{\mathscr{U}}^+$ (or $\tilde{\mathscr{U}}^-$) but not from $\mathscr{U}^+ \setminus \tilde{\mathscr{U}}^+$ (or $\mathscr{U}^- \setminus \tilde{\mathscr{U}}^-$). On the other hand, our definition of contours is such that on the set $\mathbb{Z}^\nu \setminus B_{\text{aff}}(x)$, any $(\delta, 1)$ correct configuration can appear, with sign_x being prescribed by $x(B_{\text{aff}})$. The first fact is quite arbitrary in the conceptual §4.2, but the second is a substantial one. No other property of contours will be used throughout §4.2.

4.1.3. *Definition*

Notice that any contour Γ is an outside correct configuration in the volume supp Γ (see Definition 3.1.4). Consider the function $\text{sign}_\Gamma(\cdot)$ (see 3.1.5). Distinguish, correspondingly, the + and − connected components of $(\operatorname{supp} \Gamma)^c$ (see 3.1.3).

Denote by $\operatorname{int}\Gamma$ the union of all finite connected components of $(\operatorname{supp}\Gamma)^c$. Say that Γ is a $\pm$ contour if the infinite ('external') connected component of $(\operatorname{supp}\Gamma)$ is a $\pm$ component. Define the set

$$\partial_\pm \operatorname{supp}\Gamma = \{t \in \partial \operatorname{supp}\Gamma : \operatorname{sign}_\Gamma(t) = \pm\}$$

and put

$$\partial_\pm \Gamma = \Gamma_{(\partial_\pm \operatorname{supp}\Gamma)}.$$

Thus,

$$\partial\Gamma = \partial_+\Gamma \cup \partial_-\Gamma. \tag{4.1.2}$$

In an analogous sense write

$$\operatorname{int}\Gamma = \operatorname{int}_+\Gamma \cup \operatorname{int}_-\Gamma. \tag{4.1.3}$$

4.1.4. *Definition*

Let $\Lambda \subset \mathbb{Z}^\nu$. Say that $x \in X$ is a diluted configuration with respect to Λ if all the contours of x satisfy the condition $\operatorname{dist}(V(\Gamma), \Lambda^c) \geqslant 2$. Denote by $X_{\text{dil}}(\Lambda)$ the set of all $x_\Lambda \in X(\Lambda)$ which can be continued to some $x \in X$ which is a diluted configuration with respect to Λ. Consider also the subset $X^+(\Lambda)$ (analogously, $X^-(\Lambda)$) of $X_{\text{dil}}(\Lambda)$ consisting of all x_Λ such that $\operatorname{sign}_{x_\Lambda}(t) = +$ everywhere in Λ^c. Clearly, $X_{\text{dil}}(\Lambda) = X^+(\Lambda) \cup X^-(\Lambda)$ if Λ is simple connected.

4.1.5. *Definition*

A *frame* of a + contour Γ is defined as a pair $(\operatorname{supp}\Gamma, \partial_+ \operatorname{supp}\Gamma)$. It will be identified with the corresponding class of all contours with fixed $\operatorname{supp}\Gamma$ and $\partial_+ \operatorname{supp}\Gamma$. (This is also true when + is replaced by −). We denote frames by the symbol $\mathbf{\Gamma}$. Define the notions of $V(\mathbf{\Gamma})$, $\operatorname{supp}\mathbf{\Gamma}$, $\partial_\pm \mathbf{\Gamma}$, $\operatorname{int}_\pm \mathbf{\Gamma}$ as $V(\Gamma)$, $\operatorname{supp}\Gamma$, $\partial_\pm \operatorname{supp}\Gamma$, $\operatorname{int}_\pm \Gamma$ where $\Gamma \in \mathbf{\Gamma}$.

4.1.6. *Definition*

Let $\Lambda \subset \mathbb{Z}^\nu$ be finite, let $\{\Gamma_i\}$ be a family of + contours. We say that $\{\Gamma_i\}$ is a + *contour system* in Λ if the following is satisfied:

(a) $\operatorname{dist}(V(\Gamma_i), \Lambda^c) \geqslant 2$ for each i,
(b) $\operatorname{dist}(\operatorname{supp}\Gamma_i, \operatorname{supp}\Gamma_{i'}) \geqslant 2$ whenever $i \neq i'$. (Similarly for + replaced by −.)

We also define the notion of a + frame system in Λ as a system of frames of contours of some +contour system in Λ.

NOTE. We do not mean, of course, that such a contour system would arise, in general, as a system of contours of some configuration. It is, on the contrary, an essential feature of the PS approach that contour systems of the type described above appear in the construction of the contour model, the aim being that the contour model should describe a behaviour of the *external* contour systems of the given 'physical' model only.

4.1.7. *Definition*

Let $\{\Gamma_i\}$ be a contour system. We say that $\Gamma \in \{\Gamma_i\}$ is an external contour of the system if there is no contour Γ_i of the system such that $V(\Gamma) \subset \operatorname{int} \Gamma_i$. We write $\Gamma < \Lambda$ if $\operatorname{dist}(V(\Gamma), \Lambda^c) \geqslant 2$, and also $\Gamma < \tilde{\Gamma}$ if $\Gamma < \operatorname{int} \tilde{\Gamma}$.

Given any finite contour system $\{\Gamma_i\}$, define its external contour system as the subsystem of all external contours of $\{\Gamma_i\}$. Define analogously the notion of an external frame system.

NOTE. For any contour Γ_i of a contour system $\{\Gamma_i\}$ in a finite volume Λ there is some external contour $\Gamma_{i'}$ such that $\Gamma_i < \Gamma_{i'}$. We omit the proof of statements of this type, which are based on the following observation: if $\Gamma, \tilde{\Gamma}$ are contours such that $\operatorname{supp} \Gamma \cap \operatorname{supp} \tilde{\Gamma} = \emptyset$ then either $\Gamma < \tilde{\Gamma}$ or $\tilde{\Gamma} < \Gamma$ or $V(\Gamma) \cap V(\tilde{\Gamma}) = \emptyset$.

4.1.8. *Definition*

Let $x \in X$ be a configuration with a finite boundary $B_{\text{aff}}(x)$, and let Γ be a contour of x. Say that Γ is an external contour of x if $\Gamma < \Gamma'$ for no contour Γ' of x. The family of all such Γ is called the external contour system of x. We say that this external contour, system lies in Λ if $\operatorname{dist}(v(\Gamma), \Lambda^c) \geqslant 2$ for each contour of the system.

NOTE. As in §3.1 we will often talk about contours or frames of diluted configurations in a finite volume.

4.2. REDUCTION TO A CONTOUR MODEL

In this section we define the contour models and derive the fundamental relations (4.2.21) determining the appropriate value of the contour weight. Our main object of study in §4.2 will be the following.

4.2.1. *Definition*

Let $\Lambda \subset \mathbb{Z}^\nu$ be finite, let $a \in (\mathscr{U}^+)^{\partial \Lambda^c}$. Consider the set $X^+(\Lambda)$ and the following probability density on $X^+(\Lambda)$:

$$P_{\lambda,a}(x_\Lambda) = (Z_\lambda)^{-1} \exp(-H_\lambda(x_\Lambda \mid a)) \tag{4.2.1}$$

where

$$Z_\lambda = Z_\lambda(\Lambda, a) = \int_{X^+(\Lambda)} \exp(-H_\lambda(x_\Lambda \mid a)) \, \mathrm{d}x_\Lambda \tag{4.2.1'}$$

The following proposition explains our ultimate use of the probability densities $P_{\lambda,a}$. Its proof is postponed to the very end of the paper (§4.5.6). We denote by the symbol $P_{\lambda,a}$ also the corresponding probability measure on $X^+(\Lambda)$.

PROPOSITION. *Take* $\lambda = \lambda(\Phi)$ *as in the Main Lemma (to be specified later as* $\lambda = \bar{\lambda}_{(4.5.29)}$*). Let* $\{\Lambda_n\}$ *be a sequence of finite, simply connected subsets of* $\mathbb{Z}^\nu$ *such that* $\operatorname{dist}(0, \Lambda_n^c) \to \infty$. *Let* $a_n \in (\tilde{\mathscr{U}}^+)^{\partial \Lambda_n^c}$. *Let* $A \subset \mathbb{Z}^\nu$ *be finite, let* ψ *be a bounded continuous* *function on $X(A)$. Then the limit*

$$\lim_{n\to\infty}\int X^+(\Lambda)\varphi(x_A)\mathrm{d}P_{\lambda,a_n}(x_{\Lambda^n})$$

exists and defines some Gibbs state on X, corresponding to the interaction Φ^λ. An analogous limit exists for $a_n \in (\tilde{\mathscr{U}}^-)^{\partial\Lambda_n^c}$, both the + and − limits being distinct.

NOTE. Even if the limit exists as a state on X it does not follow from the usual limit theorems of the theory of Gibbs states that the limit state is a Gibbs one. (The condition $x_\Lambda \in X^+(\Lambda)$ cannot be formulated in terms of $x_{\partial_r\Lambda}$ for some fixed r.)

According to the PS model we now construct some auxiliary contour models and reduce the study of the densities $P_{\lambda,a}$ to the study of these contour models. The following definitions are crucial for our construction. (There is some arbitrariness in some of our constructions, to be explained later. In fact, the only essential requirement of our approach is that we require (4.2.11) to hold.)

4.2.2. *Definition*

Let $\Lambda \subset \mathbb{Z}^\nu$ be finite. A pair $r_\Lambda = (x_\Lambda, \mathscr{D})$ where $x_\Lambda \in X(\Lambda)$ and $\mathscr{D}$ is a +frame system in Λ (see 4.1.6) is called a *realization of the +contour ensemble in* Λ (or simply a realization) if the following is satisfied:

(a) $x_t \in \mathscr{U}^+$ for each $t \in \Lambda \setminus \operatorname{supp} \mathscr{D}$
(b) $x_t \in \tilde{\mathscr{U}}^+$ for each $t \in \partial_+ \mathscr{D}$ where

$$\partial_+ \mathscr{D} = \bigcup_{\Gamma\in\mathscr{D}} \partial_+ \Gamma \text{ and } \operatorname{supp} \mathscr{D} = \bigcup_{\Gamma\in\mathscr{D}} \operatorname{supp} \Gamma. \tag{4.2.2}$$

Realizations of the − contour ensemble are defined analogously. Denote by $\mathscr{R}^\pm(\Lambda)$ the set of all realizations of the $\pm$ contour ensemble in Λ.

Introduce the Borel structure and also the *basic measure* on $\mathscr{R}^+(\Lambda)$ (analogously on $\mathscr{R}^-(\Lambda)$) by using the natural injection of $(\tilde{\mathscr{U}}^+)^{\partial_+\mathscr{D}} \times (\mathscr{U}^+)^{\Lambda\setminus\operatorname{supp}\mathscr{D}} \times (\mathscr{R}^k)^{\operatorname{supp}\mathscr{D}\setminus\partial_+\mathscr{D}}$ onto the subset $\mathscr{R}^+_{\mathscr{D}}(\Lambda)$ of $\mathscr{R}^+(\Lambda)$ consisting of those realizations $r_\Lambda = (x_\Lambda, \mathscr{D})$ which have a fixed set $\mathscr{D}$ of frames. We take the Lebesgue measure on $\mathbb{R}^k, \mathscr{U}^+, \tilde{\mathscr{U}}^+$.

NOTE. The condition (a) is natural for $t \in \Lambda \setminus \cup_{\Gamma\in\mathscr{D}} V(\Gamma)$ if we look at how the external configurations behave. Condition (b) respects the definition of contours. To continue with the restriction (a) 'inside $\mathscr{D}$' is the characteristic feature of PS approach. (That we do not also require (a) for $t \in \operatorname{supp} \mathscr{D} \setminus \partial_+ \mathscr{D}$ is an arbitrary feature of our construction. See Note 6, §4.2.6, where the choice (4.2.3) of the Hamiltonian $H_\lambda(r_\Lambda | a)$ is also commented on.)

4.2.3. *Definition*

By a *frame weight* (or, also, by a *contour weight*) we mean a measurable, non-negative functional F acting on all (Γ, a) where Γ is a frame of a $\pm$ contour and $a \in (\tilde{\mathscr{U}}^\pm)^{\partial_\pm\Gamma}$.

NOTE. We often omit the $\pm$ sign when talking about a $\pm$ contour weight. Sometimes, by the term contour weight, we mean a pair of both + and − contour weights. (This will be clear from the context.)

The assumption of non-negativity is a temporary condition which will be strengthened later.

4.2.4. *Definition*

Let $\Lambda \subset \mathbb{Z}^\nu$ be finite, and let F be a contour weight. For any boundary condition $x_{\partial\Lambda^c} = a \in (\mathscr{U}^+)^{\partial\Lambda^c}$ consider the Hamiltonian, acting on $r_\Lambda \in \mathscr{R}^+(\Lambda)$,

$$H_\lambda(r_\Lambda | a) = \sum_{\substack{A \notin \mathrm{Supp}^*(\mathscr{D}), \\ A \cap \Lambda \neq \emptyset}} \Phi_A^\lambda(x_A) + \sum_{A \in \mathrm{Supp}^*(\mathscr{D})} \Phi_A^{+,\lambda}(x_A) + \sum_{\Gamma \in \mathscr{D}} F(\Gamma, x_{\partial_+\Gamma}) \tag{4.2.3}$$

where

$$\mathrm{Supp}^*(\mathscr{D}) = \{A : A \cap \mathrm{supp}\,\mathscr{D} \setminus \partial_+\mathscr{D} \neq \emptyset\} \tag{4.2.4}$$

(see (4.2.2)). Consider a probability density $P_{\lambda,a,F}$ on $\mathscr{R}^+(\Lambda)$ defined as follows:

$$P_{\lambda,a,F}(r_\Lambda) = Z_\lambda^{-1} \exp(-H_\lambda(r_\Lambda | a)) \tag{4.2.5}$$

where

$$Z_\lambda = Z_\lambda(\Lambda, a, F) = \int_{\mathscr{R}^+(\Lambda)} \exp(-H_\lambda(r_\Lambda | a))\, \mathrm{d}r_\Lambda. \tag{4.2.5'}$$

(The symbol $\mathrm{d}r_\Lambda$ means integration with respect to the basic measure.) The probability measure on $\mathscr{R}^+(\Lambda)$ defined by this density is denoted by the same symbol $P_{\lambda,a,F}$ and called the *Gibbs measure on the contour ensemble* $\mathscr{R}^+(\Lambda)$. (Analogously for the $-$contour model.)

4.2.5. *Equivalence of Ensembles*

DEFINITION 1. Let $\Lambda \subset \mathbb{Z}^\nu$ be finite, and let $r \equiv (x, \mathscr{D}) \in \mathscr{R}^+(\Lambda)$. Denote by r_{ext} the pair $(x_{\mathrm{ext}}, \mathscr{D}_{\mathrm{ext}})$ where $\mathscr{D}_{\mathrm{ext}}$ is the external frame system corresponding to the frame system $\mathscr{D}$ and x_{ext} denotes the restriction of x to the set

$$\mathrm{ext}^\partial \mathscr{D} = \Lambda \setminus \bigcup_{\Gamma \in \mathscr{D}_{\mathrm{ext}}} (V(\Gamma) \setminus \partial V(\Gamma)).$$

Denote by $\mathscr{R}^+_{\mathrm{ext}}(\Lambda)$ the set of all r_{ext} corresponding to some $r \in \mathscr{R}^+(\Lambda)$. Consider the mapping

$$\{r \rightsquigarrow r_{\mathrm{ext}}\} : \mathscr{R}^+(\Lambda) \rightsquigarrow \mathscr{R}^+_{\mathrm{ext}}(\Lambda).$$

Denote by $P^{\mathrm{ext}}_{\lambda,a,F}$ the image of $P_{\lambda,a,F}$ under this map.

Analogously, for any $x \in X^+(\Lambda)$ consider a pair $(x_{\mathrm{ext}}, \mathscr{D}_{\mathrm{ext}})$ where $\mathscr{D}_{\mathrm{ext}}$ is the external frame system corresponding to the external contour system of x (see 4.1.8), and x_{ext} is the restriction of x to the set $\mathrm{ext}^\partial \mathscr{D} (\equiv \mathrm{ext}^\partial \mathscr{D}_{\mathrm{ext}})$. The set of all such pairs $(x_{\mathrm{ext}}, \mathscr{D}_{\mathrm{ext}})$ will be denoted as $X^+_{\mathrm{ext}}(\Lambda)$ in what follows

Consider the mapping

$$\{x \rightsquigarrow (x_{\mathrm{ext}}, \mathscr{D}_{\mathrm{ext}})\} : X^+(\Lambda) \to X^+_{\mathrm{ext}}(\Lambda)$$

and denote by $P^{\text{ext}}_{\lambda,a}$ the image of $P_{\lambda,a}$ under this map. (Analogously for + replaced by −.)

The following is another crucial notion.

DEFINITION 2. We identify $X^{+}_{\text{ext}}(\Lambda)$ with $\mathscr{R}^{+}_{\text{ext}}(\Lambda)$. Say that a contour weight F defines a contour ensemble which is *equivalent* to the Gibbsian one if for any finite $\Lambda \subset \mathbb{Z}^{\nu}$ and any boundary condition $a \in (\mathscr{U}^{+})^{\partial\Lambda^c}$ (analogously, $(\mathscr{U}^{-})^{\partial\Lambda^c}$) the probability distributions $P^{\text{ext}}_{\lambda,a}$ and $P^{\text{ext}}_{\lambda,a,F}$ coincide.

NOTE. This is the central idea of PS theory: to substitute a study of the original Gibbs ensemble by a study of the equivalent contour ensemble. (We have in mind the case when a phase transition is expected – otherwise one must modify this idea, which is beyond the scope of this paper.) Actually, our contour models will turn to be equivalent in some more precise sense – see the note after (4.2.15).

To find more explicit criteria of equivalence we must define further types of partition functions.

4.2.6. *Criterion of Equivalence*

DEFINITION 1. Let Γ be a + contour, and let $\mathbf{\Gamma}$ be its frame. Denote by $V^*(\mathbf{\Gamma})$ the set (compare with $V(\Gamma)$ and $V(\Gamma)\setminus\partial V(\Gamma)$)

$$V^*(\Gamma) = V^*(\mathbf{\Gamma}) = (\operatorname{supp}\mathbf{\Gamma} \cup \operatorname{int}_-\mathbf{\Gamma})\setminus\partial_+\mathbf{\Gamma}.$$

Analogously, for a − contour,

$$V^*(\Gamma) = V^*(\mathbf{\Gamma}) = (\operatorname{supp}\mathbf{\Gamma} \cup \operatorname{int}_+\mathbf{\Gamma})\setminus\partial_-\mathbf{\Gamma}.$$

Suppose again that Γ is a + contour. Denote by $X(\Gamma)$ the set of all $x \in X\,(\operatorname{int}_-\mathbf{\Gamma})$ which are 'compatible' with Γ in the following sense: Γ is a contour of the (outside correct) configuration $\Gamma \cup x$.

For any $a \in (\tilde{\mathscr{U}}^{+})^{\partial_+\mathbf{\Gamma}}$ denote by $\mathscr{C}(\mathbf{\Gamma}, a)$ the family of all + contours Γ with the same $\mathbf{\Gamma}$ and the same $(\Gamma)_{\partial_+\mathbf{\Gamma}} = a$. Denote further by $\mathscr{C}'(\mathbf{\Gamma},a)$ the family of all $x \in X\,(\operatorname{supp}\mathbf{\Gamma}\setminus\partial_+\mathbf{\Gamma})$ such that $x \cup a \in \mathscr{C}(\mathbf{\Gamma},a)$. Denote by

$$Z_\lambda(\Gamma) = \int_{X(\Gamma)} \exp(-H_\lambda(x_{V^*(\Gamma)}|\partial_+\Gamma))\,\mathrm{d}x_{\operatorname{int}_-\mathbf{\Gamma}} \tag{4.2.6}$$

and

$$Z_\lambda(\mathbf{\Gamma},a) = \int_{\mathscr{C}'(\mathbf{\Gamma},a)} Z_\lambda(\Gamma)\,\mathrm{d}\Gamma_{(\operatorname{supp}\mathbf{\Gamma}\setminus\partial_+\mathbf{\Gamma})}, \tag{4.2.6'}$$

the integrals being taken with respect to the Lebesgue measure of the corresponding subsets of $(\mathbb{R}^k)^{\operatorname{int}_-\mathbf{\Gamma}}$ (or $(\mathbb{R}^k)^{\operatorname{supp}\mathbf{\Gamma}\setminus\partial_+\mathbf{\Gamma}}$, respectively).

In addition to (4.2.1) consider also the following more general partition functions. Let $M \subset \Lambda \subset \mathbb{Z}^\nu$ be finite and let $x_{\partial\Lambda^c} = a \in (\mathscr{U}^{+})^{\partial\Lambda^c}$. Put

$$Z_\lambda(\Lambda, M, a) = \int_{X(M)\times X^{+}(\Lambda\setminus M)} \exp(-H^{(M)}_\lambda(x_\Lambda|a))\,\mathrm{d}x_\Lambda \tag{4.2.7}$$

where

$$H_\lambda^{(M)}(x_\Lambda|a) = H_\lambda(x_{\Lambda\setminus M}|a) + H_\lambda^+(x_M|a \cup x_{\Lambda\setminus M}). \tag{4.2.7'}$$

NOTE 1. We use the notation

$$H_\lambda(x_\Lambda|x_{\Lambda'}) = \sum_{A \subset \Lambda\cup\Lambda' : A\cap\Lambda \neq \emptyset} \Phi_A^\lambda(x_A)$$

in the general case when Λ,Λ' are arbitrary (Λ finite) sets such that $\Lambda \cap \Lambda' \neq \emptyset$. Notice that then $H_\lambda(x_\Lambda|x_{\Lambda'}) = H_\lambda(x_\Lambda|x_{\partial\Lambda^c})$ if $\Lambda' \supset \partial\Lambda^c$. (This is the case which is generally discussed below.)

Analogously, we will use the notion of $H_\lambda(r_\Lambda|x_\Lambda)$ for these Λ,Λ'. Consider the following analogues of the partition functions from Definition 1.

DEFINITION 2. Let $M = \Lambda \subset \mathbb{Z}^\nu$ be finite and let $x_{\partial\Lambda^c} = a \in (\mathcal{U}^+)^{\partial\Lambda^c}$. Put

$$Z_\lambda(\Lambda, M, a, F) = \int_{X(M)\times\mathcal{R}^+(\Lambda\setminus M)} \exp(-H_\lambda^{(M)}(r_{\Lambda\setminus M} \cup x_M|a))\,\mathrm{d}r_{\Lambda\setminus M}\,\mathrm{d}x_M \tag{4.2.8}$$

where

$$H_\lambda^{(M)}(r_{\Lambda\setminus M} \cup x_M|a) = H_\lambda(r_{\Lambda\setminus M}|a) + H_\lambda^+(x_M|x_{\Lambda\setminus M} \cup a) \tag{4.2.8'}$$

(compare (4.2.7')).

For any + frame Γ and any $a \in (\tilde{\mathcal{U}}^+)^{\partial_+\Gamma}$ put

$$Z_\lambda(\Gamma, a, F) = \exp(-F(\Gamma, a))Z_\lambda(V^*(\Gamma), \operatorname{supp}\Gamma, a, F) \tag{4.2.9}$$

NOTE 2. We use the notation

$$Z_\lambda(\Lambda, M, a, F) = Z_\lambda(\Lambda, M \cap \Lambda, a, F)$$

(similarly for $Z_\lambda(\Lambda, M, a)$) also in the case when $M \not\subset \Lambda$.

NOTE 3. The partition functions $Z_\lambda(\Gamma), Z_\lambda(\Gamma, a), Z_\lambda(\Gamma, a, F)$ are the 'crystalline' partition functions (according to PS terminology), whereas $Z_\lambda(\Lambda, a)$, $Z_\lambda(\Lambda, M, a)$ and $Z_\lambda(\Lambda, a, F)$, $Z_\lambda(\Lambda, M, a, F)$ are the 'diluted' ones.

Concerning the definition of $Z_\lambda(\Lambda, M, a, F)$ and $Z_\lambda(\Lambda, M, a)$ we notice that a typical application is $\Lambda = V^*(\Gamma)$ and $M = \operatorname{supp}\Gamma$ in what follows.

NOTE 4. Throughout this section and the following sections, the analogous constructions and relations hold also in the case when + is replaced by −, and vice versa.

THEOREM. *For any + frame* Γ *and any* $a \in (\tilde{\mathcal{U}}^+)^{\partial_+\Gamma}$ *put*

$$F_\lambda(\Gamma, a) = \ln Z_\lambda(V^*(\Gamma), \operatorname{supp}\Gamma, a) - \ln Z_\lambda(\Gamma, a). \tag{4.2.10}$$

Then for any finite Λ, $M \subset \mathbb{Z}^\nu$ *and any* $a \in (\mathcal{U}^+)^{\partial\Lambda^c}$,

$$Z_\lambda(\Lambda, M, a) = Z_\lambda(\Lambda, M, a, F_\lambda) \tag{4.2.11}$$

and also, for any + frame Γ *and any* $a \in (\tilde{\mathcal{U}}^+)^{\partial_+\Gamma}$,

$$Z_\lambda(\Gamma, a) = Z_\lambda(\Gamma, a, F_\lambda). \tag{4.2.12}$$

In particular, the Gibbsian ensemble and the contour ensemble are equivalent.

Proof. This follows from the following lemmas.

LEMMA 1 *Say that a +frame system* $\mathscr{D}$ *is an *-external one if the sets* $V^*(\Gamma), \Gamma \in \mathscr{D}$ *are mutually disjoint. The following formula holds for any* Λ *and* $a \in (\mathscr{U}^+)^{\partial\Lambda^c}$:

$$Z_\lambda(\Lambda, a) = \sum_{\mathscr{D}} \int_{(\mathscr{U}^+)^{\mathrm{ext}^*\mathscr{D}\setminus\partial_+\mathscr{D}} \times (\tilde{\mathscr{U}}^+)^{\partial_+\mathscr{D}}} \exp(-H_\lambda(x^*_{\mathrm{ext}}|a))$$

$$\Pi Z_\lambda(\Gamma, x_{\partial_+\Gamma}) \mathrm{d}x^*_{\mathrm{ext}}, \quad \Gamma \in \mathscr{D}, \tag{4.2.13}$$

where $\mathrm{ext}^*\mathscr{D} = \Lambda \setminus \cup_{\Gamma\in\mathscr{D}} V^*(\Gamma)$ *and* $x^*_{\mathrm{ext}} = x_{\mathrm{ext}^*\mathscr{D}}$, *the sum being taken over all possible *-external frame systems* $\mathscr{D}$ in Λ. *Analogously, for the contour ensemble,*

$$Z_\lambda(\Lambda, a, F_\lambda) = \sum_{\mathscr{D}} \int_{(\mathscr{U}^+)^{\mathrm{ext}^*\mathscr{D}\setminus\partial_+\mathscr{D}} \times (\tilde{\mathscr{U}}^+)^{\partial_+\mathscr{D}}} \exp(-H_\lambda(X^*_{\mathrm{ext}}|a))$$

$$\Pi Z_\lambda(\Gamma, x_{\partial_+\Gamma}, F_\lambda) \mathrm{d}x^*_{\mathrm{ext}}, \quad .\Gamma \in \mathscr{D}. \tag{4.2.13'}$$

An intermediate step in the proof of Lemma 1 is the following.

LEMMA 2. *Let* $\mathscr{D}$ *be an external frame system (in the usual sense) of a configuration* $x_\Lambda \in X^+(\Lambda)$. *Denote by* $\Delta = \partial_+\mathscr{D} \cup (\cup_{\Gamma\in\mathscr{D}} \partial V(\Gamma))$ *and* $\widehat{\mathrm{int}}_+\Gamma = \mathrm{int}_+\Gamma \setminus \Delta, \Gamma \in \mathscr{D}$ *(see* (4.2.5)). *Denote by* $\hat{Z}_\lambda(\widehat{\mathrm{int}}_+\Gamma, x_\Delta)$ *the analogue of* $Z_\lambda(\widehat{\mathrm{int}}_+\Gamma, x_\Delta)$ *obtained if in Definition* 4.1.4 *of* $X^+(\widehat{\mathrm{int}}_+\Gamma)$ *the condition* $\mathrm{dist}(V(\Gamma), (\widehat{\mathrm{int}}_+\Gamma)^c) \geq 2$ *is replaced by* $\mathrm{dist}(V(\Gamma), (\widehat{\mathrm{int}}_+\Gamma)^c) \geq 2$. *We have the relation* (see 4.2.5)

$$Z_\lambda(\Lambda, a) = \sum_{\mathscr{D}} \int_{(\mathscr{U}^+)^{\mathrm{ext}^\partial\mathscr{D}\setminus\partial_+\mathscr{D}} \times (\tilde{\mathscr{U}}^+)^{\partial_+\mathscr{D}}} \exp(-H_\lambda(x_{\mathrm{ext}^\partial\mathscr{D}}|a)) \times$$

$$\times \prod_{\Gamma\in\mathscr{D}} Z_\lambda(\Gamma, x_{\partial_+\Gamma}) \prod_{\Gamma\in\mathscr{D}} \hat{Z}_\lambda(\widehat{\mathrm{int}}_+\Gamma, x_\Delta) \mathrm{d}x_{\mathrm{ext}^\partial\mathscr{D}} \tag{4.2.14}$$

and similarly for $Z_\lambda(\Lambda, a, F_\lambda)$.

NOTE 5. In contrast to the obvious relation $\mathrm{dist}\,(\mathrm{int}_-\Gamma, V(\Gamma)^c) > r$ we cannot exclude the possibility that $t \in \mathrm{int}_+\Gamma$ for some $t \in \partial(V(\Gamma))$. Hence the notion of $\hat{Z}(\widehat{\mathrm{int}}_+\Gamma, x_\Delta)$.

Proof of Lemma 2. This is obvious, if we use the Fubini theorem and the definition of the external contour system. □

Proof of Lemma 1. It suffices repeatedly to express $\hat{Z}_\lambda(\widehat{\mathrm{int}}_+\Gamma, x_\Delta)$ by Lemma 2, with obvious modifications if $\hat{Z}_\lambda \neq Z_\lambda$. □

Now we can finish the proof of the Theorem. The relations (4.2.11), (4.2.12) will be proved by induction with respect to the ordering $<$ (see 4.1.7).

Comparing (4.2.13) with (4.2.13′) we see that (4.2.11) follows from (4.2.12) if the latter relation holds for each $\Gamma < \Lambda$. It is straightforward to generalize the expressions (4.2.13) and (4.2.13′) to the case of $M \neq \emptyset$ and therefore to prove the induction step (4.2.12) ⇒ (4.2.11) in full generality. On the other hand if we choose $\Lambda = V^*(\Gamma)$, $M = \mathrm{supp}\,\Gamma$ then obviously (4.2.11) implies (4.2.12) because of the relation (4.2.9) and the very definition (4.2.10) of $F_\lambda(\Gamma, a)$. Finally, to start the induction argument (with respect to $<$) notice that (4.2.11) is obvious from (4.2.14), and the corresponding relation for $Z_\lambda(\Lambda, a, F_\lambda)$ if there are no contours Γ such that $\Gamma < \Lambda$.

It is now easy to see that the probability measures $P^{\text{ext}}_{\lambda,a}$ and $P^{\text{ext}}_{\lambda,a,F_\lambda}$ are the same. In fact, we obtain more information: even the probability densities for x^*_{ext} are the same in both the Gibbs and the contour model. They are given by the formula

$$(Z_\lambda(\Lambda,a))^{-1}\exp(-H_\lambda(x^*_{\text{ext}}|a))\prod_{\Gamma\in\mathscr{D}} Z_\lambda(\Gamma,x_{\partial_+\Gamma}). \tag{4.2.15}$$

NOTE 6. As we yet noted in §4.2.2 there is some arbitrariness in our construction of contour models. Requiring (4.2.11) and (4.2.12) to hold, it is still possible to change the interactions $\Phi^{+,\lambda}_A(x_A)$, $A\in\text{Supp}^*(\mathscr{D})$ in (4.2.3), in many ways. For example one can take, instead of (4.2.3), the following choice of $H_\lambda(r_\Lambda\,|\,a)$:

$$H_\lambda(r_\Lambda\,|\,a)=\sum_{A:A\cap\Lambda\neq\emptyset}\Phi^\lambda_A(x_A)+\sum_{\Gamma\in\mathscr{D}}F(\Gamma,x_{\partial_+\Gamma}) \tag{4.2.16}$$

with the additional (quite arbitrary from the point of view of §4.2) restriction $x_t\in\mathscr{U}^+$ for $t\in\text{supp}\,\mathscr{D}\setminus\partial_+\mathscr{D}$. Any such change must be properly reflected by a corresponding change in F_λ. Our choice of (4.2.3) is technically suitable from the point of view of §4.4.

4.2.7. *Another Expression of* $F_\lambda(\Gamma,a)$.

We will investigate in more detail the expression on the right-hand side of (4.2.10). It is helpful to introduce, in addition to (4.2.10), the following quantity.

DEFINITION 1. For any + contour Γ define (see (4.2.7))

$$F_\lambda(\Gamma)=\ln Z_\lambda(V^*(\Gamma),\,\text{supp}\,\Gamma,\partial_+\Gamma)-\ln Z_\lambda(\Gamma). \tag{4.2.17}$$

DEFINITION 2. A non-negative measurable functional acting on + contours will be called a + *contour weight*. Given such a weight $F(\Gamma)$, define the corresponding *frame weight* $\mathbb{F}(\Gamma,a)$, $a\in(\tilde{\mathscr{U}}^+)^{\partial_+\Gamma}$ by

$$\mathbb{F}(\Gamma,a)=-\ln\int_{\mathscr{C}'(\Gamma,a)}\exp(-F(\Gamma))\,d\Gamma_{(\text{supp}\Gamma\setminus\partial_+\Gamma)} \tag{4.2.18}$$

(see Definition 4.2.6.1). This is called the factorization of $F(\Gamma)$.

The reader will readily recognize, that (4.2.10) is a factorization of (4.2.17).)

NOTE. In all cases when both contour and frame weights are considered, they will be related by (4.2.18). (This is the case in §§4.2 and 4.5, whereas in §4.4 and where ever partition functions $Z_\lambda(\Lambda,M,a,F)$ are investigated we deal exclusively with frame weights $F(\Gamma,a)$.) Thus, the occasional use of the term contour weight instead of frame weight will cause no confusion.

We now investigate the expression (4.2.17) in more detail. Denote by $W(\Gamma)$ the set $\text{int}_-\,\Gamma$ (for a + contour Γ). (We write simply W and V^* instead of $W(\Gamma)$ and $V^*(\Gamma)$ if this causes no ambiguity.)

Express $Z_\lambda(\Gamma)$ as follows:

$$Z_\lambda(\Gamma)=\exp(-H_\lambda(\Gamma)+H_\lambda(\partial_+\Gamma))Z_\lambda(W,\partial_-\Gamma). \tag{4.2.19}$$

Therefore, by (4.2.17),

$$F_\lambda(\Gamma) = (H_\lambda(\Gamma) - H_\lambda(\partial_+\Gamma)) + \ln Z_\lambda(V^*, \operatorname{supp}\Gamma, \partial_+\Gamma) - \\ - \ln Z_\lambda(W, \partial_-\Gamma) \tag{4.2.20}$$

$$F_\lambda(\Gamma) = H_\lambda(\Gamma) - H_\lambda(\partial_+\Gamma) + \ln Z_\lambda(V^*, \operatorname{supp}\Gamma, \partial_+\Gamma, F_\lambda) - \\ - \ln Z_\lambda(W, \partial_-\Gamma, F_\lambda). \tag{4.2.21}$$

NOTE. This is really a recurrent relation for $F_\lambda(\Gamma)$ because the right-hand side of (4.2.21) depends on $F_\lambda(\tilde{\Gamma})$ only for those $\tilde{\Gamma}$ satisfying the condition $\tilde{\Gamma} < \Gamma$ (see §4.1.7 for the definition of $<$).

DEFINITION 2. For any $\Lambda \subset \mathbb{Z}^\nu$ finite and any $a \in (\mathscr{U}^+)^{\partial\Lambda^c}$ denote by $Z_\lambda^+(\Lambda, a)$ the Gaussian partition function

$$Z_\lambda^+(\Lambda, a) = \int_{X(\Lambda)} \exp(-H_\lambda^+(x_\Lambda \mid a))\,\mathrm{d}x_\Lambda. \tag{4.2.22}$$

(Notice that $Z_\lambda^+(\Lambda, a) = Z_\lambda(\Lambda, \Lambda, a)$ in the notation of (4.2.8).) Denote by h_λ^+ the free energy of the Hamiltonian given by interactions $\{\Phi_A^{+,\lambda}\}$.

NOTE. (4.2.22) is a partition function of the type (2.2.6), but with a (generally non-standardized) Hamiltonian

$$H_\lambda^+(x_\Lambda \mid x_{\Lambda^c}) = \sum_{A \cap \Lambda \neq \emptyset} \Phi^{+,\lambda}(x_A)$$

(see §2.2.6).

Having the idea of the Gaussian approximation in mind, write (4.2.21) as

$$F_\lambda(\Gamma) = G_\lambda(\Gamma) + \Delta_\lambda(\Gamma, F_\lambda) \tag{4.2.23}$$

where (for a + contour Γ)

$$G_\lambda(\Gamma) = H_\lambda(\Gamma) - H_\lambda(\partial_+\Gamma) + \ln Z_\lambda^+(V^*, \partial_+\Gamma) - \ln Z_\lambda^-(W, \partial_-\Gamma) + \\ + h_\lambda^- |W| - h_\lambda^+ |W| \tag{4.2.24}$$

and the second term is given by the formula

$$\Delta_\lambda(\Gamma, F_\lambda) = \ln Z_\lambda(V^*, \operatorname{supp}\Gamma, \partial_+\Gamma, F_\lambda) - \ln Z_\lambda^+(V^*, \partial_+\Gamma) - \\ - \ln Z_\lambda(W, \partial_-\Gamma, F_\lambda) + \ln Z_\lambda^-(W, \partial_-\Gamma) + h_\lambda^+ |W| - h_\lambda^- |W| \tag{4.2.25}$$

(and analogously for $-$ contour Γ).

We emphasize that the contour weight F_λ of the equivalent contour ensemble, originally given by (4.2.17), is now expressed as a solution of the system of integral equations (4.2.23).

Such a reformulation is of no use until we establish some additional properties of F_λ (see the next section). These additional properties will be shown to hold in the case when a phase transition takes place.

4.2.8. *Free Energy of the Contour Model*

ASSUMPTION 1. Suppose the existence of a limit

$$h_\lambda^+(F_\lambda) = \lim_{\Lambda \uparrow \mathbb{Z}^\nu; a \in (\mathscr{U}^+)^{\partial\Lambda^c}} |\Lambda|^{-1} \ln Z_\lambda(\Lambda, a, F_\lambda) \tag{4.2.26}$$

(in the Van Hove sense).

PROPOSITION 1. (4.2.25) *can be written as*

$$\Delta_\lambda(\Gamma, F_\lambda) = \tilde{\Delta}_\lambda(\Gamma, F_\lambda) + (h_\lambda^+(F_\lambda) - h_\lambda^-(F_\lambda))|W| \tag{4.2.27}$$

where for any frame weight F we define

$$\begin{aligned}\tilde{\Delta}_\lambda(\Gamma, F) = {} & \ln Z_\lambda(V^*, \operatorname{supp}\Gamma, \partial_+\Gamma, F) - \ln Z_\lambda^+(V^*, \partial_+\Gamma) - \\ & - (h_\lambda^+(F) - h_\lambda^+)|V^*| - (\ln Z_\lambda(W, \partial_-\Gamma, F) - \ln Z_\lambda^-(W, \partial_-\Gamma) - \\ & -(h_\lambda^-(F) - h_\lambda^-)|W|) + (h_\lambda^+(F) - h_\lambda^+)|V^* \setminus W|.\end{aligned} \tag{4.2.28}$$

Proof. Immediate.

ASSUMPTION 2.

$$h_\lambda^+(F_\lambda) = h_\lambda^-(F_\lambda). \tag{4.2.29}$$

NOTE. If F_λ is given by (4.2.10), i.e. if it satisfies (4.2.11), then (4.2.29) obviously holds whenever both the quantities $h_\lambda^+(F_\lambda)$, $h_\lambda^-(F_\lambda)$ are defined. (This is an expression, through (4.2.11), of the well-known fact that the free energy of the 'physical' model does not depend on the boundary conditions.) Later we will express F_λ as a solution of a system of equations other than that of (4.2.10). The equivalence between both systems of equations will then be established only by use of the assumption (4.2.29). In this new formulation, the validity of (4.2.29) will indicate that we are in the situation when two Gibbs states (the + one and the − one) appear in the infinite limit.

Thus, the situation when there is only one limiting Gibbs state is not studied here. It is not difficult, but the more general case when there are more than two 'ground states' of the model, some of them being 'unstable', requires some further constructions. (It requires either the introduction of contour models with a parameter, or another alternative methods developed in [10]. The construction of a phase diagram outside the point of a maximal number of phases, and the proof of its completeness, will be the subject of a forthcoming paper of one of the authors.)

4.2.9. *Strategy of the Solution of* (4.2.23)

COROLLARY. *Under Assumptions* 1 *and* 2 *of the preceding section, a contour weight* F_λ *solves the system* (4.2.23) *iff it solves the system of equations*

$$F_\lambda(\Gamma) = G_\lambda(\Gamma) + \tilde{\Delta}_\lambda(\Gamma, F_\lambda) \tag{4.2.30}$$

with $\tilde{\Delta}_\lambda$ *given by* (4.2.28).

Proof. Immediate, by (4.2.27), (4.2.28). □

NOTE. Compared to the explicit formula (4.2.17) and the recurrent equation (4.2.21), the equation (4.2.30) has a more complicated structure, but it will be seen to be more useful because, as we will explain in detail in the §§4.3 and 4.4 below, it is possible to obtain a precise estimates of both G_λ and $\tilde{\Delta}_\lambda$.

The terms G_λ and $\tilde{\Delta}_\lambda$ are of a different nature: the 'main' term $G_\lambda(\Gamma)$ does not depend on F_λ, whereas $\tilde{\Delta}_\lambda(\Gamma, F_\lambda)$ is small and depends on F_λ (and $\partial_+\Gamma$). We now introduce the following important notion.

DEFINITION. A contour weight F is a *κ-functional* ($\kappa > 0$) if for each frame $\mathbf{\Gamma}$ and each $a \in (\tilde{\mathscr{U}}^+)^{\partial_+\Gamma}$ $(a \in (\tilde{\mathscr{U}}^-)^{\partial_-\Gamma})$,

$$F(\mathbf{\Gamma}, a) \geqslant \kappa |\operatorname{supp} \mathbf{\Gamma}|. \tag{4.2.31}$$

It will be shown in §4.3 below that such a condition holds, with large κ, for the frame weights $G_\lambda(\mathbf{\Gamma}, a)$, defined by (4.2.18) using the contour weight $G_\lambda(\Gamma)$, assuming that (3.2.11) is satisfied for large τ.

On the other hand, we will show in ξ4.4 that under the assumption (4.2.31), κ large, the terms $\tilde{\Delta}_\lambda(\Gamma, F)$ are estimated as

$$|\tilde{\Delta}_\lambda(\Gamma, F)| \leqslant c|\operatorname{supp} \Gamma|$$

with a small constant c not depending on F. These facts enable us to solve (4.2.30) in the class of contour weights $F_\lambda(\Gamma)$ whose factorization $F_\lambda(\mathbf{\Gamma}, a)$ is a κ – functional.

Finally, having solved (4.2.30) we take only those solutions which satisfy the condition (4.2.29). The interpretation of the corresponding contour weights and the construction of Gibbs states is briefly given in §4.5.

Section 4.3 uses some elementary estimates of §2.2 (notably that of §§2.2.3 and 2.2.7).

Section 4.4 is technically much more involved. It uses heavily the estimates of §2.3, in particular Theorem 2.3.8.

4.3. ESTIMATES OF THE MAIN TERM $G_\lambda(\Gamma)$. DECOMPOSITION OF THE CONTOUR ENERGY

In this section, we use the correlation decay results of §2.2 to show that $G_\lambda(\Gamma)$ can be estimated from below by the value

$$H^\lambda_{\mathrm{rel}}(\Gamma_{B_r}) + c\delta^2|\tilde{B}| - c'|\operatorname{supp} \Gamma|$$

(see (3.2.10)) where B_r is the set of all (δ, r) incorrect points (in the sense of §3.1 for this particular choice of r, δ) of Γ, $\tilde{B}(=\tilde{B}^+ \cup \tilde{B}^-)$ is the set of all $(\tilde{\delta}, 0)$ incorrect points of Γ not contained in the $\tilde{r}$-neighbourhood of B_r and $c, c' > 0$ are arbitrary constants. We then use the condition (3.2.11). We find that $G_\lambda(\cdot)$ is a κ-functional with a large value of κ.

4.3.1. *Notation*

Given a + contour Γ denote by $y = y_{V^* \cup \partial_+\Gamma}$ (see Definition 4.2.6) the configuration on $V^* \cup \partial_+\Gamma$ minimizing the standardized Hamiltonian $H^+_{0,\mathrm{st}}(x_{V^* \cup \partial_+\Gamma})$ (see 2.2.6) at the condition $x_{\partial_+\Gamma} = \partial_+\Gamma$. Analogously, denote by $z = z_{W \cup \partial_-\Gamma}$ the configuration on $W \cup \partial_- \Gamma$ (see 4.2.7) minimizing the Hamiltonian $H^-_{0,\mathrm{st}}(x_{W \cup \partial_- \Gamma})$ at the condition $x_{\partial_- \Gamma} = \partial_- \Gamma$. Notice the following formula, which holds for any $x \in X(V^* \cup \partial_+ \Gamma)$ such that $x_{\partial_+ \Gamma} = \partial_+ \Gamma$:

$$H^+_\lambda(x) = H^+_\lambda(y) + H^+_{0,\mathrm{qst}}((x - y)_{V^*}) \tag{4.3.1}$$

where $H^+_{0,\mathrm{qst}}$ is the quadratic standardized Hamiltonian corresponding to H^+_0. In fact, this is formula (2.2.71). (Note that $H^+_{\lambda,\mathrm{qst}} = H^+_{0,\mathrm{qst}}$ for each λ.) Therefore,

$$Z^+_\lambda(V^*, \partial_+ \Gamma) = \exp(-H^+_\lambda(y) + H^+_\lambda(\partial_+ \Gamma)) \int_{X(V^*)} \exp(-H^+_{0,\mathrm{qst}}(x))\,\mathrm{d}x. \tag{4.3.2}$$

The free energies h^+_λ, h^+_{qst} corresponding to the Hamiltonians H^+_λ, $H^+_{0,\mathrm{qst}}$ are therefore related by the formula

$$h^+_\lambda = h^+_{\mathrm{qst}} - \lambda u^+ \tag{4.3.3}$$

(see (3.2.5) and (2.2.69)).

Using this notation we can express the terms on the right-hand side of (4.2.24) as follows:

$$\ln Z^+_\lambda(V^*, \partial_+ \Gamma) - h^+_\lambda |V^*| = H^+_\lambda(\partial_+ \Gamma) - H^+_\lambda(y) + Q^+(V^*) + u^+ |V^*| \tag{4.3.4}$$

where

$$Q^+(\Lambda) = \ln \int_{X(\Lambda)} \exp(-H^+_{0,\mathrm{qst}}(x))\,\mathrm{d}x - h^+_{\mathrm{qst}} |\Lambda| + \lambda u^- |W|.$$

Similarly,

$$\ln Z^-_\lambda(W, \partial_- \Gamma) - h^-_\lambda |W| = H^-_\lambda(\partial_- \Gamma) - H^-_\lambda(z) + Q^-(W). \tag{4.3.4'}$$

with an analogously defined Q^-.

Recall from §2.2.4 that the terms $Q^\pm$ can be estimated as

$$|Q^\pm(\Lambda)| \leqslant c |\partial \Lambda^c| \tag{4.3.5}$$

where c is some constant depending only on $c_{(3.2.3)}$.

NOTE. Analogous considerations can be applied to the case of a $-$ contour Γ. Such a comment refers to all the considerations of this section and will be usually omitted below

4.3.2. *Basic Expression for $G_\lambda(\Gamma)$.*

COROLLARY. *For any + contour Γ, the term $G_\lambda(\Gamma)$ can be expressed as*

$$\begin{aligned} G_\lambda(\Gamma) = {} & \lambda(u^+ - u^-)|W| + Q^+(V^*) - Q^-(W) + H_\lambda(\Gamma) - H_\lambda(\partial_+ \Gamma) + \\ & + H^+_\lambda(\partial_+ \Gamma) - H^+_\lambda(y) - H^-_\lambda(\partial_- \Gamma) + H^-_\lambda(z) + \\ & + h^+_0 |V^* \setminus W|. \end{aligned} \tag{4.3.6}$$

Proof. This is an immediate consequence of (4.2.24), (4.3.3), (4.3.4) and (4.3.4').

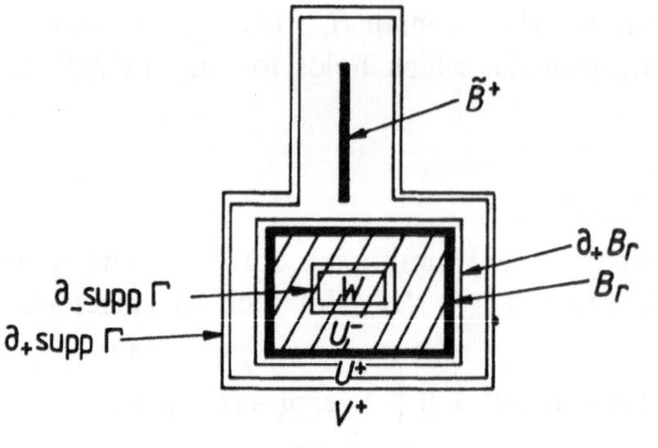

Fig. 7.

4.3.3. *Investigation of the Expression* (4.3.6).

We need some additional notations: See Figure 7.

DEFINITION 1. Denote by $B_r = B_r(\Gamma)$ the set of all (δ, r) incorrect points of Γ. Denote by Γ_r the restriction of Γ on B_r. Denote by

$$U^{\pm} = \{t \in \operatorname{supp} \Gamma \setminus B_r : \operatorname{sign}_\Gamma(t) = \pm\}$$

and analogously define $\partial_\pm B_r$. Denote by $\partial_\pm \Gamma_r$ the restriction of Γ on $\partial_\pm B_r$.

PROPOSITION 1. *Using this notation we have the expressions*

$$H_\lambda(\Gamma) = H_\lambda(\Gamma_r) + H_\lambda^+(\Gamma_{U^+} | \partial_+ \Gamma_r) + H_\lambda^-(\Gamma_{U^-} | \partial \Gamma_r) + R_\lambda(\Gamma) \tag{4.3.7}$$

and

$$H_\lambda(\partial_+ \Gamma) = H_\lambda^+(\partial_+ \Gamma) + R_\lambda^+(\partial_+ \Gamma) \tag{4.3.7'}$$

(*we use the notation of Note* 1, §4.2.6), *where* $R_\lambda(\Gamma)$, $R_\lambda^+(\partial_+ \Gamma)$ *satisfy the estimates*

$$|R_\lambda(\Gamma)| \leqslant c\,\varepsilon |\operatorname{supp} \Gamma| \tag{4.3.8}$$

$$|R_\lambda^+(\partial_+ \Gamma)| \leqslant c\,\varepsilon |\operatorname{supp} \Gamma|, \tag{4.3.8'}$$

with a suitable constant $c = c(r)$,

Proof. This is an easy consequence of the approximations (3.2.6). □

DEFINITION 2. Denote further by V^+, V^- the sets (for a +contour Γ)

$$V^+ = U^+ \cup (V^*)^c,$$
$$V^- = U^- \cup W.$$

Note that $\partial(V^\pm)^c \subset \partial_\pm B_r$ and also $\partial_\pm \Gamma \cap \partial_\pm \Gamma_r = \emptyset$ (because of the condition $\tilde{r} > 2r$).
Using (4.3.6), (4.3.7) now write $G_\lambda(\Gamma)$ as follows:

$$G_\lambda(\Gamma) = \tilde{G}_\lambda(\Gamma) + Q^+(V^*) + Q^-(W) + R_\lambda(\Gamma) - R_\lambda^+(\partial_+ \Gamma) + h_0^+ |V^* \setminus W| \tag{4.3.9}$$

where

$$\tilde{G}_\lambda(\Gamma) = H_\lambda(\Gamma_r) + H_\lambda^-(\Gamma_{U^- \setminus \partial_- \Gamma} \cup z | \partial_- \Gamma_r) + \\ + H_\lambda^+(\Gamma_{U^+} | \partial_+ \Gamma_r) - H_\lambda^+(y) + \lambda(u^+ - u^-)|W|. \tag{4.3.10}$$

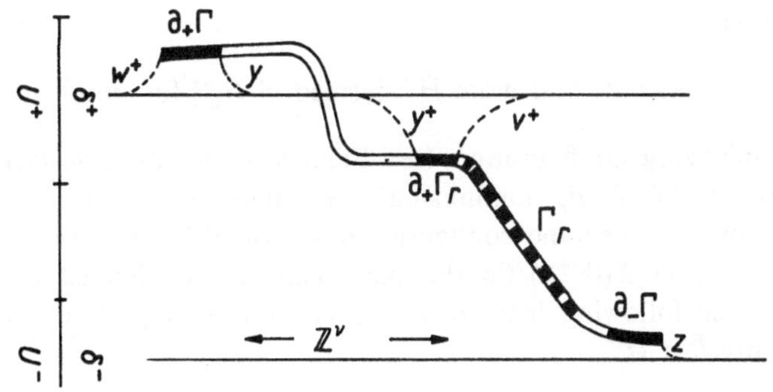

Fig. 8.

Investigate $\tilde{G}_\lambda(\Gamma)$ in more detail. Introduce the following notions (see also Figures 7 and 8). In analogy with the minimizing configurations y and z define the configuration y^+ on V^+ minimizing the Hamiltonian $H_0^+(x_{V^+}|\partial_+\Gamma_r)$, $x_{V^+} \in X(V^+)$, in the following sense (note that because V^+ is infinite we must be more precise in the definition of y^+ and $H_0^+(y^+ \mid \partial_+\Gamma_r)$). Recall the notion of $\mathring{H}^W(x_\Lambda)$ from (2.2.72), §2.2.7.

Assume that $\sigma^+ = 0$ (in (3.2.1)) for simplicity of notation (with obvious modifications in the general case). Thus, $H_{0,st}^+ = H_{0,\mathrm{qst}}^+$. We can now define y^+ as we defined $\bar{x}$ in (2.2.73), with $H = H_{0,\mathrm{st}}^+$ and with the choice of $\Lambda = \partial_+ B_r$, $W = V^+ \cup \partial_+ B_r$, $\bar{x}_\Lambda = \partial_+\Gamma_r$ in (2.2.73). Further, we can define $H_{0,\mathrm{st}}^+(y^+ \mid \partial_+\Gamma_r)$ as

$$H(\bar{x}_{W\setminus\Lambda} \mid \bar{x}_\Lambda) = \mathring{H}^W(x_\Lambda) - H(\bar{x}_\Lambda) \tag{4.3.11}$$

in this notation. It is now natural to define $H_0^+(y^+ \mid \partial_+\Gamma_r)$ from the relation

$$H_0^+(y^+ \mid \partial_+\Gamma_r) + H_0^+(x_{(V^+)^c}) = H_{0,\mathrm{st}}^+(y^+ \mid \partial_+\Gamma_r) + H_{0,\mathrm{st}}^+(x_{(V^+)^c}) \tag{4.3.12}$$

where $x_{(V^+)^c} \in X((V^+)^c)$ is arbitrary but such that $x_{\partial_+ B_r} = \partial_+\Gamma_r$ (notice that $(V^+)^c$ is finite).

Using (3.2.9) and (3.2.10), we can see that

$$\begin{aligned} &H_\lambda(\Gamma_r) + H_0^-(\Gamma_{U^-\setminus\partial_-\Gamma} \cup z \mid \partial_-\Gamma_r) + H_0^+(y^+ \mid \partial_+\Gamma_r) \\ &\quad = H_{\mathrm{rel}}^{\lambda,B_r,V^+,V^-}(\Gamma_r \cup \Gamma_{U^-\setminus\partial_-\Gamma} \cup z \cup y^+) \\ &\quad H_{\mathrm{rel}}^\lambda(\Gamma_r) + M(\Gamma) \end{aligned} \tag{4.3.13}$$

where

$$M(\Gamma) \geqslant 0. \tag{4.3.13$'$}$$

(We do not specify $M(\Gamma)$ at the moment. Obviously, it does not depend on λ.)

Therefore, using also (3.2.5), one can write $G_\lambda(\Gamma)$ as

$$\begin{aligned} \tilde{G}_\lambda(\Gamma) = H_{\mathrm{rel}}^\lambda(\Gamma_r) - H_0^+(y^+ \mid \partial_+\Gamma_r) - H_0^+(y) + H_0^+(\Gamma_{U^+} \mid \partial_+\Gamma_r) + \\ + M(\Gamma) + \lambda L(\Gamma) \end{aligned} \tag{4.3.14}$$

where $L(\Gamma) = L(\Gamma, \{u_A^\pm\})$ (we omit the explicit formula for $L(\Gamma)$) satisfies, with some $c' = c'(\{u_A^\pm\})$, the inequality

$$|L(\Gamma)| \leqslant c'|\text{supp}\,\Gamma|. \tag{4.3.14'}$$

Use the notation (2.2.72) again, and write $\mathring{H}^+$ instead of $\mathring{H}^{+z^v}_{0,\text{st}}$ for simplicity.

Define another minimizing configuration (see Figures 7 and 8). Consider the configuration v^+ on $(V^+)^c\setminus\partial_+B_r$ minimizing the Hamiltonian $H_0^+(x|\partial_+\Gamma_r)$, $x\in X((V^+)^c\setminus\partial_+B_r)$. Consider also the configuration w^+ on $(V^*)^c$ minimizing the Hamiltonian $H_0^+(x|\partial_+\Gamma)$, $x\in X((V^*)^c)$ (in the same sense as we defined y^+ and $H_0^+(y^+|\partial_+\Gamma_r)$). Then the following inequality follows immediately from the definition of $\mathring{H}^+$. For any $\tilde{B}\subset U^+$,

$$\mathring{H}^+(\partial_+\Gamma\cup\partial_+\Gamma_r\cup\Gamma_{\tilde{B}}) \leqslant H_0^+(\Gamma_{U^+}|\partial_+\Gamma_r) + H_0^+(v^+|\partial_+\Gamma_r) + H_0^+(W^+|\partial_+\Gamma). \tag{4.3.15}$$

We also obtain the relations

$$\mathring{H}^+(\partial_+\Gamma) = H_0^+(y) + H_0^+(w^+|\partial_+\Gamma)$$

and

$$\mathring{H}^+(\partial_+\Gamma_r) = H_0^+(v^+\cup\partial_+\Gamma_r) + H_0^+(y^+|\partial_+\Gamma_r).$$

Summarizing these considerations, we obtain the following result.

COROLLARY. *The following expression holds:*

$$\tilde{G}_\lambda(\Gamma) = H^\lambda_{\text{rel}}(\Gamma_r) + S(\Gamma) + M(\Gamma) + \lambda L(\Gamma) \tag{4.3.16}$$

where (for a suitable $\tilde{B}\subset U^+$ *to be specified later)*

$$S(\Gamma) = \mathring{H}^+(\partial_+\Gamma\cup\partial_+\Gamma_r\cup\Gamma_{\tilde{B}}) - \mathring{H}^+(\partial_+\Gamma) - \mathring{H}^+(\partial_+\Gamma_r) \tag{4.3.16'}$$

with the same $L(\Gamma)$ *as in* (4.3.14) *and with a new positive* $M(\Gamma)$ *(including also the difference between the right-hand side and left-hand side of* (4.3.15)*, but not depending on* λ*).*

4.3.4. *Investigation of* $S(\Gamma)$

This is the part of the proof of the main lemma where we need our notion of 'thick' contour, with some assumptions on $\tilde{r} > 2r$. (Thick contours will turn out to be useful again in §4.4 and 4.5, where we also discover another peculiar feature of contours – the condition $\partial_\pm\Gamma\in(\tilde{\mathcal{U}}^\pm)^{\pm\Gamma}$. The latter property has little relevance to §4.3, and is not used there.)

PROPOSITION. *Denote by* $\tilde{B}^+$ *the set of those* $t\in U^+$ *which satisfy the condition* $x_t\in\mathcal{U}^+\setminus\tilde{\mathcal{U}}^+$ *and which do not belong to the* $\tilde{r}$*-neighbourhood of* B_r*. (Notice also that* $\text{dist}(\tilde{B}^+,\partial_+\Gamma)\geqslant\tilde{r}-r$*.) There are some constants* c, c' *and some* $0<q<1$ *depending only on* $c_{(3.2.3)}$ *(and* r*) such that for each contour* Γ,

$$S(\Gamma) \geqslant c\delta^2|\tilde{B}^+| - c'\delta^2 q^{\tilde{r}}|\partial_+B_r\cup\partial_+\Gamma|. \tag{4.3.17}$$

Proof. Using (2.2.76) twice we have, with some $0<q<1$ and c depending on

$c_{(3.2.3)}$, the estimates

$$\begin{aligned}&\mathring{H}^+(\partial_+\Gamma \cup \partial_+\Gamma_r \cup \Gamma_{\tilde{B}^+})\\ &\quad \geqslant \mathring{H}^+(\partial_+\Gamma \cup \partial_+\Gamma_r) + \mathring{H}^+(\Gamma_{\tilde{B}^+}) - c\delta^2 q^{\tilde{r}-r}|\partial B_r \cup \partial_+\mathbf{\Gamma} \cup \tilde{B}^+|\\ &\quad \geqslant \mathring{H}^+(\partial_+\Gamma) + \mathring{H}^+(\partial_+\Gamma_r) + \mathring{H}^+(\Gamma_{\tilde{B}^+}) - c\delta^2(q^{\tilde{r}-r}|\partial B_r \cup \partial_+\mathbf{\Gamma} \cup \tilde{B}^+| +\\ &\qquad + q^{\tilde{r}}|\partial B_r \cup \partial_+\mathbf{\Gamma}|).\end{aligned}$$

On the other hand, by (2.2.75) and (3.2.3) we have the estimate

$$\mathring{H}^+(\Gamma_{\tilde{B}^+}) \geqslant c\delta^2|\tilde{B}^+| \tag{4.3.17'}$$

with some $c = c(c_{(3.2.3)})$. This clearly gives (4.3.17). □

It is now time to summarize the investigation of $G_\lambda(\Gamma)$ that we have carried out so far.

COROLLARY. *Let Γ be a + contour. Denote by $\tilde{B}^\pm$ the set of those $t \in U^+ \cap B_0^{\delta}(\Gamma)$ which do not belong to the $\tilde{r}$ neighbourhood of B_r. Then*

$$G_\lambda(\Gamma) \geqslant H^\lambda_{\text{rel}}(\Gamma_r) + c\delta^2|\tilde{B}^+ \cup \tilde{B}^-| - (c'\delta^2 q^{\tilde{r}} + c'' + c'''\lambda)|\operatorname{supp}\Gamma| \tag{4.3.18}$$

where q, c, c', c'', c''' are suitable constants depending only on $c_{(3.2.3)}$; $0 < q < 1$.

Proof. Collecting (4.3.4), (4.3.7), (4.3.13), (4.3.14), (4.3.16) and the corresponding estimates of $Q^\pm$, R, R^+, M, L, S we immediately obtain the result for the special case $\tilde{B}^- = \emptyset$. The case $\tilde{B}^- \neq \emptyset$ requires a more precise estimate of $M(\Gamma)$ in (4.3.13). That is, we can write the estimate in (4.3.13) more precisely as

$$\begin{aligned}&H^{\lambda, B_r, V^+, V^-}_{\text{rel}}(\Gamma_r \cup \Gamma_{\mathscr{U}^- \setminus \partial_-\Gamma} \cup z \cup y^+) = H^\lambda_{\text{rel}}(\Gamma_r) +\\ &\quad + \mathring{H}^-(\partial_-\Gamma_r \cup \partial_-\Gamma \cup \Gamma_{\tilde{B}^-}) - \mathring{H}^-(\partial_-\Gamma_r \cup \partial_-\Gamma) + M'(\Gamma) \text{ with } M'(\Gamma) \geqslant 0.\end{aligned}$$

Thus

$$M(\Gamma) \geqslant \mathring{H}^-(\Gamma_{\tilde{B}^-}) - c\delta^2 q^{\tilde{r}}|\tilde{B}^- \cup \partial B_r \cup \partial_-\mathbf{\Gamma}) \tag{4.3.18'}$$

with a suitable $c > 0$ and $0 < q < 1$ depending only on $c_{(3.2.3)}$. Using an analogy of (4.3.17′) we obtain the desired result (4.3.18).

At this stage we use the assumption (3.2.11) of the Main Lemma. Define the factorized weight $G_\lambda(\mathbf{\Gamma}, x_{\partial_+\Gamma})$ as in (4.2.18).

4.3.5. *Theorem*

$G_\lambda(\mathbf{\Gamma}, x_{\partial_+\Gamma})$ is a κ-functional with a suitable $\kappa = \kappa(\tau, \delta, \tilde{r}, c_{(3.2.3)})$ such that $\kappa \to \infty$ if $\delta^2(\tilde{r})^{-\nu} \to \infty$, $\tau(\tilde{r})^{-\nu} \to \infty$, and $\tilde{r}\,(\ln \delta)^{-1} \to \infty$.

Proof. This follows from (4.3.18) and the following observations:

(a) supp $\mathbf{\Gamma}$ is contained in the $\tilde{r}$-neighbourhood of the set $B_r \cup \tilde{B}^+ \cup \tilde{B}^-$. Therefore,

$$|\operatorname{supp}\mathbf{\Gamma}| \leqslant V(\tilde{r})|B_r \cup \tilde{B}^+ \cup \tilde{B}^-| \tag{4.3.19}$$

where $V(\tilde{r})$ is the cardinality of the set $\{t \in \mathbb{Z}: |t| \leqslant \tilde{r}\}$.

(b) Given a frame $\mathbf{\Gamma}$ there are no more than $2^{\operatorname{supp}\mathbf{\Gamma}}$ possibilities for B_r, and the same is true for $\tilde{B}^+$ and $\tilde{B}^-$.

(c) Given Γ_r, $\partial_+\Gamma$ and $\tilde{B}^+$, $\tilde{B}^-$ denote by Γ^{opt} the value of Γ minimizing $\tilde{G}_\lambda(\Gamma)$ (see (4.3.10)). Notice that for a general Γ, an increment in the value of $G_\lambda(\Gamma)$ is at least $-c\varepsilon|\text{supp}\,\Gamma| + c'\Sigma|\Gamma_t - \Gamma_t^{\text{opt}}|^2$ where $c = c(r)$ and c' depends only on $c_{(3.2.3)}$. Thus, integrating over all possible Γ gives an additional factor $\tilde{c}^{|\text{supp}\,\Gamma|}$ where $\tilde{c} = \tilde{c}(c_{(3.2.3)})$. Take $\tilde{c} > 1$.

Substituting all this into (4.2.18) we obtain the bound

$$\exp(-G_\lambda(\mathbf{\Gamma}, x_{\partial_+}\mathbf{\Gamma}) \leqslant \exp((c'q^{\tilde{r}}\delta^2 + c'' + c'''\lambda)^{|\text{supp}\,\Gamma|}) \times$$

$$\times \sum_{\{B_r, \tilde{B}^+, \tilde{B}^-\}} (\exp(-c\delta^2|\tilde{B}^+ \cup \tilde{B}^-|)\tilde{c})^{|\text{supp}\,\Gamma|} \times$$

$$\times \int_{\Gamma_r:\, B(\Gamma_r) = B_r} \exp(-H^\lambda_{\text{rel}}(\Gamma_r))\text{d}\Gamma_r$$

$$\leqslant (8\tilde{c}\exp(c'q^{\tilde{r}}\delta^2 + c'' + c'''\lambda - \tau_0 V(\tilde{r})^{-1}))^{|\text{supp}\,\Gamma|} \tag{4.3.20}$$

where

$$\tau_0 = \min(c\delta^2, \tau).$$

This gives the desired result. More precisely, we obtain the following bound for κ:

$$\kappa \geqslant \tau_0 V(\tilde{r})^{-1} - c - \tilde{c}\lambda - \tilde{\tilde{c}}q^{\tilde{r}}\delta^2 \tag{4.3.20'}$$

where q, c, $\tilde{c}$, $\tilde{\tilde{c}}$ are suitable constants depending only on $c_{(3.2.3)}$, $0 < q < 1$.

4.3.6. *Estimates of* $\text{d}/\text{d}\lambda\, G_\lambda(\mathbf{\Gamma}, x_{\partial_+\Gamma})$

As before, using the assumption (3.2.12) instead of (3.2.11), we can estimate the value of $\text{d}/\text{d}\lambda\, G_\lambda(\mathbf{\Gamma}, x_{\partial_+\Gamma})$. From (4.3.9) and (4.3.14) we have the relation

$$\frac{\text{d}}{\text{d}\lambda}G_\lambda(\Gamma) = \frac{\text{d}}{\text{d}\lambda}H^\lambda_{\text{rel}}(\Gamma_r) + \frac{\text{d}}{\text{d}\lambda}(R_\lambda(\Gamma) - R^+_\lambda(\partial_+\Gamma)) + L(\Gamma). \tag{4.3.21}$$

It is clear from (3.2.7) that

$$\left|\frac{\text{d}}{\text{d}\lambda}(R_\lambda(\Gamma) - R^+_\lambda(\partial_+\Gamma))\right| \leqslant c\varepsilon'|\text{supp}\,\Gamma| \tag{4.3.22}$$

where $c = c(r)$, $\varepsilon' = \varepsilon'_{(3.2.7)}$. (See (4.3.7) for the definition of R_λ and R^+_λ.) Using (4.3.14') also, we then obtain the inequality

$$\left|\frac{\text{d}}{\text{d}\lambda}G_\lambda(\Gamma) - \frac{\text{d}}{\text{d}\lambda}H^\lambda_{\text{rel}}(\Gamma_r)\right| \leqslant c'|\text{supp}\,\Gamma| \tag{4.3.23}$$

with some $c' = c'(\{u^\pm_{\lambda}\}, r)$.

THEOREM. *There is some* $\kappa' = \kappa'(\tau, \tau', \delta, \tilde{r}, c_{(3.2.3)})$ *such that* $\kappa' \to \infty$ *if* $\delta^2(\tilde{r})^{-\nu} \to \infty$, $\tau(\tilde{r})^{-\nu} \to \infty$ *and* $\tilde{r}(\ln\delta)^{-1} \to \infty$ *and such that*

$$\frac{\text{d}}{\text{d}\lambda}\exp(-G_\lambda(\mathbf{\Gamma}, x_{\partial_+\Gamma}) \leqslant \exp(-\kappa'|\text{supp}\,\mathbf{\Gamma}|). \tag{4.3.24}$$

Proof. Using (4.2.18) we see that

$$\frac{d}{d\lambda}\exp(-G_\lambda(\Gamma, x_{\partial_+\Gamma}))$$

$$= \frac{d}{d\lambda}\int_{\mathscr{C}'(\Gamma, x_{\partial_+\Gamma})} \exp(-H^\lambda_{\text{rel}}(\Gamma_r))\exp(H^\lambda_{\text{rel}}(\Gamma_r) - G_\lambda(\Gamma))\mathrm{d}\Gamma_{(\operatorname{supp}\Gamma\setminus\partial_+\Gamma)}$$

$$= \int \exp(-G_\lambda(\Gamma))\frac{d}{d\lambda}(H^\lambda_{\text{rel}}(\Gamma_r) - G_\lambda(\Gamma))d\Gamma_{(\operatorname{supp}\Gamma\setminus\partial_+\Gamma)} -$$

$$- \int\left(\frac{d}{d\lambda}\exp(-H^\lambda_{\text{rel}}(\Gamma_r))\right)\exp(-H^\lambda_{\text{rel}}(\Gamma_r) - G_\lambda(\Gamma))\mathrm{d}\Gamma_{(\operatorname{supp}\Gamma\setminus\partial_+\Gamma)}.$$

Using (4.3.23) and Theorem 4.3.5 we can estimate the first integral as $\exp(-\kappa|\operatorname{supp}\Gamma| + c|\operatorname{supp}\Gamma|)$. The second integral can be estimated using (3.2.12), (4.3.18), (4.3.23) and some considerations similar to those used in §4.3.5, as $\exp(-\kappa'|\operatorname{supp}\Gamma|)$ where κ' is given as in (4.3.20'), but with τ' instead of τ.

4.4. BOUNDARY TERMS OF PARTITION FUNCTIONS OF CONTOUR MODELS

The aim of this section is to estimate the terms $\tilde{\Delta}_\lambda(\Gamma, F)$ in (4.2.28). It is an essential part of the proof of the Main Lemma and is based on the results of §§2.2.3 and 2.3.8.

4.4.1. *Main Result*

Suppose that the conditions of the Main Lemma are satisfied (see §3.2). Fix some family of interaction $\Phi = \{\Phi^\lambda\}$ from the given class $\mathscr{F}(\Phi^+, \Phi^-, \delta, \varepsilon, \varepsilon', \tau, \tau')$, and fix also some integer $\tilde{r}$, used in the definition of contours in §4.1.

THEOREM. *There are some $\varepsilon_0 > 0$, $\delta_0 > 0$, $\kappa_0 > 0$, $r_0 \in \mathbb{N}$ depending only on $c_{(3.2.3)}$ such that if $\varepsilon \leqslant \varepsilon_0$, $\delta \geqslant \delta_0$, $\kappa \geqslant \kappa_0$, $\tilde{r} \geqslant r_0 \ln\delta$ and if F is a translation-invariant κ-functional then*

(a) *The limit (compare Assumption 4.2.8.1)*

$$h^+_\lambda(F) = \lim|\Lambda|^{-1}\ln Z_\lambda(\Lambda, \emptyset, a, F), a \in (\mathscr{U}^+)^{\partial\Lambda^c} \tag{4.4.1}$$

exists, uniformly in $a \in (\mathscr{U}^+)^{\partial\Lambda^c}$, if $\Lambda \uparrow \mathbb{Z}^\nu$ in the Van Hove sense.

(b) *There is some $\tilde{\varepsilon} = \tilde{\varepsilon}(\varepsilon_0, \kappa_0, \delta_0, c_{(3.2.3)})$ such that $\lim_{\varepsilon_0\to 0, \kappa_0\to\infty, \delta_0\to\infty} \tilde{\varepsilon} = 0$ and such that for each λ,*

$$|h^+_\lambda(F) - h^+_\lambda| \leqslant \tilde{\varepsilon}. \tag{4.4.2}$$

(c) *There is some $\tilde{\tilde{\varepsilon}} = \tilde{\tilde{\varepsilon}}(\varepsilon_0, \kappa_0, \delta_0, r_0, c_{(3.2.3)})$ such that*

$$\lim_{\varepsilon_0\to 0, \kappa_0\to\infty, \delta_0\to\infty, r_0\to\infty} \tilde{\tilde{\varepsilon}} = 0 \tag{4.4.3}$$

and such that for each + contour Γ,

$$|\ln Z_\lambda(V^*, \operatorname{supp}\Gamma, \partial_+\Gamma, F) - \ln Z^+_\lambda(V^*, \partial_+\Gamma) - (h^+_\lambda(F) - h^+_\lambda)|V^*|| \leqslant \tilde{\tilde{\varepsilon}}|\operatorname{supp}\Gamma|. \tag{4.4.4}$$

NOTATION. We will use the abbreviated notation

$$Z_\lambda^{\rm rel}(\Lambda, M, a, F) = Z_\lambda(\Lambda, M, a, F)(Z_\lambda^+(\Lambda, a))^{-1} \tag{4.4.5}$$

where $a \in (\mathscr{U}^+)^{\partial\Lambda^c}$ (see (4.2.8) and (4.2.22)).

(Analogously for $a \in (\mathscr{U}^-)^{\partial\Lambda^c}$ and $Z^-(\Lambda, a)$.) Introducing the notion

$$h_\lambda^{+\rm rel}(F) = h_\lambda^+(F) - h_\lambda^+ \tag{4.4.5'}$$

we can write (4.4.4) as

$$|\ln Z_\lambda^{\rm rel}(V^*, \operatorname{supp}\Gamma, \partial_+\Gamma, F) - h_\lambda^{+\rm rel}(F)|V^*|| \leqslant \tilde{\varepsilon}|\operatorname{supp}\Gamma|. \tag{4.4.6}$$

NOTES. (i) We will concentrate on the proof of (c). Its more general version, Proposition 4.4.3, trivially implies (a) also, and in the course of the proof, we will also obtain (b) as a simple by-product of our considerations.

(ii) An estimate similar to (4.4.4) is also needed for $Z_\lambda(W, \emptyset, \partial_-\Gamma, F)$. This case is similar to that of $Z_\lambda(V^*, \operatorname{supp}\Gamma, \partial_+\Gamma, F)$, but some technical difficulties arise (see §4.4.5).

(iii) Analogous estimates are true for any $-$ contour Γ. (Again, this remark refers to the whole section.)

(iv) We recall that $Z_\lambda(\Lambda, M, a, F)$ was defined as an integral over $X(M) \times \mathscr{R}^+(\Lambda \setminus M)$. The definition of $\mathscr{R}^+(\Lambda \setminus M)$ imposed the condition

$$\operatorname{dist}(V(\mathbf{\Gamma}), \Lambda^c \cup M) \geqslant 2 \tag{4.4.7a}$$

on any frame $\mathbf{\Gamma}$ of $r_{\Lambda\setminus M} \in \mathscr{R}^+(\Lambda \setminus M)$. This was natural in the inductive considerations of Theorem 4.2.6. On the other hand, no $V(A)$ were defined in §2.3.8, and it turns out that the estimates of §2.3.8 are applicable to the study of $Z_\lambda(\Lambda, M, a, F)$ if we replace (4.4.7a) by the condition

$$\operatorname{dist}(\operatorname{supp}\mathbf{\Gamma}, \Lambda^c \cup M) \geqslant 2. \tag{4.4.7b}$$

We do not attempt to generalize §2.3.8 such that it would include the partition functions with (4.4.7a) (which is possible). Instead, we note that there is no difference between (4.4.7a,b) if $\Lambda \setminus M$ has simply connected components (and this is the case for $\Lambda = V^*(\mathbf{\Gamma})$ and $M = \operatorname{supp}\mathbf{\Gamma}$ which we have in mind). We introduce the convention that $Z_\lambda(\Lambda, M, a, F)$ will always be defined with respect to (4.4.7b) in the rest of the paper (unless specified otherwise).

The proof of the theorem will be given in the forthcoming sections. Section 4.4.2 contains some preparatory constructions relating the problem to the situation of Theorem 2.3.8. The main estimate generalizing (4.4.4) is then stated in §4.4.3 and proved in §4.4.4.

We now introduce another convention to be used throughout §§4.4 and 4.5. We assume that

$$x^+ \equiv 0$$

for simplicity notation. Thus $H_{0,\rm st}^+ = H_{0,\rm qst}^+$ in the sense of §2.2.6. We also omit the +

sign in some notation (of $\mathscr{U}^+$ for example). This causes no confusion, because all of this section is a study of one fixed type (the + one) of a contour model.

It is now reasonable to introduce a more general object than that of (4.4.5). Let $\tilde{\Lambda} \subset \mathbb{Z}^\nu$ be finite, let $\tilde{a} \in (\mathscr{U}^+)^{\partial\tilde{\Lambda}^c}$. Denote by $\mu_{\tilde{a}}$ the conditioned Gibbs measure on $X(\tilde{\Lambda})$, defined with respect to the Hamiltonian $H_\lambda^+(x_{\tilde{\Lambda}} | x_{\tilde{\Lambda}^c})$ and the condition $x_{\partial\tilde{\Lambda}^c} = \tilde{a}$. Clearly $\mu_{\tilde{a}}$ does not depend on λ, and because of (2.2.66) it coincides with the measure $\mu_{\tilde{a}}$ defined in §2.2.1 for the standardized Hamiltonian $H_{0,\mathrm{st}}^+$. By §2.2.3 we may consider the measure $\mu_{\tilde{a}}$ even for infinite $\tilde{\Lambda}$. In the special case $\tilde{\Lambda} = \mathbb{Z}^\nu$ we denote this measure by μ.

DEFINITION. Let $\Lambda \subset \tilde{\Lambda} \subset \mathbb{Z}^\nu$, Λ finite, let $\tilde{a} \in \mathscr{U}^{\partial\tilde{\Lambda}^c}$. Denote by $\mu_{\partial\Lambda^c,\tilde{a}}$ the projection of the measure $\mu_{\tilde{a}} \otimes \delta_{\tilde{a}}$ (where $\delta_{\tilde{a}}$ is the Dirac measure concentrated on $\tilde{a}$, and $\otimes$ denotes the direct product) on $X(\partial\Lambda^c)$. Consider the quantity (see (4.4.5)), for any M,

$$Z_\lambda^{\mathrm{rel}}(\Lambda, M, \tilde{a}, F) = \int_{\mathscr{U}^{\partial\Lambda^c}} d\mu_{\partial\Lambda^c,\tilde{a}}(a) Z_\lambda^{\mathrm{rel}}(\Lambda, M, a, F). \tag{4.4.8}$$

NOTE. We are interested mainly in two special cases: $\Lambda = \tilde{\Lambda}$ and $\mathrm{dist}(\Lambda, \tilde{\Lambda}^c) \geqslant \tilde{r}$, $M = \emptyset$ where $\tilde{r}$ is the number used in the definition of contours (§4.1). In the case $\tilde{\Lambda} = \mathbb{Z}^\nu$ we will write (4.4.8) as

$$Z_\lambda^{\mathrm{rel}}(\Lambda, M, F) = \int_{\mathscr{U}^{\partial\Lambda^c}} \mathrm{d}\mu_{\partial\Lambda^c}(a) Z^{\mathrm{rel}}(\Lambda, M, a, F) \tag{4.4.8'}$$

where $\mu_{\partial\Lambda^c}$ is the projection of μ on $X(\partial\Lambda^c)$.

4.4.2. *Expression of Z_λ^{rel} as an Integral of the Type* (2.3.102)

DEFINITION 1. Denote by χ (or $\tilde{\chi}$) the indicator of $\mathscr{U}(=\mathscr{U}^+)(\tilde{\mathscr{U}}(=\tilde{\mathscr{U}}^+)$, respectively). For any interaction Φ_A^λ denote by f_A the function

$$x_A \in X(A) \rightsquigarrow \prod_{t \in A} \chi(x_t) \exp(\Phi_A^{t;\lambda}(x_A) - \Phi_A^\lambda(x_A)) - 1. \tag{4.4.9}$$

For any + frame $\mathbf{\Gamma}$ denote by $f_{\mathbf{\Gamma}}$ the function

$$x_{\mathrm{supp}\mathbf{\Gamma}} \in X(\mathrm{supp}\,\mathbf{\Gamma}) \rightsquigarrow \prod_{t \in \partial_+\mathbf{\Gamma}} \tilde{\chi}(x_t) \exp(-F(\mathbf{\Gamma}, x_{\partial_+\mathbf{\Gamma}})). \tag{4.4.9'}$$

(Because λ is fixed throughout §4.4 we do not write it in f_A, $f_{\mathbf{\Gamma}}$, etc.)

DEFINITION 2. Consider a family $\mathfrak{A}$ whose elements, denoted by E, are the following: either

(a) sets $A \subset \mathbb{Z}^\nu$ such that $\mathrm{diam}\, A \leqslant r$ or

(b) frames $\mathbf{\Gamma}$.

We take supp $A = A$ and supp Γ as in §4.1.5. Consider the compatibility relation)(on $\mathfrak{A}$ defined as follows:

(i) all the sets A are compatible;

(ii) A is compatible with $\mathbf{\Gamma}$ iff $A \cap (\operatorname{supp}\mathbf{\Gamma} \setminus \partial_+\mathbf{\Gamma}) = \emptyset$ (compare (4.2.4));
(iii) two frames $\mathbf{\Gamma}, \mathbf{\Gamma}'$ are compatible iff $\operatorname{dist}(\operatorname{supp}\mathbf{\Gamma}, \operatorname{supp}\mathbf{\Gamma}') \geqslant 2$.

Notice that)(is translation-invariant, in an obvious sense.

Given a finite $\Lambda, M \subset \mathbb{Z}^\nu$ denote by $\mathfrak{A}(\Lambda, M)$ the subsystem of $\mathfrak{A}$ consisting of

(i) sets A such that $A \cap \Lambda \neq \emptyset$, $A \cap M = \emptyset$;
(ii) frames $\mathbf{\Gamma}$ such that $\operatorname{dist}(\operatorname{supp}\mathbf{\Gamma}, \Lambda^c \cup M) \geqslant 2$.

Using this notation we obtain the following relation.

PROPOSITION 1. *In the notation of* (2.3.102) *and* (4.4.8) *we have*

$$Z_{\tilde{\lambda}}^{\text{rel}}(\Lambda, M, \tilde{a}, F) = Z(\Lambda, \tilde{a}, \{f_E, E \in \mathfrak{A}(\Lambda, M)\}). \tag{4.4.10}$$

Proof. Immediate. □

This equation has, however, only a heuristic value at the moment. In what follows we will derive a suitable modification of (4.4.10) (i.e. (4.4.17)) which will be more useful for our purposes.

DEFINITION 3. Denote by $\mu_{\Lambda,\tilde{a}}$ the projection of $\mu_{\tilde{a}}$ on $X(\Lambda)$. Write it as μ_Λ if $\tilde{\Lambda} = \mathbb{Z}^\nu$. Consider the Radon–Nikodym derivative

$$f_{\Lambda,\tilde{a}}(x_\Lambda) = \frac{\mathrm{d}\ \mu_{\Lambda,\tilde{a}}}{\mathrm{d}\ \mu_\Lambda}(x_\Lambda). \tag{4.4.11}$$

Notice that $f_{\Lambda,\tilde{a}}$ depends on $x_{\partial\Lambda}$ (and $\tilde{a}$) only.

Using the notation (2.2.14) and (2.2.44) we can write $f_{\Lambda,\tilde{a}}$ more precisely as

$$f_{\Lambda,\tilde{a}}(x_{\partial\Lambda}) = P_{\tilde{a}}^{\partial\Lambda}(x_{\partial\Lambda})(P^{\partial\Lambda}(x_{\partial\Lambda}))^{-1} \tag{4.4.11'}$$

where we take $\tilde{\Lambda}$, $\partial\Lambda$, $\tilde{a}$, $H_{0,\text{st}}$ instead of Λ, Λ', $x_{\partial\Lambda^c}$, H in (2.2.14). (We write $P^{\partial\Lambda}_{x_{\partial\tilde{\Lambda}^c}}$ instead of $P^{\partial\Lambda}_{x_{\tilde{\Lambda}^c}}$, and $P^{\partial\Lambda}$ denotes the limit $\tilde{\Lambda} \uparrow \mathbb{Z}^\nu$, $\tilde{a} \in \mathscr{U}^{\partial\tilde{\Lambda}^c}$.) Substituting (2.2.11) into (2.2.14), we have the expression (using this new notation)

$$f_{\Lambda,\tilde{a}}(x_\Lambda) = c\exp\left(-\sum_{t\in\partial\Lambda} ((\varphi_t, x_t) + (\tilde{\varphi}_{\{t\}}x_t, x_t)) - \sum_{\{s,t\}\subset\partial\Lambda, s\neq t} (\tilde{\varphi}_{\{s,t\}}x_t, x_s)\right) \tag{4.4.12}$$

where $\varphi_t \in \mathbb{R}^k$ and $\tilde{\varphi}_{\{t\}}$, $\tilde{\varphi}_{\{s,t\}}$ are symmetric real $k \times k$ matrices. c is the normalizing constant.

The estimates of §2.2.2 give the following information about φ_t and $\tilde{\varphi}_{\{s,t\}}$.

PROPOSITION 2. *There are some constants* K, $0 < q < 1$ *depending only on* $c_{(3.2.3)}$ *such that*

(a) $|\varphi_t| \leqslant K\delta q^{\operatorname{dist}(t,\tilde{\Lambda}^c)}$ (4.4.13)

(b) $|\tilde{\varphi}_{\{s,t\}}| \leqslant K q^{\operatorname{dist}(t,\tilde{\Lambda}^c) + \operatorname{dist}(s,\tilde{\Lambda}^c)}$ (4.4.14)

(c) $|\tilde{\varphi}_{\{s,t\}}| \leqslant K q^{|t-s|}$. (4.4.15)

Proof. This follows from (2.2.15), (2.2.17), (2.2.20) and the corresponding estimates of §2.2.3. The proof of (a) uses the condition $\tilde{a} \in \mathscr{U}^{\partial\tilde{\Lambda}^c}$. On the other hand, $\tilde{\varphi}_{\{s,t\}}$ does not depend on δ. □

DEFINITION 4. Extend the family $\mathfrak{A}$ introduced in Definition 2 by sets B of the type

(d) $B = \{s, t\}; \quad s, t \in \mathbb{Z}^\nu, \quad s \neq t$

and

(e) $B = \{t\}, \quad t \in \mathbb{Z}^\nu$.

We take supp $B = B$ and assume that all the sets B are mutually compatible and also compatible with any element of $\mathfrak{A}$. Denote by $\bar{\mathfrak{A}}(\Lambda, M)$ the extension of $\mathfrak{A}(\Lambda, M)$ by all objects B such that $B \subset \Lambda$. Define the functions

$$f_B(x_t \cup x_s) = \exp(-(\tilde{\varphi}_{\{s,t\}} x_s, x_t)) - 1 \quad \text{if} \quad B = \{s,t\}, \quad s, t \in \partial\Lambda \tag{4.4.16}$$

$$f_B(x_t) = \exp(-(\varphi_t, x_t) - (\tilde{\varphi}_{\{t\}} x_t, x_t)) - 1 \quad \text{if} \quad B = \{t\}, \quad t \in \partial\Lambda \tag{4.4.16'}$$

$f_B \equiv 0$ otherwise.

NOTE. Thus, for some $S \in \bar{\mathfrak{A}}(\Lambda, M)$, $|S| \leqslant 2$ there are two exemplars of E (of the type A (or B)) such that $S = \operatorname{supp} A = \operatorname{supp} B$.

The following modification of the expression (4.4.10) will play a basic role in the forthcoming investigation.

PROPOSITION 3. *Using the notation* (2.3.113) *we have the relation*

$$Z_\lambda^{\mathrm{rel}}(\Lambda, M, \tilde{a}, F) = c_{(4.4.12)} Z(\{f_E, E \in \bar{\mathfrak{A}}(\Lambda, M)\}). \tag{4.4.17}$$

Proof. Immediate. □

4.4.3. *Reformulation of Theorem* 4.4.1. *Outline of the Proof*

It will be convenient to write $Z_\lambda^{\mathrm{rel}}(V^*, \operatorname{supp}\Gamma, \partial_+\Gamma, F)$ as $Z_\lambda^{\mathrm{rel}}(\operatorname{int}_-\Gamma \cup \partial_-\Gamma, \partial_-\Gamma, \partial_+\Gamma, F)$, using the obvious relation

$$Z_\lambda^{\mathrm{rel}}(\tilde{\Lambda}, \tilde{M}, \tilde{a}, F) = Z_\lambda^{\mathrm{rel}}(\Lambda, M, \tilde{a}, F) \tag{4.4.18}$$

where $\Lambda = (\tilde{\Lambda} \setminus \tilde{M}) \cup M$, $M = \tilde{M} \cap \partial(\tilde{\Lambda} \setminus \tilde{M})^c$. Our basic strategy will be to use Corollary 2.3.8 for the estimation of $Z(\{f_E, E \in \bar{\mathfrak{A}}(\Lambda, M)\})$ from (4.4.17).

This is possible because $\operatorname{dist}(\Lambda, \tilde{\Lambda}^c)$ is large compared to $\ln \delta$, enabling us to handle $\mu_{\Lambda, \tilde{a}}$ as a small perturbation of μ_Λ. Note that the expression (4.4.10) is not so useful as (4.4.17) because we have no precise estimate of the decay of $(v_t - \bar{u})$ (if we use the expression (2.3.106)). The statements (c)–(e) of Theorem 2.3.8 are too weak for this purpose. The more powerful estimate (2.3.109) controls the value $(v_t - u_t)$, and we can profit from this fact only in the case when $(u_t - \bar{u})$ can be estimated with comparable precision (which is the case for the partition function $Z(\Lambda, \{\hat{f}_E, E \in \bar{\mathfrak{A}}(\Lambda, M)\})$.

NOTATION. Let q be as in Proposition 2, §4.4.2. Write

$$\tilde{\omega} = \delta q^{\bar{r}}. \tag{4.4.19}$$

PROPOSITION. *Suppose that F is a translation-invariant κ-functional, with large enough κ. Suppose that $\tilde{\omega}_{(4.4.19)}$ is sufficiently small. Then there is some*

$\tilde{\varepsilon} = \tilde{\varepsilon}(\varepsilon, \tilde{\omega}, \kappa, c_{(3.2.3)}, \delta)$ *such that*

$$\lim_{\varepsilon \to 0, \tilde{\omega} \to 0, \kappa \to \infty, \delta \to \infty} \tilde{\varepsilon} = 0 \tag{4.4.20}$$

and such that the following estimate holds: whenever Λ, M, $\tilde{\Lambda} \subset \mathbb{Z}^\nu$, Λ *finite are such that*

$$\operatorname{dist}(\Lambda \setminus M, \tilde{\Lambda}^c) \geqslant \tilde{r}$$

then for each $\tilde{a} \in \mathcal{U}^{\partial \tilde{\Lambda}^c}$ *the following estimate holds:*

$$|\ln Z_{\tilde{\lambda}}^{\mathrm{rel}}(\Lambda, M, \tilde{a}, F) - h_{\tilde{\lambda}}^{\mathrm{rel}}(F)|\Lambda|| \leqslant \tilde{\varepsilon}|\partial \Lambda^c \cup M| \tag{4.4.21}$$

where (*see also* (4.4.5′))

$$h_{\tilde{\lambda}}^{\mathrm{rel}}(F) = \bar{u}$$

is defined by (2.3.111′) *for the system of functions* $\{f_E, E \in \mathfrak{A}\}$ (*notice that it does not depend on* f_B). *We have also the formula*

$$h_{\tilde{\lambda}}^{\mathrm{rel}}(F) = \lim_{\Lambda \uparrow \mathbb{Z}^\nu} |\Lambda|^{-1} \ln Z_{\tilde{\lambda}}^{\mathrm{rel}}(\Lambda, \emptyset, a, F) \tag{4.4.22}$$

(*in the Van Hove sense, uniformly in* $a \in \mathcal{U}^{\partial \Lambda^c}$), *and the estimate*

$$|h_{\tilde{\lambda}}^{\mathrm{rel}}(F)| \leqslant \tilde{\varepsilon}. \tag{4.4.22′}$$

NOTES

(a) This is the second point (the first one was in §4.3) where we use our very definition of 'thick' contours (more precisely, the assumption that $\tilde{r}$ is sufficiently large). The condition of smallness of $\tilde{\omega}_{(4.4.19)}$ is of a similar nature to the condition of smallness of $\delta^2 q^{\tilde{r}}$ in (4.3.17). The second special feature of contours, the condition $\partial_{\pm}\Gamma \in (\tilde{\mathcal{U}}^{\pm})^{\mathscr{P}_{\pm}\Gamma}$, will be used only in §4.4.5).

(b) We will first replace $Z_{\tilde{\lambda}}^{\mathrm{rel}}(\Lambda, M, \tilde{a}, F)$ by $(c_{(4.4.12)})^{-1} Z_{\tilde{\lambda}}^{\mathrm{rel}}(\Lambda, M, \tilde{a}, F)$ for brevity in the proof. The estimate of $c_{(4.4.12)}$ is simple, and will be postponed to the end of the proof.

4.4.4. *Proof of Proposition* 4.4.3

The proof is based on Corollary 2.3.8. We will show that the condition of the type (2.3.105) holds for any family of indices

$$\{A_i\} \cup \{\Gamma_j\} \cup \{B_k\} \in \tilde{\mathfrak{A}}(\Lambda, M).$$

First, concerning the numbers $N(\mathscr{A})$ in (2.3.105) we notice that for any given $S \subset \mathbb{Z}^\nu$ there are no more than $2^{|S|}$ frames Γ with the same $\operatorname{supp}\Gamma = S$, and no more than two different elements $E \in \tilde{\mathfrak{A}}(\Lambda, M)$ such that $\operatorname{supp} E = S$ if $|S| \leqslant 2$. Estimate first

$$\left\langle \prod_i f_{A_i}^2 \prod_j f_{\Gamma_j}^2 \prod_k f_{B_k}^2 \right\rangle \leqslant \exp\left(-2\sum_j \kappa|\operatorname{supp}\Gamma_j|\right)\left\langle \prod_i f_{A_i}^2 \prod_j f_{\Gamma_j}^2 \right\rangle. \tag{4.4.23}$$

The symbol $\langle\,\rangle$ denotes, here and everywhere, the expectation with respect to μ.

We would like to use (2.3.155), but a straightforward application of this relation is impossible (because the condition of uniform boundedness of diam E is violated for some sets $E = B$ of the type (c) of Definition 4, §4.4.2). We therefore use first the Cauchy inequality

$$\left\langle \prod_i f_{A_i}^2 \prod_k f_{B_k}^2 \right\rangle \leqslant \left(\left\langle \prod_i f_{A_i}^4 \right\rangle \left\langle \prod_k f_{B_k}^4 \right\rangle \right)^{1/2}. \tag{4.4.24}$$

Clearly, $\langle \Pi_i f_{A_i}^4 \rangle$ can be estimated using Proposition 2.3.13. We get (with $p = p_{(2.3.155)}$)

$$\left\langle \prod_i f_{A_i}^4 \right\rangle \leqslant c^{\Sigma_i |A_i|} \prod_i (\langle |f_{A_i}|^{4p} \rangle)^{1/p}. \tag{4.4.25}$$

This estimate must be combined with the following simple fact.

LEMMA 1. *There is some* $\omega = \omega(\delta, \varepsilon, p)$ *such that*

$$\lim_{\varepsilon \to 0, \delta \to \infty} \omega = 0 \tag{4.4.26}$$

and such that for each A, diam $A \leqslant r$, $A \cap \Lambda \neq \emptyset$,

$$\langle |f_A|^{4p} \rangle \leqslant \omega. \tag{4.4.27}$$

Proof. Using the definition of f_A (see (4.4.9)) and the condition (3.2.6) we have the inequality

$$|f_A(x_A)| \leqslant \exp(\varepsilon) - 1 + \left(1 - \prod_{t \in A} \chi(x_t)\right). \tag{4.4.28}$$

Thus it is sufficient to show that $\langle 1 - \chi(x_t) \rangle \to 0$ for $\delta \to \infty$, but this is clear because the mean value of μ is zero and its variance does not depend on t. This implies (4.4.27) with (4.4.26). □

LEMMA 2. *There is some* $\tilde{\tilde{\omega}} = \tilde{\tilde{\omega}}(\tilde{\omega}_{(4.4.19)})$ *such that*

$$\lim_{\tilde{\omega} \to 0} \tilde{\tilde{\omega}} = 0 \tag{4.4.29}$$

and such that for each $B = \{t\}$, $t \in \partial\Lambda$ (*of the type* (d), *Definition* 4, §4.4.2)

$$\langle |f_B|^{4p} \rangle \leqslant \tilde{\tilde{\omega}}. \tag{4.4.30}$$

Proof. This is an analogy of Lemma 1, which uses (4.4.14), (4.4.13) instead of (3.2.6). □

What now remains is to find a sufficient bound also for $\langle \Pi_k f_{B_k}^4 k \in \mathscr{K} \rangle$ where $\{B_k\}$ is some family of indices of the type (d), of Definition 4, §4.4.2. Using the obvious estimate

$$|\exp x - 1| \leqslant |x| \exp|x|$$

we can write

$$|f_B(x_t \cup x_s)| \leqslant |(\tilde{\varphi}_B x_t, x_s)| \exp(|\tilde{\varphi}_B x_t, x_s|), \quad B = \{s, t\}. \tag{4.4.31}$$

Writing $B_k = \{t_k, s_k\}$ we then have

$$\left|\prod_k f^4_{B_k}(x_{B_k})\right| \leqslant \prod_k (|(\tilde{\varphi}_{B_k} x_{t_k}, x_{s_k})|) \exp(|(\tilde{\varphi}_{B_k} x_{t_k}, x_{s_k})|)^4.$$

We show below that

$$\left\langle \prod_k (|(\tilde{\varphi}_{B_k} x_{t_k}, x_{s_k})|) \exp(|(\tilde{\varphi}_{B_k} x_{t_k}, x_{s_k})|)^4 \right\rangle \leqslant q^{\Sigma_k |s_k - t_k|} \tilde{\varepsilon}^{|\mathscr{K}|} \tag{4.4.32}$$

for a suitable $0 < q < 1$ depending on $q_{(4.4.15)}$ and a suitable $\tilde{\varepsilon}$, depending on q, $\tilde{\omega}_{(4.4.19)}$ such that

$$\lim_{\tilde{\omega} \to 0} \tilde{\varepsilon} = 0. \tag{4.4.33}$$

This estimate follows from the following considerations. By (4.4.14), (4.4.15) we have

$$|\tilde{\varphi}_{\{s,t\}}| \leqslant \min\{2\tilde{\omega}K, Kq^{|t-s|}\} \leqslant (2\tilde{\omega})^{1/2} Kq^{|t-s|/2}.$$

Writing $\tilde{\varphi}_{\{s,t\}} = \tilde{\tilde{\varphi}}_{\{s,t\}} q^{|s-t|/4}$ we easily obtain (4.4.32) as a consequence of the following lemma (used for $\varphi \equiv \tilde{\tilde{\varphi}}$).

LEMMA 3. *Let μ be a Gaussian measure on $X(\Lambda)$, $\Lambda \subset \mathbb{Z}^\nu$, with zero mean values and with a covariance matrix B satisfying the bound*

$$(Bx_\Lambda, x_\Lambda) \leqslant c|x_\Lambda|^2, \qquad x_\Lambda \in X(\Lambda)$$

for some $c > 0$ not depending on x_Λ. Let $\mathscr{K} = \{\{s,t\};\ s,t \in \Lambda,\ s \neq t\}$ be such that for each $t \in \Lambda$ there is some $s \in \Lambda$ with $\{s,t\} \in \mathscr{K}$. Let $0 < q < 1$. Let $\{\varphi_{\{s,t\}}, \{s,t\} \in \mathscr{K}\}$ be a family of $k \times k$ matrices such that $|\varphi_{\{s,t\}}| \leqslant \tilde{c}q^{|t-s|}$, $\{s,t\} \in \mathscr{K}$, for a sufficiently small $\tilde{c}$ (depending on c and q). Then there is some $\tilde{K} = \tilde{K}(c,q)$ such that $\lim_{\tilde{c} \to 0} \tilde{K} = 0$ and such that

$$\int \left(\prod_{\{s,t\} \in \mathscr{K}} |(\varphi_{\{s,t\}} x_s, x_t)| \exp(|\varphi_{\{s,t\}} x_s, x_t)|) \right) d\mu \leqslant \tilde{K}^{|\mathscr{K}|}. \tag{4.4.34}$$

Proof of Lemma 3. Consider the function φ as a function on the whole $\{\{s,t\}; s,t \in \Lambda\}$ by taking $\varphi_{\{s,t\}} = 0$ if $\{s,t\} \in \mathscr{K}$. Because

$$\sum_{\{s,t\} \in \mathscr{K}} |(\varphi_{\{s,t\}} x_s, x_t)| \leqslant \frac{1}{2} \sum_{s,t \in \Lambda} |\varphi_{\{s,t\}}| (|x_s|^2 + |x_t|^2) \leqslant K\tilde{C}|x_\Lambda|^2$$

with a suitable $K = K(q)$ we may also assume that an inequality

$$\sum_{\{s,t\} \in \mathscr{K}} |(\varphi_{\{s,t\}} x_s, x_t)| \leqslant K\tilde{C}(B^{-1} x_\Lambda, x_\Lambda) \tag{4.4.35}$$

holds, with some other K.

Applying the elementary estimate $u \leqslant c(\alpha) \exp(\alpha u)$, $u \geqslant 0$ where $\lim_{\alpha \to \infty} c(\alpha) = 0$ and using (4.4.35) we see that

$$\prod_{\{s,t\} \in \mathscr{K}} |(\varphi_{s,t} x_s, x_t)| \leqslant c(\alpha)^{|\mathscr{K}|} \exp(K\tilde{C}\alpha(B^{-1} x_\Lambda, x_\Lambda))$$

So the left-hand side of (4.4.34) is bounded from above by

$$(c(\alpha))^{|\mathscr{K}|}\int \exp(K\tilde{C}(1+\alpha))(B^{-1}x_\Lambda,x_\Lambda)\,\mathrm{d}\mu(x_\Lambda)$$

Choose $\alpha=\alpha(\tilde{c})$ such that $k\tilde{c}(1+\alpha)=\frac{1}{4}$. Then the last integral is equal to $2^{|\Lambda|}\leqslant 2^{|\mathscr{K}|}$ and therefore (4.4.34) is true for $\tilde{K}=2c(\alpha(\tilde{c}))$. Because $\alpha(\tilde{c})\to\infty$ for $\tilde{c}\to 0$ we have $\tilde{K}\to 0$ if $\tilde{c}\to 0$. □

It is now easy to see that by collecting the estimates (4.4.23), (4.4.24), (4.4.25), (4.4.27), (4.4.30), (4.4.33) we obtain the desired estimate (2.3.105), for any family $\mathscr{A}=\{A_i\}\,\&\,\{\Gamma_j\}\,\&\,\{B_k\}$. (We recall the obvious upper bound $2^{\Sigma \operatorname{supp} E}$ for the number of possible $\mathscr{A}$ with the same $\{\operatorname{supp} E, E\in\mathscr{A}\}$.)

Using Corollary 2.3.8 we obtain the estimate

$$|\ln((c_{(4.4.12)})^{-1}Z^{\mathrm{rel}}_{\lambda}(\Lambda,M,\tilde{a},F))-h^{\mathrm{rel}}_{\lambda}(F)|\Lambda||\leqslant\varepsilon'|\partial\Lambda^c\cup M| \tag{4.4.36}$$

with some ε' such that

$$\lim_{\varepsilon\to 0,\tilde{\omega}\to 0,\delta\to\infty}\varepsilon'=0.$$

To finish the proof of (4.4.22) it remains to prove a bound

$$|\ln c_{(4.4.12)}|\leqslant\varepsilon'|\partial\Lambda^c| \tag{4.4.37}$$

(with some new ε'). From Definition (4.4.11) we see that $\langle f_{\Lambda,\tilde{a}}\rangle=1$ and therefore

$$c_{(4.4.12)}=Z(\Lambda,\{f_B\})^{-1}. \tag{4.4.38}$$

where by $\{f_B\}$ we denote the system of functions (4.4.16), (4.4.16′). So we can apply a simplified variant of the previous considerations to obtain, instead of (4.4.36), the estimate (4.4.37). (We leave it to the reader to elaborate the details of the proof of (4.4.37).)

Thus, the proof of Proposition 4.4.3 is complete. □

NOTE. The choice of q in (2.3.105) is taken with respect to $q_{(4.4.15)}$ only. Because $\operatorname{supp}\Gamma$ is connected and diam $A\leqslant r$, the functions f_Γ and f_A affect the choice of q only in a trivial way.

Finally, few words about the proof of statement (b) of Theorem 4.4.1. This is a consequence of (2.3.108), exploiting again the very condition (2.3.105) which we have established before. The independence of $\tilde{\varepsilon}_{(4.4.2)}$ on $\tilde{r}$ is clear from (2.3.109). Statement (a) of Theorem 4.4.1 is a trivial consequence of (c).

4.4.5. *Estimate of* $Z^{\mathrm{rel}}_{\lambda}(W,\emptyset,\partial_-\Gamma,F)$

The result of this section is similar to Theorem 4.4.1.

THEOREM. *There are some* $\varepsilon_0>0,\delta_0>0,\kappa_0>0$, $r_0\in\mathbb{N}$ *depending only on* $c_{(3.2.3)}$ *such that if* $\varepsilon\leqslant\varepsilon_0,\delta\geqslant\delta_0\kappa\geqslant\kappa_0,\tilde{r}\geqslant r_0\ln\delta$ *and* F *is a translation-invariant* κ*-functional then*

(a) *The limit* $h_\lambda^{-}(F) = \lim|\Lambda|^{-1}\ln Z_\lambda(\Lambda,\phi,a,F), a\in(\mathcal{U}^{-})^{\partial\Lambda^c}$ *exists, uniformly in* $a\in(\mathcal{U}^{-})^{\partial\Lambda^c}$, if $\Lambda\uparrow\mathbb{Z}^\nu$ *in the Van Hove sense.*

(b) *There is some* $\tilde{\varepsilon} = \tilde{\varepsilon}(\varepsilon_0,\kappa_0,\delta_0,c_{(3.2.3)})$ *such that*

$$\lim_{\varepsilon_0\to 0,\kappa_0\to\infty,\delta_0\to\infty,r_0\to\infty} \tilde{\varepsilon} = 0 \tag{4.4.40}$$

and such that for each + *contour* Γ,

$$|\ln Z_\lambda(W,\emptyset,\partial_-\Gamma,F) - \ln Z_\lambda^{-}(W,\partial_-\Gamma) - (h_\lambda^{-}(F) - h_\lambda^{-})|W|| \leqslant \tilde{\varepsilon}|\operatorname{supp}\Gamma|. \tag{4.4.41}$$

where $W = W(\Gamma)$.

NOTE. This case differs from that of Proposition 4.4.3 and Theorem 4.4.1 by the absence of the 'intermediate zone' M or $\operatorname{supp}\Gamma$, enabling us to look on μ_a as a small perturbation of μ. On the other hand, the case of the partition function $Z_\lambda^{\text{rel}}(W,\partial_{\bar{r}} W,\partial_-\Gamma,F)$ causes no difficulties and is completely analogous to the situation of §4.4.3. What is new in this section is the estimate of a difference

$$\ln Z_\lambda^{\text{rel}}(W,\emptyset,\partial_-\Gamma,F) - \ln Z_\lambda^{\text{rel}}(W,\partial_{\bar{r}} W,\partial_-\Gamma,F).$$

(see Lemma 2 below).

NOTATION. Theorem 4.4.5 is a result about the − contour ensemble for a + contour Γ. It is convenient, however, to revert to the previous situation of the +contour ensemble, with all the previous conventions and notation. Thus we will prove a variant of Proposition 4.4.3, designed to prove Theorem 4.4.5 but formulated in the language of the +contour ensemble.

As before, we assume that any statement of §4.4 has its counterpart formulated by replacing + by −, and vice versa (which will be usually omitted).

This is the first part of the proof of the Main Lemma where we will profit from our definition of contours requiring that $\partial_+\Gamma\in(\tilde{\mathcal{U}}^{+})^{\partial_+\Gamma}$ instead of $\partial_+\Gamma\in(\mathcal{U}^{+})^{\partial_+\Gamma}$. We will specify our choice of $\tilde{\mathcal{U}} = \tilde{\mathcal{U}}^{+}$: assume that $K>0$ is chosen such that for any finite $\Lambda\subset\mathbb{Z}^\nu$ and any $\bar{x}_{\partial\Lambda^c}\in\mathcal{U}^{\partial\Lambda^c}$, the mean value $\bar{x}_\Lambda$ of the measure $\mu_{\bar{x}_{\partial\Lambda^c}}$ satisfies the condition

$$\bar{x}_t\in K\mathcal{U}, \qquad t\in\Lambda. \tag{4.4.42}$$

(We recall that we take $\sigma^{+}\equiv 0$.)

Such a choice of K is possible as follows, from (2.2.15) for example, and it depends only on $c_{(3.2.3)}$. We may assume $K\geqslant 1$.

Choose

$$\tilde{\mathcal{U}} = (2K)^{-1}\mathcal{U}. \tag{4.4.43}$$

We will prove the following result:

PROPOSITION. *Suppose that* F *is a translation-invariant* κ *functional, with large enough* κ. *Suppose that* $\tilde{\omega}_{(4.4.19)}$ *is sufficiently small. Then there is some* $\tilde{\varepsilon} = \tilde{\varepsilon}(\varepsilon,\tilde{\omega},\kappa,c_{(3.2.3)},\delta)$ *such that* (4.4.20) *holds and such that for each finite* Λ *and each* $a\in(\tilde{\mathcal{U}})^{\partial\Lambda^c}$,

$$|\ln Z_\lambda^{\text{rel}}(\Lambda,\emptyset,a,F) - h_\lambda^{\text{rel}}(F)|\Lambda|| \leqslant \tilde{\varepsilon}|\partial_{\bar{r}}\Lambda|. \tag{4.4.44}$$

NOTE. This clearly proves the theorem, if we also use the following geometrical lemma.

LEMMA 1. *There is some constant $c = c(\nu)$ such that for each $+$contour Γ,*

$$|\partial_{\bar r} W| \leqslant c|\operatorname{supp}\Gamma|. \tag{4.4.45}$$

Proof. Denote by $S(\Gamma)$ the set of all $t \in \operatorname{supp}\Gamma$ such that $x_t \notin \tilde{\mathcal{U}}^-$. We will treat $\mathbb{Z}^\nu$ as a subset of $\mathbb{R}^\nu$, with the Euclidean metric ρ. For any $t \in \partial_{\bar r} W$ denote by t' a point of $S(\Gamma)$ which is the closest possible one to t (in the metric ρ). Take $\tilde t = t + \frac{2}{3}(t' - t)$ and find some $\hat t \in \mathbb{Z}^\nu$ which is a closest possible one to $\tilde t$. The definition of contours (see §4.1.1) implies that $\hat t \in \operatorname{supp}\Gamma$. We will use the simple but important observation that for any $t_1, t_2 \in \partial W$, $t_1 \neq t_2$ we also have $\tilde t_1 \neq \tilde t_2$. Moreover, all $\tilde t$ belong to the lattice $\frac{1}{3}\mathbb{Z}^\nu$. Consider the mapping $\{t \rightsquigarrow \hat t\}: \partial_{\bar r} W \to \operatorname{supp}\Gamma$. It is easy to see that any $\hat t \in \operatorname{supp}\Gamma$ has no more than 3^ν preimages with respect to this mapping. □

Proof of Proposition. We will pick out the points where the proof differs from that of Proposition 4.4.3. We have previously explained the disadvantage of the expression (4.4.10), but here (4.4.17), which was so useful in §4.4.4, cannot be used directly because the measure μ_a cannot be viewed as a small perturbation of μ on Λ.

We now combine both these expressions. As we noted previously, the estimate

$$|\ln Z^{\mathrm{rel}}_{\lambda}(\Lambda, \partial_{\bar r}\Lambda, a, F) - h^{\mathrm{rel}}_{\lambda}(F)|\Lambda|| \leqslant \bar\varepsilon|\partial_{\bar r}\Lambda| \tag{4.4.46}$$

follows immediately from Proposition 4.4.3, for any $a \in \mathcal{U}^{\partial\Lambda^c}$. (We note that $|\partial\Lambda^c| \leqslant c|\partial_{\bar r}\Lambda|$ where $c = c(r, \nu)$.) We need another estimate (with another $\tilde\varepsilon$)

$$|\ln Z^{\mathrm{rel}}_{\lambda}(\Lambda, \emptyset, a, F) - \ln Z^{\mathrm{rel}}_{\lambda}(\Lambda, \partial_{\bar r}\Lambda, a, F)| \leqslant \tilde\varepsilon|\partial_{\bar r}\Lambda| \tag{4.4.47}$$

and this will be proved using the expression (4.4.10).

LEMMA 2. *The estimate* (4.4.47) *holds, for any finite $\Lambda^c \subset \mathbb{Z}^\nu$ and any $a \in (\tilde{\mathcal{U}})^{\partial\Lambda^c}$, with a suitable $\tilde\varepsilon = \tilde\varepsilon(\varepsilon, \delta, \kappa)$ such that*

$$\lim_{\varepsilon \to 0, \delta \to \infty, \kappa \to \infty} \tilde\varepsilon = 0.$$

Proof. As in Proposition 4.4.3 we must check a condition of the type (2.3.105), but with respect to the expression (4.4.10) (rather than (4.4.17)). This is a much simpler task here than it was in Proposition 4.4.3, because there are no functions f_B in (4.4.10). Because of the condition $a \in \tilde{\mathcal{U}}^{\partial\Lambda^c}$, the mean value $\bar a_\Lambda$ of μ_a satisfies the condition $\bar a_\Lambda \in (\frac{1}{2}\mathcal{U})^\Lambda$ and therefore, using (4.4.27), the smallness of $\langle f^p_A \rangle^\Lambda$ is verified. This gives (2.3.105) for the family $\{f_E, E \in \mathfrak{A}(\Lambda, \emptyset)\}$ (with a trivial choice of q).

Thus, using the estimate (2.3.109) we obtain

$$\begin{aligned} &|\ln Z^{\mathrm{rel}}_{\lambda}(\Lambda, \emptyset, a, F) - \ln Z^{\mathrm{rel}}_{\lambda}(\Lambda, \partial_{\bar r}\Lambda, a, F)| \\ &= |\ln Z(\Lambda, a, \{f_E, E \in \mathfrak{A}(\Lambda, \emptyset)\})| - \ln Z(\Lambda, a, \{f_E, E \in \mathfrak{A}(\Lambda, \partial_{\bar r}\Lambda)\})| \\ &\leqslant \sum_{t \in \Lambda} |v_t - \tilde v_t| \leqslant c\varepsilon' \sum_{t \in \Lambda} q'^{\operatorname{dist}(t, \partial_{\bar r}\Lambda)} \end{aligned} \tag{4.4.48}$$

where v_t $(\tilde v_t)$ is taken with respect to the families $\{f_E, E \in \mathfrak{A}(\Lambda, \emptyset)\}$ (or $\{f_E, E \in \mathfrak{A}(\Lambda, \partial_{\bar r}\Lambda)\}$, respectively). This obviously proves the desired estimate (4.4.47).

Finally, (4.4.46) and (4.4.47) prove the proposition. □

NOTE 2. We recall that this is the first place where the condition $a \in \tilde{\mathscr{U}}^{\partial\Lambda^c}$ is essential. In fact the proof may seem to be unnecessarily complicated because an elementary estimate of the type

$$|Z_\lambda(\Lambda, \phi, a, F)| \leqslant \exp(\tilde{\varepsilon}|M|) Z_\lambda(\Lambda, M, a, F) \tag{4.4.49}$$

holds. (This is easily seen from the inequality

$$H_\lambda(r_\Lambda | a) \leqslant H_\lambda^{(M)}(r_\Lambda | a) + \varepsilon c|M| - \text{see (4.2.8').)}$$

The reason for our choice of the more complicated proof is that we will consider also, in §4.5, the case of complex Hamiltonians. It is then not at all obvious how to prove (4.4.49) without the use of Theorem 2.3.8. Our proof, based on (4.4.48), still works in the complex case.

The following statement summarizes the main results of §4.4.

COROLLARY. *The term* $\tilde{\Delta}_\lambda(\Gamma, F)$ *(see* (4.2.28)) *satisfies, under the assumption that* F *is a* κ*-functional, the bound*

$$|\tilde{\Delta}_\lambda(\Gamma, F)| \leqslant \tilde{\varepsilon}|\operatorname{supp} \Gamma| \tag{4.4.50}$$

where

$$\tilde{\varepsilon} = \tilde{\varepsilon}(\varepsilon, \delta, \tilde{r}, \kappa, c_{(3.2.3)})$$

is such that

$$\lim_{\varepsilon \to 0, \tilde{\omega} \to 0, \delta \to \infty, \kappa \to \infty} \tilde{\varepsilon} = 0$$

(with $\tilde{\omega} = \tilde{\omega}_{(4.4.19)}$*).*

Proof. Theorems 4.4.1 and 4.4.5. □

4.5. CONCLUSION OF THE PROOF OF THE MAIN LEMMA

In §4.2 we showed that if equation (4.2.30) has a solution satisfying (4.2.29) then this solution also satisfies (4.2.17)((4.2.10) in a factorized form) and therefore, by Theorem 4.2.6, the contour ensemble is equivalent to the Gibbsian one. In this section, we prove the existence of such a solution and also its analyticity with respect to λ. We will also explain in more detail the relation between the contour model and the real model, and construct the limit Gibbs states. This will complete the proof of the Main Lemma.

4.5.1. *Summary of* §§4.3 *and* 4.4

For the reader's convenience we now sum up the main results of §§4.3, 4.4, estimating the terms $G_\lambda(\Gamma, a), \tilde{\Delta}_\lambda(\Gamma, F)$ in (4.2.30).

COROLLARY. *Suppose that the assumptions of the Main Lemma are satisfied. Then it is possible to choose some* $\tilde{r} = \tilde{r}(\delta, \tau, \tau')$ *(see* (4.5.22) *below for its more precise specification) such that the following is true, if we use this particular value of* $\tilde{r}$ *in the definition of contours.*

(a) *There is some* $\kappa = \kappa(\delta, \tau, \tilde{r})$ *such that*

$$\lim_{\tilde{r}(\ln\sigma)^{-1}\to\infty,\, \tau(\tilde{r})^{-\nu}\to\infty,\, \delta^2(\tilde{r})^{-\nu}\to\infty} \kappa = \infty \tag{4.5.1}$$

and such that

$$G_\lambda(\mathbf{\Gamma}, a) \geqslant \kappa|\operatorname{supp}\mathbf{\Gamma}| \tag{4.5.1'}$$

for each $\pm$ *frame* $\mathbf{\Gamma}$ *and each* $a \in (\tilde{\mathscr{U}}^{\pm})^{\partial_\pm \mathbf{\Gamma}}$. *Similarly, there is some* $\kappa' = \kappa'(\delta, \tau', \tilde{r})$ *such that*

$$\lim_{\tilde{r}(\ln\sigma)^{-1}\to\infty,\, (\tilde{r})^{-\nu}\to\infty,\, \delta 2(\tilde{r})^{-\nu}\to\infty} \kappa' = \infty \tag{4.5.2}$$

and such that

$$\left|\frac{\mathrm{d}}{\mathrm{d}\lambda}\exp(-G_\lambda(\mathbf{\Gamma}, a))\right| \leqslant \exp(-\kappa'|\operatorname{supp}\mathbf{\Gamma}|). \tag{4.5.2'}$$

(b) *There is some* $\tilde{\varepsilon} = \tilde{\varepsilon}(\varepsilon, \delta, \kappa, \tilde{r})$ *such that*

$$\lim_{\varepsilon\to 0,\, \delta\to\infty,\, \kappa\to\infty,\, \tilde{r}(\ln\delta)^{-1}\to\infty} \tilde{\varepsilon} = 0 \tag{4.5.3}$$

and such that the following estimates hold:

(i) $|h^{\pm}_{\lambda}(E) - h^{\pm}_{\lambda}| \leqslant \tilde{\varepsilon}$ (4.5.4)

for any translation-invariant frame weight F which is a κ*-functional.*

(ii) *For any such frame weight E and for any contour* Γ,

$$|\tilde{\Delta}_\lambda(\Gamma, F)| \leqslant \tilde{\varepsilon}|\operatorname{supp}\Gamma|. \tag{4.5.5}$$

PROOF. This is simply a recapitulation of Theorems 4.3.5 and 4.3.6 and Corollary 4.4.5.

4.5.2. *Solution of* (4.2.30) *in the Class of* κ*-Functionals*

It will prove useful to introduce the notion of a contour functional as a functional of contours which is not necessarily positive (in contrast to the notion of contour weight).

DEFINITION 1. For any contour functional F write

$$|F| = \operatorname{supp}_{\Gamma} |F(\Gamma)|V^*(\Gamma)|^{-1}|. \tag{4.5.6}$$

Given a contour functional G denote by $\mathscr{F}_G$ the metric space of contour functionals which are of the type $G + F$ where $|F| < \infty$, the metric being chosen as follows:

$$\rho(G + F, G + F') = |F - F'|. \tag{4.5.6'}$$

Write $\mathscr{F}$ instead of $\mathscr{F}_0$. Given $d \in \mathbb{R}$, denote by $\mathscr{F}_{G,d}$ the closed d-neighbourhood of G.

NOTE. We do not suppose that $G \in \mathscr{F}$. (This is not satisfied for the typical application $G = G_\lambda$.)

DEFINITION 2. Denote by $\mathscr{F}_a^b$ the set

$$\{F \in \mathscr{F} : a|\operatorname{supp}\Gamma| \leqslant F(\Gamma) \leqslant b|\operatorname{supp}\Gamma| \text{ for each } \Gamma\}.$$

(This is a closed subset of the complete metric space $\mathscr{F}$.)

In what follows, we will also study the factorized functionals given by (4.2.18).

NOTATION. Given a contour functional F such that the factorized functional (4.2.18) exists (in the sense of existence of the Lebesgue integral) we denote this factorized functional (acting on $(\boldsymbol{\Gamma}, a)$ where $a \in (\tilde{\mathscr{U}}^+)^{\partial_+\boldsymbol{\Gamma}}$ or $a \in (\tilde{\mathscr{U}}^-)^{\partial_-\boldsymbol{\Gamma}}$) by the symbol $\mathbb{F}$. We call the functionals acting on $(\boldsymbol{\Gamma}, a)$ *frame functionals*, and denote by $\mathscr{F}^*$ the set of all frame functionals.

DEFINITION 3. Given $\kappa \in \mathbb{R}$ we denote by $\mathscr{F}_\kappa^*$ the set of all frame functionals $\mathbb{F}$ such that for each frame $\boldsymbol{\Gamma}$ and each $a \in (\tilde{\mathscr{U}}^+)^{\partial_+\boldsymbol{\Gamma}}$ (or $a \in (\tilde{\mathscr{U}}^-)^{\partial_-\boldsymbol{\Gamma}}$),

$$\mathbb{F}(\boldsymbol{\Gamma}, a) \geqslant \kappa|\operatorname{supp}\boldsymbol{\Gamma}|$$

We call the elements of $\mathscr{F}_\kappa^*$ as κ-functionals. We also write $F \in \mathscr{F}_\kappa^*$ and say that F is a κ-functional if $\mathbb{F} \in \mathscr{F}_\kappa^*$ for the factorized functional $\mathbb{F}$.

NOTE. The condition $F \in \mathscr{F}_\kappa^*$ does not mean that $F \in \mathscr{F}_\kappa^\infty$. On the contrary, the following is obviously true:

$$\text{if} \quad G \in \mathscr{F}_\kappa^* \quad \text{and if} \quad G' \in \mathscr{F}_\mu^\infty \quad \text{then} \quad G + G' \in \mathscr{F}_{\kappa+\mu}^*. \tag{4.5.7}$$

(A typical application is $G = G_\lambda$ and $G' = \tilde{\Delta}_\lambda$.)

We will now study, as it is usual in PS theory, the mapping

$$\{F(\Gamma) \rightsquigarrow G_\lambda(\Gamma) + \tilde{\Delta}_\lambda(\Gamma, F)\} \tag{4.5.8}$$

(with translation-invariant F). As a first step we show its contractivity properties.

PROPOSITION 1. *Denote by* $\mathscr{F}_{G,d}^{\mathrm{tr}}$ *the set of all translation-invariant functionals from* $\mathscr{F}_{G,d}$. *There is some* $\kappa_0 = \kappa_0(c_{(3.2.3)}, d)$ *such that for all* $\kappa \geqslant \kappa_0$, *the mapping*

$$\{F \rightsquigarrow \tilde{\Delta}_\lambda(\cdot, F)\} : \mathscr{F}_{G,d}^{\mathrm{tr}} \to \mathscr{F} \tag{4.5.9}$$

is a contraction, for any contour functional $G \in \mathscr{F}_\kappa^*$.

Proof. For any $F \in \mathscr{F}_G$ and $F' \in \mathscr{F}$ denote by $[F; F']$ the segment

$$[F; F'] = \{F_t = F + tF', t \in [0, 1]\}. \tag{4.5.10}$$

By (4.5.7) we can claim that if $F \in \mathscr{F}_\kappa^*$ and $F' \in \mathscr{F}_\mu^\infty$ then $[F; F'] \subset \mathscr{F}_\kappa^* \cup \mathscr{F}_{\kappa+\mu}^*$. We shall see (from (4.5.20), (4.5.21)) that it suffices to prove the following lemma.

LEMMA 1. *Let* Λ, $M \subset \mathbb{Z}^\nu$ *be finite, let* $[F; F'] \in \mathscr{F}_\kappa^*$. *Then for any* $a \in \tilde{\mathscr{U}}^{\partial\Lambda^c}$ *the following estimate holds, for large enough* κ:

$$\left|\frac{\mathrm{d}}{\mathrm{d}t} \ln Z_\lambda(\Lambda, M, a, F_t)\right| \leqslant \omega|\Lambda||F'| \tag{4.5.11}$$

where $\omega = \omega(\kappa)$ *is such that* $\lim_{\kappa\to\infty} \omega = 0$. (*We write* $\tilde{\mathscr{U}}^+ = \tilde{\mathscr{U}}$, *as in* §4.4.)

NOTE. We investigate the case $\tilde{\mathscr{U}} = \tilde{\mathscr{U}}^+$ only. The analogous case arising if $\pm$ is replaced by $\mp$ is omitted, here and throughout §4.5. We use the simplified notation of §4.4 if no ambiguity arises.

Proof of Lemma 1. By the definition (4.2.8) we get the formula

$$\frac{\mathrm{d}}{\mathrm{d}t}\ln Z_\lambda(\Lambda, M, a, \mathbb{F}_t) = -\int\Big(\sum_{\Gamma\in\mathscr{D}}\frac{\mathrm{d}}{\mathrm{d}t}\mathbb{F}_t(\boldsymbol{\Gamma}, x_{\partial_+\Gamma})\Big)P(r_{\Lambda\setminus M}\cup x_M)\,\mathrm{d}r_{\Lambda\setminus M}\,\mathrm{d}x_M \qquad (4.5.12)$$

where $r_{\Lambda\setminus M} = (x_{\Lambda\setminus M}, \mathscr{D})$ and where $P = P_{\lambda, M, a, F_t}$ denotes the probability density

$$P_{\lambda, M, a, F_t}(r_{\Lambda\setminus M}\cup x_M) = Z_\lambda^{-1}(\Lambda, M, a, \mathbb{F}_t)\exp(-H_\lambda^{(M)}(r_{\Lambda\setminus M}\cup x_M \,|\, a)) \qquad (4.5.12')$$

(see (4.2.8) with $\mathbb{F}_t$ instead of F).

From (4.5.6) we have

$$\left|\frac{\mathrm{d}}{\mathrm{d}t}F_t(\Gamma)\right| \leqslant |F'(\Gamma)| \leqslant |F'|\,|V^*(F)|$$

and therefore

$$\left|\frac{\mathrm{d}}{\mathrm{d}t}\mathbb{F}_t(\boldsymbol{\Gamma}, x_{\partial_+\Gamma})\right| \leqslant \left|\frac{\mathrm{d}}{\mathrm{d}t}\int\exp(-F_t(\Gamma))\,\mathrm{d}\Gamma_{\operatorname{supp}\Gamma\setminus\partial_+\Gamma} \times\right.$$
$$\left.\times\left(\int\exp(-F_t(\Gamma))\,\mathrm{d}\Gamma_{\operatorname{supp}\Gamma\setminus\partial_+\Gamma}\right)^{-1}\right|$$
$$\leqslant |F'|\,|V^*(\Gamma)| \qquad (4.5.13)$$

(see (4.2.18)).

Denote by $\hat{P}_\Gamma$ the projection, on $(\tilde{\mathscr{U}})^{\partial_+\Gamma}$, of the measure $P(r_{\Lambda\setminus M}\cup x_M)\,\mathrm{d}r_{\Lambda\setminus M}\,\mathrm{d}x_M$ restricted to all $r_{\Lambda\setminus M} = (x_{\Lambda\setminus M}, \mathscr{D})$ satisfying the condition $\mathscr{D}\supset\boldsymbol{\Gamma}$. Then, using (4.5.12) and (4.5.13) we can write

$$\left|\frac{\mathrm{d}}{\mathrm{d}t}\ln Z_\lambda(\Lambda, M, a, F_t)\right| \leqslant |F'| \sum_{\boldsymbol{\Gamma}:\,\operatorname{dist}(\operatorname{supp}\boldsymbol{\Gamma}, \Lambda^c\cup M)\geqslant 2}\int V^*(\boldsymbol{\Gamma})\,\mathrm{d}\hat{P}_\Gamma(x_{\partial_+\Gamma}). \qquad (4.5.14)$$

Now we use the elementary inequality

$$|V^*(\boldsymbol{\Gamma})| \leqslant c\exp\left(\frac{\kappa}{2}|\operatorname{supp}\boldsymbol{\Gamma}|\right) \qquad (4.5.15)$$

where $c = c(\kappa)$ is such that $\lim_{\kappa\to\infty} c = 0$ and also the following important estimate, which is an analogy of the well-known Peierls estimate. (We formulate it more precisely than it is necessary there; this will be useful later.)

LEMMA 2. *There is some* $\tilde{\varepsilon} = \tilde{\varepsilon}(\varepsilon_{(3.2.6)}, \kappa, c_{(3.2.3)})$ *such that* $\lim_{\varepsilon\to 0, \kappa\to\infty}\tilde{\varepsilon} = 0$ *and such that for any frame weight* $\mathbb{F}$ *which is a* κ *functional,*

$$\frac{\mathrm{d}\hat{P}_\Gamma}{\mathrm{d}\hat{P}}(x_{\partial_+\Gamma}) \leqslant \exp(\tilde{\varepsilon}\,|\operatorname{supp}\boldsymbol{\Gamma}| - \mathbb{F}(\boldsymbol{\Gamma}, x_{\partial_+\Gamma})) \qquad (4.5.16)$$

where $\hat{P}$ *denotes the projection, on* $x(\partial_+\Gamma)$, *of the measure* $P(r_{\Lambda\setminus M}\cup x_M)\mathrm{d}r_{\Lambda\setminus M}\,\mathrm{d}x_M$ (see (4.5.12')).

Let us show, at first, that the estimates (4.5.15) and (4.5.16) imply (4.5.11). By (4.5.14) we have, if we also use the condition $\mathbb{F}_t(\mathbf{\Gamma}, x_{\partial_+\Gamma}) \geqslant \kappa\,|\mathrm{supp}\,\Gamma|$, the inequality

$$\left|\frac{\mathrm{d}}{\mathrm{d}t}\ln Z_\lambda(\Lambda, M, a, \mathbb{F}_t)\right| \leqslant |F'| \sum_{\Gamma:\,\mathrm{dist}(\mathrm{supp}\,\Gamma,\Lambda^c\cup M)\geqslant 2} c\exp\left(\left(\tilde{\varepsilon}-\frac{\kappa}{2}\right)|\mathrm{supp}\,\mathbf{\Gamma}|\right)$$

$$\leqslant |F'|c\left(\sum_{\Gamma:\,\mathrm{supp}\,\Gamma\ni 0}\exp\left(\left(\tilde{\varepsilon}-\frac{\kappa}{2}\right)|\mathrm{supp}\,\Gamma|\right)\right)|\Lambda|, \tag{4.5.17}$$

□

Proof of Lemma 2. Instead of $\hat{P}$ consider the measure $\hat{P}_{\mathrm{non}\,\Gamma}$ defined as the projection, on $\mathcal{U}^{\partial_+\Gamma}$, of the measure $P(r_{\Lambda\setminus M}\cup x_M)\mathrm{d}r_{\Lambda\setminus M}\mathrm{d}x_M$, $r_{\Lambda\setminus M} = (x_{\Lambda\setminus M}, \mathcal{D})$ restricted to the set of all $r_{\Lambda\setminus M}$ such that $\mathrm{dist}(\mathrm{supp}\,\mathbf{\Gamma}, \mathrm{supp}\,\tilde{\mathbf{\Gamma}}) \geqslant 2$ for any $\tilde{\mathbf{\Gamma}} \in \mathcal{D}$. Clearly,

$$\frac{\mathrm{d}\hat{P}_\Gamma}{\mathrm{d}\hat{P}_{\mathrm{non}\Gamma}}(x_{\partial_+\Gamma}) = \exp(-\mathbb{F}(\mathbf{\Gamma}, x_{\partial_+\Gamma}))Z_\lambda(\Lambda_\Gamma, M_\Gamma, a_\Gamma, \mathbb{F})Z_\lambda^{-1}(\Lambda_\Gamma, M, a_\Gamma, \tilde{\mathbb{F}}) \tag{4.5.18}$$

where

$$\Lambda_\Gamma = \Lambda\setminus\partial_+\mathbf{\Gamma}$$
$$M_\Gamma = M\cup\mathrm{supp}\,\mathbf{\Gamma}$$
$$a_\Gamma = a\cup x_{\partial_+\Gamma}$$

and where

$$\tilde{\mathbb{F}}(\tilde{\mathbf{\Gamma}}, x_{\partial_+\tilde{\Gamma}}) = \mathbb{F}(\tilde{\mathbf{\Gamma}}, x_{\partial_+\Gamma}) \text{ if } \mathrm{dist}(\mathrm{supp}\,\mathbf{\Gamma}, \mathrm{supp}\,\tilde{\mathbf{\Gamma}}) \geqslant 2$$
$$\tilde{\mathbb{F}}(\tilde{\mathbf{\Gamma}}, x_{\partial_+\tilde{\Gamma}}) = \infty \text{ otherwise.}$$

Expressing both $Z_\lambda(\Lambda_\Gamma, M_\Gamma, a_\Gamma, \mathbb{F})$ and $Z_\lambda(\Lambda_\Gamma, M, a_\Gamma, \tilde{\mathbb{F}})$ by (4.4.10), we easily see from Theorem 2.3.8 that

$$|\ln Z_\lambda(\Lambda_\Gamma, M_\Gamma, a_\Gamma, \mathbb{F}) - \ln Z_\lambda(\Lambda_\Gamma, M, a_\Gamma, \tilde{\mathbb{F}})| \leqslant \tilde{\varepsilon}\,|\mathrm{supp}\,\mathbf{\Gamma}| \tag{4.5.19}$$

where $\tilde{\varepsilon} = \tilde{\varepsilon}(\varepsilon, \delta, c_{(3.2.3)}, \kappa)$ is such that $\lim_{\varepsilon\to 0,\delta\to\infty,\kappa\to\infty}\tilde{\varepsilon} = 0$. Now it suffices to use the obvious fact that

$$\hat{P}_{\mathrm{non}\Gamma} \leqslant \hat{P}$$

which concludes the proof of Lemma 2. □

NOTE. This is a similar estimate to that of (4.4.48). The condition $a \in \mathcal{U}^{\partial\Lambda^c}$ is again rather substantial there. A typical application is $\Lambda = W(\Gamma)$, $M = \emptyset$, $a = \partial_-\Gamma$.

COROLLARY. *If F_t is translation-invariant, then $h_\lambda(F_t)$ satisfies the bound*

$$\left|\frac{\mathrm{d}}{\mathrm{d}t}h_\lambda(F_t)\right| \leqslant \omega\,|F'| \tag{4.5.20}$$

with the same ω as in (4.5.11).

Proof. Obvious from (4.5.11) if we assume the existence of $\mathrm{d}/\mathrm{d}t\, h_\lambda(\mathbb{F}_t)$. Without this assumption one must interpret (4.5.20) and also some formulas (to be given below) in the sense of a derivative of an absolutely continuous function. (Nevertheless $d/dt\, h_\lambda(\mathbb{F}_t)$ exists and we will in fact prove this in §4.5.6, where we will show that $\hat{P}$ converges, sufficiently fast, to some translation-invariant measure.)

Substituting the bounds (4.5.11), (4.5.20) into (4.2.28) we obtain

$$\left| \frac{d}{dt} \tilde{\Delta}_\lambda(\Gamma, \mathbb{F}_t) \right| \leqslant 4\omega V^*(\Gamma). \tag{4.5.21}$$

Thus, the mapping (4.5.9) is a contraction if $\omega < \frac{1}{4}$. □

PROPOSITION 2. *There is some* $\tilde{\varepsilon} = \tilde{\varepsilon}(\varepsilon, \delta, \kappa, \tilde{r})$ *such that* $\lim_{\varepsilon \to 0, \delta \to \infty, \kappa \to \infty, \tilde{r}(\ln \delta)^{-1} \to \infty} \tilde{\varepsilon} = 0$ *and such that the image of the mapping* (4.5.9) *is a subset of* $\mathscr{F}^{\tilde{\varepsilon}}_{-\tilde{\varepsilon}}$.

Proof. This is simply a reformulation of (4.5.5). □

Now we will solve the equation (4.2.30). First, let us specify more precisely our assumptions on the value of κ (and, therefore, also of τ, δ, $\tilde{r}$ – see (4.5.1)). Consider the value $\tilde{\varepsilon}$ from Proposition 2 and the value κ_0 from Proposition 1. Choose a suitable $\hat{\kappa}$ such that

$$\kappa_{(4.5.1)} - \tilde{\varepsilon}(\varepsilon, \delta, \hat{\kappa}, \tilde{r}) \geqslant \hat{\kappa} \geqslant \kappa_0(c_{(3.2.3)}, \tilde{\varepsilon}). \tag{4.5.22}$$

Such a choice of $\hat{\kappa}$ is possible if the conditions of the Main Lemma are satisfied. We can then choose sufficiently large $\tilde{r}(\ln \delta)^{-1}$ such that $\tau(\tilde{r})^{-\nu}$ and $\delta^2(\tilde{r})^{-\nu}$ are also sufficiently large.

Consider the sequence

$$F^{(n)}_\lambda(\Gamma) = G_\lambda(\Gamma) + \tilde{\Delta}_\lambda(\Gamma, F^{(n-1)}_\lambda); F^{(0)}_\lambda = G_\lambda. \tag{4.5.23}$$

THEOREM. *Under the condition* (4.5.22), *there exists a limit*

$$F_\lambda(\Gamma) = \lim_{n \to \infty} F^{(n)}_\lambda(\Gamma) \tag{4.5.23'}$$

satisfying (4.2.30).

Proof. It suffices to combine Propositions 1 and 2. By (4.5.22), the mapping (4.5.8) maps the set $\mathscr{F}_{G_\lambda, \tilde{\varepsilon}}$, $G_\lambda \in \mathscr{F}^*_{\hat{\kappa}}$ into itself and it is a contraction on this complete metric space. □

4.5.3. *Solution of* (4.2.29)

Using the continuity (even analyticity) of any mapping

$$\{\lambda \rightsquigarrow G_\lambda(\Gamma) \colon [\lambda_1, \lambda_2] \to \mathbb{R} \tag{4.5.24}$$

and using the contractivity of (4.5.9) and also the condition (3.2.7) one obtains the equicontinuity of all the mappings (for a fixed Γ)

$$\{\lambda \rightsquigarrow F^{(n)}_\lambda(\Gamma)\} \colon [\lambda_1, \lambda_2] \to \mathbb{R} \tag{4.5.24'}$$

and also of all the factorized mappings (for a fixed $\mathbf{\Gamma}$, $x_{\partial_+ \mathbf{\Gamma}}$)

$$\lambda \rightsquigarrow \mathbb{F}^{(n)}_\lambda(\mathbf{\Gamma}, x_{\partial_+ \mathbf{\Gamma}})\} \colon [\lambda_1, \lambda_2] \to \mathbb{R}. \tag{4.5.25}$$

To understand this investigate the difference

$$F_{\lambda}^{(n)}(\Gamma) - F_{\lambda'}^{(n)}(\Gamma) = (\tilde{\Delta}_{\lambda}(\Gamma, F_{\lambda}^{(n-1)}) - \tilde{\Delta}_{\lambda'}(\Gamma, F_{\lambda}^{(n-1)})) + \\ + (\tilde{\Delta}_{\lambda'}(\Gamma, F_{\lambda}^{(n-1)}) - \tilde{\Delta}_{\lambda'}(\Gamma, F_{\lambda'}^{(n-1)})). \tag{4.5.26}$$

The first term on the right-hand side is estimated as $c\varepsilon'_{(3.2.7)}|V(\Gamma)||\lambda - \lambda'|$, whereas the second term is estimated as $|qV(\Gamma)||F_{\lambda}^{(n-1)} - F_{\lambda'}^{(n-1)}|$ with $0 < q < 1$. (See the next section for a more detailed study of $\{\lambda \leadsto F_{\lambda}^{n}\}$.)

Thus we obtain the continuity of

$$\{\lambda \leadsto h_{\lambda}^{\pm}(\mathbb{F}_{\lambda})\}: [\lambda_1, \lambda_2] \to \mathbb{R}. \tag{4.5.27}$$

where

$$\mathbb{F}_{\lambda} = \lim \mathbb{F}_{\lambda}^{(n)}.$$

Assume now that

$$\tilde{\varepsilon}_{(4.5.3)} < \Delta_{(3.2.8)}. \tag{4.5.28}$$

Then

$$h_{\lambda_1}^{+}(\mathbb{F}_{\lambda_1}) < h_{\lambda}^{-}(\mathbb{F}_{\lambda_1}), \qquad h_{\lambda_2}^{+}(\mathbb{F}_{\lambda_2}) > h_{\lambda_2}^{-}(\mathbb{F}_{\lambda_2})$$

and so the solution of equation (4.2.29) exists. We denote it by

$$\bar{\lambda} = \bar{\lambda}(\Phi), \qquad \Phi = \{\Phi^{\lambda}, \lambda \in [\lambda_1, \lambda_2]\} \tag{4.5.29}$$

(we mean one of these solutions, at the moment; the conditions assuring its unicity will be formulated below). We expect that for this value of $\bar{\lambda}$ there are two translation-invariant Gibbs states, described by an appropriate contour model. This will be explained in more detail in §4.5.6.

4.5.4. *Investigation of the Phase Diagram*

In this section we will prove some results about the unicity (in $[\lambda_1, \lambda_2]$) of the solution (4.5.29), exploiting the conditions (3.2.7) and (3.2.12).

Our basic estimate will be the proof of a uniform smallness of $\mathrm{d}/\mathrm{d}\lambda(h_{\lambda}^{\pm}(\mathbb{F}_{\lambda}) - h_{\lambda}^{\pm})$. This will guarantee the strict monotonicity of the mapping

$$\{\lambda \leadsto h_{\lambda}^{+}(\mathbb{F}_{\lambda}) - h_{\lambda}^{-}(\mathbb{F}_{\lambda})\}: [\lambda_1, \lambda_2] \to \mathbb{R} \tag{4.5.30}$$

and therefore the unicity of $\bar{\lambda}(\Phi)$. (We use the same convention as in (4.5.20), the existence of a derivative will be clear only later in §4.5.5.) To get some bounds for $\mathrm{d}/\mathrm{d}\lambda(h_{\lambda}^{\pm}(\mathbb{F}_{\lambda}) - h_{\lambda}^{+})$ we need, at the same time, appropriate bounds also for $\mathrm{d}/\mathrm{d}\lambda\, G_{\lambda}(\cdot)$, $\mathrm{d}/\mathrm{d}\lambda \tilde{\Delta}_{\lambda}(\cdot, F_{\lambda})$ and $\mathrm{d}/\mathrm{d}\lambda\, F_{\lambda}(\cdot)$. So it is reasonable to investigate the general expression (say for the + contour model)

$$\frac{\mathrm{d}}{\mathrm{d}\lambda} \ln Z_{\lambda}^{\mathrm{rel}}(\Lambda, M, a, F_{\lambda}) \tag{4.5.31}$$

$$= \sum_{\Gamma:\, \mathrm{dist}(\mathrm{supp}\,\Gamma, \Lambda^c \cup M) > 2} \int \frac{\mathrm{d}}{\mathrm{d}\lambda} \mathbb{F}_{\lambda}(\Gamma, x_{\partial_+\Gamma}) \mathrm{d}\hat{P}_{\Gamma}(x_{\partial_+\Gamma}) +$$

$$+ \sum_{A \notin \operatorname{supp} \mathscr{D}, A \cap \Lambda \neq \emptyset} \int \frac{\mathrm{d}}{\mathrm{d}\lambda} (\Phi_A^\lambda(x_A) - \Phi_A^{+,\lambda}(x_A))\, \mathrm{d}\hat{P}_A(x_A)$$

(see 4.2.3) where $\hat{P}_\Gamma$ has the same meaning as in (4.5.14) and similarly $\hat{P}_A$ is the projection on $\mathscr{U}^{\partial A}$ of the measure

$$P(r_{\Lambda\setminus M} \cup x_M)\,\mathrm{d}r_{\Lambda\setminus M}\,\mathrm{d}x_M,\; r_{\Lambda\setminus M} = (x_{\Lambda\setminus M}, \mathscr{D}),\; A \notin \operatorname{supp} \mathscr{D}.$$

It is easy to estimate, using the assumption (3.2.7) of the Main Lemma, the second term on the right-hand side of (4.5.27):

$$\sum_A \int \left| \frac{\mathrm{d}}{\mathrm{d}\lambda} (\Phi_A^\lambda(x_A) - \Phi_A^{+,\lambda}(x_A)) \right| \mathrm{d}\hat{P}_A(x_A) \leqslant \varepsilon' |\Lambda| \tag{4.5.32}$$

with $\varepsilon' = c\varepsilon'_{(3.2.7)}$ and some $c = c(r)$. To find analogous bounds for the first term in (4.5.31) we have also to investigate similar expressions for all frame functionals $F_\lambda^{(n)}$, $n = 0, 1, \ldots$.

Investigate first the case $n = 0$. Using (4.5.16) and (4.5.2′) it is easy to see that

$$\sum_\Gamma \int \left| \frac{\mathrm{d}}{\mathrm{d}\lambda} \mathbb{G}_\lambda(\Gamma, x_{\partial_+\Gamma}) \right| \mathrm{d}\hat{P}_\Gamma(x_{\partial_+\Gamma}) \leqslant \varepsilon'' |\Lambda| \tag{4.5.33}$$

where $\varepsilon'' = \varepsilon''(\kappa', c_{(3.2.3)})$ is such that $\lim_{\kappa' \to \infty} \varepsilon'' = 0$. The estimates (4.5.32) and (4.5.33), collected together and generalized to any $F_\lambda^{(n)}$, give the following result.

PROPOSITION. *There is some* $\varepsilon = \varepsilon(\varepsilon', c_{(3.2.3)}, \kappa')$ *such that* $\lim_{\varepsilon' \to 0, \kappa' \to \infty} \varepsilon = 0$ (*where* $\varepsilon' = \varepsilon'_{(3.2.7)}$, $\kappa' = \kappa'_{(4.5.2')}$) *and such that the following inequalities are valid for each* $n = 0, 1, \ldots$, *each finite* Λ, $M \subset \mathbb{Z}^\nu$ *and such* $a \in \tilde{\mathscr{U}}^{\partial\Lambda^c}$:

$$\left| \frac{\mathrm{d}}{\mathrm{d}\lambda} \ln Z_\lambda^{\mathrm{rel}}(\Lambda, M, a, F_\lambda^{(n)}) \right| \leqslant \varepsilon |\Lambda|, \tag{4.5.34}$$

$$\left| \frac{\mathrm{d}}{\mathrm{d}\lambda} h_\lambda^{\mathrm{rel}}(F_\lambda^{(n)}) \right| \leqslant \varepsilon, \tag{4.5.35}$$

$$\left| \frac{\mathrm{d}}{\mathrm{d}\lambda} \tilde{\Delta}_\lambda(\Gamma, F_\lambda^{(n)}) \right| \leqslant 4\varepsilon |V^*(\Gamma)|. \tag{4.5.36}$$

Proof. We will proceed by induction on n. First notice that for $n = 0$ we obtain (4.5.34) by summing (4.5.32) and (4.5.33). Taking the limit $\Lambda \uparrow \mathbb{Z}^\nu$ we obtain (4.5.35). (We again omit the proof of existence of $\mathrm{d}/\mathrm{d}\lambda\, h_\lambda^{\mathrm{rel}}(F_\lambda^{(n)})$.) Using the very definition (4.2.28) of $\tilde{\Delta}_\lambda(\Gamma, F_\lambda^{(n)})$ we finally obtain (4.5.36) from the preceding two relations (4.5.34), (4.5.35).

Clearly, the last two steps of the reasoning remain true also for general n and it therefore suffices to prove the induction step only for (4.5.34). We will prove it with

$$\varepsilon = 2(\varepsilon'_{(4.5.32)} + \varepsilon''_{(4.5.33)})$$

(and with a reasonably taken value of ε', ε'' – see below).

Assuming that (4.5.36) holds for $n - 1$ and considering also (4.5.31), (4.5.32),

(4.5.33) we can estimate $\mathrm{d}/\mathrm{d}\lambda \ln Z_\lambda^{\mathrm{rel}}$ as follows:

$$\ln Z_\lambda^{\mathrm{rel}}(\Lambda, M, a, F_\lambda^{(n)}) \leqslant (\varepsilon' + \varepsilon'')|\Lambda| + \sum_{\Gamma:\, \mathrm{dist}(\mathrm{supp}\,\Gamma, \Lambda^c \cup M) \geqslant 2} 4\varepsilon |V^*(\Gamma)| \int \mathrm{d}\hat{P}_\Gamma(x_{\partial_+\Gamma})$$
$$\leqslant (\varepsilon' + \varepsilon'' + 4c\varepsilon)|\Lambda|. \tag{4.5.37}$$

Concerning the choice of ε', ε'', notice that ε'' is evaluated by using the Peierls estimate (4.5.16) for $\hat{P}_\Gamma$, depending on $F_\lambda^{(n)}$. We replace $F_\lambda^{(n)}(\Gamma)$ by the lower bound $G_\lambda(\Gamma) - \varepsilon|\mathrm{supp}\,\Gamma|(\varepsilon = \varepsilon_{(4.5.5)})$ in this estimate. This gives the 'proper' value of ε'.

The constant $c_{(4.5.37)}$ is taken from the obvious inequality

$$\sum_{\Gamma:\, \mathrm{dist}(\mathrm{supp}\,\Gamma, \Lambda^c \cup M) \geqslant 2} |V^*(\Gamma)| \mathrm{d}\hat{P}_\Gamma(\Gamma) \leqslant c|\Lambda|. \tag{4.5.38}$$

It satisfies the relation $\lim_{\kappa \to \infty} c = 0$.

Suppose that κ is so large that $c < \frac{1}{8}$, and therefore $(\varepsilon' + \varepsilon'' + 4c\varepsilon) \leqslant \varepsilon$. The induction step for (4.5.34) is then proved. □

COROLLARY. *The inequalities* (4.5.34,35,36) *remain true (with the same* ε*) for the contour weight* F_λ. *If moreover*

$$\varepsilon_{(4.5.35)} < \frac{1}{2}(u^+_{(3.2.5)} - u^-_{(3.2.5)})$$

then the mapping

$$\{\lambda \rightsquigarrow (h_\lambda^+(F_\lambda) - h_\lambda^-(F_\lambda))\} : [\lambda_1, \lambda_2] \to \mathbb{R} \tag{4.5.39}$$

is strictly monotone and therefore the solution (4.5.29) *is unique.*

We now study the analytical features of this solution,

4.5.5. *Analyticity of the Phase Diagram*

Choose some neighbourhood $\mathscr{V}(1)$ of $1 \in \mathbb{R}$ and assume that all the assumptions needed for the construction of $\tilde{\lambda}(\beta\Phi)$ are satisfied for all $\beta \in \mathscr{V}(1)$. (This is obviously true if $\mathrm{diam}(\mathscr{V}(1))$ is sufficiently small.) Consider the mapping

$$\{\beta \rightsquigarrow \tilde{\lambda}(\beta)\}\mathscr{V}(1) \to [\lambda_1, \lambda_2] \tag{4.5.40}$$

where $\tilde{\lambda}(\beta) = \tilde{\lambda}(\beta\Phi)$ (see (4.5.29)).

We now prove the analyticity of this mapping. The main step is the proof of the following statement.

PROPOSITION 1. *The mapping* (4.5.25) *is analytical, for each* Γ.

Proof. First note the following.

LEMMA 1. *The mapping* (4.5.24) *is analytic, for each* Γ.

Proof of Lemma 1. This is obvious from the definition (4.2.24) of $G_\lambda(\Gamma)$ (or, better, from (4.3.6)) and the assumption 4 of the Main Lemma. □

Proof of Proposition 1. We use the considerations of §2.3.14. We need also to extend the constructions of §4.2 to the case of a complex λ. Assume that the set $\mathscr{V}$ mentioned in the assumption 4 of Main Lemma is a parallelipiped

$$\mathscr{V} = \{\lambda \in \mathbb{C}: \operatorname{Re}\lambda \in (\lambda_1 - \omega, \lambda_2 + \omega), \operatorname{Im}\lambda \in (-\omega, \omega)\} \tag{4.5.41}$$

with a sufficiently small ω. For each $\lambda \in \mathscr{V}$ define the approximations

$$\Phi_A^{+,\lambda} \equiv \Phi_A^{+,\operatorname{Re}\lambda}. \tag{4.5.42}$$

(This means that we handle the imaginary part of the quadratic Hamiltonian as an additional perturbation.) It is obvious that all the considerations of §4.2 which do not use the notion of $h_\lambda^\pm(F_\lambda)$ (i.e. §§4.2.1–4.2.7) can be reformulated to the case of complex interactions Φ_A^λ, $\lambda \in \mathscr{V}$. Because of the condition (3.2.3) and the because ε is required to be small there are no problems with the convergence of the finite-dimensional integrals (4.2.1), (4.2.9) etc., if we also use Assumption 5 of the Main Lemma in the case of (4.2.1) and smallness of ω in both cases.

From (4.2.26) onwards we must be more cautious. For example, the formula (4.5.12) is of no use for $\Lambda \uparrow \mathbb{Z}^\nu$ because then the Feynman measure P ceases to exist. On the other hand, all the considerations of §2.3.8 are still applicable to the right-hand side of (4.4.10) and we obtain the existence of $h_\lambda^\pm(F)$ as an analytical function of F (see Proposition 2.3.14).

LEMMA 2. *Let $\{\tilde{F}_\lambda\}$ be a family of translation-invariant contour functionals (with values in $\mathbb{C}$) depending on $\lambda \in \mathscr{V} \subset \mathbb{C}$. Suppose that $\operatorname{Re} \mathbb{F}_\lambda \in \mathscr{F}_\kappa^*$ with a sufficiently large $\kappa > 0$. Suppose that for each frame Γ, the functions $\tilde{F}_{(\cdot)}(\Gamma)$, $\Gamma \in \boldsymbol{\Gamma}$ are analytic in $\lambda \in \mathscr{V}$, and there is an integrable majorant for all $\exp(-\tilde{F}_\lambda(\cdot))$ on $\boldsymbol{\Gamma}$. Then $h_\lambda^\pm(\tilde{F}_\lambda)$ is an analytic function of $\lambda \in \mathscr{V}$.*

Proof. This is an immediate consequence of Proposition 2.3.14. □

Now we apply Lemmas 1 and 2 to the recurrently defined functionals $F_\lambda^{(n)}$ (see (4.5.23)). From the definition of $F_\lambda^{(n)}$ and $\tilde{\Delta}_\lambda$ we obtain, by induction on n, the following statement.

LEMMA 3. *All the functions*

$$\{\lambda \rightsquigarrow F_\lambda^{(n)}(\Gamma)\}: \mathscr{V} \to \mathbb{C} \tag{4.5.44}$$

are analytic, as well as the function

$$\{\lambda \rightsquigarrow h_\lambda(F_\lambda^{(n)})\}: \mathscr{V} \to \mathbb{C}, \tag{4.5.45}$$

for each $n \in \mathbb{N}$.

Proof. The induction step is carried out using Lemma 2. In the case of (4.5.44) we further use the obvious fact that $Z_\lambda(\Lambda, M, a, F_\lambda)$ is an analytical function of λ (in the same sense as in Lemma 2).

We now apply Lemma 4 in the limit case $n \to \infty$. Notice that we have not yet proved the existence of the limit (4.5.23′) for complex λ. This follows, however, from the well-known properties of analytical functions if we prove (for example) the uniform boundedness of all $h_\lambda^\pm(F_\lambda^{(n)})$, with $\lambda \in \mathscr{V}$ and $n \in \mathbb{N}$. The latter fact is clear from Assumption 5 of the Main Lemma for the contour weights G_λ (so that

$\mathbb{G}_\lambda \in \mathscr{F}^*_\kappa$ for a large κ) and for a general n is proven by induction analogously to the case of a real λ (taking κ such that (4.5.20) holds and such that $\operatorname{Re}\mathbb{F}^{(n)}_\lambda \in \mathscr{F}^*_\kappa$).

Thus we have shown that the mapping (4.5.27) can be extended to an analytical function on $\mathscr{V}$ if ω is small, and the same is true for any function (4.5.25). Finally we can apply our considerations to the case of the function (4.5.40).

THEOREM. *The function* (4.5.40) *can be extended to some analytic function on some open* $\mathscr{V}(1) \subset \mathbb{C}$, $\mathscr{V}(1) \ni 1$.

Proof. Denote by $F_{\lambda,\beta}$, $h^\pm_{\lambda,\beta}(\mathbb{F}_{\lambda,\beta})$ the corresponding values of F_λ, $h^\pm_\lambda(\mathbb{F}_\lambda)$ if the potential Φ is replaced by $\beta\Phi$. (Clearly, the considerations of §4.4 and §4.5 can be also applied to the potential $\beta\Phi$ if β is very close to 1.) By Lemma 2, the function

$$\{(\lambda, \beta) \in \mathscr{V}' \times \mathscr{V}(1) \to h^\pm_{\lambda,\beta}(\mathbb{F}_{\lambda,\beta})\} \tag{4.5.46}$$

is analytical both in $\lambda \in \mathscr{V}'$ and $\beta \in \mathscr{V}(1)$ where $\mathscr{V}' = \{\lambda \in \mathscr{V} : \beta\lambda \in \mathscr{V}$ for each $\beta \in \mathscr{V}(1)\}$. Therefore, the manifold $\{(\lambda, \beta);\ h^+_{\lambda,\beta}(\mathbb{F}_{\lambda,\beta}) = h^-_{\lambda,\beta}(\mathbb{F}_{\lambda,\beta})\}$ is analytical, (the conditions of the implicit function theorem being clearly satisfied). □

4.5.6. *Construction of Gibbs States*

Let $\lambda = \bar{\lambda}_{(4.5.29)}$. We recall that F_λ then satisfies the relations (4.2.23) and therefore also the equivalent relations (4.2.17) and (4.2.10). As we explained at the end of §4.2.6, the relation (4.2.10) implies that there are the same probabilities for

$$x^*_{\text{ext}} \equiv (x_\Lambda)_{\text{ext}^*\mathscr{D}}$$

both in the case of the Gibbsian probability $P_{\lambda,a}$ on $X^\pm(\Lambda)$ and in the case of the probability P_{λ,a,F_λ} on the contour ensemble $\mathscr{R}^\pm(\Lambda)$, for any $a \in (\mathscr{U}^+)^{\partial\Lambda^c}$ $(a \in (\mathscr{U}^-)^{\partial\Lambda^c})$.

Using this observation we now describe in more detail the limit Gibbs states which arise in this case.

The results are presented as follows.

(a) Because we have worked, throughout §4, almost exclusively with the contour ensemble and with the 'diluted' partition functions $Z_\lambda(\Lambda, a, F_\lambda)$, it is reasonable to start with the proof of some limit properties of these contour models. This will be done in Lemma 1 (where we will show that x^*_{ext} exists also in the infinite ensemble) and in Lemma 2 (where the uniqueness of the limit contour ensemble is proven).

(b) We must relate the limit thus obtained to the limit of the 'physical' ensemble which we are investigating (see Lemmas 3 and 4). In particular, we must show that the limit state we construct is a Gibbs state. See the forthcoming theorem.

NOTATION. All the sets Λ, Λ'', etc. are assumed to have simply connected components for the of §4.5 (compare Note 4 in §4.4.1).

LEMMA 1. *There is some* ρ *such that* $\rho \to \infty$ *for* $\tilde{r}(\ln\delta)^{-1} \to \infty$, $\tau(\tilde{r})^{-\nu} \to \infty$, $\delta^2(\tilde{r})^{-\nu} \to \infty$ *and such that for each finite* $\Lambda \subset \mathbb{Z}^\nu$, *each* $a \in (\tilde{\mathscr{U}})^{\partial\Lambda^c}$ *and each* $t \in \Lambda$, $L > 0$,

$$P_{\lambda,a}\{(x^*_{\text{ext}}, \mathscr{D}) : t \in V^*(\Gamma),\quad \operatorname{diam}(\operatorname{supp}\Gamma) \geqslant L \text{ for some } \Gamma \in \mathscr{D}\} \leqslant \exp(-\rho L). \tag{4.5.47}$$

Proof. We can write P_{λ,a,F_λ} instead of $P_{\lambda,a}$ in (4.5.47). We will combine Lemma

4.5.2.2 with the usual estimates for the number of possible frames Γ such that $t \in V^*(\Gamma)$ and $\text{diam}(\text{supp}\,\Gamma) \leqslant L$. We obtain an upper bound c^L, $c = c(\nu)$ for the number of these frames. Noting that $\mathbb{F}_\lambda$ is an κ-functional, with $\kappa = \hat{\kappa}_{(4.5.22)}$ we obtain that the left-hand side of (4.5.47) is no greater than

$$\sum_{N=L}^{\infty} \exp((\tilde{\varepsilon} - \kappa)N)c^N \quad (\tilde{\varepsilon} = \tilde{\varepsilon}_{(4.5.16)}) \qquad \square$$

LEMMA 2. *Let $\Lambda \subset \mathbb{Z}^\nu$ be finite, let $\mathscr{L}$ be a family of all frame systems in Λ. Let f be a measurable function of the pair $(x_\Lambda, \mathscr{D})$ where $x_\Lambda \in X(\Lambda)$ and $\mathscr{D} \in \mathscr{L}$. Suppose that $\{\Lambda^n\}$ is a sequence of finite subsets of $\mathbb{Z}^\nu$ such that $\text{dist}(\Lambda, \Lambda^{nc}) \to \infty$. Let $a_n \in (\tilde{\mathscr{U}})^{\partial \Lambda^{nc}}$. Then the limit*

$$\lim_n \int dP_{\lambda, a_n, \mathbb{F}_\lambda}(r_n)f(r_n) \tag{4.5.48}$$

exists. Here, we interpret f as function on $\mathscr{R}^+(\Lambda_n)$ given as follows: for $r_n = (x^n, \mathscr{D}^n)$,

$$f(r_n) = f(x_\Lambda^n, \mathscr{D}_\Lambda^n)$$

where $\mathscr{D}_\Lambda^n = \{\Gamma \in \mathscr{D}^n : \text{dist}(V^*(\Gamma), \Lambda^c) \geqslant 2\}$.

Proof. We can suppose that f differs only slightly from 1, i.e. the function $\Phi^f(r_n) = \ln f(r_n)$ is small (we used this argument previously in §2.3.15). Considering Φ^f as another 'interaction' we can write $\int dP_{\lambda, a_n, \mathbb{F}_\lambda}(r_n)f(r_n)$ as

$$\int dP_{\lambda, a_n, \mathbb{F}_\lambda}(r_n)f(r_n) = Z_\lambda^{-1}(\Lambda^n, a_n, \mathbb{F}_\lambda)Z_\lambda^f(\Lambda^n, a_n, \mathbb{F}_\lambda) \tag{4.5.49}$$

where $Z_\lambda^f(\Lambda^n, a_n, \mathbb{F}_\lambda)$ is defined analogously to $Z_\lambda(\Lambda^n, a_n, \mathbb{F}_\lambda)$ but including the interaction Φ^f. Both Z_λ and Z_λ^f can be expressed, using Theorem 2.3.8, as follows (considerations of this type were used in §4.4.5 and also in §4.5.2):

$$\ln Z_\lambda(\Lambda^n, a_n, \mathbb{F}_\lambda) = \sum_{t \in \Lambda^n} v_{t,n} + \ln Z_\lambda^+(\Lambda^n, a_n)$$

$$\ln Z_\lambda^f(\Lambda^n, a_n, \mathbb{F}_\lambda) = \sum_{t \in \Lambda^n} v_{t,n}^f + \ln Z_\lambda^+(\Lambda^n, a_n) \tag{4.5.50}$$

where $v_{t,n}$, $v_{t,n}^f$ are such that (see (2.3.109))

$$|v_{t,n} - v_{t,n}^f| \leqslant cq^{\text{dist}(t, \Lambda)}. \tag{4.5.51}$$

By (2.3.111), both $\lim_n v_{t,n}$ and $\lim_n v_{t,n}^f$ do exist. Together with (4.5.51) this proves the convergence of (4.5.49) ((4.5.50) with (4.5.51) deteriorate if $a_n \notin \mathscr{U}^{\partial \Lambda^{nc}}$).

NOTE. Because we are interested in the 'physical ensemble' X with the limit probability $\lim_n P_{\lambda, a_n}$ we do not formulate the 'infinite volume contour ensemble' on which the limit probability $\lim_n P_{\lambda, a_n, \mathbb{F}_\lambda}$ lives. Thus, (4.5.48) becomes interesting for us only after translation into the 'physical ensemble'.

LEMMA 3. *Let $\{\varphi_t\}$ be a collection of bounded measurable functions depending on*

$x_t \in \mathbb{R}^k$. For any + frame $\mathbf{\Gamma}$ and any $x_{\partial_+\mathbf{\Gamma}} \in (\tilde{\mathcal{U}})^{\partial_+\mathbf{\Gamma}}$ define the function

$$\varphi_\Lambda^*(\mathbf{\Gamma}, x_{\partial_+\mathbf{\Gamma}}) = \int \prod_{t \in \Lambda \cap V^*(\mathbf{\Gamma})} \varphi_t(x_t)\, \mathrm{d}P_{\lambda, x_{\partial_+\mathbf{\Gamma}}}(x_{V^*(\mathbf{\Gamma})}) \tag{4.5.52}$$

where $P_{\lambda, x_{\partial_+\mathbf{\Gamma}}}(\cdot)$ is the Gibbs measure on $\cup_{\Gamma \in \mathbf{\Gamma}: \partial_+\Gamma = x_{\partial_+\mathbf{\Gamma}}} X(\Gamma)$ taken with respect to the boundary condition $x_{\partial_+\mathbf{\Gamma}}$. Then we can write

$$\int \prod_{t \in \Lambda} \varphi_t(x_t)\, \mathrm{d}P_{\lambda, a_n}(x_{\Lambda^n})$$

$$= \int \prod_{t \in \mathrm{ext}^*\mathcal{D} \cap \Lambda} \varphi_t(x_t) \prod_{\mathbf{\Gamma} \in \mathcal{D}: V^*(\mathbf{\Gamma}) \cap \Lambda \neq \emptyset} \varphi_\Lambda^*(\mathbf{\Gamma}, x_{\partial_+\mathbf{\Gamma}})\, \mathrm{d}P_{\lambda, a_n}(x_{\Lambda^n}) \tag{4.5.53}$$

where $\mathcal{D}$ is the collection of all $*$-external contours of x_{Λ^n}, and where $\mathrm{ext}^*\mathcal{D} = \Lambda^n \setminus \cup_{\mathbf{\Gamma} \in \mathcal{D}} V^*(\mathbf{\Gamma})$. Assuming that all φ_t differ only slightly from 1, we have the estimate

$$\varphi_\Lambda^*(\mathbf{\Gamma}, x_{\partial_+\mathbf{\Gamma}}) \leqslant \exp(c|\Lambda|) \tag{4.5.54}$$

with a small $c > 0$.

Proof. Immediate, using the well-known properties of Gibbs states. □

Denote the function within the integral on the right-hand side of (4.5.53) by $\varphi_\Lambda^*(x_{\Lambda^n})$. Notice that using Theorem 4.2.6 this can also be viewed as a function of $\mathcal{R}^+(\Lambda^n)$, with

$$\int \varphi_\Lambda^*(x_{\Lambda^n})\, \mathrm{d}P_{\lambda, a_n} = \int \varphi_\Lambda^*(r_{\Lambda^n})\, \mathrm{d}P_{\lambda, a_n, \mathrm{F}_\lambda}. \tag{4.5.55}$$

Of course, φ_Λ^* is not of the type considered in Lemma 2, but it can be approximated by those functions, except of some set of a small measure.

Given $m \in \mathbb{N}$ denote by $\mathcal{R}_m^+(\Lambda^n)$ the system of all $(x_{\Lambda^m}, \mathcal{D}^n) \in \mathcal{R}^+(\Lambda^n)$ which satisfy the condition

$$\mathbf{\Gamma} \in \mathcal{D}^n, V^*(\mathbf{\Gamma}) \cap \Lambda \neq \emptyset \Rightarrow \mathrm{dist}(t, \Lambda) \leqslant m\, \forall\, t \in V^*(\mathbf{\Gamma}). \tag{4.5.56}$$

For any realization $r_{\Lambda^n} = (x_{\Lambda^n}, \mathcal{D}) \in \mathcal{R}^+(\Lambda^n)$ define the approximating function

$$\int \prod_{t \in \mathrm{ext}^*\mathcal{D}^m \cap \Lambda} \varphi_t(x_t) \prod_{\mathbf{\Gamma} \in \mathcal{D}^m: V(\mathbf{\Gamma}) \cap \Lambda \neq \emptyset} \varphi_\Lambda^*(\mathbf{\Gamma}, x_{\partial_+\mathbf{\Gamma}})\, \mathrm{d}P_{\lambda, a_n \mathrm{F}_\lambda}(r_{\Lambda^n}) \tag{4.5.57}$$

where $\mathcal{D}^m$ is the collection of all $\mathbf{\Gamma} \in \mathcal{D}$ which satisfy the condition $\mathrm{dist}(t, \Lambda) \leqslant m$ for each $t \in V^*(\mathbf{\Gamma})$.

LEMMA 4. *Take Λ^n and a_n as in Lemma 2. Then*

$$\lim_{n \to \infty} P_{\lambda, a_n}(\varphi_\Lambda^*) \tag{4.5.58}$$

exists and does not depend on $\{\Lambda^n\}$ and $\{a_n\}$.

Proof. If we approximate

$$\int \varphi_\Lambda^*(x_{\Lambda_n})\,\mathrm{d}P_{\lambda,a_n}\left(= \int \varphi_\Lambda^*(r_{\Lambda^n})\,\mathrm{d}P_{\lambda,a_n,\mathbb{F}\lambda}\right)$$

by $\int \varphi_\Lambda^{*m}(r_{\Lambda^n})\,\mathrm{d}P_{\lambda,a_n,\mathbb{F}\lambda}$ then the existence of the limit follows from Lemma 2. On the other hand, Lemma 1 easily yields an inequality of the type

$$P_{\lambda,a_n,\mathbb{F}\lambda}(\mathscr{R}^+(\Lambda^n)\setminus\mathscr{R}_m^+(\Lambda^n)) \leqslant \exp((c-\rho)m) \tag{4.5.59}$$

where $c = c(\Lambda)$ is some constant. This proves Lemma 4. □

It is now clear from Lemma 4 and the Stone–Weierstrass theorem that the limit

$$\lim_n P_{\lambda,a_n}(f) \tag{4.5.60}$$

exists for each bounded measurable cylindrical function on X.

What remains now is to show that (4.5.60) really corresponds to some measure on X, and that the limit measure is a Gibbs one. Concerning the former property, it suffices to check some standard 'compactness criteria' for the existence of the Radon probability corresponding to a given 'cylindrical measure' (see [25]).

LEMMA 5. *For any $t\in\mathbb{Z}^\nu$ and any $\varepsilon > 0$ there is a compact set $\mathscr{V}\subset\mathbb{R}^k$ such that for any $\Lambda^n\ni t$, $a_n\in\tilde{\mathscr{U}}^{\partial\Lambda^{nc}}$*

$$P_{\lambda,a_n}\{x_{\Lambda^n}\in X^+(\Lambda^n): x_t\in\mathscr{V}\} \geqslant 1-\varepsilon. \tag{4.5.61}$$

Proof. (We will take $\mathscr{V}\supset\mathscr{U}$.) By Assumption 5 of the Main Lemma, we can find, for each frame Γ and each $x_{\partial_+\Gamma}\in(\tilde{\mathscr{U}})^{\partial_+\Gamma}$, a compact set $\mathscr{V} = \mathscr{V}(\Gamma, x_{\partial_+\Gamma})$ such that the event $x_t\notin\mathscr{V}$ conditioned with respect to a given $x_{\partial_+\Gamma}$ and with respect to the condition 'Γ is an $*$external frame of x_{Λ^n} and $t\in V^*(\Gamma)$, has a probability (in the conditioned Gibbs ensemble) smaller than ε. Using the uniform continuity of these conditioned Gibbs probabilities in $x_{\partial_+\Gamma}$. $\mathscr{V}$ can be chosen uniformly for all $x_{\partial_+\Gamma}$. By Lemma 1, only those Γ which do not 'exceed' t very much have to be checked. □

To prove that the limit measure (4.5.60) is a Gibbs measure we will use the following method. In addition to the ('diluted') ensemble $X^+(\Lambda)$ we will also consider the restricted ensemble ($\subset X^+(\Lambda)$)

$$\tilde{X}^+(\Lambda) = \{x_\Lambda\in X(\Lambda): x_t\in\tilde{\mathscr{U}} \text{ for each } t\in\partial_{\bar{r}}\Lambda^c\}. \tag{4.5.62}$$

We denote by $\tilde{P}_{\lambda,a}$ the measure $P_{\lambda,a}$ conditioned on $\tilde{X}^+(\Lambda)$.

THEOREM. *Let $\{\Lambda^n\}$ be a sequence of cubes*

$$\Lambda^n = \{t\in\mathbb{Z}^\nu: |t_i|\leqslant n; i = 1,2,\ldots,\nu\}.$$

Let $a_n = (x^+)_{\partial\Lambda^{nc}}$. Then the limit

$$\lim_{n\to\infty} \tilde{P}_{\lambda,a_n}(f) \tag{4.5.63}$$

exists for each cylindrical bounded measurable function f and equals the limit (4.5.60).

NOTE. It is a standard consequence of the theory of limit Gibbs states that the state (4.5.63), if it exists, is a Gibbs one. This completes also the proof of Proposition 4.2.1.

Proof of Theorem. Notice first that we have in fact proved the existence of the limit (4.5.60) in a rather strong sense: for dist $(0,(\Lambda^n)^c) \to \infty, a_n \in \tilde{\mathscr{U}}^{\partial(\Lambda^n)^c}$. Consider the following decomposition of $\tilde{X}^+(\Lambda)$. Given $x_\Lambda \in \tilde{X}^+(\Lambda)$ denote by $\tilde{\mathscr{D}}$ the collection of all external frames of x_Λ which satisfy the condition dist$(V(\Gamma), \partial_{\tilde{r}}\Lambda) \leqslant 1$. Denote by

$$\tilde{\Lambda}(x_\Lambda) = \Lambda \setminus (\partial_{\tilde{r}}\Lambda \cup \bigcup_{\Gamma \in \tilde{\mathscr{D}}} V(\Gamma)).$$

Denote by $\tilde{X}(\Lambda, \tilde{\Lambda}, \tilde{a})$ the collection of all $x_\Lambda \in \tilde{X}^+(\Lambda)$ with the same $\tilde{\Lambda}(x_\Lambda) = \tilde{\Lambda}$, such that $x_{\Lambda \setminus \tilde{\Lambda}} = \tilde{a}$ (where $\tilde{a} \in (\tilde{\mathscr{U}})^{\Lambda \setminus \tilde{\Lambda}}$). Conditioning $\tilde{P}_{\lambda,a}$ on $\tilde{X}^+(\Lambda, \tilde{\Lambda}, \tilde{a})$, we obtain $P_{\lambda,\tilde{a}}$. It suffices, therefore, to show that for any finite $M \subset \mathbb{Z}^\nu$,

$$\lim_n \tilde{P}_{\lambda, a_n}\Big(\bigcup_{\tilde{\Lambda} \supset M} \tilde{X}^+(\Lambda, \tilde{\Lambda}, \tilde{a}) \Big) = 1. \tag{4.5.64}$$

In other words, if we denote by $\bar{\mathscr{E}}_{M,n}$ the event $\{x_{\Lambda^n} \in \tilde{X}^+(\Lambda^n)$ iff there is some external frame of x_{Λ^n} such that $V(\Gamma) \cap M \neq \emptyset$, dist $(V(\Gamma), \partial_{\tilde{r}}\Lambda^n) \leqslant 1\}$, it suffices to prove that

$$\lim_n \tilde{\mathrm{P}}_{\lambda, a_n}(\bar{\mathscr{E}}_{M,n}) = 0. \tag{4.5.64$'$}$$

Clearly, some analogy of Lemma 2 of §4.5.2 is useful here. Denote by $\tilde{\mathscr{R}}^+(\Lambda^n) = \{r^n = (x^n, \mathscr{D}^n) \in \mathscr{R}^+(\Lambda^n), \; x_t \in \tilde{\mathscr{U}}^+$ for each $t \in \partial_{\tilde{r}}\Lambda^n \cap (\Lambda^n \setminus \cup_{\Gamma \in \mathscr{D}^n} V^*(\Gamma))$. One has to define a suitable frame with $\tilde{\mathbb{F}}_\lambda$ such that the probabilities of external frames both in $\tilde{X}^+(\Lambda^n)$ and $\tilde{\mathscr{R}}^+(\Lambda^n)$ would be the same. (A mere conditioning of $P_{\lambda, a_n, \mathbb{F}_\lambda}$ on $\mathscr{R}^+(\Lambda^n)$ does not give an equivalent ensemble!)

DEFINITION. Say that a frame Γ is *critical* if dist(supp $\Gamma, \Lambda^{nc}) \leqslant \tilde{r}$. Define a modified weight $\tilde{\mathbb{F}}_\lambda(\Gamma, a)$ as in (4.2.10), (4.2.17) but with

$$\tilde{Z}_\lambda(\Gamma, a) = -\ln \int_{\tilde{\mathscr{C}}'(\Gamma, a)} \exp(-F(\Gamma))\, d\Gamma_{(\operatorname{supp}\Gamma \setminus \partial_+\Gamma)}$$

instead of $Z_\lambda(\Gamma; a)$, where $\tilde{\mathscr{C}}(\Gamma, a)$ is the collection of those $\Gamma \in \mathscr{C}(\Gamma, a)$ only which can appear in $\tilde{X}^+(\Lambda^n)$. Clearly, $\tilde{\mathbb{F}} = \mathbb{F}$ for non-critical Γ.

It is easily seen (as in Theorem 4.2.6) that the ensembles $(\tilde{X}^+(\Lambda^n), \tilde{P}_{\lambda, a_n})$ and $(\tilde{\mathscr{R}}^+(\Lambda^n), \tilde{P}_{\lambda, a_n, \tilde{\mathbb{F}}_\lambda})$ are equivalent (where $\tilde{P}_{\lambda, a_n, \tilde{\mathbb{F}}_\lambda}$ is defined using the frame weight $\tilde{\mathbb{F}}_\lambda$).

One can analyse $\tilde{\mathbb{F}}(\Gamma, a)$ as in (4.2.23). There are no changes in the estimate of Δ_λ (notice for example that $\tilde{\mathbb{F}} = \mathbb{F}$ within $W(\Gamma)$). The lower bound obtained for $G_\lambda(\Gamma, a)$ is even better in this case because $\tilde{\mathscr{C}}(\Gamma, a) \subset \mathscr{C}(\Gamma, a)$. Thus, the condition that G_λ is a κ-functional can be established. The following lemma estimates the probability of a given frame in the ensemble $\tilde{\mathscr{R}}^+(\Lambda^n)$ (but we are really interested only in the external frames in the ensemble $\tilde{X}^+(\Lambda^n)$).

LEMMA 6. *Let Λ^n, $a_n = (x^+)_{\partial\Lambda^{nc}}$ be as in the theorem. Let Γ be a frame, denote by $\tilde{P}_{\lambda,n}(\Gamma)$ the probability (in $\tilde{P}_{\lambda, a_n, \tilde{\mathbb{F}}_\lambda}$) of the event '$\Gamma$ is a frame of $r_{\Lambda^n} \in \tilde{\mathscr{R}}^+(\Lambda^n)$'. Assume that $\tilde{\mathbb{F}}_\lambda$ is a κ-functional. Then, with some small $c > 0$,*

$$\tilde{P}_{\lambda,n}(\Gamma) \leqslant \exp((c - \kappa)|\operatorname{supp}\Gamma|). \tag{4.5.65}$$

Proof. (This is a coarser analogy of (4.5.16); the same is true of the proof.) As in

(4.5.18) we have

$$\tilde{P}_{\lambda,n}(\Gamma) \leqslant \exp(-\kappa|\operatorname{supp}\Gamma|)\tilde{Z}_\lambda(\Lambda^n, \operatorname{supp}\Gamma, a_n, \tilde{\mathbb{F}}_t)(\tilde{Z}_\lambda(\Lambda^n, a_n, \tilde{\mathbb{F}}_t))^{-1} \tag{4.5.66}$$

where $\tilde{Z}_\lambda$ are defined analogously to Z_λ but with the additional condition $x_t \in \tilde{\mathcal{U}}$ for $t \in \partial_{\tilde{r}}\Lambda^n \cap (\Lambda^n \setminus \cup_{\Gamma\in\mathcal{D}^n} V^*(\Gamma))$. We have also an analogy of (4.5.19):

$$|\ln \tilde{Z}_\lambda(\Lambda^n, \operatorname{supp}\Gamma, a_n, \tilde{\mathbb{F}}_\lambda) - \ln \tilde{Z}(\Lambda^n, a_n, \tilde{\mathbb{F}}_\lambda)| \leqslant \varepsilon|\operatorname{supp}\Gamma| \tag{4.5.67}$$

(the condition $a_n = (x^+)_{\partial(\Lambda^n)^c}$ is useful there; a mere condition $a_n \in (\tilde{\mathcal{U}})^{\partial(\Lambda^n)^c}$ would cause some difficulties). This proves (4.5.65). Clearly, for a large κ, the estimate (4.5.65) implies (4.5.64′). This concludes the proof of the theorem. □

4.5.7. *Concluding Notes; Unicity of the Construction*

It is also necessary to prove that the limit Gibbs states constructed in §4.5.6, which can be denoted by P_λ^+, P_λ^-, are really distinct.

Let Λ be a (large) cube, and let d be a (large) integer. Consider a cylindrical function f defined as follows: if there is some $\bar{\Lambda} \supset \Lambda$ such that $\operatorname{dist}(t, \Lambda) \leqslant d$ for each $t \in \bar{\Lambda}$ and if $x_{\bar{\Lambda}} \in X^\pm(\bar{\Lambda})$ then $f(x) = \pm 1$. In the opposite case, or if there is an ambiguity in the preceding prescription, $f(x) = 0$. It follows easily from Lemma 1, §4.5.6 that

$$\lim_{\Lambda \uparrow \mathbb{Z}^\nu} \lim_{d \to \infty} \int_X f(x)\,\mathrm{d}P_\lambda^\pm(x) = \pm 1 \qquad \square$$

It still remains to prove the statements (d) and (e) of the Main Lemma. The former statement follows easily from (4.4.2). The latter one is a consequence of the following result.

PROPOSITION. *Consider the limit Gibbs state* P_λ^+. *Assuming that* $\tau(\ln\delta)^{-\nu} \to \infty$, $\delta \to \infty, \varepsilon \to 0$ *(in the formulation of the Main Lemma), we have the asymptotic relation*

$$\lim_{\tau,\delta,\varepsilon} \int_X f(x)\,\mathrm{d}P_\lambda^\pm(x) = \int_X f(x)\,\mathrm{d}\mu^\pm(x) \tag{4.5.68}$$

for any bounded continuous cylindrical function f. ($\mu^\pm = \mu$ was defined in §4.4.1.)

Proof. We assume that $\tilde{r}$ is chosen such that $\tilde{r}\ (\ln\delta)^{-1} \to \infty$, $\tau(\tilde{r})^{-\nu} \to \infty$, $\delta^2(\tilde{r})^{-\nu} \to \infty$. It follows from (4.5.47) that instead of the left-hand side of (4.5.68) we can study the limit of

$$\lim_{\tau,\delta,\varepsilon}\left(\lim_n \int_{\mathcal{R}_+(\Lambda^n)} f(r_{\Lambda^n})\,\mathrm{d}P^\pm_{\lambda,a_n,\mathbb{F}_\lambda}(r_{\Lambda^n})\right) \tag{4.5.69}$$

where we identify f as a function

$$\{r_{\Lambda^n} = (x_{\Lambda^n}, \mathcal{D}^n) \rightsquigarrow f(x_{\Lambda^n})$$

on $\mathcal{R}^\pm(\Lambda^n)$. (The difference between $\int\Pi\varphi_t(x_t)$ and (4.5.57) disappears in the limit.) But the right-hand side of (4.5.49) gives $\int_{x(\Lambda^n)} f(x)\,\mathrm{d}\mu^\pm_{a_n}$ if $\varepsilon \to 0$, $\tilde{r}(\ln\delta)^{-1} \to \infty$, $\tau(\tilde{r})^{-\nu} \to \infty$, $\delta^2(\tilde{r})^{-\nu} \to \infty$ (see 4.4.1 for the definition of $\mu^\pm_{a_n} = \mu_{a_n}$). □

We now give some remarks about the unicity of the construction. It is clear that our construction of the solution (4.5.29) does not depend on the concrete choice of δ, $\tilde{\delta}$, $\tilde{r}$ or on the choice of the approximating potentials $\Phi^{\pm}$. To see the latter, note that our constructions in §4 can be carried over to the more general case when the balls $\mathscr{U}^{\pm}$, $\tilde{\mathscr{U}}^{\pm}$ are not centred exactly in $x^{\pm}$ (or more generally, when $\mathscr{U}^{\pm}$, $\tilde{\mathscr{U}}^{\pm}$ are measurable sets not differeing very much from the former balls). From (4.2.11) it is clear that for $\mathscr{U}^{\pm}$, $\tilde{\mathscr{U}}^{\pm}$ fixed, the contour models change only insignificantly (inside the external frames) while 'moving' $\Phi^{\pm}$.

What is by no means so obvious is the fact that there are no translation-invariant limit Gibbs states other than those corresponding to our contour models. It is reasonable to study this question in a more general setting as a question of the completeness of the PS construction, even in the case when there are more than two ground states, and also outside of the point of a maximal number of coexisting phases. This requires some new concepts and goes beyond the scope of this paper. (One must combine the methods of this paper and of [10].)

Another question which arises in the study of the limit

$$\lim_n P_{\lambda, a_n}, a_n \in (\tilde{\mathscr{U}}^{\pm})^{\partial\Lambda^{nc}}$$

is what happens if we replace $\tilde{\mathscr{U}}^{\pm}$ by a general bounded set $\hat{\mathscr{U}}$. If $\hat{\mathscr{U}} = \mathscr{U}^{+}$ or $\hat{\mathscr{U}} = \mathscr{U}^{-}$ then one expects the same limit Gibbs states as those arising for $\tilde{\mathscr{U}}^{+}, \tilde{\mathscr{U}}^{-}$. (The proof is omitted there.) This is not, however, the case if $\hat{\mathscr{U}}$ is a general bounded set. We do not investigate this situation. While it reasonable to expect the existence of limit Gibbs states (at least by taking a suitable subsequence of $\{n\}$) it is apparently possible that the limit Gibbs state is not translation-invariant. (If the condition of boundedness of $\hat{\mathscr{U}}$ is relaxed then many further non-translation invariant states can appear even in the gaussian approximation – see for example [17]). The study of non-translation-invariant Gibbs states obtained for a general bounded set $\hat{\mathscr{U}}$ is the subject of further research.

CONCLUDING NOTE (on the geometrical description of configurations in the limit Gibbs state). We obtain the following picture, characteristic of PS theory. In the limit Gibbs state, arising if the boundary condition is taken from $\tilde{\mathscr{U}}^{+}$, almost any configuration has the following behavior: it attains values from $\mathscr{U}^{+}$ except for some 'islands' which are scattered scarsely but uniformly throughout $\mathbb{Z}^{\nu}$. The distribution of these islands is determined by the distribution of external frames in the $+$contour model.

Acknowledgements

The authors would like to express their heartiest thanks to V. A. Malyshev, R. A. Minlos, S. A. Pirogov and Ya. G. Sinai for very useful discussions concerning the subject of this paper.

References

1. Dobrushin, R. L. and Shlosman, S. B.: 'Phases Corresponding to Minima of the Local Energy', *Select. Math. Sov.* 1, (1981), 317–338.

2. Glimm, I., Jaffe, A. and Spencer, T.: 'Phase Transitions for Quantum Field', *Commun. Math. Phys.* **45** (1975), 203–216.
Fröhlich, J., Israel, R., Lieb, E. and Simon, B.: 'Phase Transitions and Reflection Positivity; I. General Theory and Long Range Interactions', *Commun. Math. Phys.* **62** (1978), 1–34; 'II. Lattice Systems with Short Range Coulomb Interactions', *J. Statist. Phys.* **22** (1980), 297–347.
Fröhlich, J. and Lieb, E.: 'Phase Transitions in Anisotropic Lattice Spin Systems', *Commun. Math. Phys.* **60** (1978), 233–267.
3. Pirogov, S. A. and Sinai, Ya. G.: 'Phase Diagrams of Classical Lattice Systems' *Theor. Math. Phys.* **25** (1975), 1185–1192; *Theor. Math. Phys.* **26** (1976), 39–49.
4. Sinai, Ya. G.: *Theory of Phase Transitions: Rigorous Results*, Pergamon Press, Oxford, 1982.
5. Dobrushin, R. L. and Pecherski, E. A.: 'Uniqueness Conditions for Finitely Dependent Random Fields, in *Random Fields*, Vol. I, eds. I. Fritz, J. L. Lebowitz and D. Szasz, North-Holland, Amsterdam, 1981, 223–262.
6. Minlos, R. A. and Sinai, Ya. G.: '"Phase Separation" Phenomenon at Low Temperatures, in Some Lattice Models of a Gas, I', *Mat. Sb.*, **73** (1967), 375–448; *Trudy Mosk. Mat. Obsch.* **19** (1968), 113–178 (in Russian).
7. Glimm, J. and Jaffe, A.: *Quantum Physics, A Functional Integral Point of View*, Springer-Verlag, Berlin, 1981.
8. Malyshev, V. A.: 'Cluster Expansions in Lattice Models of Statistical Physics and the Quantum Theory of Fields', *Russian Math. Surveys* **35** (1980), 1–62.
9. Seiler, E.: *Gauge Theories as a Problem of Constructive Quantum Field Theory and Statistical Mechanics*. Lecture Notes in Physics **159**, Springer-Verlag, Berlin, 1982.
10. Zahradník, M.: 'An Alternate Version of Pirogov Sinai Theory', *Commun. Math. Phys.* **93** (1984) 559–581.
11. Imbrie, J. Z.: 'Phase Diagrams and Cluster Expansions for Low Temperature $P(\varphi)_2$ Models. I. The Phase Diagram', *Commun. Math. Phys.* **82** (1981), 261–304; 'II. The Schwinger Functions', *Commun. Math. Phys.* **82**, (1981), 305–344.
12. Malyshev, V. A. and Minlos, R. A.: *Gibbsian Fields: the Method of Cluster Expansions*, Nauka, Moscow, 1985 (in Russian).
13. Malyshev, V. A., Minlos, R. A., Petrova, E. N. and Terleckij, Yu. A.: 'Generalized Contour Models', in *Itogi Nauk. Techn.*, tom 19, VINITI, Moscow, 1982, 3–54 (in Russian).
14. Dinaburg, E. I. and Sinai, Ya. G.: 'Contour Models with Interactions and their Applications', *Proc. Conf. Math. Phys. Dubna*, 1984.
15. Bricmont, J., Kuroda, K. and Lebowitz, J. L.: 'First Order Phase Transitions in Lattice and Continuous Systems: Extension of Pirogov Sinai Theory', Preprint, 1984/85.
16. Pechersky, E. A.: 'The Peierls Condition (G.P.S. Condition) is not Always Satisfied', *Select. Math. Sov.* **3** (1983/84), 87–92.
17. Dobrushin, R. L.: 'Gaussian Random Fields – Gibbsian Point of View'; in *Multicomponent Random Systems*, eds. R. L. Dobrushin and Ya. G. Sinai, Marcel Dekker, New York, 1980, 119–152.
18. Doob, J. L.: *Stochastic Processes*, Wiley, New York, 1953.
19. Bellman, R.: *Introduction to Matrix Analysis*, McGraw-Hill, New York, 1960.
20. Simon, B.: *The $P(\varphi)_2$ Euclidean (Quantum) Field Theory*, Princeton University Press, Princeton, NJ, 1974.
21. Künsch, H.: 'Thermodynamics and Statistical Analysis of Gaussian random fields', *Z. Wahrsch. verw. Geb.* **58** (1981), 407–421.
22. Grenander, U. and Szegö, G.: *Toeplitz Forms and their Applications*, University of California Press, Berkeley, 1958.
23. Dobrushin, R. L. and Minlos, R. A.: Polynomes of Random Functions', *Uspehi Mat. Nauk.* **32** (1977), 67–122 (in Russian).
24. Dunea, M., Iagolnitzer, D. and Souillard, B.: 'Decay Properties of Truncated Correlation Functions and Analyticity Properties for Classical Lattice and Continuous Systems', *Commun. Math. Phys.* **31** (1973), 191–208.
25. Bourbaki, N. *Integration*, chapitre 9, livre VI, Hermann, Paris, 1969.

N. I. CHERNOV

Space-Time Entropy of Infinite Classical Systems

1. Introduction

We study a class of dynamical systems generated by the motion of infinitely many identical particles in $\mathbb{R}^d$, $d \geqslant 1$. The particles interact via a pair, finite-range, hard-core potential $U(r)$ which is more fully specified below. We require the mean density of particles to be sufficiently small. The existence of time evolution for such systems under various conditions on the function $U(r)$ has been proved by Lanford [1], Sinai [2, 3] and Presutti *et al.* [4]. We consider the entropy of the systems described.

Infinite classical systems possess the time-space group $\mathfrak{R} = \{S^u \circ T^t\}$ generated by the space translations S^u to a vector $u \in \mathbb{R}^d$ and the time evolution T^t, $t \in \mathbb{R}$ (this group acts only on the subset of the phase space where the time evolution exists). The translations S^u commute with T^t and the group $\mathfrak{R}$ is a $(d + 1)$-dimensional flow of automorphisms on the phase space (see below). The entropy of this flow was studied in [5] for an infinite ideal gas and in [6] for a system of infinitely many hard spheres. The time evolutions of infinite systems of particles typically have infinite (K–S) entropy (see [5]), as do their space translations. For this reason, in the study of infinite systems of particles we consider the entropy of the multi-parameter group $\mathfrak{R}$. The main result of the present chapter is that the space-time entropy of infinite classical systems is finite.

First we give some definitions and notation. The potential $U(q, q')$ depends on the distance between particles only: $U(q, q') = U(|q - q'|)$. We require that for the function $U(r)$ the following conditions are satisfied:

(U1) It has a 'hard core': $U(r) \equiv \infty$ for $0 < r \leqslant r_0$.
(U2) $U(r)$ is a c^2 function in (r_0, ∞).
(U3) It is of finite range: $U(r) \equiv 0$ for $r \geqslant r_1 > r_0$.
(U4) The following restrictions on the increase of $U(r)$ hold as $r \to r_0 + 0$:

$$\kappa^{(-)}(r - r_0)^{-\lambda(-)} < U(r) < \kappa^{(+)}(r - r_0)^{-\lambda(+)}$$

for some $\kappa^{(\pm)} > 0$ and $\lambda^{(\pm)} > 0$.

(U5) The following restrictions on the increasing of the first and the second derivatives of $U(r)$ hold as $r \to r_0 + 0$:

$$|U'(r)| < \kappa_1 |U(r)|^{\lambda_1}, |U''(r)| < \kappa_2 |U(r)|^{\lambda_2}$$

for some $\lambda_{1,2} > 0$.

We denote a particle $x = (q, v)$, where $q = (q_1, \ldots, q_d)$ is its position and $v = (v_1, \ldots, v_d)$ its momentum, i.e. the particle x is a point from $\mathbb{R}^{2d}$. The configuration of the system is an infinite subset $X \subset \mathbb{R}^{2d}$ such that for any two different

R. L. Dobrushin (ed.), Mathematical Problems of Statistical Mechanics and Dynamics, 125–137.

points $x = (q, v) \in X$ and $x' = (q', v') \in X$ holds $|q - q'| > r_0$, where r_0 is the hard-core diameter – see (U1). The set of the positions of all particles $x \in X$ is an infinite subset $Q = Q(X) \subset \mathbb{R}^d$. The phase space $\mathfrak{M}$ is the set of all configurations $\{X\}$.

The time evolution of a phase point $X \in \mathfrak{M}$ is formally described by the usual equations of motion:

$$\frac{dq}{dt} = v, \quad \frac{dv}{dt} = -\sum_{x' \neq x} \text{grad}_q\, U(|q - q'|) \tag{1}$$

(the masses of particles are equal to 1). The existence of their solutions will be discussed below.

A measurable structure is introduced onto $\mathfrak{M}$ in the usual way [7]. The equilibrium measure $\mathscr{P}_{\beta,\mu}$ on $\mathfrak{M}$ is called the Gibbs distribution (μ is the chemical potential and $\beta = (kT)^{-1}$, where T is the temperature). For every bounded region $V \subset \mathbb{R}^d$ denote by $\mathfrak{M}_V$ the set of particle configurations in V. Then $\mathfrak{M}_V = \mathfrak{M}_{V,0} \cup \mathfrak{M}_{V,1} \cup \cdots \cup \mathfrak{M}_{V,N(V)}$, where $\mathfrak{M}_{V,n}$ is the set of configurations in V containing n particles ($N(V) < \infty$ because of (U1)). For each fixed configuration $X(\bar{V})$ of particles outside V the conditional distribution on $\mathfrak{M}_{V,n}$ has the density

$$p(x_1, \ldots, x_n) = \Xi^{-1} \frac{1}{n!} \exp(-\beta H_V(X) + \beta\mu n), \tag{2}$$

where

$$H_V(X) = \sum_{x \in X(V)} \left(\frac{|v|^2}{2} + \sum_{\substack{x' \in X(V) \\ x' \neq x}} \tfrac{1}{2} U(|q - q'|) + \sum_{x' \in X(\bar{V})} U(|q - q'|) \right) \tag{3}$$

is the total energy of the configuration $X(V)$ under the 'boundary condition' $X(\bar{V})$ and

$$\Xi = 1 + \sum_{n=1}^{N(V)} \frac{1}{n!} \int_{V^n} \exp(-\beta H_V(X) + \beta\mu n)\, dx_1 \cdots dx_n$$

is the grand partition function.

We suppose that the mean particle density $\rho = \rho(\beta, \mu)$ is sufficiently small: $\rho < \rho_0(\beta)$. Sinai [3] has proved that under this condition the equations (1) define the time evolution $\{T^t\}$ on a subset $\mathfrak{M}' \subset \mathfrak{M}$ of full $\mathscr{P}_{\beta,\mu}$-measure for every β and μ. The time evolution and the space translations preserve the Gibbs measure $\mathscr{P}_{\beta,\mu}$. The subset $\mathfrak{M}' \subset \mathfrak{M}$ is explicitly described in [3]. More precisely, for every $X \in \mathfrak{M}'$ the particles fall into finite clusters and for a finite time clusters do not interact with each other.

Our main result is the following.

THEOREM. *The entropy $h(\mathfrak{R})$ of the group $\mathfrak{R}$ is finite and satisfies the following estimate:* $h(\mathfrak{R}) < \rho\, \text{const}(\beta)$.

We prove this theorem in §§2–4.

2. Statistical Estimates of the Gibbs Distribution

For every $L > 0$, $\Lambda_L = \{q \in \mathbb{R}^d : |q_i| < L, i = 1, 2, \ldots, d\}$ is the cube in $\mathbb{R}^d$. For each phase point X and $L > 0$ consider the time evolution $\{T_L^t\}$ in which the particles outside Λ_L are frozen, those inside Λ_L move pairwise interacting in the field of external particles and when they reach the boundaries of Λ_L they are elastically reflected. These groups $\{T_L^t\}$ are called partial flows (see [4]). For every $L > 0$ the partial flow $\{T_L^t\}$ preserves the Gibbs distribution $\mathscr{P}_{\beta,\mu}$ ([3]).

DEFINITION. For every $X \in \mathfrak{M}$ and $D > 0$ let $V_D(X) \subset \mathbb{R}^d$ be the union of all open balls of radius D centred in the positions $\{q\}$ of particles $x = (q, v) \in X$. The connectedness components of the set $V_D(X)$ are called *D-clusters* of the configuration X. The number of points $q \in Q(X)$ belonging to the D-cluster of X is called the *size* of this cluster.

For every particle $x = (q, v)$ denote by $x_t = (q_t, v_t)$ its T^t-evolution and by $x_t^{(n)} = (q_t^{(n)}, v_t^{(n)})$ its T_n^t-evolution. Denote

$$\mathfrak{M}'_{\tau,n,c} = \left\{ \sup_{|t| \leqslant \tau} \sup_{x \in X(\Lambda_n)} |v_t^{(n)}| \leqslant c\sqrt{\ln n} \right\}.$$

LEMMA 1. $1 - \mathscr{P}_{\beta,\mu}(\mathfrak{M}'_{\tau,n,c}) \leqslant \text{const}(\beta)\tau n^{-1/2\beta c^2 + d}$

This lemma is proved in [4]. □

Denote

$$\mathfrak{M}''_{n,c} = \left\{ \inf_{|t| \leqslant 1} \inf_{\substack{x \in X(\Lambda_n) \\ x' \in X, x' \neq x}} |q_t^{(n)} - q_t'^{(n)}| \geqslant r_0 + (c \ln n)^{-1/\lambda(-)} \right\},$$

where the particle x' is either moving or frozen.

LEMMA 2. $1 - \mathscr{P}_{\beta,\mu}(\mathfrak{M}''_{n,c}) \leqslant \text{const}(\beta) n^{-\beta c + 2d}$.

This lemma is proved in [4]. □

The inequality $|q - q'| \geqslant r_0 + (c \ln n)^{-1/\lambda(-)}$ together with condition (U4) implies $U(|q - q'|) \leqslant \kappa^{(+)} (c \ln n)^{\lambda(+)/\lambda(-)}$.

LEMMA 3. $1 - \mathscr{P}_{\beta,\mu} \{H_{\Lambda_n}(X) \leqslant n^{d+1}\} \leqslant \tilde{c}_0 n^d \rho \exp(-\tilde{c}_1 n)$, *where* $\tilde{c}_0$ *and* $\tilde{c}_1$ *depend on* $U(r)$ *and* β *only.*

Proof. Divide the cube Λ_n into small identical cubes with edge l, where l satisfies two conditions: $l \in (10r_1, 20r_1)$ and $2n/l$ is an integer. If $H_{\Lambda_n}(X) > n^{d+1}$ then for at least one of these cubes $\Lambda^{(i)}$ we have $H_{\Lambda^{(i)}}(X) > \text{const}(r_1)n$. The property (2) of the Gibbs distribution implies that for every cube V of fixed volume ($l < 20r_1$) and $A > 0$

$$\mathscr{P}_{\beta,\mu}\{H_V(X) > A\} \leqslant \text{const}(\beta)\exp(\mu)\exp(-\beta A).$$

This together with the obvious inequality $\exp(\mu) < \text{const}\,\rho$ proves the lemma. □

Let $r_2 > r_1/2$ be a fixed constant. Consider the set $\mathfrak{M}_\varepsilon = \{X \in \mathfrak{M}: Q(X) \cap \Lambda_\varepsilon \neq \emptyset\}$, where $\Lambda_\varepsilon \subset \mathbb{R}^d$, $\varepsilon = r_0/d$ is the small cube containing no more than one particle. For every point $X \in \mathfrak{M}_\varepsilon$ let K be the r_2-cluster containing this particle and N be the size of this cluster.

LEMMA 4. *If the mean density of particles* ρ *is small enough* ($\rho < \rho_1(\beta, r_2)$) *then for each* $m \geqslant 1$

$$\mathscr{P}_{\beta,\mu}(\{N = m+1\}/\mathfrak{M}_\varepsilon) \leqslant \bar{c}\exp(-\gamma m),$$

where $\bar{c} = \bar{c}(\beta, r_2) > 0$ *and* $\gamma = \gamma(\beta, r_2) > 0$.

Proof. We use the so-called correlation function (see for instance [8]). The m-particle correlation function $f_m(q_1, \ldots, q_m)$, $m \geqslant 1$, $q_1 \in \mathbb{R}^d, \ldots, q_m \in \mathbb{R}^d$ is the 'probability density' for finding m different particles at points $q_1, \ldots, q_m$. Ruelle [8] and Minlos [9] have proved the following estimate:

$$f_m(q_1, \ldots, q_m) \leqslant \zeta^m, \tag{4}$$

where ζ is a constant independent of m, $q_1, \ldots, q_m$ and $\zeta \to 0$ as $\rho \to 0$ and β is constant.

Let the cluster K contain $m+1$ particles. Join every pair of particles q, q' in the cluster K by a line segment if $|q - q'| < 2r_2$. We obtain a connected graph G with $m+1$ vertexes. Consider an arbitrarily chosen connected tree with $m+1$ vertexes being a subgraph of G. The number of non-isomorphic connected trees with $m+1$ vertexes is less than ψ^m, where ψ is a constant (see [10]). Integrating the correlation function f_{m+1} and using (4) we obtain

$$\mathscr{P}_{\beta,\mu}(\{N = m+1\}/\mathfrak{M}_\varepsilon) \leqslant [c(r_0)(r_2 - r_0/2)]^m \psi^m \zeta^m.$$

We can define the function $\rho_1(\beta, r_2)$ so that if $\rho < \rho_1$ then $c(r_2 - r_0/2)\psi\zeta < 1$, proving the lemma. □

For our purpose it is sufficient to fix an arbitrary constant $r_2 > r_1/2$, i.e. function $\rho_0(\beta)$ (see the Introduction) can be defined as $\rho_0(\beta) = \rho_1(\beta, r_1)$.

In the following lemma we estimate the velocities of all particles moving under the global flow $\{T^t\}$. Let $\mathfrak{M}'''_{t,n,c}$ be a set of configurations $X \in \mathfrak{M}$ such that

(a) $\sup\limits_{|t| \leqslant \tau} \sup\limits_{x: q_t \in \Lambda_n} |v_t| \leqslant c\sqrt{\ln n}$;

(b) $\sup\limits_{|t| \leqslant \tau} \sup\limits_{x: q_t \notin \Lambda_n} \dfrac{|v_t|}{\sqrt{\ln |q_t|}} \leqslant c.$

LEMMA 5. *Let* $\alpha \in (0,1)$ *be a fixed constant. Then*

$$1 - \mathscr{P}_{\beta,\mu}(\mathfrak{M}'''_{n^\alpha,n,c}) \leqslant \mathrm{const}(\beta,\mu,\alpha) n^{-1}$$

for some $c = c(\beta, \mu, \alpha) > 0$.

Proof. For every $m \geqslant 1$ let $\mathfrak{M}_{\tau,m,a,b}$ be a set of the configurations $X \in \mathfrak{M}$ such that

(a) $\sup\limits_{|t| \leqslant \tau} \sup\limits_{x \in X(\Lambda_m)} |v_t^{(m)}| \leqslant a\sqrt{\ln m}$;

(b) at every instant

$$t_i = \frac{r_2 - r_1/2}{a\sqrt{\ln m}}, \qquad |i| \leqslant \frac{\tau a\sqrt{\ln m}}{r_2 - r_1/2}$$

the size of each r_2-cluster of the configuration $T_n^{t_i} X$ intersecting the cube Λ_m does not exceed $b \ln m$.

From Lemmas 1 and 4 it follows that

$$1 - \mathscr{P}_{\beta,\mu}(\mathfrak{M}_{\tau,m,a,b}) \leqslant \text{const}(\beta,\mu)\tau(m^{-1/2\beta a2+2d} + \ln m \cdot m^{-\gamma b+d}).$$

Let $\tau = m^\alpha$ and a,b be chosen so that

$$1 - \mathscr{P}_{\beta,\mu}(\mathfrak{M}_{m^\alpha,m,a,b}) \leqslant \text{const}(\beta,\mu)m^{-2}.$$

From the Borel–Cantelli lemma, for almost every phase point X we have $X \in \mathfrak{M}_{m^\alpha,m,a,b}$ for all $m \geqslant m_0(X)$. It is easy to show that $\mathscr{P}_{\beta,\mu}\{m_0(X) \geqslant p\} \leqslant \text{const}\, p^{-1}$ for every $p \geqslant 1$. Let $m \geqslant \max\{m_0(X), c'\}$, where the constant $c' = c'(a,b,\alpha)$ is defined below. We want to show that for every configuration $X \in \mathfrak{M}_{m^\alpha,2m,a,b} \cap \mathfrak{M}_{m^\alpha,4m,a,b}$ the motion of particles inside Λ_m is the same under the partial flows $\{T_{2m}^t\}$ and $\{T_{4m}^t\}$ within the time interval $|t| \leqslant m^\alpha$. Let us assume that the motion of a particle $y \in \Lambda_{2m}$ is not the same under these flows. Then we can find a sequence of particles $y_1, \ldots, y_k$ and a sequence of time moments $0 < t_1 < t_2 < \cdots < t_k < m^\alpha$ with the following property. The particle y_1 at the moment $t = 0$ belongs to an r_2-cluster intersecting the boundary $\partial\Lambda_{2m}$, the particles y_i and y_{i+1} $(1 \leqslant i \leqslant k-1)$ belong to the same r_2-cluster at the moment t_i and the particles y_k and y belong to the same r_2-cluster at the moment t_k. According to the conditions (a) and (b) we can assume that $t_i = [(r_2 - r_1/2)/a\sqrt{\ln m}]j$ for some integer j. Therefore $k \leqslant \text{const}(a,b) \cdot m^\alpha \cdot \sqrt{\ln m}$ and the sizes of all above r_2-clusters do not exceed $b \ln m$. Then the trajectory of the particle y lies in the L-neighbourhood of the boundary $\partial\Lambda_{2m}$ within the time interval $|t| \leqslant m^\alpha$, where $L = \text{const}(a,b) \cdot m^\alpha \cdot (\ln m)^{3/2}$.

We might say that the freezing of particles outside Λ_{2m} spreads its influence within the time interval $|t| \leqslant m^\alpha$ only into the L-neighbourhood of the boundary $\partial\Lambda_{2m}$. Let us choose the constant c' so that $L < m/2$ for $m \geqslant c'$. Therefore for all $m \geqslant \max\{m_0(X), c'\}$ the motion of particles inside the cube Λ_m is the same under the partial flows $\{T_{2m}^t\}$ and $\{T_{4m}^t\}$. Then we have $\{m_0(X) \leqslant n\} \subset \mathfrak{M}'''_{n^\alpha,n,4a}$ for all $n > c'$, and the proof of the lemma is complete. □

3. Reduction to Partial Flows

We now begin the study of the entropy of the group $\mathfrak{R}$ (see the Introduction). First we give the necessary definitions and results, see [11]. Let Γ be a group of automorphisms on a measurable space M, $\Gamma \simeq \mathbb{Z}^p$, $p \geqslant 1$. Therefore the group Γ is generated by p commuting automorphisms $T_1, \ldots, T_p$. Let ξ be a measurable partition of M having finite entropy. Then by definition

$$h(\xi,\Gamma) = \lim_{i_1,\ldots,i_p \to \infty} \frac{1}{(2i_1)\cdot\cdots\cdot(2i_p)} H(\xi_{i_1,\ldots,i_p}), \tag{5}$$

where we denote

$$\xi_{i_1,\ldots,i_p} = \bigvee_{\substack{|j_1| \leqslant i_1 \\ |j_p| \leqslant i_p}} T_1^{j_1} \circ \cdots \circ T_p^{j_p} \xi$$

and $H(\varphi)$ is the entropy of the partition φ. Let $(i_1^{(k)}, \ldots, i_p^{(k)})$, $k = 1, 2, \ldots$ be a sequence of multi-indexes such that $i_m^{(k)} \to \infty$ as $k \to \infty$ for every $m = 1, \ldots, p$. Then the limit (5) taken over this sequence exists and is equal to the same quantity $h(\xi, \Gamma)$ (see [11]). The entropy of the group Γ is defined as $h(\Gamma) = \sup_\xi h(\xi, \Gamma)$. Let $G = \{F^u\}$, $u \in \mathbb{R}^p$ $(p \geqslant 1)$ be a continuous flow of automorphisms on M. Let $\{T^{u_1}, \ldots, T^{u_p}\}$ be a fixed basis in G. It generates a subgroup $\Gamma \subset G$, $\Gamma \simeq \mathbb{Z}^p$ and the entropy of the flow G is defined as $h(G) = h(\Gamma)$.

We fix the basis $\{S_1^1, \ldots, S_d^1, T^1\}$ in the flow $\mathfrak{R}$. Here S_i^1 is the unit space translation along the ith coordinate axis of $\mathfrak{R}^d$ and T^1 is the unit time evolution. This basis generates a subgroup $\Gamma_1 \subset \mathfrak{R}$ and we define $h(\mathfrak{R}) = h(\Gamma_1)$.

We use the following result from entropy theory (see [11, 12]). Let Γ be a group of automorphisms on M, $\Gamma \simeq \mathbb{Z}^p$, $p \geqslant 1$.

LEMMA 6. *Let $\xi_1 \leqslant \xi_2 \leqslant \ldots$ be an increasing sequence of measurable partitions of finite entropy such that $\xi_n \to \varepsilon$ as $n \to \infty$ (where ε is the partition into individual points). Then*

$$h(\Gamma) = \lim_{n \to \infty} h(\xi_n, \Gamma).$$

Now we introduce a special family of partitions of the phase space $\mathfrak{M}$. For each $\varepsilon \in (0, r_0/d)$

$$\Lambda_{0,\varepsilon} = \{q \in \mathbb{R}^d : 0 \leqslant q_i < \varepsilon, i = 1, 2, \ldots, d\}$$

is the cube in $\mathbb{R}^d$. Let v_k be an arbitrary partition of the momentum space into $(k-1)$ subsets $V_1, \ldots, V_{k-1}$. Let $A = \{X \in \mathfrak{M} : \Lambda_{0,\varepsilon} \cap Q(X) = \emptyset\}$ and for each $i = 1, 2, \ldots, k-1$ $B_i = \{X \in \mathfrak{M} : \Lambda_{0,\varepsilon} \cap Q(X) = \{q\}$ for some particle $x = (q, v)$ and $v \in V_i\}$. Then we have the partition $\zeta^{(0)}(\varepsilon, v_k)$ of $\mathfrak{M}$ into the subsets $\{A, B_1, B_2, \ldots, B_{k-1}\}$. Let for each vector $(i_1, \ldots, i_d) \in \mathbb{Z}^d$ $C_{i_1, \ldots, i_d} = \{X \in \mathfrak{M} : \Lambda_{0,\varepsilon} \cap Q(X) = \{q\}$ for some particle $x = (q, v)$ and $i_p\varepsilon \leqslant v_p \leqslant (i_p + 1)\varepsilon$ for each $p = 1, \ldots, d\}$. Then we have the partition $\zeta^{(0)}(\varepsilon)$ of $\mathfrak{M}$ into the subsets A and $C_{i_1, \ldots, i_d}$.

Suppose that $1/\varepsilon = l$ is an integer. For every $n \geqslant 1$ denote

$$\xi_n(\varepsilon, v_k) = \bigvee_{i_1, \ldots, i_d = -D}^{D-1} S_1^{i_1\varepsilon} \circ \cdots \circ S_d^{i_d\varepsilon} \xi^{(0)}(\varepsilon, v_k)$$

and

$$\xi_n(\varepsilon) = \bigvee_{i_1, \ldots, i_d = -D}^{D-1} S_1^{i_1\varepsilon} \circ \cdots \circ S_d^{i_d\varepsilon} \xi^{(0)}(\varepsilon),$$

where we denote $D = nl + [r_1 l]$ (r_1 is the range of the potential – see (U3)). By Lemma 6 we have

$$h = \sup_{\varepsilon, k, |v_k|} \lim_{m, n \to \infty} \frac{1}{2m(2n)^d} H\left(\bigvee_{|j| \leqslant m} T^j \xi_n(\varepsilon, v_k) \right). \tag{6}$$

Let us fix a constant $\alpha \in (0, 1)$ and denote in what follows $m = m(n) = [n^\alpha]$. For every $n \geqslant 1$ consider the cube Λ_n and inside that the smaller cube Δ_{n_0} for $n_0 = n - m \ln^2 n$. Denote by $\mathfrak{M}_n'$ the set of configurations $X \in \mathfrak{M}$ for which the motion of particles inside Λ_{n_0} under the partial flow $\{T_n^t\}$ and the global one $\{T^t\}$ within the time interval $|t| \leqslant m$ is the same.

LEMMA 7. *$\mathscr{P}_{\beta,\mu}(\mathfrak{M}'_n) \to 1$ as $n \to \infty$.*

Proof. Let $\mathfrak{M}''_n$ be the set of configurations $X \in \mathfrak{M}$ such that

(a) $$\sup_{|t| \leqslant m} \sup_{x: q_t \in \Lambda_n} |v_t| < a\sqrt{\ln n},$$

$$\sup_{|t| \leqslant m} \sup_{x \in X(\Lambda_n)} |v_t^{(n)}| < a\sqrt{\ln n},$$

$$\sup_{|t| \leqslant m} \sup_{x: q_t \notin \Lambda_n} \frac{|v_t|}{\sqrt{\ln |q_t|}} < a,$$

where a is the constant defined in Lemma 5;

(b) at each instant $t_i = i\delta$ ($|i| \leqslant m\delta^{-1}$), the sizes of all r_2-clusters of the configurations $T^{t_i}X$ and $T_n^{t_i}X$ intersecting the cube Λ_{2n} do not exceed $b \ln n$.

By virtue of Lemmas 4 and 5 there exists $b > 0$ such that $\mathscr{P}_{\beta,\mu}(\mathfrak{M}''_n)_{n\to\infty} \to 1$.

It remains to show that $\mathfrak{M}''_n \subset \mathfrak{M}'_n$ if n is large enough. Indeed, the conditions (a) and (b) imply in a manner completely analogous to the proof of Lemma 5 that the freezing of particles outside Λ_n spreads its influence within the time interval $|t| \leqslant m = [n^\alpha]$ only into the L-neighbourhood of the boundary $\partial\Lambda_n$ for $L = \text{const}(a, b)\cdot m\cdot(\ln n)^{3/2}$. Since $L < m\cdot(\ln n)^2$ for large enough n, the lemma is proved. □

Let $\zeta_n = \bigvee_{|j| \leqslant m} T^j \xi_n(\varepsilon, v_k)$ and $\zeta'_n = \bigvee_{|j| \leqslant m} T_n^j \xi_n(\varepsilon, v_k)$.

LEMMA 8. *If ε and v_k are fixed then $|H(\zeta_n) - H(\zeta'_n)| = o(mn^d)$ as $n \to \infty$.*

Proof. Let ψ be the partition of $\mathfrak{M}$ into two subsets: $\mathfrak{M}'_n$ and $\mathfrak{M}\setminus\mathfrak{M}'_n$ (see Lemma 7). Denote $\varphi_n = \zeta_n \vee \psi$ and $\varphi'_n = \zeta'_n \vee \psi$. It is clear that $|H(\zeta_n) - H(\varphi_n)| < \ln 2$, $|H(\zeta'_n) - H(\varphi'_n)| < \ln 2$ then Lemma 8 is equivalent to the relation $|H(\varphi_n) - H(\varphi'_n)| = o(mn^d)$ as $n \to \infty$. Denote $P = \mathscr{P}_{\beta,\mu}$ for convenience and write

$$H(\varphi_n) = -\sum_{\substack{\Delta\in\varphi_n\\ \Delta\subset\mathfrak{M}_n}} P(\Delta)\ln P(\Delta) - \sum_{\substack{\Delta\in\varphi_n\\ \Delta\not\subset\mathfrak{M}_n}} P(\Delta)\ln P(\Delta) = H_0 + H_1$$

and

$$H(\varphi'_n) = -\sum_{\substack{\Delta\in\varphi_n\\ \Delta\subset\mathfrak{M}_n}} P(\Delta)\ln P(\Delta) - \sum_{\substack{\Delta\in\varphi_n\\ \Delta\not\subset\mathfrak{M}_n}} P(\Delta)\ln P(\Delta) = H'_0 + H'_1.$$

Denote $p_n = 1 - \mathscr{P}_{\beta,\mu}(\mathfrak{M}'_n)$. Since the partitions φ_n and φ'_n contain no more than $k^{\text{const}\cdot m\cdot n^d}$ elements, then from Lemma 7 we have $H_0 \leqslant p_n \ln k^{\text{const}\cdot m\cdot n^d} = o(mn^d)$, Analogously, $H'_0 = o(mn^d)$. Consider the partitions

$$\varphi_{n,n_0} = \left(\bigvee_{|j|\leqslant m} T^j \xi_{n_0}(\varepsilon, v_k)\right) \vee \psi$$

and

$$\varphi'_{n,n_0} = \left(\bigvee_{|j| \leqslant m} T^j_n \xi_{n_0}(\varepsilon, v_k)\right) \vee \psi$$

for $n_0 = n - m \ln^2 n$. These partitions are identical within the set $\mathfrak{M}'_n$. For each element $\Delta_0 \in \varphi_{n,n_0}, \Delta_0 \subset \mathfrak{M}'_n$ the number of elements of the partitions φ_n and φ'_n within Δ_0 is not more than $k^{\text{const.}\, D_n}$, where we write $D_n = |\Lambda_n| - |\Lambda_{n_0}|$. Then we have $|H_1 - H'_1| \leqslant \text{const}\, D_n m = o(mn^d)$, proving the lemma. □

From Lemma 8 and (6) we have the relation

$$h = \sup_{\varepsilon, k, v_k} \lim_{n \to \infty} \frac{1}{2m(2n)^d} H\left(\bigvee_{|j| \leqslant m} T^j_n \xi_n(\varepsilon, v_k)\right), \tag{7}$$

i.e. in order to estimate h we may consider the partial flows $\{T^t_n\}$ only.

4. Estimate of Space-Time Entropy

For every $X \in \mathfrak{M}$ let $H_{\Lambda_n}(X)$ be the total energy (3) of the configuration $X(\Lambda_n)$ under the boundary condition $X(\bar{\Lambda}_n)$ and $N_{\Lambda_n}(X)$ be the number of particles inside Λ_n. These quantities are invariant under the partial flow $\{T^t_n\}$. Note that $N_{\Lambda_n}(X) \leqslant N_0 \cdot n^d$ where N_0 depends on r_0 and d only. Denote $\hat{\mathfrak{M}}_n = \{X \in \mathfrak{M} : H_{\Lambda_n}(X) \leqslant n^{d+1}\}$. By virtue of Lemma 3 $\mathscr{P}_{\beta,\mu}(\hat{\mathfrak{M}}_n) \to 1$ as $n \to \infty$. Let $\eta_n(\varepsilon)$ and $\eta_n(\varepsilon, v_k)$ be the partitions of $\mathfrak{M}$ coinciding with $\xi_n(\varepsilon)$ and $\xi_n(\varepsilon, v_k)$, respectively, within the set $\mathfrak{M}_n$ and containing the set $\mathfrak{M} \setminus \hat{\mathfrak{M}}_n$ as an element. Since the partition $\xi_n(\varepsilon, v_k)$ contains no more than $k^{\text{const}\cdot n^d}$ elements then

$$\begin{aligned} H\left(\bigvee_{|j| \leqslant m} T^j_n \xi_n(\varepsilon, v_k)\right) \\ = H\left(\bigvee_{|j| \leqslant m} T^j_n \eta_n(\varepsilon, v_k)\right) + o(mn^d). \end{aligned} \tag{8}$$

We can construct the sets $V_i (i = 1, \ldots, k-1)$ so that $\xi_n(\varepsilon) \geqslant \xi_n(\varepsilon, v_k)$. Setting $\varepsilon_n = \varepsilon/[\exp(n^\theta)]$ for some fixed $\theta \in (0, \alpha)$ we obtain $\eta_n(\varepsilon, v_k) \leqslant \eta_n(\varepsilon) \leqslant \eta_n(\varepsilon_n)$ and then

$$H\left(\bigvee_{|j| \leqslant m} T^j_n \eta_n(\varepsilon, v_k)\right) \leqslant H\left(\bigvee_{|j| \leqslant m} T^j_n \eta_n(\varepsilon_n)\right). \tag{9}$$

The direct calculation of the number of elements z_n of the partition $\eta_n(\varepsilon_n)$ gives an estimate

$$z_n \leqslant \exp(\text{const} \cdot n^{d+\theta}). \tag{10}$$

We need one known formula from entropy theory: if ξ and η are measurable partitions of finite entropy then we have (see for instance [12]) $H(\xi \vee \eta) = H(\eta) + H(\xi/\eta)$, where $H(\alpha/\beta)$ is the conditional entropy of the partition α relative to the partition β. Applying this formula n times we obtain $H(\xi \vee T\xi \vee \cdots \vee T^n\xi) = H(\xi) + \Sigma^n_{k=1} H(\xi/T\xi \vee \cdots \vee T^k\xi) \leqslant H(\xi) + nH(\xi/T\xi)$. This result together with (10) gives

$$H\left(\bigvee_{|j|\leqslant m} T_n^j \eta_n(\varepsilon_n)\right) \leqslant 2mH(\eta_n(\varepsilon_n)/T_n^1\eta_n(\varepsilon_n)) + o(mn^d). \tag{11}$$

For every element $\Delta \in \eta_n(\varepsilon_n)$ denote by N_Δ the number of elements of $\eta_n(\varepsilon_n)$ intersecting the set $T_n^1 \Delta$. The formulae (7)–(9) give the following estimate:

$$k \leqslant \sup_{\varepsilon} \lim_{n\to\infty} \sup (2n)^{-d} \sum_{\Delta\in\eta_n(\varepsilon_n)} \mathscr{P}_{\beta,\mu}(\Delta) \ln N_\Delta. \tag{12}$$

Let $\mathfrak{M}^{(0)}_{n,a,b}$ be the set of configurations $X \in \mathfrak{M}$ such that

(a) $\sup_{x:q\in\Lambda_n} |v| \leqslant a\sqrt{\ln n}$;

(b) $\sup_{x:q\in\Lambda_n} \sup_{x\neq x} U(|q-q'|) \leqslant b(\ln n)^\lambda$ for $\lambda = \lambda^{(+)}/\lambda^{(-)}$ (see the Condition (U4)).

Denote by $\mathfrak{M}_{n,a,b} = \cap_{t\in[-1,0]} T_n^t \mathfrak{M}^{(0)}_{n,a,b}$. By virtue of Lemmas 1 and 2

$$1 - \mathscr{P}_{\beta,\mu}(\mathfrak{M}_{n,a,b}) \leqslant \text{const}\cdot n^{-1} \tag{13}$$

for some $a > 0$ and $b > 0$. Note that $\mathfrak{M}_{n,a,b} \subset \hat{\mathfrak{M}}_n$ for all sufficiently large n. The set of elements of the partition $\eta_n(\varepsilon_n)$ can be divided into two groups: $\zeta_0 = \{\Delta\in\eta_n(\varepsilon_n): \Delta\cap\mathfrak{M}_{n,a,b} = \emptyset\}$ and $\zeta_1 = \eta_n(\varepsilon_n)\setminus\zeta_0$. The relations (10) and (13) imply the estimate

$$-\sum_{\Delta\in\zeta_0} \mathscr{P}_{\beta,\mu}(\Delta)\ln N_\Delta \leqslant \text{const.}\, n^{-1} n^{d+\theta} = o(n^d). \tag{14}$$

Now we study the time evolution of a set $\Delta\in\zeta_1$ under the partial flow $\{T_n^t\}$. Let K_Δ be the number of particles inside Λ_{n+r_1} (it is obviously the same for all configurations $X\in\Delta$ and its images $T_n^t X$). Taking the coordinates of these particles only we can conceive a set $\Delta\in\zeta_1$ as a cube in $\mathbb{R}^{2dk_\Delta}$ with edge ε_n. Let us introduce a special metric dist on $\mathfrak{M}$. For every $X\in\mathfrak{M}$ the configuration $X(\Lambda_{n+r_1})$ is a finite subset of $\mathbb{R}^{2d}$ (we consider each particle as a point of $\mathbb{R}^{2d}$). The distance between $X\in\mathfrak{M}$ and $X'\in\mathfrak{M}$ is defined as the Hausdorf distance between the subsets $X(\Lambda_{n+r_1})$ and $X'(\Lambda_{n+r_1})$ in $\mathbb{R}^{2d}$, generated by the metric $r(x,y) = \max_{1\leqslant i\leqslant 2d}|x_i - y_i|$ on $\mathbb{R}^{2d}$. Then the diameter of every set $\Delta\in\zeta_1$ is equal to ε_n. Our aim is to estimate the diameter of its image $T_n^1\Delta$.

Consider an arbitrary closed system of N particles moving in accordance with the equations (1). Its phase space M is a subset of $\mathbb{R}^{2dN}$ and the time evolution $\{T_M^t\}$ is a smooth flow on M. The total energy U of this system is invariant under $\{T_M^t\}$.

LEMMA 9. *Let $dT_M^t(X)$ be the differential of the mapping $T_M^t: M\to M$ at the point $X\in M$. Then*

$$\|dT_M^t(X)\| \leqslant 1 + t\hat{c}\max\{1, U^{\lambda_3}\} + o(t)$$

as $t\to 0$. Here we denote

$$\lambda_3 = \max\{\lambda_1,\lambda_2\} \quad \textit{and} \quad \hat{c} = \max\left\{1, \left(\frac{1}{r_0}\kappa_1 + \frac{1}{2}\kappa_2\right)d^2\left(\frac{2r_1}{r_0}\right)^{2d}\right\}.$$

Proof. Let $(q_1^{(i)},\ldots,q_d^{(i)})$ be the position of the ith particle and $(v_1^{(i)},\ldots,v_d^{(i)})$ its momentum. Then $q_j^{(i)}$ and $v_j^{(i)}$ are $2dN$ coordinates of the point $X\in M$. Setting $Q = (q_1^{(1)},\ldots,q_d^{(N)})$ and $V = (v_1^{(1)},\ldots,v_d^{(N)})$ we can write $X = (Q,V)$. Let $(Q^t,V^t) =$

$F(Q,V,t)$ be a solution of (1) with initial conditions $F(Q,V,0) = (Q,V)$. The equations (1) can be written in the form

$$\frac{\partial F(Q,V,t)}{\partial t} = \left(V, -\frac{\partial H}{\partial Q}\right)$$

(here $\partial H/\partial Q = \text{grad}_Q H(Q,V)$), where

$$H(X) = \sum_{x \in X} \frac{|v|^2}{2} + \sum_{q' \neq q''} U(|q' - q''|)$$

is the Hamiltonian of the system. Then we have

$$F(Q,V,t) = (Q,V) + \left(V, -\frac{\partial H}{\partial Q}\right)t + o(t)$$

and, consequently,

$$dT_M^t(X) = \left(\frac{\partial F(Q,V,t)}{\partial Q}, \frac{\partial F(Q,V,t)}{\partial V}\right) = E + At + o(t),$$ where E is the unit matrix and

$$A = \begin{pmatrix} 0 & E \\ -\dfrac{\partial^2 H}{\partial Q^2} & 0 \end{pmatrix}.$$

Therefore

$$\|dT_M^t(X)\| = 1 + t \max\left\{1, \left\| \frac{\partial^2 H}{\partial Q^2} \right\|\right\} + o(t).$$

Every particle interacts with no more than $K_0 = (2r_1/r_0)^d$ other particles at the same time. Then each line and each column of the matrix $\partial^2 H/\partial Q^2$ contains no more than $K_0 d$ non-zero elements.

The direct calculations of the elements of this matrix together with (U5) give

$$\left\| \frac{\partial^2 H}{\partial Q^2} \right\| \leqslant K_0^2 \left(\frac{1}{2}\kappa_2 + \frac{1}{r_0}\kappa_1\right) U^{\lambda_3}.$$

This proves the lemma. □

Consider the evolution of a set $\Delta \in \zeta_1$ under the partial flow $\{T_n^t\}$ for $t \in [0,1]$. We suppose in what follows that n is large enough: $n \geqslant n_0(a,b,\beta,\mu)$. Let $X_0 \in \Delta \cap \mathfrak{M}_{n,a,b}$ be a fixed configuration (it exists because $\Delta \in \zeta_1$). It is clear that $T_n^t X_0 \in \mathfrak{M}_{n,a,b}^{(0)}$ for all $t \in [0,1]$. We want to show that for all $t \in [0,1]$

$$T_n^t \Delta \subset \mathfrak{M}_{n,2a,2b}^{(0)}, \qquad \text{diam}\, T_n^t \Delta \leqslant \varepsilon \exp(-\tfrac{1}{2} n^\theta). \tag{15}$$

If $t = 0$ then it follows from the definition of $\mathfrak{M}_{n,2a,2b}^{(0)}$ and from the relation $\text{diam}\,\Delta = \varepsilon_n$. Suppose that the relations (15) hold for all $t \in [0,t_0]$, $t_0 < 1$. Then Lemma 9 implies

$$\|dT_n^{t_0}(X)\| \leqslant \exp(t_0 \hat{c}(b(\ln n)^\lambda)^{\lambda_3})$$

for every $X \in \Delta$. Because of the connectedness of the set $T_n^t \Delta$ we have

$$\operatorname{diam} T_n^t \Delta \leqslant \varepsilon_n \exp(t_0 \hat{c}(b(\ln n)^{\lambda})^{\lambda_3}) \leqslant \varepsilon \exp(-\tfrac{1}{4} n^{\theta}).$$

This implies (15) for some $t > t_0$. Suppose that the relations (15) hold for all $t \in [0, t_0)$. Then we obtain (15) for $t = t_0$ due to the continuity of the partial flow $\{T_n^t\}$. Thus the relations (15) are proved for all $t \in [0,1]$.

Let $X \in \Delta$ be an arbitrary configuration. According to (15) there exists a bijection $F_{X_0,X}: X_0(\Lambda_{n+r_1}) \to X(\Lambda_{n+r_1})$ such that for every $t \in [0,1]$ and $x \in X(\Lambda_{n+r_1})$

$$r(T_n^t x, T_n^t(F_{X_0,X} x)) \leqslant \varepsilon \exp(-\tfrac{1}{2} n^{\theta}) \tag{16}$$

(here $T_n^t x$ is the trajectory of the particle x in $\mathbb{R}^{2d}$).

Consider the time instants $t_i = i\delta$ for $i = 0, 1, \dots I = [\delta^{-1}]$ and $\delta = (r_2 - r_1/2)/8a\sqrt{\ln n}$. For each $i = 0, 1, \dots, I$ divide all particles of the configuration $T_n^{t_i} X_0(\Lambda_{n+r_1})$ into r_2-clusters. The bijection $F_{X_0,X}$ induces a partition into some groups of the particles of the configuration $T_n^{t_i} X(\Lambda_{n+r_1})$ for every $X \in \Delta$. By virtue of (16) these groups do not interact with each other within the time interval $(i\delta, i\delta + \delta)$. Consequently, the evolution $\{T_n^t\}$ within this time interval is the direct product of the evolutions of individual groups.

Let $y \in X_0(\Lambda_{n+r_1})$ be an arbitrary particle. Denote by $K_{y,i}$ the r_2-cluster of the configuration $T_n^{t_i} X_0$ containing the particle y and denote by $U_{y,i}$ the total energy of this cluster ($i = 0, 1, \dots, I$). For each $X \in \Delta$ consider the group of particles of configuration $T_n^{t_i} X$ corresponding to the cluster $K_{y,i}$ under the bijection $F_{X_0,X}$. Its total energy is less than $2U_{y,i}$ because of (15). Denote $\hat{U}_{y,i} = \max\{1, U_{y,i}\}$. Using Lemma 9 we get

$$r(T_n^1 y, T_n^1(F_{X_0,X} y)) \leqslant \varepsilon_n \exp\Big(2\hat{c}\delta \sum_{i=1}^{I} (2\hat{U}_{y,i})^{\lambda_3}\Big).$$

Therefore

$$N_\Delta \leqslant \exp\left(\text{const} \cdot \delta \cdot \sum_{y \in X_0(\Lambda_{n+r_1})} \sum_{i=0}^{\mathcal{T}} \hat{U}_{y,i}^{\lambda_3}\right). \tag{17}$$

Let $K_{\Delta,i,p}$, $p = 1, \dots, P_{\Delta,i}$ be all the r_2-clusters of the configuration $T_n^{t_i} X_0 (X_0 \in \Delta, \Delta \in \zeta_1)$. Here $P_{\Delta,i}$ is the number of these clusters and the numeration is arbitrary. Let $N_{\Delta,i,p}$ be the number of particles in the cluster $K_{\Delta,i,p}$ and $U_{\Delta,i,p}$ be its total energy (the latter quantity depends on the choice of the configuration $X_0 \in \Delta$ but this dependence is negligible because of (16)). Combining (12), (14) and (17) we get

$$k \leqslant \sup_{\varepsilon} \limsup_{n \to \infty} \Big[\text{const} \cdot \sum_{\Delta \in \zeta_1} \mathscr{P}_{\beta,\mu}(\Delta) \delta \times \\ \times \Big(\sum_{i=0}^{I} \sum_{p=1}^{P_{\Delta,i}} N_{\Delta,i,p} \hat{U}_{\Delta,i,p}^{\lambda_3} \Big) \Big]. \tag{18}$$

For every configuration $X \in \mathfrak{M}$ denote by $K_p (p = 1, \dots, P(X))$ all r_2-clusters intersecting the cube Λ_{n+r_1}. Let N_p be the number of particles in the cluster K_p, and

U_p be its total energy. Denote $\hat{U}_p = \max\{1, U_p\}$. The inequality (18) implies

$$h \leqslant \text{const} \cdot \lim_{n \to \infty} (2n)^{-d} \left\langle \sum_{p=1}^{P(X)} N_p \hat{U}_p^{\lambda_3} \right\rangle, \tag{19}$$

where $\langle\ \rangle$ means the expectation with respect to the Gibbs measure $\mathscr{P}_{\beta,\mu}$. The constant in (19) depends on the function $U(r)$ and do not on β and μ.

Consider the small cube $\Lambda_{\varepsilon_0}, \varepsilon_0 = r_0/d$ in $\mathbb{R}^d$. Let $N = N(X)$ be the number of particles in the r_2-cluster containing the particle $x = (q, v)$ with $q \in \Lambda_{\varepsilon_0}$ and $U = U(X)$ be the total energy of this cluster (if $\Lambda_{\varepsilon_0} \cap Q(X) = \emptyset$ we set $N(X) = U(X) = 0$). Let

$$\hat{U}(X) = \begin{cases} \max\ \{1, U(X)\} & \text{if} \quad U(X) \neq 0 \\ 0 \ \text{otherwise.} \end{cases}$$

Then (19) implies the following estimate: $h \leqslant \text{const} \langle N \hat{U}^{\lambda_3} \rangle$. The proof of the main theorem is completed by the following result.

LEMMA 10. *The quantity* $\langle N \hat{U}^{\lambda_3} \rangle$ *is finite and satisfies the estimate* $\langle N \hat{U}^{\lambda_3} \rangle \leqslant \text{const} \cdot (\beta) \cdot \rho$.

Proof. First we write $\langle N \hat{U}^{\lambda_3} \rangle \leqslant \langle N^2 \rangle + \langle \hat{U}^{2\lambda_3} \rangle$. Denote $\mathfrak{M}_{\varepsilon_0} = \{Q(X) \cap \Lambda_{\varepsilon_0} \neq \emptyset\}$ and $p_0 = \mathscr{P}_{\beta,\mu}(\mathfrak{M}_{\varepsilon_0})$. Obviously $p_0 \leqslant \text{const}(\beta) \cdot \rho$. Then Lemma 4 implies the estimate $\langle N^2 \rangle \leqslant \text{const}(\beta) \cdot p_0 \leqslant \text{const}(\beta) \cdot \rho$.

It remains to obtain the same estimate for the quantity $\langle \hat{U}^{2\lambda_3} \rangle$. For every $A > 1$ let $\mathfrak{M}_A$ be the set of all configurations $X \in \mathfrak{M}$ such that (a) $N(X) \leqslant n$ and (b) $H_{\Lambda_\mu}(X) \leqslant A$, where $u = A^{1/(d+1)}$ and $n = u/2r_1$. According to Lemmas 3 and 4

$$1 - \mathscr{P}_{\beta,\mu}(\mathfrak{M}_A / \mathfrak{M}_{\varepsilon_0}) \leqslant \text{const}(\beta) \cdot A \exp(-\text{const}(\beta) \cdot A^{1/(d+1)}).$$

If $X \in \mathfrak{M}_A$ then the size of the r_2-cluster of the configuration X containing the particle $x = (q, v)$ with $q \in \Lambda_{\varepsilon_0}$ is not greater than n because of (a). Then this cluster lies in the cube Λ_u and its total energy is less than A because of (b). Consequently,

$$1 - \mathscr{P}_{\beta,\mu}(\{U(X) \geqslant A\} / \mathfrak{M}_{\varepsilon_0}) \leqslant \text{const}(\beta) \cdot A \exp(-\text{const}(\beta) \cdot A^{1/(d+1)}).$$

Therefore we have $\langle \hat{U}^{2\lambda_3} \rangle \leqslant \text{const}(\beta) \cdot p_0 \leqslant \text{const}(\beta) \cdot \rho$, completing the proof. □

Acknowledgements

I would like to thank Professor Ya.G.Sinai for very valuable discussions and helpful suggestions. I also wish to thank V. A. Chulaevskij for very helpful discussions.

References

1. Lanford O. E. III, 'Time Evolution of Large Classical Systems.' In *Lecture Notes in Physics.* **38**, Springer-Verlag, Berlin, pp. 1–111.
2. Sinai, Ya. G.: 'The Construction of Dynamics in One Dimensional Systems of Statistical Mechanics', *Sov. Theor. Math. Phys.* **12** (1972), 487–497.
3. Sinai, Ya. G.: 'The Construction of Cluster Dynamics for Dynamical Systems of Statistical Mechanics,' *Vest. Moscow Univ.* **1** (1974), 152–158.
4. Presutti, E., Pulvirenti, M., and Tirozzi, B.: 'Time Evolution of Infinite Classical Systems with Singular, Long Range, Two Body Interactions', *Commun. Math. Phys.* **47** (1976), 81–94.

5. Goldstein, S.: 'Space-Time Ergodic Properties of Systems of Infitely Many Independing Particles', *Commun. Math. Phys.* **39** (1975), 303–328.
6. Sinai, Ya. G. and Chernov, N. I.: 'Entropy of Gas of Hard Spheres with Respect to the Group of Time-Space Translations. *Trudy Sem. im. I. G. Petrovsky* **8** (1982), 218–238 (in Russian).
7. Minlos, R. A.: 'Gibbs' Limit Distribution', *Funkts. Analiz.*, **1** (2) (1967), 60–73 (in Russian).
8. Ruelle, D.: 'Superstable Interactions in Classical Statistical Mechanics', *Commun. Math. Phys.* **18** (1970), 127–159.
9. Minlos, R. A.: 'The Regularity of Gibbs' Limit Distribution', *Funkts. Analiz.*, **1** (3) (1967), 40–53 (in Russian).
10. Otter, R.: 'The Number of Trees', *Ann. Math.* **49** (1948), 583–599.
11. Conze, J. P.: 'Entropie d'un Groupe abelien de Transformations', *Z. Wahrsch. Verw. Geb.* **25** (1972), 11–30.
12. Billingsley, P.: *Ergodic Theory and Information*, Wiley, New York; 1965.

R. A. MINLOS and A. I. MOGILNER

Spectrum Analysis and Scattering Theory for a Three-Particle Cluster Operator

1. Introduction. A General Definition of the Cluster Operator

Consider the symmetric Fock space $F = \mathrm{Exp}^{\mathrm{sym}}\{L_2(T^\nu, d\lambda)\}$ over the space $L_2(T^\nu, d\lambda)$, i.e. the symmetrized tensor exponent [1] of the space $L_2(T^\nu, d\lambda)$ where T^ν is a ν-dimensional torus and $d\lambda$ is the Haar measure on T^ν which is normalized in the following way: $\lambda(T) = (2\pi)^\nu$. Consider the following standard representation:

$$F = \bigoplus_{h=0}^{\infty} L_2^n \tag{1.1}$$

where $L_2^0 = \mathbb{C}^1$ and $L_2^n = L_2^{\mathrm{sym}}((T^\nu)^n, (d\lambda)^n)$, $n \geqslant 1$ is the Hilbert space of symmetric functions $f(\lambda_1, \ldots, \lambda_n)$ of n variables $\lambda_i \in T^\nu$, $i = 1, \ldots, n$. Then any operator A acting in F is represented by a matrix

$$A = \|A_{m,n}\|,\ A_{m,n}\colon L_2^n \to L_2^m,\ m, n = 0, 1, \ldots$$

Suppose that each operator $A_{m,n}$ is given by the formula

$$(A_{m,n}f)(\lambda_1, \ldots, \lambda_m) = \int_{(T^\nu)^n} K_{m,n}(\lambda_1, \ldots, \lambda_m, \mu_1, \ldots, \mu_n) \cdot f(\mu_1, \ldots, \mu_n)\, d\mu_1 \cdots d\mu_n \tag{1.2}$$

where $f \in L_2^n$ and the kernel $K_{m,n}$ is a distribution of the variables $\{\lambda_1, \ldots, \lambda_m\}$ and $\mu_1, \ldots, \mu_n$ and which is symmetric in $\lambda - s$ and $\mu - s$.

Further, we assume that $K_{m,n}$ can be represented in the form

$$K_{m,n}(\lambda_1, \ldots, \lambda_m; \mu_1, \ldots, \mu_n) = \sum_{\gamma \in \mathscr{I}_{m,n}} a_\gamma(\lambda_1, \ldots, \lambda_m; \mu_1, \ldots, \mu_n)\delta_\gamma \tag{1.3}$$

where the summation is over the set $\mathscr{I}_{m,n}$ of all partitions $\gamma = (\alpha_1, \ldots, \alpha_s)$ of the pair of sets $M = (1, \ldots, m)$ and $N = (1, \ldots, n)$ into the non-empty pairs

$$\alpha_i = (\beta_i, \beta_i'),\ \beta_i \subseteq M,\ \beta_i' \subseteq N,\ \beta, \beta_i' \neq \emptyset, \qquad i = 1, \ldots, s \tag{1.4}$$

and

$$\delta_\gamma = \prod_{i=1}^{s} \delta\Big(\sum_{j \in \beta_i} \lambda_j - \sum_{j' \in \beta_i'} \mu_{j'}\Big)$$

($\delta(\cdot)$ is a δ-function on the torus T^ν).

R. L. Dobrushin (ed.), Mathematical Problems of Statistical Mechanics and Dynamics, 139–160.

We suppose further that $a_\gamma(\lambda_1,\dots,\lambda_m;\ \mu_1,\dots,\mu_n)$ is from the class $C^p(\Gamma_\gamma)$ (p is fixed) where the manifold Γ_γ is defined in the following way:

$$\Gamma_\gamma = \left\{(\lambda_1,\dots,\lambda_m;\mu_1,\dots,\mu_n): \sum_{j\in\beta_i}\lambda_j = \sum_{j'\in\beta_i'}\mu_{j'}, i=1,\dots,s\right\}. \tag{1.5}$$

Let $\tau(\sigma,\sigma')$ be the map of $\mathcal{I}_{m,n}$ into itself induced by the maps $\sigma: M\to M$, $\sigma': N\to N$. The symmetry of the kernel $K_{m,n}$ in the variables $\lambda_1,\dots,\lambda_m$ and $\mu_1,\dots,\mu_n$ means that

$$\begin{aligned} &a_\gamma(\lambda_{\sigma(1)},\dots,\lambda_{\sigma(m}; \mu_{\sigma'(1)},\dots,\mu_{\sigma'(n)}) = \\ &= a_{\tau^{-1}(\sigma,\sigma')\gamma}(\lambda_1,\dots,\lambda_m;\mu_1,\dots,\mu_n). \end{aligned} \tag{1.6}$$

LEMMA 1. *The operator $A_{m,n}: L_2^n\to L_2^m$ given by the formulas* (1.2)–(1.4) *is bounded. Its norm can be estimated by the inequality*

$$\|A_{m,n}\| \leqslant C_{m,n} \max_{\substack{\{\lambda_1,\dots,\lambda_m,\mu_1,\dots,\mu_n\}\\ \gamma\in\mathcal{I}_{m,n}}} |a_\gamma(\lambda_1,\dots,\lambda_m;\mu_1,\dots,\mu_n) \tag{1.7}$$

where $C_{m,n}$ is an absolute constant.

The proof follows from the Causchy–Bynakovsky inequality. A bounded operator acting in F in such a way that its blocks $A_{m,n}$ have the form (1.2)–(1.4), is called a *cluster operator*, and the functions $\{a_\gamma, \gamma\in\mathcal{I}_{m,n}, m,n=0,1\dots\}$ are called its *cluster functions*. Note that the representation of the kernel $K_{m,n}$ in the form (1.3) is unique and so the cluster functions a_γ are uniquely defined. Note also that each cluster operator commutes with the group U_t of unitary operators acting in F:

$$\begin{aligned} (U_t f)_m(\lambda_1,\dots,\lambda_m) &= \exp\{i(t,\lambda_1+\cdots+\lambda_m)\}.\ f(\lambda_1,\dots,\lambda_m), m>0,\\ U_t f_0 &= f_0,\quad t\in\mathbb{Z}^\nu. \end{aligned} \tag{1.8}$$

In the case when the operator A is self-adjoint then for all $\gamma\in\mathcal{G}_{m,n}$,

$$a_\gamma(\lambda_1,\dots,\lambda_m;\mu_1,\dots,\mu_n) = \overline{a_{\gamma^1}(\mu_1,\dots,\mu_n;\lambda_1,\dots,\lambda_m)} \tag{1.9}$$

where the partition $\gamma'\in\mathcal{G}_{n,m}$ is obtained from the partition $\gamma\in\mathcal{G}_{m,n}$ by the transposition of all pairs $\alpha_i\in\gamma$. The cluster operators acting in the space $L_2^k\subset F$ (which will be called k-particle cluster operators) are defined in the same way. The cluster operators (or k-particle cluster operators) appear in many problems of quantum physics [2]; hence the spectral analysis of these operators is of great importance, as is scattering theory for them.

In the present work this problem is solved for the case of a three-particle cluster operator in the general position.

2. Three-Particle Cluster Operators

Consider the cluster functions for a three-particle cluster operator. Note that the set $\mathcal{G}_{3,3}$ of the partitions of the pair $\mathcal{N}_3=(1,2,3)$, $\mathcal{N}'_3=(1',2',3')$ can be decomposed into the following orbits with respect to the acting of the group of all $\tau(\sigma,\sigma')$, σ, $\sigma'\in S_3$:

(a) Six partitions of the form

$$\gamma_\sigma = (\{1\}, \{\sigma(1)\}), \qquad (\{2\}, \{\sigma(2)\}), \qquad (\{3\}, \{\sigma(3)\})$$

where σ is any permutation of three elements. For any σ the cluster function $a_{\gamma_\sigma} \equiv a_\sigma$ defined on the manifold $\Gamma_{\gamma_\sigma} \equiv \Gamma_\sigma$ is equal to

$$a_\sigma(\lambda_1, \lambda_2, \lambda_3; \mu_1, \mu_2, \mu_3) = \tfrac{1}{6}a_0(\lambda_1, \lambda_2, \lambda_3) \tag{2.1}$$

where a_0 is a p-smooth symmetric real function with three variables $\lambda_1, \lambda_2, \lambda_3 \in T^\vee$.

(b) Nine partitions of the form

$$\gamma_{i,j} = (\{i\}, \{j\}), (\mathcal{N}_3 \setminus \{i\}, \mathcal{N}'_3 \setminus \{j\}), \quad i = 1, 2, 3; j = 1', 2', 3'.$$

The cluster function $a_{\gamma_{i,j}} \equiv a_{i,j}$ defined on the manifold $\Gamma_{\gamma_{i,j}} \equiv \Gamma_{i,j}$ is equal to

$$a_{i,j}(\lambda_2, \lambda_2, \lambda_3; \mu_1, \mu_2, \mu_3) = a_1(\pi_{\{i\}}, \pi_{\mathcal{N}_3 \setminus \{i\}}; \lambda_{(i+1)} \mu_{j+1}) \tag{2.2}$$

where $\pi_{\{i\}} = \lambda_i = \mu_j$; $\pi_{\mathcal{N}_3 \setminus \{i\}} = \lambda_{i+1} + \lambda_{i+2} = \mu_{j+2} + \mu_{j+2}$ and $+1$ in the subscripts $i + 1, j + 1$, denotes cyclic shift of the sets (1, 2, 3) and (1′, 2′, 3′); $a_1(\pi_1, \pi_2; \lambda, \mu)$ is a p-smooth function of four variables such that

$$a_1(\pi_1, \pi_2; \lambda, \mu) = a_1(\pi_1, \pi_2; \pi_2 - \lambda, \mu) = a_1(\pi_1, \pi_2; \lambda, \pi_2 - \mu). \tag{2.3}$$

In the case when the operator A is self-adjoint

$$a_1(\pi_1, \pi_2; \lambda, \mu) = \overline{a_1(\pi_1, \pi_2, \mu, \lambda)}. \tag{2.4}$$

(c) Nine partitions of the form

$$\hat{\gamma}_{i,j} = (\{i\}, \mathcal{N}'_3 \setminus \{j\}), (\mathcal{N}_3 \setminus \{i\}, \{j\}), i = 1, 2, 3, j = 1', 2', 3'.$$

The cluster function $a_{\hat{\gamma}_{i,j}} \equiv \hat{a}_{i,j}$ is defined on the manifold $\Gamma_{\hat{\gamma}_{i,j}} \equiv \hat{\Gamma}_{i,j}$ and is equal to

$$\hat{a}_{i,j}(\lambda_2, \lambda_2, \lambda_3; \mu_1, \mu_2, \mu_3) = \hat{a}_1(\pi_{\{i\}}, \pi_{N_3 - \{i\}}; \lambda_{i+1}, \mu_{j+1}) \tag{2.5}$$

where $\hat{a}_1(\pi_1, \pi_2; \lambda, \mu)$ is a p-smooth function of four variables such that

$$\hat{a}_1(\pi_1, \pi_2; \lambda, \mu) = \hat{a}_2(\pi_1, \pi_2; \pi_2 - \lambda, \mu) = \hat{a}_1(\pi_1, \pi_2, \lambda, \pi_2 - \mu). \tag{2.6}$$

In the case when the operator A is self-adjoint

$$\hat{a}_1(\pi_1, \pi_2; \lambda, \mu) = \overline{\hat{a}_1(\pi_2, \pi_1; \mu, \lambda)} \tag{2.7}$$

(d) One partition of the form $\gamma_0 = (\mathcal{N}_3, \mathcal{N}'_3)$. The cluster function a_{γ_0} defined on the manifold $\Gamma_{\gamma_0} \equiv \Gamma_0$ is equal to

$$a_{\gamma_0}(\lambda_1, \lambda_2, \lambda_3; \mu_1, \mu_2, \mu_3) = a_2(\pi; \lambda_1, \lambda_2; \mu_1, \mu_2) \tag{2.8}$$

where $\pi = \lambda_1 + \lambda_2 + \lambda_3 = \mu_1 + \mu_2 + \mu_3$ and a_2 is a p-smooth function of five variables which is separately symmetric in λ_1, λ_2 and in μ_1, μ_2. Moreover,

$$\begin{aligned} a_2(\pi; \lambda_1, \lambda_2; \mu_1, \mu_2) &= a_2(\pi; \pi - \lambda_1 - \lambda_2, \lambda_2; \mu_1, \mu_2) = \\ &= a_2(\pi; \lambda_1, \lambda_2; \pi - \mu_1 - \mu_2, \mu_2). \end{aligned} \tag{2.9}$$

When the operator A is self-adjoint

$$a_2(\pi; \lambda_1, \lambda_2; \mu_1, \mu_2) = a_2(\pi, \mu_1, \mu_2; \lambda_1, \lambda_2). \tag{2.10}$$

The functions a_0, a_1, $\hat{a}_1$ and a_2 will also be called the cluster functions of the three-particle cluster operator. With regard to their smoothness, we suppose that $p \geqslant 2\nu + 2$. The set of cluster operators forms an algebra; the sum $A + B$ and the product $A \cdot B$ of two cluster operators are again cluster operators. We shall give the proof of this fact only for the case of three-particle cluster operators.

LEMMA 2 (the multiplication table). *Let A and B be the three-particle cluster operators with cluster functions a_0^A, a_1^A, $\hat{a}_1^A$, a_2^A and a_0^B, a_1^B, $\hat{a}_1^B$, a_2^B. Then their product $C = AB$ is also a three-particle cluster operator, and its cluster functions are calculated in the following way:*

1. $a_0^C(\lambda_1 \lambda_2 \lambda_3) = a_0^A(\lambda_1, \lambda_2, \lambda_3) a_0^B(\lambda_1, \lambda_2, \lambda_3).$ $\tag{2.11}$

2. $a_1^C(\pi_1, \pi_2; \lambda, \mu)$

$$= a_0^A(\pi_1, \lambda, \pi_2 - \lambda) a_1^B(\pi_1, \pi_2; \lambda, \mu) +$$

$$+ a_1^A(\pi_1, \pi_2, \lambda, \mu) a_0^B(\pi_1, \mu, \pi_2 - \mu) + 3 \int_{T^\nu} a_1^A(\pi_2, \pi_2; \lambda, \kappa) a_1^B(\pi_1, \pi_2; \kappa, \mu) \, d\kappa +$$

$$+ 3 \int_{T^\nu} \hat{a}_1^A(\pi_1, \pi_2; \lambda, \kappa) \hat{a}_1^B(\pi_1, \pi_2; \kappa, \mu) \, d\kappa. \tag{2.12}$$

3. $\hat{a}_1^C(\pi_1, \pi_2; \lambda, \mu)$

$$= a_0^A(\pi_1, \lambda, \pi_2 - \lambda) \hat{a}_1^B(\pi_1, \pi_2, \lambda, \mu) +$$

$$+ \hat{a}_1^A(\pi_1, \pi_2, \lambda, \mu) a_0^B(\mu, \pi_2 - \mu, \pi_2) + 3 \int_{T^\nu} \hat{a}_1^A(\pi_1, \pi_2; \lambda, \kappa) a_1^B(\pi_2, \pi_1, \kappa, \mu) \, d\kappa +$$

$$+ 3 \int_{T^\nu} a_1^A(\pi_1, \pi_2; \lambda, \kappa) \cdot \hat{a}_1^B(\pi_1, \pi_2; \kappa, \mu) \, d\kappa. \tag{2.13}$$

4. $a_2^C(\pi; \lambda_1, \lambda_2; \mu_1, \mu_2)$

$$= a_0^A(\lambda_1, \lambda_2; \pi - \lambda_1 - \lambda_2) \cdot a_2^B(\pi; \lambda_1, \lambda_2; \mu_1, \mu_2) +$$

$$+ a_2^A(\pi; \lambda_1, \lambda_2; \mu_1, \mu_2) \cdot a_0^B(\mu_1, \mu_2, \pi - \mu_1 - \mu_2) +$$

$$+ 3 \left[\int_{T^\nu} a_1^A(\lambda_1, \lambda_2, \kappa) \cdot a_2^B(\pi, \lambda_1, \kappa, \mu_1, \mu_2) \, dx \right. \tag{2.14}$$

+ *two analogous integrals obtained by the following change of variables:*

$$\lambda_1 \to \lambda_2 \to \pi - \lambda_1 - \lambda_2, \lambda_2 \to \pi - \lambda_1 - \lambda_2 \to \lambda_1, +$$

$$+ \int_{T^\nu} \hat{a}_1^A(\lambda_1, \pi - \lambda_1, \lambda_2, \kappa) a_2^B(\pi, \kappa, \lambda_1 - \kappa, \mu_1, \mu_2) \, d\kappa$$

+ *two analogous integrals obtained by the same change of variables as above,*

$$+ \int_{T^\nu} a_2^A(\pi; \lambda_1, \lambda_2; \mu_1, \kappa) a_1^B(\mu_1, \pi - \mu_1, \kappa, \mu_2) \, dx$$

+ *two analogous integrals obtained by the change of variables* $\mu_1 \to \mu_2 \to \pi - \mu_1 - \mu_2$; $\mu_2 \to \pi - \mu_1 - \mu_2 \to \mu_1$

$$+ \int_{T^\nu} a_2^A(\pi; \lambda_1, \lambda_2; \mu_1, \kappa)\hat{a}_1^B(\mu_1, \pi - \mu_1; \mu_2, \kappa)\, d\kappa$$

+ *two analogous integrals obtained by the same change as above* $\Big] +$

$$+ \int_{T^\nu}\int_{T^\nu} a_2^A(\pi; \lambda_1\lambda_2; \kappa_1, \kappa_2)\, a_2^B(\pi; \kappa_1, \kappa_2; \mu_1, \mu_2)\, d\kappa_1 d\kappa_2 +$$

$$+ \Bigg\{ \int_{T^\nu} a_1^A(\lambda_1, \pi - \lambda_1; \lambda_2, \kappa)\,[a_1^B(\mu_2, \pi - \mu_2; \kappa, \mu_1) +$$

$$+ a_1^B(\pi - \mu_1 - \mu_2; \mu_1 + \mu_2; \kappa, \mu_1)]\, d\kappa +$$

$$+ \int_{T^\nu} a_1^A(\lambda_2, \pi - \lambda_2; \lambda_1, \kappa)[a_1^B(\mu_1, \pi - \mu_1; \kappa, \mu_2) +$$

$$+ a_1^B(\pi - \mu_1 - \mu_2, \mu_1 + \mu_2; \kappa, \mu_2)]\, d\kappa +$$

$$+ \int_{T^\nu} a_1^A(\pi - \lambda_1 - \lambda_2, \lambda_1 + \lambda_2; \lambda_1, \kappa)\,[a_1^B(\mu_1, \pi - \mu_1, \kappa, \mu_2) +$$

$$+ a_1^B(\mu_2, \pi - \mu_2; \kappa, \mu_1)]\, d\kappa \tag{2.14}$$

+ 3×3 *analogous terms with integrals of the products*

$$a_1^A \cdot \hat{a}_1^B, \qquad \hat{a}_1^A \cdot a_1^B, \qquad \hat{a}_1^A \cdot \hat{a}_1^B \Big\}.$$

The proof is based on direct calculation of cluster functions using the following formula:

$$\int_{(T^\nu)^n} a_{\gamma_1}^{(1)}(\lambda, \kappa)\, a_{\gamma_2}^{(2)}(\kappa, \mu)\, \delta_{\gamma_1}(\lambda, \kappa)\delta_{\gamma_2}(\kappa, \mu)\, d\kappa$$

$$= a_{\tilde{\gamma}}(\lambda, \mu)\delta_{\tilde{\gamma}}(\lambda, \mu) \qquad \gamma_1, \gamma_2, \tilde{\gamma} \in \mathcal{G}_{n,n} \tag{2.15}$$

where $\lambda = (\lambda_1, \ldots, \lambda_n)$, $\mu = (\mu_1, \ldots, \mu_n)$, $\kappa = (\kappa_1, \ldots, \kappa_n)$ and partition $\tilde{\gamma}$ is obtained in the following way. The partition $\gamma_1 = (\alpha_1^{(1)}, \ldots, \alpha_{s_1}^{(1)})$ can be viewed as the partition of the pair $\{\mathcal{N}, \mathcal{N}'\}$ while $\gamma_2 = (\alpha_1^{(2)}, \ldots, \alpha_{s_2}^{(2)})$ can be viewed as a partition of the pair $\{\mathcal{N}', \mathcal{N}''\}$ where $\alpha_i^{(1)} = (\beta_i, \beta_i')$, $\alpha_j^{(2)} = (\bar{\beta}_j, \beta_j'')$, $\beta_i \subset \mathcal{N}$, $\beta_i', \beta_j^{-1} \subset \mathcal{N}'$, $\beta_j'' \subset \mathcal{N}''$ and $(\beta_1', \ldots, \beta_{s_1}') \vee (\beta_1^{-1}, \ldots, \bar{\beta}_{s_2}') = (\varepsilon_1, \ldots, \varepsilon_l)$ is the lower bound of two partitions of the set $\mathcal{N}'$. Then, the pairs

$$\tilde{\alpha}_m = \left\{ \bigcup_{i;\beta_i' \subseteq \varepsilon_m} \beta_i, \bigcup_{j;\bar{\beta}_j' \subseteq \varepsilon_m} \beta_j'' \right\}, \qquad m = 1, \ldots, l$$

generate a partition of the pair of the sets $(\mathcal{N}, \mathcal{N}'')$ denoted by $\tilde{\gamma}$.

3. Equations for the Resolvent of a Self-Adjoint Three-Particle Cluster Operator

We assume that for all regular values z the resolvent $R(z) = (A - zE)^{-1}$ of a self-adjoint three-particle cluster operator A is again a cluster operator with cluster functions $r_0, r_1, \hat{r}_1, r_2$. This assumption will be justified later. Now we obtain a set of equations for the functions $r_0, r_1, \hat{r}_1, r_2$. Substituting the cluster representation (1.3) for the kernels of the operators A and $R(z)$ in the identity

$$(A - zE)\,R(z) = E \tag{3.1}$$

and using the multiplication table for cluster operators and the fact that the cluster functions of the operator E are equal to

$$a_0^E \equiv 1, \quad a_1^E = \hat{a}_1^E \equiv 0, \quad a_2^E \equiv 0, \tag{3.2}$$

we obtain the following equations for the functions r_0, r_1, $\hat{r}_1$ and r_2. In order to write these equations we introduce the following notation:

$$\begin{aligned}
&\alpha^0_{\pi_1,\pi_2,\pi_3} = a_0(\pi_1, \pi_2, \pi_3),\\
&\alpha^0_{\pi_1,\pi_2}(\kappa) = a_0(\pi_1, \kappa, \pi_2 - \kappa),\\
&\alpha^0_{\pi}(\kappa_1, \kappa_2) = a_0(\kappa_1, \kappa_2, \pi - \kappa_1 - \kappa_2),\\
&K_{\pi_1,\pi_2}(\kappa, \omega) = 3a_1(\pi_1, \pi_2; \kappa, \omega),\\
&\hat{K}_{\pi_1,\pi_2}(\kappa, \omega) = 3\hat{a}_1(\pi_1, \pi_2; \kappa, \omega),\\
&F_\pi(\kappa_1, \kappa_2; \omega_1, \omega_2) = a_2(\pi; \kappa_1\, \kappa_2; \omega_1, \omega_2),\\
&M_\pi(\kappa_1, \kappa_2, \omega) = 3a_1(\kappa_1, \pi - \kappa_1; \kappa_2, \omega),\\
&\hat{M}_\pi(\kappa_1, \kappa_2; \omega) = 3\hat{a}_1(\kappa_1, \pi - \kappa_1, \kappa_2, \omega),\\
&\rho^{(0)}_{\pi_1,\pi_2,\pi_3} = r_0(\pi_1, \pi_2, \pi_3),\\
&\rho^{(1)}_{\pi_1,\pi_2;\mu}(\kappa) = r_1(\pi_1, \pi_2; \kappa, \mu),\\
&\hat{\rho}^{(1)}_{\pi_1,\pi_2;\mu}(\kappa) = \hat{r}_1(\pi_1, \pi_2; \kappa, \mu),\\
&\rho^{(2)}_{\pi_s,\mu_1,\mu_2}(\kappa_1, \kappa_2) = r_2(\pi; \kappa_1, \kappa_2; \mu_1, \mu_2)
\end{aligned}$$

With this notation the set of equations for the cluster functions of the resolvent take of the form

$$\text{(a)}\quad (\alpha^0_{\pi_1,\pi_2,\pi_3} - z)\,\rho^{(0)}_{\pi_1,\pi_2,\pi_3} = 1. \tag{3.4}$$

$$\begin{aligned}
\text{(b)}\quad &(\alpha^0_{\pi_1,\pi_2}(\kappa) - z)\,\rho^{(1)}_{\pi_1,\pi_2;\mu}(\kappa) +\\
&+ \int_{T^\nu} K_{\pi_1,\pi_2}(\kappa, \omega)\,\rho^{(1)}_{\pi_1,\pi_2;\mu}(\omega)\,d\omega +\\
&+ \int_{T^\nu} \hat{K}_{\pi_1,\pi_2}(\kappa, \omega)\,\hat{\rho}^{(1)}_{\pi_2,\pi_1;\mu}(\omega)\,d\omega) = \delta^{(1)}_{\pi_1,\pi_2,\mu}(\kappa),\\
&(\alpha^0_{\pi_2,\pi_1}(\kappa) - z)\,\hat{\rho}^{(1)}_{\pi_2,\pi_1;\mu}(\kappa) +\\
&+ \int_{T^\nu} K_{\pi_2,\pi_1}(\kappa, \omega)\,\hat{\rho}^{(1)}_{\pi_2,\pi_1;\mu}(\omega)\,d\omega +\\
&+ \int \hat{K}_{\pi_2,\pi_1}(\kappa, \omega)\,\rho^{(1)}_{\pi_1,\pi_2}(\omega)\,d\omega = \hat{\delta}^{(1)}_{\pi_2,\pi_1,\mu}(\kappa).
\end{aligned} \tag{3.5}$$

(c) $$(\alpha^0_\pi(\kappa_1,\kappa_2)-z)\,\rho^{(2)}_{\pi,\mu_1,\mu_2}(\kappa_1,\kappa_2)+$$

$$+\int M_\pi(\kappa_1,\kappa_2,\kappa)\rho^{(2)}_{\pi,\mu_1,\mu_2}(\kappa_1,\kappa)\,\mathrm{d}\kappa+$$

$$+\int M_\pi(\kappa_2,\pi-\kappa_1-\kappa_2,\kappa)\,\rho^{(2)}_{\pi,\mu_1,\mu_2}(\kappa_2,\kappa)\,\mathrm{d}\kappa+$$

$$+\int M_\pi(\pi-\kappa_1-\kappa_2,\kappa_1,\kappa)\,\rho^{(2)}_{\pi,\mu_1,\mu_2}(\pi-\kappa_1-\kappa_2,\kappa)\,\mathrm{d}\kappa+$$

$$+\int \hat{M}_\pi(\kappa_1,\kappa_2,\kappa)\,\rho^{(2)}_{\pi,\mu_1,\mu_2}(\kappa_1-\kappa,\kappa)\,\mathrm{d}\kappa+$$

$$+\int \hat{M}_\pi(\kappa_2,\pi-\kappa_1-\kappa_2,\kappa)\,\rho^{(2)}_{\pi,\mu_1,\mu_2}(\kappa_2-\kappa,\kappa)\,\mathrm{d}\kappa+$$

$$+\int \hat{M}_\pi(\pi-\kappa_1-\kappa_2,\kappa_1,\kappa)\,\rho^{(2)}_{\pi,\mu_1,\mu_2}(\pi-\kappa_1-\kappa_2-\kappa,\kappa)\,\mathrm{d}\kappa+$$

$$+\iint F_\pi(\kappa_1,\kappa_2;\kappa,\kappa')\,\rho^{(2)}_{\pi,\mu_1,\mu_2}(\kappa,\kappa')\,\mathrm{d}\kappa\,\mathrm{d}\kappa'=\delta^{(2)}_{\pi,\mu_1,\mu_2}(\kappa_1,\kappa_2). \tag{3.6}$$

Here

$$\begin{aligned}\delta^{(1)}_{\pi_1,\pi_2,\mu}(\kappa)&=-a_1(\pi_1,\pi_2;\kappa,\mu)\,\rho^{(0)}_{\pi_1,\mu,\pi_2-\mu}\\ \hat{\delta}^{(1)}_{\pi_1,\pi_2,\mu}(\kappa)&=-\hat{a}_1(\pi_1,\pi_2;\kappa,\mu)\,\rho^{(0)}_{\pi_2,\mu,\pi_2-\mu}\end{aligned} \tag{3.6}$$

and the function $\delta^{(2)}_{\pi,\mu,\mu_2}(\kappa_1,\kappa_2)$ is equal to

$$\delta^{(2)}_{\pi,\mu_1,\mu_2}(\kappa_1,\kappa_2)=a_2(\pi,\kappa_1,\kappa_2,\mu_1,\mu_2)\cdot$$

$$\cdot\rho^0_{\mu_1,\mu_2,\pi-\mu_1-\mu_2}+3\left[\left[\int_{T^\nu}a_2(\pi,\kappa_1,\kappa_2,\mu_1,\kappa)\cdot\right.\right.$$

$$\left.\cdot\rho^{(1)}_{\mu_1,\pi-\mu_1,\mu_2}(\kappa)\,\mathrm{d}\kappa\right]$$

+2 analogous terms obtained by the change of variables

$$\mu_1\to\mu_2\to\pi-\mu_1-\mu_2+\int_{T^\nu}a_2(\pi,\kappa_1,\kappa_2,\mu_1,\kappa)\hat{\rho}^{(1)}_{\mu_1,\pi-\mu_1-\mu_2}(\kappa)\,\mathrm{d}\kappa$$

+2 analogous terms obtained as above +

$$+\int_{T^\nu}a_1(\kappa_1,\pi-\kappa_1,\kappa_2,\kappa)[\rho^{(1)}_{\mu_2,\pi-\mu_2,\mu_1}(\kappa)+$$

$$+\hat{\rho}^{(1)}_{\mu_2,\pi-\mu_2,\mu_1}(\kappa)]\,\mathrm{d}\kappa+$$

$$+\int\hat{a}_1(\kappa_1,\pi-\kappa_1,\kappa_2,\kappa)[\rho^{(1)}_{\mu_2,\pi-\mu_2,\mu_1}(\kappa)+$$

$$+\hat{\rho}^{(1)}_{\mu_2,\pi-\mu_2,\mu_1}(\kappa)]\,\mathrm{d}\kappa \tag{3.6}$$

$+4\times 5$ analogous terms.

LEMMA 3. *Let z be such that there exist p-smooth solutions of the equations (3.4)–(3.6): $\rho^{(0)}_{\pi_1,\pi_2,\pi_3}$, $\rho^{(1)}_{\pi_1,\pi_2,\mu}(\kappa)$, $\hat{\rho}^{(1)}_{\pi_1,\pi_2,\mu}(\kappa)$, $\rho^{(2)}_{\pi,\mu_1,\mu_2}(\kappa_1,\kappa_2)$ defined for all values of the parameters (π_1,π_2,π_3), (π_1,π_2,μ) and (π,μ_1,μ_2) and p-smoothly depending on these parameters. Then z is a regular value for the operator A, and its resolvent $(A-zE)^{-1}$ is a cluster operator with the cluster functions $r_0,r_1,\hat{r}_1,r_2$ defined by the formulas (3.3).*

The proof is obvious.

Note that the set of equations (a), (b) and (c) has a hierarchical structure: the first function $\rho^{(0)}_{\pi_1,\pi_2,\pi_3}$ and the domain $\Phi^{(0)}\subset R^1$ of values z for which such a solution does not exist for some values π_1,π_2,π_3 are defined by the equation (3.4). Then the functions $\rho^{(1)}_{\pi_1,\pi_2,\mu}(\kappa)$ and $\hat{\rho}^{(1)}_{\pi_1,\pi_2,\mu}(\kappa)$ and the domain $\Phi^{(1)}\subset R^1\setminus\Phi^{(0)}$ of z for which such solutions do not exist for some π_1,π_2 are defined from the equations (3.5). At last from the equation (3.6) we find the function $\rho^{(2)}_{\pi,\mu_1,\mu_2}(\kappa_1,\kappa_2)$ and the set of $z\notin\Phi^{(0)}\cup\Phi^{(1)}$ for which such solutions do not exist for some π. Now we turn to the detailed realization of this plan.

4. Study of Equations (3.4)–(3.6)

(a) The solution of equation (3.4) has the form

$$\rho^{(0)}_{\pi_1,\pi_2,\pi_3}=\frac{1}{\alpha^{(0)}_{\pi_1,\pi_2,\pi_3}-z} \tag{4.1}$$

and exists for all z such that

$$z\notin \operatorname{Im} a_0(\pi_1,\pi_2,\pi_3)$$

(b) Let us investigate the set of equations (3.5).

We introduce the auxiliary Hilbert space

$$\mathscr{H}^{(1)}=L_2(T^\nu)\oplus L_2(T^\nu)$$

consisting of the pairs of functions $r=(\rho,\hat{\rho}),\rho,\hat{\rho}\in L_2(T^\nu)$. Let $\{\mathscr{A}^{(2)}_{\pi_1,\pi_2},\pi_1,\pi_2\in T^\nu\}$ be the family of operators in $\mathscr{H}^{(1)}$ given by the formula

$$\mathscr{A}^{(1)}_{\pi,\pi_2}r=d,\qquad d=(\delta,\hat{\delta})\in\mathscr{H}^{(1)} \tag{4.2}$$

where

$$\delta(\kappa)=\alpha^0_{\pi_1,\pi_2}(\kappa)\rho(\kappa)+\int_{T^\nu}K_{\pi_1,\pi_2}(\kappa,\kappa')\cdot \rho(\kappa')\,d\kappa'+\int_{T^\nu}\hat{K}_{\pi_1,\pi_2}(\kappa,\kappa')\hat{\rho}(\kappa')\,d\kappa' \tag{4.3}$$

$$\hat{\delta}(\kappa)=\alpha^0_{\pi_2,\pi_1}(\kappa)\hat{\rho}(\kappa)+\int_{T^\nu}K_{\pi_2,\pi_1}(\kappa,\kappa')\cdot \hat{\rho}(\kappa')\,d\kappa'+\int_{T^\nu}\hat{K}_{\pi_2,\pi_1}(\kappa,\kappa')\rho(\kappa')\,d\kappa'. \tag{4.4}$$

It follows from (2.7) and (2.10) that for all π_1,π_2 the operator $A^{(1)}_{\pi_1,\pi_2}$ is self-adjoint.

Then the subspace $C_p^{(2)} = C_p \oplus C_p \subset \mathscr{H}^{(1)}$ is invariant with respect to A_{π_1,π_2} and its restriction to this subspace is a bounded operator in $C_p^{(2)}$ (for which we retain the same notation). Finally, the operator

$$\mathscr{A}^{(1)}_{\pi_2,\pi_1} = \mathscr{I} A^{(1)}_{\pi_1,\pi_2} \mathscr{I}^{-1}$$

where $\mathscr{I}(\rho,\hat{\rho}) = (\hat{\rho},\rho)$, so $A^{(1)}_{\pi_1,\pi_2}$ and $\mathscr{A}^{(1)}_{\pi_2,\pi_2}$ are unitary equivalent.

The system (3.5) can be represented in the form

$$(\mathscr{A}^{(1)}_{\pi_1,\pi_2} - zE) r_{\pi_1,\pi_2,\mu} = d_{\pi_1,\pi_2,\mu} \tag{4.5}$$

where

$$r_{\pi_1,\pi_2,\mu} = (\rho^{(1)}_{\pi_1,\pi_2,\mu}, \hat{\rho}^{(1)}_{\pi_1,\pi_2,\mu})$$

and

$$d_{\pi_1,\pi_2,\mu} = (\delta^{(1)}_{\pi_1,\pi_2,\mu}, \hat{\delta}^{(1)}_{\pi_1,\pi_2,\mu})$$

Denote by $\Phi^0_{\pi_1,\pi_2} = \Phi^0_{\pi_2,\pi_1} \subset \Phi^{(0)}$ the set

$$\Phi^0_{\pi_1,\pi_2,} = \operatorname{Im}\alpha^0_{\pi_1,\pi_2} \cup \operatorname{Im}\alpha^0_{\pi_2,\pi_1}$$

In what follows, we shall assume that critical points of the function $\alpha^0_{\pi_1,\pi_2}(\kappa)$ are not degenerated for all π_1,π_2. Note that this assumption is strong enough. In the general case the non-degenerate critical points lie outside the union of smooth surfaces of codimension 1. For the values π_1,π_2 lying on these surfaces, the critical points of $\alpha^0_{\pi_1,\pi_2}(\kappa)$ are now singular points in a general position [3]. This case is considered separately.

THEOREM 4. *The operator* $A^{(1)}_{\pi_1,\pi_2}$ for all $\pi_1,\pi_2 \in T^\nu$ *has a finite number of eigenvalues*

$$\varepsilon_1^{(1)}(\pi_1,\pi_2),\ldots,\varepsilon_s^{(1)}(\pi_1,\pi_2),\ s = s(\pi_1,\pi_2) \tag{4.6}$$

not belonging to the set $\Phi^0_{\pi_1,\pi_2}$ *(each eigenvalue is repeated in* (4.6) *according to its multiplicity). The components of the corresponding eigenvectors*

$$r_i(\pi_1,\pi_2) = (\rho_i(\kappa',\pi_1,\pi_2), \hat{\rho}_i(\kappa',\pi_1,\pi_2))\ i = 1,\ldots,s$$

are the p-smooth functions of $\kappa \in T^\nu: r_i(\pi_1,\pi_2) \in c_p^2$. *Moreover, there exist such a covering of* $T^\nu \times T^\nu$ *by the domains* G_j *and such a continuous piecewise p-smooth functions* $\varepsilon_j(\pi_1,\pi_2),(\pi_1,\pi_2) \in G_j$ *and* $r_j(\pi_1,\pi_2),(\pi_1,\pi_2) \in G_j$ *defined on corresponding domains* G_j *that for any point* $(\pi_1,\pi_2) \in T^\nu \times T^\nu$ *the set of the values* $\{\varepsilon_j(\pi_1,\pi_2)\}$ *and* $\{r_j(\pi_1,\pi_2)\}$ *of these functions coincides with the sequence of the eigenvalues* (4.6) *and of the eigenvectors of the operator* $\mathscr{A}^{(1)}_{\pi_1,\pi_2}$.

Proof. Denote by

$$f^\lambda = ((\alpha^0_{\pi_1,\pi_2}(\kappa) - \lambda)\rho(\kappa), (\alpha^0_{\pi_2,\pi_1}(\kappa) - \lambda)\hat{\rho}(\kappa)) = (\varphi^\lambda, \hat{\varphi}^\lambda). \tag{4.7}$$

Then the equation

$$(A^{(1)}_{\pi_1,\pi_2} - \lambda E) r = 0$$

for the eigenvector takes the form

$$f^{\lambda} + B^{\lambda}_{\pi_1,\pi_2} f^{\lambda} = 0 \tag{4.8}$$

where

$$B^{\lambda}_{\pi_1,\pi_2} f^{\lambda} = \Bigg(\Bigg(\int_{T^{\nu}} \Bigg[K_{\pi_1,\pi_2}(\kappa,\kappa') \frac{\varphi^{\lambda}(\kappa')}{\alpha^0_{\pi_1,\pi_2}(\kappa') - \lambda} + \\ + \hat{K}_{\pi_1,\pi_2}(\kappa,\kappa') \frac{\hat{\varphi}^{\lambda}(\kappa')}{\alpha^0_{\pi_2,\pi_1}(\kappa') - \lambda}\Bigg] d\kappa'\Bigg), \Bigg(\int_{T^{\nu}} \Bigg[\hat{K}_{\pi_2,\pi_1}(\kappa,\kappa') \times \\ \times \frac{\hat{\varphi}^{\lambda}(\kappa')}{\alpha^0_{\pi_2,\pi_1}(\kappa') - \lambda} + K_{\pi_2,\pi_1}(\kappa,\kappa') \frac{\varphi^{\lambda}(\kappa')}{\alpha^0_{\pi_1,\pi_2}(\kappa') - \lambda}\Bigg] \cdot d\kappa'\Bigg)\Bigg). \tag{4.9}$$

The following Fredholm determinant is defined for $\lambda \notin \Phi^0_{\pi_1,\pi_2}$:

$$\det(E + B^{\lambda}_{\pi_1,\pi_2}) \equiv \mathscr{D}_{\pi_1,\pi_2}(\lambda) \tag{4.9^a}$$

and zeros of the function $\mathscr{D}_{\pi_1,\pi_2}(\lambda)$ are eigenvalues of the operator $A^{(1)}_{\pi_1,\pi_2}$. Since $\mathscr{D}_{\pi_1,\pi_2}(\lambda) \to 1$ when $\lambda \to \infty$ and the function $\mathscr{D}_{\pi_1,\pi_2}(\lambda)$ is analytic in the domain $\Pi_{\pi_1,\pi_2} = \mathbb{C}^1 \setminus \Phi^0_{\pi_1,\pi_2}$ only a finite number of zeros of $\mathscr{D}_{\pi_1,\pi_2}(\lambda)$ lie outside any neighbourhood of the set $\Phi^0_{\pi_1,\pi_2}$. The set $\Phi^0_{\pi_1,\pi_2}$ is either a segment $[a_1, b_1]$ or consists of two segments: $[a_1, b_1] \cup [a_2, b_2]$, $a_1 < b_1 < a_2 < b_2$.

Under the assumption made above on the function $\alpha^0_{\pi_1,\pi_2}(\kappa)$ it can be shown that asymptotically $\mathscr{D}_{\pi_1,\pi_2}(\lambda)$ in a neighbourhood of any endpoints of these segments has the following form [4].

$$\mathscr{D}_{\pi_1,\pi_2} = \begin{cases} \dfrac{K}{|c-\lambda|^{1/2}}(1 + o(1)), & \nu = 1 \\ K \ln|c-\lambda|\,(1 + o(1)), & \nu = 2 \\ K + o(1) & \nu \geqslant 3, \end{cases} \tag{4.10}$$

where $c = a_i$ or $c = b_i$, $i = 1,2$ and $K \neq 0$. It follows from the asymptotic form (4.10) that the function $\mathscr{D}_{\pi_1,\pi_2}(\lambda)$ in a small neighbourhood of the set $\Phi^{(0)}_{\pi_1,\pi_2}$ and so the number of its zeros, i.e. the eigenvalues of the operator $\mathscr{A}^{(1)}_{\pi_1,\pi_2}$, is finite. Smoothness of the eigenvectors $r_i(\pi_1, \pi_2)$ follows, for example, from the representation

$$r_i(\pi_1, \pi_2) = B^{\lambda}_{\pi_1,\pi_2}\, r_i(\pi_1, \pi_2), \qquad \lambda = \varepsilon_i(\pi_1, \pi_2) \tag{4.11}$$

The possibility of choosing branches of the eigenvalues $\varepsilon_i(\pi_1, \pi_2)$ which are continuous with respect to π_1, π_2 and also of the continuous family of corresponding eigenvectors $r_i(\pi_1, \pi_2)$ follows from general theorems of the perturbation theory of operators [5].

In the case when the eigenvalue $\delta_i(\pi_1, \pi_2)$ is simple in some point (π^0_1, π^0_2), this eigenvalue and the corresponding eigenvector $r_i(\pi_1, \pi_2)$ are differentiable with respect to π_1, π_2 at this point. So a singularity of $\varepsilon_i(\pi_1, \pi_2)$ and of $r_i(\pi_1, \pi_2)$ can appear only on hypersurfaces in $T^{\nu} \times T^{\nu}$ where at least two eigenvalues $\varepsilon_i(\pi_1, \pi_2)$ coincide. The theorem is thus proven. □

We now introduce some additional assumptions about the functions $\varepsilon_i(\pi_1,\pi_2)$ and $r_i(\pi_1,\pi_2)$ for later convenience.

(a) For all π_1,π_2 the eigenvalues of the operator $A^{(1)}_{\pi_1,\pi_2}$ do not lie in the set $\Phi^0_{\pi_1,\pi_2}$.
(b) The eigenvalues $\varepsilon_i(\pi_1,\pi_2)$ lie at a finite distance from the set $\Phi^{(0)}_{\pi_1,\pi_2}$

$$\inf_{i,\pi_1,\pi_2} d(\varepsilon_i(\pi_1,\pi_2), \Phi^0_{\pi_1,\pi_2}) > 0$$

where $d(x,A)$ is the distance of the point $x\in R^1$ from the set $\mathscr{A}\subset R^1$. It follows from (a) and (b) that the number $s(\pi_1,\pi_2)$ of the eigenvalues of the operator $A^{(1)}_{\pi_1,\pi_2}$ is constant and each of the functions $\varepsilon_i(\pi_1,\pi_2)$ (and also $r_i(\pi_1,\pi_2)$ is defined everywhere on $T^\nu\times T^\nu$.

(c) The functions $\varepsilon_i(\pi_1,\pi_2)$ and $r_i(\pi_1,\pi_2)$ are differentiable in π_1 and π_2.

Let $\varepsilon^\pi_i(\lambda)=\varepsilon_i(\lambda,\pi-\lambda)$. We shall assume that for all π the critical points of $\varepsilon^\pi_i(\lambda)$ are non-degenerate. The case of the arbitrary functions $\varepsilon^\pi_i(\lambda)$ (such as the general case for the $\alpha^0_{\pi_1,\pi_2}(\lambda)$) will be considered separately.

Denote by $\Phi^{(1)}_{\pi_1,\pi_2}$ the set of the eigenvalues of the operator $A^{(1)}_{\pi_1,\pi_2}$ and let $\Phi^{(1)}=\cup_{\pi_1,\pi_2}\Phi^{(1)}_{\pi_1,\pi_2}=\cup_i \operatorname{Im}\varepsilon_i(\pi_1,\pi_2)$.

LEMMA 4.5. *The solution of the equations* (3.5) *exists, is unique and can be represented for all pairs* π_1,π_2 *and any* $z\in\mathbb{C}^1\setminus(\Phi^{(0)}_{\pi_1,\pi_2}\cup\Phi^{(1)}_{\pi_1,\pi_2})=\Pi^{(1)}_{\pi_1,\pi_2}$ *in the form*

$$\begin{aligned}\rho^{(1)}_{\pi_1,\pi_2,\mu}(\kappa) &= \frac{t^{(1)}_{\pi_1,\pi_2}(\kappa,\mu',z)}{(\alpha^0_{\pi_1,\pi_2}(\kappa)-z)(\alpha^0_{\pi_1,\pi_2}(\mu)-z)}+\\ &+\frac{t^{(2)}_{\pi_1,\pi_2}(\kappa,\mu;z)}{(\alpha^0_{\pi_2,\pi_1}(\kappa)-z)(\alpha^0_{\pi_2,\pi_1}(\mu)-z)}+\\ &+\sum_{i=1}^{s}\frac{\varphi^{(i)}_{\pi_1,\pi_2}(\kappa)\varphi^{(i)}_{\pi_1,\pi_2}(\mu)}{\varepsilon_i(\pi_1,\pi_2)-z},\\ \hat\rho^{(1)}_{\pi_1,\pi_2,\mu}(\kappa) &= \frac{\hat t^{(1)}_{\pi_1,\pi_2}(\kappa,\mu;z)}{(\alpha^0_{\pi_2,\pi_1}(\kappa)-z)(\alpha^0_{\pi_2,\pi_1}(\mu)-z)}+\\ &+\frac{\hat t^{(2)}_{\pi_1,\pi_2}(\kappa,\mu;z)}{(\alpha^0_{\pi_1,\pi_2}(\kappa)-z)(\alpha^0_{\pi_1,\pi_2}(\mu)-z)}+\\ &+\sum_{i=1}^{S}\frac{\hat\varphi^{(i)}_{\pi_1,\pi_2}(\kappa)\hat\varphi^{(i)}_{\pi_1,\pi_2}(\mu)}{\varepsilon_i(\pi_1,\pi_2)-z}\end{aligned}\tag{4.12}$$

where $\varphi^{(i)}_{\pi_1,\pi_2}(\kappa)$ *and* $\hat\varphi^{(i)}_{\pi_1,\pi_2}(\kappa)$ *are components of the eigenvector* $r^{(i)}_{\pi_1,\pi_2}$ *and* $t^{(1),(2)}_{\pi_1,\pi_2}(\kappa,\mu;z)$ *and* $\hat t^{(1),(2)}_{\pi_1,\pi_2}(\kappa,\mu;z)$ *are analytic functions of* κ,μ defined for all π_1,π_2 *in the domain* $z\in\mathbb{C}^1\setminus\Phi^{(0)}_{\pi_1,\pi_2}$. *If* $\nu\geqslant 3$ *the limits*

$$t^{(1),(2)}_{\pi_1,\pi_2}(\kappa,\mu;\lambda\pm i0)=\lim_{\varepsilon\to+0}t^{(1),(2)}_{\pi_1,\pi_2}(\kappa,\mu;\lambda\pm i\varepsilon)\tag{4.13}$$

$$\hat t^{(1),(2)}_{\pi_1,\pi_2}(\kappa,\mu;\lambda\pm i0)=\lim_{\varepsilon\to+0}t^{(1),(2)}_{\pi_1,\pi_2}(\kappa,\mu;\lambda\pm i\varepsilon)$$

exist for all $\lambda\in\Phi^{(0)}_{\pi_1,\pi_2}$. *When* $\nu=1,2$ *the limits* (4.13) *exist only for values* $\lambda\in\Phi_{\pi_1,\pi_2}$ *different from the values of the function* $\alpha^0_{\pi_1,\pi_2}(\kappa)$ *at their critical points. (We will call these values regular.) The limit functions* $t^{(1),(2)}_{\pi_1,\pi_2}$ *and* $\hat{t}^{(1),(2)}_{\pi_1,\pi_2}$ *are differentiable on* λ *in all regular points. Derivatives of the functions* $t^{(1),(2)}_{\pi_1,\pi_2}$ *and* $\hat{t}^{(1),(2)}_{\pi_1,\pi_2}$ *have power singularities* $|\lambda-\lambda_0|^{(\nu-1)/2}$ *with any* $\nu \geqslant 3$ *in singular values* $\lambda=\lambda_0$.

Proof. Substituting for (4.7) in equation (4.5) we obtain

$$f_{\pi_1,\pi_2,\mu} + B^z_{\pi_1,\pi_2} f_{\pi_1,\pi_2,\mu} = d_{\pi_1,\pi_2,\mu}. \tag{4.14}$$

The solution of this equation can be written in the form

$$f_{\pi_1,\pi_2,\mu}(\kappa) = d_{\pi_1,\pi_2,\mu}(\kappa) + \frac{1}{\mathscr{D}_{\pi_1,\pi_2}(z)}\int_{T^\nu} \mathscr{D}^z_{\pi_1,\pi_2}(\kappa,\kappa')d_{\pi_1,\pi_2,\mu}(\kappa')d\kappa'$$

where $\mathscr{D}_{\pi_1,\pi_2}(z)$ is the Fredholm determinant (4.9a), and matrix elements of the kernel

$$\mathscr{D}^z_{\pi_1,\pi_2}(\kappa,\kappa') = \|\Delta^{(j,k)}_{\pi_1,\pi_2}(\kappa,\kappa';z)\| \; j,k=1,2$$

can be represented with the help of the well-known Fredholm formulae [6] in the form

$$\Delta^{(j,k)}_{\pi_1,\pi_2}(\kappa,\kappa';z)$$

$$= \sum_n \frac{1}{n!} \sum_{(i_2,\dots,i_n) i_s=1,2} \overbrace{\int \cdots \int}^{n} \frac{s^{(j,k)}_{i_1,\dots,i_n}(\kappa,\kappa'';\omega_1,\dots,\omega_n)}{\prod_s (\alpha^0_{i_s,i'_s}(\omega_s)-z)} \prod_s d\omega_s \tag{4.15}$$

where $i'_s = 2,1$ if $i_s = 1,2$, and the functions $s^{(j,K)}_{i_1,\dots,i_n}(\kappa,\kappa';\omega_1,\dots,\omega_n)$ are represented in the form of the determinants corresponding to the kernels $k_{\pi_{i_s},\pi_{i'_s}}(\omega_s,\omega'_{s'})$ and $\hat{k}_{\pi_{i_s},\pi_{i'_s}}(\omega_s,\omega_{s'})$

Introducing functions

$$\varphi^{(j,k)}_{i_1,\dots,i_n}(\kappa,\kappa';\, t_1,\dots,t_n) = \int_{\alpha^0_{\pi_{i_1},\pi_{i'_1}}=t_1} \cdots \int_{\alpha^0_{\pi_{i_n},\pi_{i'_n}}=t_n} s^{(j,K)}_{i_1,\dots,i_n}(\kappa,\kappa';\omega_1,\dots,\omega_n) \prod_1^n d\omega_i$$

we will write the integrals in (4.15) after integration by parts in the form

$$(-1)^n \int_{t'}^{t''} \cdots \int_{t'}^{t''} \frac{\partial^n \varphi^{(j,K}_{i_1,\dots,i_n}(\kappa,\kappa';t_1,\dots,t_n)}{\partial t_1\cdots\partial t_n} \prod_{s=1}^n \ln(t_s-z)dt_1\cdots dt_n \tag{4.16}$$

where $t' = \min \alpha^0_{\pi_1,\pi_2}$, $t'' = \max \alpha^0_{\pi_1,\pi_2}$

As follows from out assumption about non-degeneracy of critical points of the function $\alpha^0_{\pi_1,\pi_2}(\kappa)$, the function $\partial^n \varphi^{(j,K)}_{i_1,\dots,i_n}/\partial t_1\cdots\partial t_n$ has singularities only if $t_i = t^0_p$ for some i (the values of the function $\alpha^0_{\pi_1,\pi_2}$ in its critical points), and these singularities have the form $|t_i - t^0_p|^{(\nu-4)/2}$, $\nu \geqslant 3$ [7]. This follows easily from this fact that the integral (4.16) for any n is an analytic function of z outside the cut (or cuts) $\Phi^0_{\pi_1,\pi_2}$. This function is bounded and continuous up to the cut. We assume also that the functions $\mathscr{D}_{\pi_1,\pi_2}(z)$ are not equal to zero on the set $\Phi^0_{\pi_1,\pi_2}$. The derivative of the limit values $\mathscr{D}^{\pm}_{\pi_1,\pi_2}(\lambda) = \lim_{\varepsilon\to+0} \mathscr{D}_{\pi_1,\pi_2}(\lambda \pm i\varepsilon)$ exists in all regular points, and its singularities in the singular points t^0_p have the form const $|\lambda - t^0_p|^\beta$, $\beta > 0$.

Then estimating the determinants in the representation of the function

$s^{(j,K)}_{i_1,\ldots,i_n}(\kappa,\kappa';\omega_1,\ldots,\omega_n)$ and their derivatives with the help of the Hadamard inequality we find that the integral (4.16) does not exceed $C^n n^{n/2}$ where c is some constant. Thus the series (4.15) converges and is an analytic function outside the cut $\Phi^{(0)}_{\pi_1,\pi_2}$ bounded and continuous up to the $\Phi^{(0)}_{\pi_1,\pi_2}$ as above. Then, using the form (3.6a) of the function $d_{\pi_1,\pi_2,\mu}$ and the Fredholm formulas, we get that the components of the vector $\int_{T^\nu}\mathscr{D}^z_{\pi_1,\pi_2}(\kappa,\kappa')\,d_{\pi_1,\pi_2,\mu}(\kappa')d\kappa'$ are equal to

$$\sum\frac{1}{n!}\sum_{(i_1,\ldots,i_n)}\int_{T^\nu}\cdots\int_{T^\nu}\frac{s_{i_1,\ldots,i_n}(\kappa,\mu;\omega,\ldots,\omega_n)}{\prod_s(\alpha^0_{\pi_{i_s}\pi_{i'_s}}(\omega_s)-z)}\cdot\Pi\,d\omega.$$

$$\frac{1}{(\alpha^{(0)}_{\pi_1,\pi_2}(\mu)-z)}\left(\text{or }\frac{1}{(\alpha^0_{\pi_2,\pi_1}(\mu)-z)}\right). \tag{4.17}$$

Repeating previous arguments with respect to the terms of the series (4.17) again and turning to the functions $\rho^{(1)}_{\pi_1,\pi_2,\mu}$, $\hat{\rho}_{\pi_1,\pi_2,\mu}$ we have finally that the components of the vector $r_{\pi_1,\pi_2,\mu}$ have the form

$$\frac{\tilde{t}_{\pi_1,\pi_2'}(\kappa,\mu;z)}{(\alpha^{(0)}_{\pi_1,\pi_2}(\kappa)-z)(\alpha^0_{\pi_1,\pi_2}(\kappa)-z)}+\frac{\tilde{t}_{\pi_1,\pi_2}(\kappa,\mu;z)}{(\alpha^0_{\pi_2,\pi_1}(\kappa)-z)(\alpha^0_{\pi_2,\pi_1}(\kappa)-z)}$$

where $\tilde{t}_{\pi_1,\pi_2}(\kappa,\mu;z)$ and $\tilde{t}_{\pi_1,\pi_2}(\kappa,\mu;z)$ are meromorphic functions outside of cut $\Phi^0_{\pi_1,\pi_2}$ and continuous up to $\Phi^{(0)}_{\pi_1,\pi_2}$, and their poles coincide with the zeros of the function $\mathscr{D}_{\pi_1,\pi_2}(z)$ (lying outside the cut $\Phi^0_{\pi_1,\pi_2}$ as we have assumed). Separating the main terms of the Lorenz expansion in these poles we find that the components of the vector $r_{\pi_1,\pi_2,\mu}(\kappa)$ are represented in the form

$$\frac{t(\kappa,\mu;z)}{(\alpha^{(0)}_{\pi_1,\pi_2}(\kappa)-z)(\alpha^0_{\pi_1,\pi_2}(\kappa)-z)}+\frac{\hat{t}(\kappa,\mu;z)}{(\alpha^0_{\pi_1,\pi_2}(\kappa)-z)(\alpha^0_{\pi_1,\pi_2}(\mu)-z)}+$$
$$+\sum_i Q^{(i)}_{\pi_1,\pi_2}(\kappa,\mu,z)$$

where $Q^{(i)}_{\pi_1,\pi_2}(\kappa,\mu;z)$ is a polynomial of $1/\varepsilon_i(\pi_1,\pi_2)-z$. Note that it follows from the equations (4.5) that

$$r_{\pi_1,\pi_2,\mu}=(A-zE)^{-1}d_{\pi_1,\pi_2,\mu}.$$

It follows from spectral theorem that any isolated eigenvalue $\varepsilon_i(\pi_1,\pi_2)$ is a first-order pole for the resolvent, and the residue at this pole is equal to the projection operator on the corresponding eigensubspace. Hence we find that the residue $r_{\pi_1,\pi_2,\mu}$ in the point $z=\varepsilon_i$ is equal to:

$$(\varphi^{(i)}_{\pi_1,\pi_2}(\kappa)\chi^{(i)}_{\pi_1,\pi_2}(\mu),\qquad \hat{\varphi}^{(i)}_{\pi_1,\pi_2}(\kappa)\cdot\hat{\chi}^{(i)}_{\pi_1,\pi_2}(\mu))$$

where $\chi^{(i)}_{\pi_1,\pi_2}(\mu)=(\varphi^{(i)}_{\pi_1,\pi_2}\delta_{\pi_1,\pi_2,\mu})$ and $\varphi^{(i)}_{\pi_1,\pi}$ is the normalized eigenfunction corresponding to the eigenvalue $\varepsilon_i(\pi_1,\pi_2)$. Then using the symmetry of $r_{\pi_1,\pi_2}(\kappa,\mu)$ with respect to the variables κ) and μ we find that $\chi^{(1)}_{\pi_1,\pi_2}(\mu)=\varphi^{(i)}_{\pi_1,\pi_2}(\mu)$. The representation (4.12) thus follows. We now turn to the investigation of equation (3.6), writing it in the form

$$(\mathscr{D}_\pi-zE)\,\rho^{(2)}_{\pi,\mu_1,\mu_2}=\delta^{(2)}_{\pi,\mu_1,\mu_2} \tag{4.18}$$

where $\mathscr{D}_\pi$ is an operator in the space $L_2^{\mathrm{sym},\pi}$ of the form

$$(\mathscr{D}_\pi\rho)(\kappa_1\kappa_2) = \alpha_\pi^{(0)}(\kappa_1,\kappa_2)\rho(\kappa_1,\kappa_2) +$$

$$+ \int_{(T^\nu)^2} F_\pi(\kappa_1,\kappa_2;\ \omega_1,\omega_2)\,\rho(\omega_1,\omega_2)\,d\omega_1\,d\omega_2 +$$

$$+ \int_{T^\nu} [M_\pi(\kappa_1,\kappa_2,\omega) + \hat{M}_\pi(\kappa_1,\kappa_2,\omega)]\,\rho(\kappa_1,\omega)\,d\omega +$$

$$+ \int_{T^\nu} [M_\pi(\kappa_2,\pi-\kappa_1-\kappa_2,\omega) + \hat{M}_\pi(\kappa_2,\pi-\kappa_1-\kappa_2,\omega)]\,\rho(\kappa_2,\omega)\,d\omega +$$

$$+ \int [M_\pi(\pi-\kappa_1-\kappa_2,\kappa_1,\omega) +$$

$$+ \hat{M}_\pi(\pi-\kappa_1-\kappa_2,\kappa_1,\omega)]\,\rho(\pi-\kappa_1-\kappa_2,\omega)\,d\omega.$$

Here by $L_2^{\mathrm{sym},\pi}$ we denote the space of symmetric functions on the manifold $\gamma_\pi = \{\kappa_1,\kappa_2,\kappa_3\colon\ \kappa_1+\kappa_2+\kappa_3 = \pi\} \subset (T^\nu)^3$ integrable with respect to the usual measure $d\kappa_1\,d\kappa_2$ on γ_π. As a rule we will use variables κ_1 and κ_2 as coordinate on γ_π.

Note that this operator is self-adjoint, and its restriction on the space of p-smooth symmetric functions $C_p^{\mathrm{sym},\pi}$ defines a bounded operator in $C_p^{\mathrm{sym},\pi}$ (we retain the previous notation for this). Denote

$$\Phi_\pi^0 = \operatorname{Im}\alpha_\pi^0 = \bigcup_{\pi_1+\pi_2=\pi} \Phi_{\pi_2,\pi_2}^{(0)}$$

$$\Phi_\pi^{(1)} = \bigcup_{\pi_2+\pi_2=\pi} \Phi_{\pi_2,\pi_2}^{(1)}.$$

LEMMA 4.6. *The operator $\mathscr{D}_\pi - zE$ in $C_p^{\mathrm{sym},\pi}$ (and also in $L_2^{\mathrm{sym},\pi}$) is a Fredholm operator for any π and any $z \notin \Phi_\pi^{(1)} \cup \Phi_\pi^{(0)}$, and its index is equal to zero.*

Proof. It is evident that for $z \notin \Phi_\pi^{(1)} \cup \Phi_\pi^{(0)}$ the functions

$$r_0^\pi(\lambda_1,\lambda_2) = \rho^0_{\lambda_1,\lambda_2,\pi-\lambda_1-\lambda_2}$$

$$r_1^\pi(\lambda_1,\lambda_2;\mu) = \rho^{(1)}_{\lambda_1,\pi-\lambda_1,\mu}(\lambda_2)$$

$$\hat{r}_1^\pi(\lambda_1,\lambda_2;\mu) = \hat{\rho}^{(1)}_{\lambda_1,\pi-\lambda_1,\mu}(\lambda_2)$$

exist. Introduce the operator $\tilde{R}^\pi(z)$ in $L_2^{\mathrm{sym},\pi}$ (and in $C_p^{\mathrm{sym},\pi}$). Its kernel has the form

$$K^\pi(\lambda_1,\lambda_2;\mu_1,\mu_2) = \tfrac{1}{6}r_0^\pi(\lambda_1,\lambda_2)\cdot \sum_{\substack{(i_1,i_2)\\ i_1,i_2=1,2,3}} \delta(\lambda_1-\mu_{i_1})\cdot\delta(\lambda_2-\mu_{i_2}) +$$

$$+ \sum_{i=1,2,3;\,i'=1,2,3} (r_1^\pi(\lambda_i,\lambda_{i+1},\mu_{i'+1})\,\delta(x_i-\mu_{i'}) +$$

$$+ \hat{r}_1(\lambda_i,\lambda_{i+1},\mu_{i'+1})\,\delta(\lambda_i-\mu_{i'}-\mu_{i'+1})). \tag{4.20}$$

Here $\lambda_3 = \pi - \lambda_1 - \lambda_2$ and $\mu_3 = \pi - \mu_1 - \mu_2$. With the help of direct calculation using the equations (3.5) and the conditions (2.2) and (2.3), we find that

$$\tilde{R}^\pi\mathscr{D}_\pi = E + N_{\mathrm{left}}^\pi \text{ and } \mathscr{D}_\pi\tilde{R}^\pi = E + N_{\mathrm{right}}^\pi \tag{4.21}$$

where N^{π}_{left} and N^{π}_{right} are integral operators in $L_2^{\text{sym},\pi}$ (and in $C_p^{\text{sym},\pi}$) with p-smooth kernels. So N^{π}_{left} and N^{π}_{right} are compact operators in $L_2^{\text{sym},\pi}$ (and in $C_p^{\text{sym},\pi}$) and, consequently, the operator $\mathcal{D}_\pi - zE$ is a Fredholm operator [8]. In the case when $\mathcal{D}_\pi - zE$ is considered as an operator in $L_2^{\text{sym},\pi}$ its index is equal to zero because of the self-adjointness of $\mathcal{D}_\pi$. We now show that $\operatorname{ind}(\mathcal{D}_\pi - zE) = 0$ also in the case when $\mathcal{D}_\pi$ is treated as an operator in $C_p^{\text{sym},\pi}$. Let $\operatorname{Im} z \neq 0$. Then $\operatorname{Ker}(\mathcal{D}_\pi - zE) = 0$. We show that $\operatorname{Ker}(\mathcal{D}'_\pi - zE) = 0$. We assume the contrary, and let $\varphi \in C_p^{\text{sym},\pi}$ be a distribution satisfying the equation

$$(\mathcal{D}'_\pi - zE)\varphi = 0. \tag{4.22}$$

Because the action of $\mathcal{D}'_\pi$ coincides formally with the action of $\mathcal{D}_\pi$, on multiplying the equation (4.22) on $\tilde{R}^\pi$ we see that

$$\varphi + N^{\pi}_{\text{left}}\varphi = 0. \tag{4.23}$$

The integral operator N^{π}_{left} transforms the distribution φ into a usual function, and so φ is a usual function. In this case $\varphi \in L_2^{\text{sym},\pi}$, which is a contradiction. Thus, $\operatorname{ind}(\mathcal{D}_\pi - zE) = 0$ when $\operatorname{Im} z \neq 0$. Because of the continuity of the index with respect to small perturbations of Fredholm operator, $\operatorname{ind}(\mathcal{D}_\pi - zE) = 0$ for all $z \notin \Phi^{(1)}_\pi \cup \Phi^{(0)}_\pi$. The lemma is proved. □

THEOREM 4.7. *The operator $\mathcal{D}_\pi$ in $L_2^{\text{sym},\pi}$ for any $\pi \in T^\nu$ has a finite number of eigenvalues $\hat{\varepsilon}_2(\pi), \ldots, \hat{\varepsilon}_s(\pi)$ outside the set $\Phi^{(0)}_\pi \cup \Phi^{(1)}_\pi$ (each value $\hat{\varepsilon}_i(\pi)$ is repeated as often as its multiplicity). Then there exists a covering of T^ν by a finite number of domains $\hat{G}_1, \ldots, \hat{G}_s$ and piecewise smooth continuous functions $\{\hat{\varepsilon}_i(\pi), \pi \in G_i\}$ and $\{\hat{\varphi}^{(i)}_\pi(\kappa_1, \kappa_2), \pi \in G_i\}$ defined in the domain G_i such that for any $\pi \in T^\nu$ the set of values $\{\varepsilon_i(\pi)\}$ coincides with the set of eigenvalues of the operator $\mathcal{D}_\pi$ and the family $\{\hat{\varphi}^{(i)}_\pi(\kappa_1, \kappa_2)\}$ coincides with the set of its eigenfunctions.*

Proof. Note that the equation for the eigenvalues

$$\mathcal{D}_\pi \varphi - \lambda\varphi = 0$$

reduces to the equation

$$\varphi + N^{\pi}_{\text{left}}\varphi = 0$$

after application of the multiplication onto $\tilde{R}^\pi(\lambda)$ from the left. So the eigenvalues are included among the zeros of the function $\Delta_\pi(\lambda) = \operatorname{Det}(E + N^{\pi}_{\text{left}})$, a number which is finite. This fact follows from the fact that $\Delta_\pi(\lambda)$ is an analytic function of λ outside the set $\varphi^{(0)}_\pi \cup \Phi^{(1)}_\pi$ consisting of the finite numbers of segments, and the behaviour of $\Delta_\pi(\lambda)$ in a neighbourhood of any end point of these segments is described by the formula (4.10). The other statements of the theorem can be obtained from general facts of the perturbation theory of operators, in the same way as in the proof of Theorem 4.4. The theorem is proved.

As above, we again introduce some simplifying assumptions about the eigenvalues and the eigenfunctions describing the case of the general position.

(a) For all $\pi \in T^\nu$ all eigenvalues of the opeator $\mathcal{D}_\pi$ lie outside the set $\Phi^{(0)\pi} \cup \varphi^{(1)}_\pi$, i.e. the eigenvectors introduced in Theorem 4.7 exhaust all the discrete spectrum of the operator $\mathcal{D}_\pi$.

(b) The set $\Phi_\pi^{(2)} = \{\hat{\varepsilon}_i(\pi)\}$ of the eigenvalues lies at a finite distance from $\Phi_\pi^{(0)} \cup \Phi_\pi^{(1)}$: $\inf_\pi d(\Phi_\pi^{(2)}, \Phi_\pi^{(0)} \cup \Phi_\pi^{(1)}) > 0$. It follows from this that the number of eigenvalues $\hat{\varepsilon}_i(\Pi)$ is constant, and each function $\hat{\varepsilon}_i(\pi)$ is defined everywhere on the torus T^ν.

LEMMA 4.8. *For all $\pi \in T^\nu$ and any $z \notin \Phi_\pi^{(0)} \cup \Phi_\pi^{(1)} \cup \Phi_\pi^{(2)}$ the solution $\rho_{\pi,\mu_1\mu_2}^{(2)}(\lambda_1,\lambda_2) \equiv \rho_\pi^{(2)}(\lambda_1,\lambda_2;\mu_1\mu_2;z)$ of equation (4.16) exists and can be represented in the form*

$$
\begin{aligned}
&[\rho_\pi^{(2)}(\lambda_1,\lambda_2;\mu_1,\mu_2;z)\\
&= \frac{S_\pi(\lambda_1,\lambda_2;\mu_1,\mu_2;z)}{(\alpha_\pi^0(\lambda_1,\lambda_2)-z)(\alpha_\pi^0(\mu_1,\mu_2)-z)} + \\
&+ \frac{1}{\alpha_\pi^0(\lambda_1,\lambda_2)-z}\left[\sum_i \left(\frac{u_\pi^{(i)}(\lambda_1,\lambda_2;\mu_1,\mu_2;z)}{\varepsilon_i(\pi-\mu_1,\mu_1)-z} + \right.\right.\\
&\left.\left.+\frac{u_\pi^{(i)}(\lambda_1,\lambda_2;\mu_2,\pi-\mu_1-\mu_2;z)}{\varepsilon_i(\pi-\mu_2,\mu_2)-z} + \frac{u_\pi^{(i)}(\lambda_1,\lambda_2;\pi-\mu_1-\mu_2;\mu_1;z)}{\varepsilon_i(\mu_1+\mu_2,\pi-\mu_1-\mu_2)-z}\right)\right] + \\
&+ \frac{1}{\alpha_\pi^0(\mu_1,\mu_2)-z)}\left[\sum_i \left(\frac{u_\pi^{(i)}(\lambda_1\lambda_2;\mu_1,\mu_2;z)}{\varepsilon_i(\pi-\lambda_1,\lambda_1)-z} + \right.\right.\\
&+\frac{u_\pi^{(i)}(\lambda_2,\pi-\lambda_1-\lambda_2;\mu_1,\mu_2;z)}{\varepsilon_i(\pi-\lambda_2,\lambda_2)-z} + \\
&\left.\left.+\frac{u_\pi^{(i)}(\pi-\lambda_1-\lambda_2,\lambda_1;\mu_1,\mu_2;z)}{\varepsilon_i(\lambda_1+\lambda_2,\pi-\lambda_1-\lambda_2)-z}\right)\right] + \\
&+ \sum_{\substack{i,j,k=1,2,3\\ K'=1,2,3}} \frac{v^{(i,j)}(\lambda_k,\lambda_{k+1},\mu_{k'},\mu_{k'+1};z)}{(\varepsilon_i(\pi-\lambda_k,\lambda_k)-z)(\varepsilon_j(\pi-\mu_{k'},\mu_{k'})-z)} + \\
&+ \sum_j \frac{\hat{\varphi}_\pi^{(j)}(\lambda_1,\lambda_2)\hat{\varphi}_\pi^{(j)}(\mu_1,\mu_2)}{\hat{\varepsilon}_j(\pi)-z}
\end{aligned}
$$

and

$$
S_\pi(\lambda_1,\lambda_2;\mu_1,\mu_2;z) = \sum_{i,j} \frac{t_\pi^{(i,j)}(\lambda_1,\lambda_2,\mu_1,\mu_2;z)}{\alpha_\pi^0(\lambda_i,\mu_j)-z} + S'_\pi(\lambda_1,\lambda_2;\mu_1,\mu_2;z).
$$

Here the functions $t_\pi^{(i,j)}(\lambda_1,\lambda_2;\mu_1,\mu_2;z)$, $S'_\pi(\lambda_1,\lambda_2;\mu_1,\mu_2;z)$, $u_\pi^{(i)}(\lambda_1,\lambda_2;\mu_1,\mu_2;z)$, $v_\pi^{(i,j)}(\lambda_1,\lambda_2;\mu_1,\mu_2;z)$ are analytic with respect to z outside the set $\Phi_\pi^{(0)} \cup \Phi_\pi^{(1)} = \Phi\diagdown_\pi$, and for $\nu \geqslant 3$ the limits

$$
\begin{aligned}
&\lim_{\varepsilon\to\pm 0} S'_\pi(\lambda_1,\lambda_2;\mu_1,\mu_2;t+i\varepsilon) = S'_\pi(\lambda_1,\lambda_2;\mu_1,\mu_2;t\pm i0)\\
&\lim_{\varepsilon\to\pm 0} u_\pi^{(i)}(\lambda_1,\lambda_2;\mu_1,\mu_2;t+i\varepsilon) = u_\pi^{(i)}(\lambda_1,\lambda_2;\mu_1,\mu_2;t+i0)\\
&\lim_{\varepsilon\to\pm 0} v_\pi^{(i,j)}(\lambda_1,\lambda_2;\mu_1,\mu_2;t+i\varepsilon) = v_\pi^{(i,j)}(\lambda_1,\lambda_2;\mu_1,\mu_2;t\pm i0)\\
&\lim_{\varepsilon\to\pm 0} t_\pi^{(i,j)}(\lambda_1,\lambda_2;\mu_1,\mu_2;t+i\varepsilon) = t_\pi^{(i,j)}(\lambda_1,\lambda_2;\mu_1,\mu_2;t\pm i0)
\end{aligned}
$$

exist in all points of the cut $\tilde{\Phi}_\pi$. *These functions are differentiable, and the singularities of their derivatives in the singular points* $t_0 \in \tilde{\Phi}_\pi$ *have the form* $\sim |t - t_0|^{(\nu-1)/2}$. *Then* $\hat{\varphi}_\pi^{(j)}$ *are the eigenvectors of the operator* $\mathcal{D}_\pi$.

Proof. The equation (4.18) for the definition of the function $\rho_\pi^{(2)}(\lambda_1, \lambda_2; \mu_1, \mu_2; z)$ is reduced to the equation

$$\rho_\pi^{(2)} + N_{\text{left}}^\pi \rho_\pi^{(2)} = \tilde{R}_\pi(z)\delta_\pi^{(2)}$$

after the application of the operator $\tilde{R}_\pi(z)$ to both parts. The kernel of the operator N_{left}^π has the form

$$N_{\text{left}}^\pi(\lambda_1, \lambda_2; \mu_1, \mu_2; z) = \frac{K_\pi(\lambda_1, \lambda_2; \mu_1, \mu_2; z)}{\alpha_\pi^0(\lambda_1, \lambda_2) - z} + \\ + \sum_{i;k=1,2,3} \frac{Q_\pi^{(i,k)}(\lambda_1, \lambda_2; \mu_1 \mu_2)}{\varepsilon_i(\pi - \lambda_k, \lambda_k) - z} \tag{4.25}$$

where k_π is the function analytical in the domain $z \in \mathbb{C}^1 \setminus \Phi_\pi^{(0)}$ and the functions $Q_\pi^{(i,k)}$ are p-smooth functions of $\lambda_1, \lambda_2; \mu_1, \mu_2$. Hence, using Fredholm formulas, we find that the solution of the equation (4.18) can be represented in the form

$$\rho_\pi^{(2)} = \tilde{R}_\pi(z)\delta_\pi + G_\pi \tilde{R}_\pi(z)\delta_\pi \tag{4.26}$$

where G_π is the integral operator, and its kernel can be represented in the form

$$G_\pi(\lambda_1, \lambda_2; \mu_1, \mu_2; z) = \frac{\tilde{S}_\pi(\lambda_1, \lambda_2; \mu_1, \mu_2; z)}{\alpha_\pi^0(\lambda_1, \lambda_2) - z} + \\ + \sum_{i;k} \frac{\tilde{u}_\pi^{(i,k)}(\lambda_1 \lambda_2; \mu_1, \mu_2)}{\varepsilon_i(\pi - \lambda_k, \lambda_k) - z} + \sum_m Q_\pi^{(m)}$$

where $\lambda_3 = \pi - \lambda_1 - \lambda_2$; the functions $\tilde{S}_\pi$ and $\tilde{u}_\pi^{(i,k)}$ are analytic outside the cut $\tilde{\Phi}_\pi$ and bounded and continuous up to the cut $\tilde{\Phi}_\pi$, their limit values at this cut are differentiable, and the derivatives have the singularities $\sim |t - t_0|^{(\nu-\lambda)/2}$ in the singular points (values of the functions $\alpha_\pi^0(\lambda_1, \lambda_2)$ and $\varepsilon_i(\pi - \lambda_k, \lambda_k)$ in their critical points). The main part of the Lorenz expansion in the pole $z = \hat{\varepsilon}_m(\pi)$ is denoted by $Q_\pi^{(m)}$. We then use the representation (4.20) for the kernel of the operator $\tilde{R}_\pi(z)$ and also the form of the function δ_π (3.6) again and get a representation for $\rho_\pi^{(2)}$. The last term corresponding to the zeros of the determinant Δ_π in this representation can be obtained in the same way as the last term in the representation (4.12). □

5. The Main Result

After carrying out the investigation of the resolvent of the operator A it is convenient to formulate the main result about the spectral expansion of this operator.

Let us introduce auxiliary subspaces

$$\mathcal{H}_0 = L_2^{\text{sym}}((T^\nu)^3) = \mathcal{H}, \qquad \mathcal{H}_i = L_2((T^\nu)^2), i = 1, \ldots, s_1$$

$$\mathcal{H}_m = L_2(T^\nu), \qquad m = 1, \ldots, s_2$$

and introduce operators in each of them:

$$(\hat{A}_0 f^{(0)})(\pi_1,\pi_2,\pi_3) = a_0(\pi_1,\pi_2,\pi_3) f^{(0)}(\pi_1,\pi_2,\pi_3),$$
$$(\hat{U}_t^{(0)} f^{(0)})(\pi_1,\pi_2,\pi_3) = \exp\{i(t,\pi_1+\pi_2+\pi_3)\} f^{(0)}(\pi_1,\pi_2,\pi_3) t \in z^\nu,$$
$$(\hat{A}_i f^{(i)})(\pi_1,\pi_2) = \varepsilon_i(\pi_1,\pi_2) f^{(i)}(\pi_1,\pi_2),$$
$$(\hat{U}_t^{(i)} f^{(i)})(\pi_1,\pi_2) = \exp\{i(t,\pi_1+\pi_2)\} f^{(i)}(\pi_1,\pi_2),$$
$$(\tilde{A}_m f^{(m)})(\pi) = \hat{\varepsilon}_m(\bar{\mu}) f^{(m)}(\bar{\mu}) t \in z^\nu,$$
$$(\bar{U}_t^{(m)} f^{(m)})(\pi) = \exp\{i(t,\pi)\} f^{(m)}(\pi).$$

Then let $\bar{\mathscr{H}}$ be the direct sum of these spaces, and $\bar{A}$ and $\bar{U}_t$ be the direct sums of the operators (5.1).

THEOREM 5.9. *The unitary mapping $W\colon \bar{\mathscr{H}} \to \mathscr{H}$ transforming the operators $\bar{A}$ and $\bar{U}_t$ into the operators A and U_t exists:*

$$A = W\bar{A}W^{-1}, \qquad U_t = W\bar{U}_t W^{-1}.$$

Note that the spaces $\mathscr{H}$ and $\bar{\mathscr{H}}$ and the operators A and $\bar{A}$, U_t and $\bar{U}_t$ can be expanded into the direct integral

$$\mathscr{H} = \int_{T^\nu} \mathscr{H}_\pi \mathrm{d}\pi, \qquad (\mathscr{H}_\pi = L_2^{\mathrm{sym},\pi}), \qquad \bar{\mathscr{H}} = \int_{T^\nu} \bar{\mathscr{H}}_\pi \mathrm{d}\pi,$$

$$A = \int_{T^\nu} \mathscr{D}_\pi \mathrm{d}\pi, \qquad \bar{A} = \int_{T^\nu} \bar{\mathscr{D}}_\pi \mathrm{d}\pi,$$

$$U_t = \int U_t^{(\pi)} \mathrm{d}\bar{\mu}, \qquad \bar{U}_t = \int \bar{U}_t^{(\pi)} \mathrm{d}t,$$

where the space $L_2^{\mathrm{sym},\pi} = L_2^{\mathrm{sym}}(\gamma_\pi^3)$ and the operator $\mathscr{D}_\pi$ are defined above. The operator $U_t^{(\pi)} = \exp(i(t,\pi))E_\pi$. The space $\bar{\mathscr{H}}_\pi$ is the direct sum of the spaces

$$\bar{\mathscr{H}}_\pi = \mathscr{H}_\pi^0 \oplus \sum_{\mathrm{sym},\pi} \mathscr{H}_\pi^{(i)} \oplus \sum \bar{\mathscr{H}}_\pi^{(m)}$$

where

$$\mathscr{H}_\pi^0 = L_2^{\mathrm{sym},\pi}, \qquad \mathscr{H}_\pi^{(i)} = L_2(\gamma_\pi^2),$$

and

$$\gamma_\pi^2 = \{\pi_1,\pi_2 \colon \pi_1+\pi_2=\pi\} \subset (T^\nu)^2, \mathscr{H}_\pi^{(m)} = \mathbb{C}^1$$

Then the operators $\bar{\mathscr{D}}_\pi$ and $U_t^{(\pi)}$ can be represented as the direct sum of the operators $\mathscr{D}_\pi^0, \mathscr{D}_\pi^{(i)}, \mathscr{D}_\pi^{(m)}$ and $\bar{U}_{t,\pi}^{(0)}, \bar{U}_{t,\pi}^{(i)}, \bar{U}_{t,\pi}^{(m)}$ acting in these spaces and defined analogously to (5.1).

Theorem 5.9 evidently follows from the following theorem.

THEOREM 5.10. *For all $\pi \in T^\nu$ the unitary mapping $W_\pi\colon \bar{\mathscr{H}}_\pi \to \mathscr{H}_\pi$ transforming the operators $\bar{\mathscr{D}}_\pi$ and $\bar{U}_t^{(\pi)}$ into the operators $\mathscr{D}_\pi$ and $U_t^{(\pi)}$ exists.*

For the proof of this theorem we will use methods of scattering theory allowing us to write the operator W_π directly in the form of a so-called wave operator. Define operators of imbedding of the spaces $\mathscr{H}_\pi^{(0)}, \mathscr{H}_\pi^{(i)}, \mathscr{H}_\pi^{(m)}$ into $\mathscr{H}_\pi$. Let

$$\mathscr{T}_\pi^{(m)}: \bar{\mathscr{H}}_\pi^{(m)} \to \mathscr{H}_\pi: \lambda \to \lambda \hat{\varphi}_\pi^{(m)}(\kappa_1, \kappa_2), \lambda \in \mathbb{C}^1 \equiv \bar{\mathscr{H}}_\pi^{(m)}$$

where $\hat{\varphi}_\pi^{(m)}$ is the eigenfunction of the operator $\mathscr{D}_\pi$ corresponding to the eigenvalue $\hat{\varepsilon}_m(\pi)$. The mapping $\mathscr{T}_\pi^{(p)} = \oplus \mathscr{T}_\pi^{(m)}$ is evidently an isometric imbedding and the space $\mathscr{T}_\pi^p(\oplus \bar{\mathscr{H}}_\pi^{(m)}) = \mathscr{H}_\pi^{(p)} \subset \mathscr{H}_\pi$ is the invariant subspace of the operator $\mathscr{D}_\pi$ where its spectrum is discrete and coincides with $\{\hat{\varepsilon}_2(\pi), \ldots, \hat{\varepsilon}_{s_2}(\pi)\}$. Let $\tilde{\mathscr{H}}_\pi = \mathscr{H}_\mu^{(0)} \oplus \Sigma_i \mathscr{H}_\pi^{(i)}$ and $\mathscr{H}_\pi^c = \mathscr{H}_\pi \ominus \mathscr{H}_\pi^{(p)}$. Denote by $\mathscr{D}_\pi^c$ and $U_{t,\pi}^c$ the parts of the operators $\mathscr{D}_\pi$ and $U_t^{(\pi)}$ acting in $\mathscr{H}_\pi^c$. Then let

$$\mathscr{T}_\pi^{(i)}: \mathscr{H}_\pi^{(i)} \to \mathscr{H}_\pi; \qquad f(\lambda) \to \frac{1}{\sqrt{6}} \sum_{k=1,2,3} f(\lambda_k) \times$$

$$\times \varphi_{\lambda_k}^{(i)}, \frac{(\lambda_{k+1})}{\pi - \lambda_k}(\lambda_{k+1}) + \frac{1}{\sqrt{2}} \sum_{k=1,2,3} f(\lambda_k) \times$$

$$\times \hat{\varphi}_{\lambda_k, \pi-\lambda_k}^{(i)}(\hat{\lambda}_{k+1}).$$

Here $\varphi_{\pi_1,\pi_2}^{(i)}$ is the eigenvector of the operator A_{π_1,π_2} corresponding to the eigenvalue $\varepsilon_i(\pi_1, \pi_2)$. At least, let

$$\mathscr{T}_\pi^{(0)}: \bar{\mathscr{H}}_\pi^0 \to \mathscr{H}_\pi: f \to f - \sum_i p_\pi^{(i)} f \tag{5.3}$$

where $p_\pi^{(i)} \equiv \mathscr{T}_\pi^{(i)}(\mathscr{T}_\pi^{(i)})^*$ is the projecting operator on the subspace $\mathscr{T}_\pi^{(i)} \bar{\mathscr{H}}_\pi^{(i)} \subset \mathscr{H}_\pi$ (the equality $p_\pi^{(i)} = \mathscr{T}_\pi^{(i)} \cdot (\mathscr{T}_\pi^{(i)})^*$ follows from the equality $(\mathscr{T}_\pi^{(i)})^* \mathscr{T}_\pi^{(i)} = E_{\mathscr{H}_\pi^{(i)}}$, verified by a direct calculation). Furthermore, $\mathscr{T}_\pi^{(i)} \bar{\mathscr{H}}_\pi^{(i)} \perp \mathscr{T}_\pi^{(j)} \bar{\mathscr{H}}_\pi^{(j)}$, $i \neq j$ and so $p_\pi^{(i)} p_\pi^{(j)} = p_\pi^{(j)} p_\pi^{(i)} = 0$. Let $\tilde{\mathscr{T}}_\pi = \mathscr{T}_\pi^{(0)} \oplus \Sigma \mathscr{T}_\pi^{(i)}: \tilde{\mathscr{H}}_\pi \to \mathscr{H}_\pi$. Construct now the unitary mapping $\tilde{\mathscr{H}}_\pi \to \mathscr{H}_\pi^c$ defining

$$\tilde{W}_\pi = s - \lim_{t \to \infty} \exp\{i \mathscr{D}_\pi^c t\} \tilde{\mathscr{T}}_\pi \exp\{-i \tilde{\mathscr{D}}_\pi t\}. \tag{5.4}$$

THEOREM 5.11. *The limit* (5.4) *exists and defines the unitary operator* $\tilde{W}_\pi$; $\tilde{\mathscr{H}}_\pi \to \mathscr{H}_\pi^c$ *reducing* $\tilde{\mathscr{D}}_\pi$ *into* $\mathscr{D}_\pi^c$ *and* $\tilde{U}_t^{(\pi)}$ *into* $U_{t,c}^{(\pi)}$.

6. Proof of Theorem 5.11 (Scattering Theory)

In the proof of Theorem 5.11 we will follow the method of [9]. Introduce the operator

$$T_\pi: \tilde{\mathscr{H}}_\pi \to \mathscr{H}_\pi:$$

$$T_\pi = \mathscr{D}_\pi \tilde{\mathscr{T}}_\pi - \tilde{\mathscr{T}}_\pi \tilde{\mathscr{D}}_\pi = T_\pi^{(0)} \oplus \sum_i T_\pi^{(i)}$$

where

$$T_\pi^{(0)} = \mathscr{D}_\pi \mathscr{T}_\pi^{(0)} - \mathscr{T}_\pi^{(0)} \tilde{\mathscr{D}}_\pi^0, \qquad T_\pi^{(i)} = \mathscr{D}_\pi \mathscr{T}_\pi^{(i)} - \mathscr{T}_\pi^{(i)} \tilde{\mathscr{D}}_\pi^c \tag{6.1}$$

It is easy to show, using the equation for the eigenfunctions $\varphi_{\pi_1,\pi_2}^{(i)}$, that the operator $T_\pi^{(i)}$ is an integral operator acting from the space $L_2(\gamma_\pi^2)$ into $L_2^{\mathrm{sym}}(\gamma_\pi^3)$ with kernel $K_\pi(\lambda_1, \lambda_2, \pi)$ p-smooth with respect to all variables. Then let us represent the operator $\mathscr{D}_\pi$ as the sum

$$\mathscr{D}_\pi = \mathscr{D}_\pi^{(0)} + \mathscr{D}_\pi^{(1)} + \mathscr{D}_\pi^{(2)}$$

where $\mathscr{D}_\pi^{(0)}$ is the operator corresponding to the multiplication on a function, $\mathscr{D}_\pi^{(1)}$ is the operator corresponding to the integration on one variable and $\mathscr{D}_\pi^{(2)}$ is an integral operator. We have

$$\mathscr{D}_\pi \mathscr{T}_\pi^0 - \mathscr{T}_\pi^0 \tilde{\mathscr{D}}_\pi^0 = \mathscr{D}_\pi^{(1)} + \mathscr{D}_\pi^{(2)} - (\mathscr{D}_\pi^{(0)} + \mathscr{D}_\pi^{(1)} + \mathscr{D}_\pi^{(2)}) \times$$
$$\times \left(\sum p_\pi^{(i)}\right) + \left(\sum p_\pi^{(i)}\right)(\mathscr{D}_\pi^{(0)} + \mathscr{D}_\pi^{(1)})$$
$$- \left(\sum_i p_\pi^{(i)}\right)\mathscr{D}_\pi^{(1)}$$

Analogously to the above it can be shown that for all $i = 1, 2, \ldots, s$.

$$(\mathscr{D}_\pi^{(0)} + \mathscr{D}_\pi^{(1)})p_\pi^{(i)} - p_\pi^{(i)}(\mathscr{D}_\pi^{(0)} + \mathscr{D}_\pi^{(1)}) = \mathscr{R}$$

where $\mathscr{R}$ is an integral operator in the space $L_2^{\mathrm{sym}}(\gamma_\pi^3)$ with a p-smooth kernel. Hence we finally get that the operator $T_\pi^{(0)}: L_2^{\mathrm{sym}}(\gamma_\pi^3) \to L_2^{\mathrm{sym}}(\gamma_\pi^3)$ has the form

$$T_\pi^{(0)} = \left(E - \sum_i p_\pi^{(i)}\right)\mathscr{D}_\pi^{(1)} + V_\pi \tag{6.1}$$

where V_π is an integral operator with a p-smooth kernel.

LEMMA 6.1 *The operators $T_\pi^{(0)}$, $T_\pi^{(i)}$ can be represented in the form*

$$T_\pi^{(i)} = (Q_\pi^{(i)})^* \cdot S_\pi^{(i)}, \qquad Q_\pi^i: \mathscr{H}_\pi \to \bar{\mathscr{H}}_\pi^{(i)}, \qquad S_\pi^{(i)}: \bar{\mathscr{H}}_\pi^{(i)} \to \mathscr{H}_\pi^{(i)} \tag{6.2}$$

The operator $Q_\pi^{(i)}$ is $\mathscr{D}_\pi$-smooth and the operator $S_\pi^{(i)}$ is $\bar{\mathscr{D}}_\pi^{(i)}$smooth. Also

$$T_\pi^0 = (E - \sum p_\pi^{(i)}) \sum_e (M_\pi^{(l)})^* N_\pi^{(l)} +$$
$$+ \sum_m (Q_{\pi,m}^{(0)})^* S_{\pi,m}^{(0)},$$

$$M_\pi^{(l)}, N_\pi^{(l)}, Q_{\pi,m}^0, S_{\pi,M}^0: \mathscr{H}_\pi \to \mathscr{H}_\pi$$

where the operators $N_\pi^{(l)}$ and $S_{\pi,m}^0$ are $\bar{\mathscr{D}}_\pi^0$-smooth and $Q_{\pi,m}^{(0)}$ and $M_\pi^{(l)} \cdot (E - \Sigma_i p_\pi^{(i)})$ are $\mathscr{D}_\pi$-smooth (see the definition of a $\mathscr{D}$-smooth operator for the given self-adjoint operator $\mathscr{D}$ in [10]).

Proof. Let us show that the operator $F: L_2(\gamma_\pi^2) \to L_2^{\mathrm{sym}}(\gamma_\pi^3)$ with $(\nu + 3)$-smooth kernel can be represented in the product form $F = F' \cdot F''$ where $F'': L_2(\gamma_\pi^2) \to L_2(\gamma_\pi^2)$ and $F': L_2(\gamma_\pi^2) \to L_2^{\mathrm{sym}}(\gamma_\pi^3)$ and both operators have the smooth kernel $F'(\lambda_1, \lambda_2, \pi)$, $F''(\pi_1, \pi_2)$. We let

$$F''(\pi_1, \pi_2) = \sum_{n \in Z^\nu} \frac{\exp\{i(\pi_1 - \pi_2), n)\}}{(n^2 + 1)^q} \tag{6.3}$$

where $q = \frac{1}{2}(\nu + 2)$. In this case the kernel $F''(\pi, \pi')$ is differentiable; the inverse operator $(F'')^{-1}$ is a differential operator of power $(\nu + 2)$. Because of the sufficient smoothness of the kernal F, the operator $F' = F(F'')^{-1}$ has a smooth kernel. Hence the representation (6.2) for the operator $T_\pi^{(i)}$ follows. Note that for $\nu \geqslant 3$ the operator $S_\pi^{(i)}$ with smooth kernel is $\bar{\mathscr{D}}_\pi^{(i)}$-smooth and the operator $Q_\pi^{(i)}$ is $\mathscr{D}_\pi$-smooth, as follows from the formula for the resolvent of the operator $\mathscr{D}_\pi$ (4.24).

Let us turn to the operator T_π^0 and consider the terms $(E - \Sigma p_i)\mathscr{D}_\pi^{(1)}$ in the representation (6.1). Note that the operator $K: L_2(\gamma_\pi^3) \to L_2(\gamma_\pi^3)$ with the kernel of the form

$$K(\lambda_1, \lambda_2; \mu_1, \mu_2)\delta(\lambda_1 - \mu_1) \quad \text{or} \quad K(\lambda_1, \lambda_2; \mu_1, \mu_2)\cdot\delta(\lambda_1 - \mu_1 - \mu_2)$$

where K is the p-smooth kernel on the manifold $\{\lambda_1, \lambda_2, \mu_1, \mu_2;\ \lambda_1 = \mu_1\}\mathscr{T} \subset (T^\nu)^4$ (or on the manifold $\{\lambda_1, \lambda_2, \mu_1, \mu_2 : \lambda_1 = \mu_1 + \mu_2\} \subset (T^\nu)^4$) can be represented as a product $K = K' \cdot K''$ where K' and K'' are operators with the kernels of the form (6.3) as above, and the functions $K'(\lambda_1, \lambda_2; \mu_1 \mu_2)$ and $K''(\lambda_1, \lambda_2; \mu_1, \mu_2)$ defined at corresponding manifolds are differentiable. Having carried out this expansion for each term in $\mathscr{D}_\pi^{(1)}$ and turning to the symmetrized product K' and K'', we find that

$$\left(E - \sum p_\pi^{(i)}\right)\mathscr{D}_\pi^{(1)} = \sum_l \left(E - \sum p_\pi^{(i)}\right)\left(M_\pi^{(l)}\right)^* N_\pi^{(l)} + V'_\pi$$

where $M_\pi^{(l)}$ and $N_\pi^{(l)}$ are operators in $L_2(\gamma_\pi^3)$ with the kernels represented as the sum of δ-functions of the form $\delta(\lambda_i - \mu_j)$ or $\delta(\lambda_i - \mu_j - \mu_k)$, $i, j, k = 1, 2, 3$ with a smooth multiplier; V'_π is an integral operator in $L_2(\gamma_\pi^3)$. Then the sum of the integral operators $V'_\pi + V_\pi$ with smooth kernels can be expanded in the product

$$V'_\pi + V_\pi = Q_\pi^{(0)} S_\pi^{(0)}$$

again, where the integral operators $Q_\pi^{(0)}$ and $S_\pi^{(0)}$ have smooth kernels. The operator $S_\pi^{(0)}$ is $\bar{\mathscr{D}}_\pi^0$-smooth and $Q_\pi^{(0)}$ is $\mathscr{D}_\pi$-smooth as follows from the representation (4.24) of the resolvent of the operator $\mathscr{D}_\pi$. Then the operators $N_\pi^{(l)}$ are also $\bar{\mathscr{D}}_\pi$-smooth. We now show the $\mathscr{D}_\pi$-smoothness of the operator $M_\pi^{(l)}(E - \Sigma p_\pi^{(i)})$. Having used the expansion of the kernel of the resolvent $\mathscr{D}_\pi$, analogously to the expansion (4.24) we get that the kernel of the operator

$$K(z) = \left(E - \sum p_\pi^{(i)}\right)(\mathscr{D}_\pi - zE)\left(E - \sum p_\pi^{(i)}\right)$$

does not contain terms of the form

$$\frac{K_1(\lambda_1, \lambda_2; \mu_1, \mu_2)}{\varepsilon_i(\pi - \lambda_j, \lambda_j) - z}\delta(\lambda_j - \mu_k),$$

$$\frac{K_2(\lambda_1, \lambda_2; \mu_1, \mu_2; z)}{(\alpha_\pi^0(\lambda_1, \lambda_2) - z)(\varepsilon_i(\pi - \mu_k, \mu_k) - z)}\cdot\delta(\lambda_j - \mu_k),$$

$$\frac{K_3(\lambda_1, \lambda_2; \mu_1, \mu_2; z)}{(\varepsilon_i(\pi - \lambda_j, \lambda_j) - z)(\alpha_\pi^0(\mu_1 \mu_2) - z)}\delta(\lambda_j - \mu_k)$$

and analogous terms with δ-function of the form $\delta(\lambda_j - \mu_k - \mu_k)$ j, k, $k' = 1, 2, 3$. The remaining terms in $K(z)$ are such that the operator $(M_\pi^{(l)})^* K(z) M_\pi^{(l)}$ is uniformly bounded in the exterior of any neighbourhood of the eigenvalues of $\mathscr{D}_\pi$. The lemma is proved. □

It follows from the results of [11] that the operator $\tilde{W}_\pi$ exists, and $\mathrm{Im}\,\tilde{W}_\pi$ coincides with the absolutely continuous subspace $\mathscr{H}^c_\pi$ of the operator $\mathscr{D}_\pi$. The inverse mapping $\tilde{W}_\pi^{-1}: \mathscr{H}^c_\pi \to \mathscr{H}_\pi$ exists. In order to prove the equality

$$\mathrm{Im}\,\tilde{W}_\pi \equiv \tilde{\mathscr{H}}^c_\pi = \mathscr{H}^c_\pi$$

we need to prove the following lemma.

LEMMA 6.2. *The singular spectrum of the operator $\mathscr{D}_\pi$ is empty.*

Proof. We will use the following criterion of the absence of the singular spectrum [10]. Let the quadratic form $(\varphi, (\mathscr{D}_\pi - zE)\varphi)$ where φ has an everywhere dense set $\mathfrak{N}$ be bounded for all fixed π as the function of z outside any neighbourhood of a finite set of points z (independent of φ). Then the singular spectrum of the operator $\mathscr{D}_\pi$ is empty. Choosing a set of sufficiently smooth functions as the set $\mathfrak{N}$ and using the representation (4.24) for the kernel of the resolvent, it is again easy to prove that the above criterion is fulfilled. So $\mathrm{Im}\tilde{W}_\pi = \mathscr{H}^c_\pi$.

LEMMA 6.3. *The operator $\tilde{W}_\pi: \tilde{\mathscr{H}}_\pi \to \mathscr{H}^c_\pi$ is unitary. The proof is analogous to the proof of a similar statement in* [9].

Acknowledgements

In the course of this work the authors benefited from useful consultations with M. Sh. Birman, D. R. Jafaev and A. F. Vakulenko on scattering theory. We wish to express our deep gratitude to them.

References

1. Dobrushin, R. L. and Minlos, R. A.: 'Polynomes of Linear Random Functions', *Uspehi Mat. Nauk* **32** (1977), 67–122 (in Russian).
2. Malyshev, V. A. and Minlos, R. A.: Cluster Operators, *Trudy Sem. im. I. G. Petrovsky* **9** (1981), 63–80 (in Russian).
3. Arnold, V. I., Varchenko, A. N. and Gusein-Zadeh, S. M.: *Singularity of Differentiable Mappings, Part I*, Nauka, Moscow, 1982 (in Russian).
4. Lakaev, S. N.: 'Some Spectral Properties of the Generalized Friedrich's Model', *Trudy Sem. im. I. G. Petrovsky* **11** (1986), 210–238 (in Russian).
5. Kato, T.: *Perturbation Theory for Linear Operators*, Springer-Verlag, Berlin, 1966.
6. Lovitt, W. V.: *Linear Integral Equations*, McGraw-Hill, New York, 1924.
7. Gelfand, I. M. and Shilov, G. E.: *Generalization Functions and Actions Over Them.* 1, Phys.-mat. Cover Pub. Moscow, 1959 (in Russian).
8. Kirillov, A. A. and Gvishiany, A. D.: *Theorems and Problems of Functional Analysis*, Nauka, Moscow, 1979 (in Russian).
9. Jafaev, D. R.: On the Theory of Multichannel Scattering in a Pair of Spaces', *Theor. Math. Phys.* **37** (1978), 48–57 (in Russian).
10. Reed, M. and Simon, B.: *Methods of Modern Mathematical Physics.* IV. *Analysis of Operators.* Academic Press, New York, 1978.
11. Lavine, R.: 'Commutators and Scattering Theory II. A Class of One-Body Problems', *Indiana Univ. Math. J.* **21** (1972), 643–656.

M. L. BLANK

Stochastic Attractors and their Small Perturbations

1. Introduction

Let X be a ν-dimensional unit cube (torus) with the Euclidean norm $|\cdot|$ on X, and assume a piecewise differentiable endomorphism $T: X \to X$. According to the definition of Sinai [1] the subset $\Lambda \subset X$ is called a *stochastic attractor* of the dynamical system (T, X) if

(a) there exists an open set $U \subset X$ such that $\Lambda \subset U$, $\overline{TU} \subset U$ and $\Lambda = \cap_{n \geq 1} T^n U$;

(b) there exists a measure μ_* such that supp $\mu_* = \Lambda$ and iterations of any smooth measure μ with supp $\mu \subset U$ by endomorphism T converge weakly to μ_*;

(c) the dynamical system $(T|\Lambda, \Lambda, \mu_*)$ is mixing.

In this paper, we understand by a *smooth measure* any Borel measure absolutely continuous with respect to the Lebesgue measure m, and $\bar{U}$ is the closure of the set U.

Of all deterministic dissipative systems, those with stochastic attractors possess the strongest stochastic properties. Examples of such systems give hyperbolic systems satisfying Axiom A of Smale [2, 3, 4] and piecewise expanding mappings of the interval [5, 6]. In [7] it was shown that the Lorenz system

$$\begin{cases} dx/dt = \sigma y - \sigma x \\ dy/dt = rx - y - xz \\ dz/dt = xy - bz \end{cases}$$

also has the stochastic attractor. Technically it is more convenient to consider a Poincaré mapping of a suitably chosen two-dimensional plane area. This plane area transversly intersects the two-dimensional stable manifold of the null fixed point. The Poincaré mapping is therefore discontinuous, which makes this problem like that of the piecewise expanding type. The main purpose of this paper is to investigate piecewise differentiable dynamical systems with stochastic attractors and their various small perturbations.

The stochastic map $G_\varepsilon: X \to X$ is a small ε-*perturbation* of the identical map if the density of the probability distribution $G_\varepsilon x$ is concentrated in the ε-neighbourhood of the point $x \in X$. Correspondingly the ε-perturbed system (T_ε, X) is a dynamical system which is a composition of the system (T, X) and of ε-perturbation, namely $T_\varepsilon x = (G_\varepsilon \cdot T)x$.

In dealing with dynamical systems with stochastic attractors one certainly cannot expect any kind of stable behaviour of separate trajectories, hence only from the

R. L. Dobrushin (ed.), Mathematical Problems of Statistical Mechanics and Dynamics, 161–197.

statistical point of view is there any hope that such systems may be stable. We therefore introduce the notion of intrinsic stability of T-invariant measures.

We say that the T-invariant measure μ is *intrinsically stable* if there exists an open subset $U \subset X$ such that iterations of any smooth measure with support in U converge weakly under the action of the endomorphism T to the measure μ. The notion of intrinsic stability for ε-perturbed systems is analogous. If an invariant measure is not intrinsically stable we shall call it *intrinsically unstable*. In this terminology the measure μ_* in (b) is the only intrinsically stable mixing T-invariant measure with support in U.

A T-invariant measure μ is called *stable* with respect to the family of ε-perturbations $\{G_\varepsilon\}_{\varepsilon>0}$ if there exists a sequence of intrinsically stable invariant measures of ε-perturbed systems which converge weakly to μ. In this paper ε-perturbed systems (T_ε, X) have as a rule only one invariant measure, μ_ε.

We shall study conditions for families of ε-perturbations under which intrinsically stable T-invariant measures are stable (regular case), or intrinsically unstable invariant measures become stable (singular case). In Section 3 we prove that if stochastic ε-perturbations satisfy some mixing condition then the only stable measure with the support on the stochastic attractor is the measure μ_*. In [1, 8, 9, 10, 11, 12, 13, 14] analogous results were obtained for some specific classes of dynamical systems (hyperbolic diffeomorphisms, piecewise monotonic mappings of interval, topological Markov chains). In these papers the properties of unperturbed systems used were such that it was not possible to transfer proofs suitable for one class of dynamical systems to another. For example, in the investigation of hyperbolic diffeomorphisms [12] the existence of the uniform approximation of ε-trajectories [3] by trajectories of the system itself was very important. In the general case of discontinuous endomorphisms this property does not hold, and it is not possible to transfer the methods of [12] to the case of piecewise monotonic mappings of the interval or the Poincaré mapping of the Lorenz system.

In this paper we have essentially used in the proofs only such properties of stochastic attractors (a)–(c) that allow us to investigate uniform stability properties of invariant measures for all the cases mentioned above, and also to examine stochastic perturbations of the Lorenz-type systems which have not been investigated previously.

Closely connected to these problems is the problem of the stochastic stabilization of dynamical systems. If there exists a sequence of perturbations $\{G_\varepsilon\}_{\varepsilon>0}$ such that a sequence of their intrinsically stable measures converges to the measure μ_*, then we call this sequence of perturbations the stabilizing one.

We now give a brief description of the results of this paper.

In Section 2 properties of dynamical systems with stochastic attractors are examined. It is proved that a system which can be represented in the form of a skew product of systems, one of which is a system with stochastic attractors while the other is a contractive one, also has a stochastic attractor (Theorem 2.1). The character of the decay of correlations is also investigated. We prove that if the basic system has the property of exponential decay of correlations then the same decay is valid for the whole system. The construction of the skew product allows us to construct new examples of systems with stochastic attractors and to investigate some dynamical systems by reducing their dimension (Lorenz-type systems). In this section we also give some definitions and prove several technical results which we shall use later during the investigation of various perturbations.

In Sections 3–5 stochastic perturbations of dynamical systems with stochastic attractors are examined. The main result of this part gives sufficient conditions for the stability of the measure μ_* on the stochastic attractor (Theorems 3.1, 3.2, 4.1). We also examine the connection between supports of the smooth invariant measure of the initial dynamical system with the stochastic attractor and the invariant measure of the stochastically ε-perturbed system for each fixed $\varepsilon > 0$ (Theorem 3.4). In Section 4 we deal with the map $T: X \to R^{\nu}$, where X is an open bounded set in R^{ν}. The results obtained in this section are equivalent to the results of Section 3 with respect to conditionally invariant measures of the dynamical system (T, X) and its small stochastic perturbations.

In Section 5 we examine the singular case in the presence of stochastic ε-perturbations, namely the case of stabilization of the intrinsically unstable invariant measures. Theorem 5.1 demonstrates that if the map T locally satisfies the Lipschitz condition and there exists a cycle of the system (T, X) then, independently of its stablity there exists such a sequence of stochastic ε-perturbations that it stabilizes the T-invariant measure concentrated on this cycle.

In Sections 6 and 7 small deterministic perturbations of dynamical systems with stochastic attractors are investigated.

In Section 6 we consider the case when the ε-perturbation itself is a deterministic dynamical system (G_{ε}, X) with a stochastic attractor. In connection with general ideas about stochasticity in deterministic dynamical systems, it is essential to assume that results in this case must be the same as in the case of stochastic perturbations. Here we obtain some results which confirm this assumption; we investigate regular and singular cases. The last case is remarkable because the stabilization effect arises in the case of interactions of strongly unstable dynamical systems (T, X) and (G_{ε}, X) with stochastic attractors. The comparison of results obtained in Sections 3–5 and in Section 6 of this paper allows us to conclude that effects which are observed to react as purely stochastic or 'quasi-stochastic' perturbations are qualitatively the same. Moreover, the conditions generating the singular case in the presence of small perturbations are equal in both cases.

In Section 7 a type of small deterministic perturbation which appears in the presence of space discretizations is considered. Typical examples of such perturbations are round-off errors, which inevitably appear in the modelling of dynamical systems on computers. We shall demonstrate that even for extremely small perturbations of this type, structurally stable dynamical systems [22] in the presence of space discretizations acquire properties qualitatively far from the initial ones. We investigate for local changes the phenomenon of *period multiplication*, changing the structure of the neighbourhood of a periodic trajectory, and for global changes the existence of the global stabilization of the unstable cycle and the influence of small discretizations on the ergodicity property of some dynamical systems. The only approach which exists in the literature (see a survey in [15]) for asymptotic properties of such perturbations consists of replacing round-off errors by independent stochastic values uniformly distributed in the interval $[-\varepsilon, \varepsilon]$. Such an approach does not explain the above effects, but in this case a statistical approach is valid (Theorems 7.5 and 7.6). In these theorems we prove the correctness of the statistical approach with respect to all round-off errors, and calculate 'probabilities' of unstable cycle stabilizations and the ergodicity of perturbed systems.

2. Dynamical Systems with Stochastic Attractors

Let $T: X \to X$ be a measurable non-singular endomorphism of a compact set $X \subset R^\nu$. This means that for any Borel subset $A \subset X$ with $m(A) = 0$ we have $m(T^{-1}A) = 0$. By the dynamical system we mean the pair (T, X).

The main aim of this paper is to investigate properties of systems with discontinuities, therefore the systems under consideration may satisfy some smoothness conditions only piecewise.

We say that the endomorphism $T: X \to X$ piecewise satisfies the property $\mathcal{F}(T \in P\mathcal{F}(X))$ if there exists a finite partition $X = \cup_{i=1}^{w} X_i$, $w = w(T)$, $m(X_i) = m(X_i(T)) > 0$, each element X_i is a connected domain with the rectifiable boundary, $\text{int}(X_i) \cap \text{int}(X_j) = \emptyset$ if $i \neq j$ and the map $T_i = T_{\text{int}(X_i)}$ satisfies the property $\mathcal{F}$ and T_i can be extended to the closure of X_i as a function satisfying the condition $\mathcal{F}$. The partition $X_i(T)$ is called an $\mathcal{F}$-partition for the map T in terms of the property $\mathcal{F}$.

We assume that T is a piecewise diffeomorphism $(T \in PC^1(X))$ with a stochastic attractor. In some important examples of such dynamical systems (systems of the Lorenz type and others) these systems can be represented in the form of skew products (see below) that allow us to reduce them to the case of a smaller dimension. Let us examine a general model of this type. Let the dynamical system $(S, X \times Y)$ be represented in the form of a skew product of systems (T, X) and (G_x, Y), namely for any $x \in X$ and $y \in Y$ $S(x, y) = (Tx, G_x y)$.

THEOREM 2.1. *Let (T, X) be a system with the stochastic attractor Λ_T, $T \in PC^1(X)$ where the system (G_x, Y) is a contraction for all $x \in X$, i.e. there exists $\lambda \in (0, 1)$ such that for all $y_1, y_2 \in Y$ $|G_x y_1 - G_x y_2| \leqslant \lambda |y_1 - y_2|$ and the function $G_x y$ satisfies the Hölder condition on $x \in X$ $(G_x y \in C^\alpha(X), \alpha \in (0, 1))$:*

$$|G_{x_1} y - G_{x_2} y| \leqslant A |x_1 - x_2|^\alpha. \tag{2.1}$$

Then $(S, X \times Y)$ also has a stochastic attractor and for any $f, g \in C^\alpha(X \times Y)$ the rate of decay of correlations in the dynamical system $(S, X \times Y)$

$$K_S^{(n)}(f, g) = |\mu_S((f \circ T^n) g) - \mu_S(f) \mu_S(g)| \tag{2.2}$$

is the same as $K_T^{(n)}(f_1, g_1)$ in the system (T, X, μ_T), where $f_1, g_1 \in C^\alpha(X)$ and μ_T is the measure from the definition of the stochastic attractor Λ_T, μ_S is the measure on Λ_S, $\mu_S(f) = \int f d\mu_S$

Let us establish some preliminary results before proving the theorem.

LEMMA 2.1. *Under the conditions of Theorem 2.1 there exists a function $\varphi \in C^\alpha(X)$ such that for all $y \in Y$*

$$|G_x^n y - \varphi(T^n x)| \leqslant \lambda^n \operatorname{diam}(X)$$

where

$$G_x^n y = G_{T^{n-1}x} \cdot G_x^{n-1} y.$$

Proof.

$$\begin{aligned}|G_x^n y - G_x^n z| &= |G_{T^{n-1}x}\cdot G_x^{n-1}y - G_{T^{n-1}x}\cdot G_x^{n-1}z|\\ &\leqslant \lambda|G_x^{n-1}y - G_x^{n-1}z| \leqslant \cdots \leqslant \lambda^n|y-z|.\end{aligned}$$

Therefore there exists a function φ such that the exponential rate of the convergence $|G_x y - \varphi(T^n x)| \to 0$ as $n \to \infty$ follows from the inequality previously obtained. Let us check that $\varphi \in C^\alpha(X)$. Define the function $G_x^{(n)}y = G_x^{(n-1)}\cdot G_x y$. Then the function $\varphi(x)$ can be obtained as a limit in the uniform norm of functions $G_x^{(n)}y$ for given $y \in Y$. In this case the rate of the convergence is obviously the same as above:

$$\begin{aligned}|\varphi(u) - \varphi(v)| &\leqslant |\varphi(u) - G_u^{(n)}y| + |G_u^{(n)}y - G_v^{(n)}y| + |G_v^{(n)}y - \varphi(v)|\\ &\leqslant 2\lambda^n \operatorname{diam}(X) + \lambda|G_u^{(n-1)}y - G_v^{(n-1)}y| + A|u-v|^\alpha \leqslant \cdots\\ &\leqslant 2\lambda^n \operatorname{diam}(X) + A|u-v|^\alpha/(1-\lambda)\end{aligned}$$

and as $|\varphi(u) - \varphi(v)|$ does not depend on $n \in Z_+$ then it follows that $\varphi(x)$ satisfies the Hölder condition (2.1). □

Proof of Theorem 2.1. From the definition of the stochastic attractor Λ_T and the results of Lemma 2.1 one can see that all trajectories of $(S, X \times Y)$ converge to the set $\varphi(\Lambda_T) = \Lambda_S$. Let us show that Λ_S is a stochastic attractor for the system $(S, X \times Y)$ and the measure μ_S, defined for all Borel sets $A \subset X$ by the formula $\mu_S(\varphi(A)) = \mu_T(A)$ is an intrinsically stable measure (see the definition of the stochastic attractor). Note that it follows from Lemma 2.1 that the hypersurface $y = \varphi(x)$ is S-invariant. From this, and also from other results of Lemma 2.1 we immediately derive all properties of the stochastic attractor.

Now let pass to the second part of the theorem. We have

$$\begin{aligned}K_S^{(n)}(f,g) &= \left|\int f(T^n x, G_x^n y) g(x,y)\,\mathrm{d}\mu_S(x,y) - \int f\,\mathrm{d}\mu_S \int g\,\mathrm{d}\mu_S\right|\\ &= \left|\int f(T^n x, \varphi(T^n x) + y_n) g(x,y)\,\mathrm{d}\mu_S(x,y) - \int f\,\mathrm{d}\mu_S \int g\,\mathrm{d}\mu_S\right|\\ &\leqslant \left|\int f(T^n x, \varphi(T^n x)) g(x,\varphi(x))\,\mathrm{d}\mu_T(x) - \int f(x,\varphi(x))\,\mathrm{d}\mu_T \times\right.\\ &\quad\times \left.\int g(x,\varphi(x))\,\mathrm{d}\mu_T\right| + \left|\int f(T^n x, \varphi(T^n x) + y_n) g(x,y)\,\mathrm{d}\mu_S(x,y) - \right.\\ &\quad - \left.\int f(T^n x, \varphi(T^n x)) g(x,y)\,\mathrm{d}\mu_S(x,y)\right|.\end{aligned}$$

By Lemma 2.1, $|y_n| \leqslant C\lambda^n$. So the second term can be estimated from the above as follows:

$$\begin{aligned}R_n &= \int |f(T^n x, \varphi(T^n x) + y_n) - f(T^n x, \varphi(T^n x))|\;|g(x,y)|\,\mathrm{d}\mu_S(x,y)\\ &\leqslant D\int |y_n|^\alpha |g(x,y)|\,\mathrm{d}\mu_S(x,y) \leqslant DC^\alpha \lambda^{n\alpha}\int |g(x,y)|\,\mathrm{d}\mu_S(x,y)\end{aligned}$$

which converges to zero exponentially with respect to n. Now to estimate $K_S^{(n)}(f, g)$ from the above we examine the function $f_1(x) = f(x, \varphi(x))$ and $g_1(x) = g(x, \varphi(x))$. Clearly $f_1, g_1 \in C^\alpha(X)$, therefore $K_S^{(n)}(f, g) \leqslant K_T^{(n)}(f_1, g_1) + R_n$. □

COROLLARY. *If the rate of decay of correlations in the dynamical system* (T, X) *is exponential, then under the conditions of Theorem 2.1 the same statement is valid for the dynamical system* $(S, X \times Y)$.

For systems satisfying Smale's Axiom A the exponential estimates of the rate of the decay of correlations were obtained in [2]. Also, almost exponential estimates of the type $\exp(-An^{2/3})$ for some smooth dynamical systems and systems of the Lorenz type are known. For systems with discontinuities, exponential estimates have previously been obtained only in the case of one-dimensional maps [16, 17].

REMARK 2.1. If the spectrum of the dynamical system (T, X) has a discrete component, i.e. the mixing condition in the definition of the stochastic attractor holds only for some power n of the endomorphism T restricted to subsets of the stochastic attractor Λ_i, the situation is more complicated. In this case iterations of the smooth measure μ do not converge to μ_* (as in the definition of the stochastic attractor), but their Cesaro means do: $n^{-1}\Sigma_{k=0}^{n-1} T^{*k}\mu \to \mu_*$ as $n \to \infty$, where μ_* does not depend on μ as before. Also several stochastic attractors may appear in dynamical systems. In both cases the results of Theorem 2.1 remain valid, because the restriction of T^n to a small neighbourhood of the stochastic attractor Λ_i pertains to the situation described in Theorem 2.1.

Another means of constructing systems with stochastic attractors is *non-singular topological conjugation.* We say that the dynamical systems (T, X) and $(\hat{T}, X)$ are non-singularly conjugated if there exists a non-singular gomeomorphism $F: X \to X$ such that $FTF^{-1} = \hat{T}$.

THEOREM 2.2. *If a dynamical system* $(\hat{T}, X)$ *is non-singularly conjugated with a dynamical system* (T, X) *with a stochastic attractor, then* $(\hat{T}, X)$ *also has a stochastic attractor.*

REMARK 2.2. To prove the existence of stochastic attractors we impose some conditions of hyperbolicity [3]. A transition to a system non-singularly conjugated with the given one allows us to weaken these conditions. For example, in [18] it was shown that some systems with critical points satisfy the conditions of Theorem 2.2.

In the one-dimensional case, in dynamical systems which are non-singularly conjugated with piecewise expanding mappings of an interval, there is also the fact of the exponential rate of decay of correlations [16]. For an example of systems in which this property can be proved by the construction described above, let us examine the one-dimensional family of mappings of the unit interval $T_a x = ax(1 - x)$, $a \in (3, 4]$. In [18] it was demonstrated that this holds for the countable family of values of the parameter a.

We now introduce some notation which we need in order to investigate the stability properties of dynamical systems with stochastic attractors. By $M(X)$ we denote the set of normalized Borel measures on X; on $M(X)$ the operator T^* acts, which any $\mu \in M(X)$ transforms to $T^*\mu$ induced by the dynamical system (T, X); that is, if A is a Borel subset in X, then $T^*\mu(A) = \mu(T^{-1}A)$. For the investigation of weak convergence of measures let us introduce a metric $\rho(.,.)$ in $M(X)$ which is equivalent to weak convergence. For this purpose we fix in $C(X)$ a dense family of functions $\{\varphi_k(x)\}_{k=1}^{\infty}$ and define

$$\rho(\mu_1, \mu_2) = \sum_{k=1}^{\infty} 2^{-k} \|\varphi_k\|_C^{-1} \left| \int \varphi_k \, d\mu_1 - \int \varphi_k \, d\mu_2 \right|. \tag{2.3}$$

The space of Lebesgue-integrable functions on X is denoted by $L_1(X, \|\cdot\|)$. According to [5] we can define for the dynamical system (T, X) the Perron–Frobenius operator (PF-operator) $P: L_1(X) \to L_1(X)$. For a smooth measure μ with the density $f(x) = d\mu/dm$ the value of this operator on f is $Pf = d(T^*\mu)/dm$, or in other terms

$$Pf(x) = \sum_{i=1}^{w} f(T_i^{-1}x) |\det(DT_i^{-1}x)| 1_{TX_i}(x). \tag{2.4}$$

The formula (2.4) is the multidimensional generalization of the description of the PF-operator for the piecewise monotonic mappings of an interval [5, 6, 8].

LEMMA 2.2. *For all* $f, g \in L_1(X)$ $\|Pf - Pg\| \leqslant \|f - g\|$.

Proof. We shall use the formula (2.4) for the operator P. Denoting $h = f - g$ we obtain

$$\|Pf - Pg\| = \|Ph\| \leqslant \int |Ph(x)| \, dx = \int \left| \sum_{i=1}^{w} h(T_i^{-1}x) |\det(DT_i^{-1}x)| 1_{TX_i}(x) \right| dx$$

$$\leqslant \int \sum_{i=1}^{w} |h(T_i^{-1}x)| \cdot |\det(DT_i^{-1}x)| \, dx = \int P(|h|)(x) dx = \|h\| = \|f - g\|.$$

In this calculation we have essentially used the fact that for non-negative functions the operator P, defined by the formula (2.4), preserves the integral; this follows immediately from its definition (see also [24]).

LEMMA 2.3. *If measures* μ_1 *and* μ_2 *are smooth and their densities are* h_1 *and* h_2 *respectively, then* $\rho(\mu_1, \mu_2) \leqslant \|h_1 - h_2\|$.

Proof.

$$\rho(\mu_1, \mu_2) = \sum_{k=1}^{\infty} 2^{-k} \|\varphi_k\|_C^{-1} \left| \int \varphi_k \, d\mu_1 - \int \varphi_k \, d\mu_2 \right|$$

$$\leqslant \sum_{k=1}^{\infty} 2^{-k} \cdot \|\varphi_k\|_C^{-1} \cdot \int |h_1(x) - h_2(x)| \, dx = \|h_1 - h_2\|. \qquad \square$$

LEMMA 2.4. *Under conditions of Lemma* 2.3,

$$\rho(T^*\mu_1, T^*\mu_2) \leqslant \|h_1 - h_2\|.$$

Proof. By Lemma 2.3,

$$\rho(T^*\mu_1, T^*\mu_2) \leqslant \|dT^*\mu_1/dm - dT^*\mu_2/dm\| = \|Ph_1 - Ph_2\| \leqslant \|h_1 - h_2\|$$

as follows from Lemma 2.2. $\square$

The PF-operator P given by the formula (2.4) transforms the set of Hölder functions $C^\alpha(X)$ to the set of functions for which the Hölder condition can be performed only piecewise. The set of such functions is denoted by $PC^\alpha(X)$. In this space we introduce the norm

$$||f||_\alpha = ||f|| + \sum_i \sup_{x,y\in X_i(f), x\neq y} |f(x) - f(y)|/|x-y|^\alpha \tag{2.5}$$

where $\{X_i(f)\}_{i=1}^{w(f)}$ is the Hölder partition for the function $f \in PC^\alpha(X)$. In the case $w(f) = 1$, $f \in C^\alpha(X)$ the norm $||f||_\alpha$ coincides with the usual norm for Hölder-continuous functions of the compact space [23].

We now describe properties of piecewise Hölder functions that we shall need in future.

LEMMA 2.5. *For piecewise Hölder functions $f \in PC^\alpha(X)$ the following properties are valid:*

(a) *if $x, y \in X_i(f)$ then $|f(x) - f(y)| \leqslant (||f||_\alpha - ||f||)|x - y|^\alpha$;*
(b) *for all $x \in X$ there exists a constant $A = A(\{X_i(f)\}_{i=1}^{w(f)})$ such that $|f(x)| \leqslant A||f||_\alpha$;*
(c) *for all $\beta \in (0, \alpha)$ $f \in PC^\beta(X)$;*
(d) *if for all $x \in X$ $f(x) \geqslant b > 0$ then $1/f \in PC^\alpha(X)$;*
(e) *if $f \in PC^\alpha(X)$, $g \in PC^\beta(X)$ and $\alpha \wedge \beta = \min(\alpha, \beta)$ then $(f + g)$, $(fg) \in PC^{\alpha\wedge\beta}(X)$ and $f \cdot g \in PC^{\alpha\beta}(X)$.*

We shall say that the endomorphism $T \in PC^{1+\alpha}(X)$, $0 < \alpha < 1$, if the endomorphism T is a piecewise diffeomorphism $(T \in PC^1(X))$ and $J_i(x) = |\det(DT_i^{-1}x)| \in PC^\alpha(TX_i(T))$ for all $i = 1, 2, \ldots, w(T)$ is bounded from above by a constant $a < \infty$.

Note that $DTx = (DT_i^{-1}(Tx))^{-1}$, and therefore among maps $T \in PC^{1+\alpha}(X)$ there exist maps with unbounded derivatives (this follows from the fact that $J_i(x)$ may be equal to zero for some $x \in X$) that allow to deal with such dynamical system as the Lorenz system.

LEMMA 2.6. *If $T \in PC^{1+\alpha}(X)$ then the* PF-*operator P of the dynamical system (T, X) is such that $P: PC^\alpha(X) \to PC^\alpha(X)$ and there exists $C < \infty$ such that for all $f \in PC^\alpha(X)$ $||Pf||_\alpha \leqslant C||f||_\alpha$.*

Proof. It follows immediately from the definition of $T \in PC^{1+\alpha}(X)$ that for all $i \leqslant w(T)$, T_i^{-1} is correctly defined and continuously differentiable in $TX_i \subset X$. Assume that points x, y belong to the same domain of the C^α-partition for $Pf(x)$. There are $w(f, T) \leqslant w(f)w(T)$ such domains. Therefore

$$|Pf(x) - Pf(y)| \leqslant \sum_j |f(T_j^{-1}x)J_j(x) - f(T_j^{-1}y)J_j(y)|$$

$$\leqslant w(f, T)a||f||_\alpha|x - y|^\alpha \leqslant aw(f, T)||f||_\alpha(\operatorname{diam}(X))^\alpha.$$

Now from Lemma 2.5(b) one can easily obtain the statement of the lemma. □

The second class of functions which we shall be interested in is the class of functions of bounded variation (BV(X)). For the investigation of functions of this class one can introduce in $L_1(X)$ equivalence classes of functions which distinguish on sets of the m-measure zero. For equivalence classes in $L_1(X)$ ($X = I_1 \times \cdots \times I_\nu$ where $I_i = [0, 1]$) we introduce the notion of variation:

$$V(f) = \inf\{\operatorname{var}(\hat{f}): \hat{f} \in f\}, \tag{2.6}$$

where $\operatorname{var}(\hat{f})$ is the Tonelly variation [19] defined by the formula

$$\operatorname{var}(\hat{f}) = \sum_{i=1}^{\nu} \int_{I_1,\ldots,I_{i-1},I_{i+1},\ldots,I_\nu} \operatorname{var}_{I_i}(\hat{f})\,dx_1 \cdots dx_{i-1}\,dx_{i+1} \cdots dx_\nu$$

and $\operatorname{var}_{I_i}(\hat{f})$ is the usual one-dimensional variation [20] through the segment I_i with fixed variables $x_1, \ldots, x_{i-1}, x_{i+1}, \ldots, x_\nu$; f is the equivalence class in $L_1(X)$. In the space $\mathrm{BV}(X)$ one can introduce the variation norm

$$\|f\|_v = \|f\| + V(f). \tag{2.7}$$

We now show that the PF-operator transforms the space $\mathrm{BV}(X)$ into itself, and study some properties of these operators in $\mathrm{BV}(X)$.

LEMMA 2.7. *If $T \in PC^1(X)$ and $V(J_i(x)) < \infty$ then $P: \mathrm{BV}(X) \to \mathrm{BV}(X)$ and there exist constants $\alpha, \beta > 0$ such that for all $f \in PC(X) \cap \mathrm{BV}(X)$ the following inequality is valid*:

$$\|Pf\|_v \leqslant \alpha\|f\|_v + \beta\|f\|. \tag{2.8}$$

Proof. As $Pf(x)$ is represented (by the formula (2.4)) in the form of the sum of multiplications of bounded functions with bounded Tonelly variation, it is enough to use properties of functions with bounded Tonelly variation. A survey of these properties can be found in [19]. □

Usually to investigate a dynamical system (T, X) we fix some properties of the endomorphism T, and all properties of the system (T, X) are determined from these. To investigate properties of intrinsically stable measures of this system it is more natural in some cases to set the properties of its PF-operator. For instance, in the one-dimensional case the following statement is valid.

PROPOSITION 2.1. *Let $X = [0, 1]$ and let the* PF-*operator P satisfy the condition* (2.8); *if there exists $N \in Z_+$ and $\gamma < \infty$ such that for all functions $f \in \mathrm{BV}(X)$ and $n \geqslant N$ $\|P^n f\|_v \leqslant \gamma\|f\|$, then for all $f \in \mathrm{BV}(X)$*

$$S_n(f) = n^{-1} \sum_{k=0}^{n-1} P^k f \to f^* \in \mathrm{BV}(X)$$

in the L_1 sense, and f^* is the density of the intrinsically stable measure if $f \geqslant 0$ and $\|f\| = 1$.

Proof. Let $f \in \mathrm{BV}(X)$, then

$$V(S_n(f)) = V\left(n^{-1} \sum_{k=0}^{n-1} P^k f\right) \leqslant V\left(n^{-1} \sum_{k=0}^{N} P^k f\right) + V\left(n^{-1} \sum_{k=N+1}^{n-1} P^k f\right)$$

$$\leqslant n^{-1} V\left(\sum_{k=0}^{N} P^k f\right) + (n - N)\gamma\|f\|/n \leqslant \gamma\|f\|$$

provided $n \in Z_+$ is large enough. Hence by the Helly theorem [20] the sequence $S_n(f)$ has the limit point $f^* \in BV(X)$. We can therefore apply the statistical ergodic theorem [21], whence it follows that the sequence $S_n(f)$ strongly converges to f^* and $Pf^* = f^*$, which is what we require. □

COROLLARY. *In the one-dimensional case the conditions of Proposition* 2.1 *are valid for piecewise-expanding maps of an interval* [6, 8].

LEMMA 2.8. *Let* $\{\mu^{(n)}\}_{n=0}$, $\{\mu_\varepsilon^{(n)}\}_{n=0}$ *be two sequences of normalized Borel measures in* $M(X)$ *with* $\mu^{(0)} = \mu_\varepsilon^{(0)} = \mu$, $\varepsilon > 0$ *and assume two sequences of finite functionals* $\Psi_n^{(1)}(\mu) \to 0$ *as* $n \to \infty$ *and* $\Psi_n^{(2)}(\mu)$. *If there exists* $\alpha \in (0, 1]$ *and measures* $\mu_*, \mu_\varepsilon \in M(X)$ *such that*

$$\max\{\rho(\mu^{(n)}, \mu_*), \rho(\mu_\varepsilon^{(n)}, \mu_\varepsilon)\} \leqslant \Psi_n^{(1)}(\mu) \tag{2.9}$$

$$\rho(\mu_\varepsilon^{(n)}, \mu^{(n)}) \leqslant \varepsilon^\alpha \Psi_n^{(2)}(\mu) \tag{2.10}$$

then $\rho(\mu_\varepsilon, \mu_*) \to 0$ *as* $\varepsilon \to 0$.

Proof. By the triangle inequality, for an arbitrary $n \in Z_+$ one has

$$\begin{aligned}\rho(\mu_\varepsilon, \mu_*) &\leqslant \rho(\mu_\varepsilon, \mu_\varepsilon^{(n)}) + \rho(\mu_\varepsilon^{(n)}, \mu^{(n)}) + \rho(\mu^{(n)}, \mu_*)\\ &\leqslant 2\Psi_n^{(1)}(\mu) + \varepsilon^\alpha \Psi_n^{(2)}(\mu).\end{aligned}$$

Fix an arbitrary $\delta > 0$; then there exists $n = n(\delta)$ such that $\Psi_n^{(1)}(\mu) \leqslant \delta/3$, then for all $0 < \varepsilon \leqslant (\delta/(3\Psi_n^{(2)}(\mu)))^{1/\alpha}$

$$\rho(\mu_\varepsilon, \mu_*) \leqslant 2\delta/3 + \delta/(3\Psi_n^{(2)}(\mu))\Psi_n^{(2)}(\mu) = \delta.$$

As the left-hand side of the last inequality does not depend on δ for sufficiently small $\varepsilon > 0$, then $\mu_\varepsilon \to \mu_*$ as $\varepsilon \to 0$ weakly, which is what is required. □

We also use technical results obtained in Lemmas 2.2–2.8 to investigate the stochastic and deterministic perturbations of dynamical systems. In the case of one-dimensional systems with stochastic attractors, stochastic perturbations have already been obtained in the same manner in [8]. In this paper we have used the fact that not only does the weak convergence of the sequence of iterations of smooth measures to an intrinsically stable measure under the action T^* take place, but so does the convergence in $L_1(X)$ of corresponding densities to the density of μ_*, which is also a smooth measure. These facts allow us to deal only with PF-operators, and essentially simplify the calculations in our proofs. However, in the general case of multidimensional systems with stochastic attractors, the measures μ_* are not smooth.

3. Stochastic Perturbations (Regular Case)

We now give a more precise definition of stochastic perturbations than the general one in §1. The stochastic function $G_\varepsilon x$ is a stochastic ε-perturbation of the identical map if the sequence $x^{(n+1)} = G_\varepsilon x^{(n)}$ for all $n \in Z_+$ is the Markov chain on X with transition probability densities $q_\varepsilon(x^{(n)}, x^{(n+1)})$. We shall examine only local perturbations which satisfy the condition $q_\varepsilon(x, y) = 0$ if $y \notin B_\varepsilon(x)$, where $B_\varepsilon(x) = \{y \in X: |x - y| \leqslant \varepsilon\}$.

An operator $Q_\varepsilon: L_1(X) \to L_1(X)$ which describes the dynamics of densities for this process is defined by the formula

$$Q_\varepsilon f(x) = \int f(u) q_\varepsilon(u, x) \mathrm{d}u. \tag{3.1}$$

The corresponding operator acting in the measure space $M(X)$ is denoted by G_ε^*. This operator satisfies the condition $\mathrm{d}(G_\varepsilon^* \mu)/\mathrm{d}m = Q_\varepsilon(\mathrm{d}\mu/\mathrm{d}m)$ for each smooth measure $\mu \in M(X)$.

Also, as in §1 the ε-perturbed system (T_ε, X) is a composition of the ε-perturbation and the initial system, i.e. $T_\varepsilon = G_\varepsilon \cdot T$. It is more natural in this case to deal not with maps themselves (one of them is deterministic and the other one is stochastic), but with operators describing the dynamics of their measures and densities. For the dynamical system (T, X) these operators are T^* and P defined in §2, and for stochastic perturbations G_ε^* and Q_ε. It is clear that the ε-perturbed dynamical system (T_ε, X) is a stochastic Markov chain with transition probability densities $p_\varepsilon(x, y) = q_\varepsilon(Tx, y)$, and operators describing the dynamics of measures and densities for this process are $T_\varepsilon^* = G_\varepsilon^* T_\varepsilon^*$ and $P_\varepsilon = Q_\varepsilon P$.

The first question which arises in the investigation of invariant measures of ε-perturbed systems is a connection between these measures and invariant measures of the initial system.

PROPOSITION 3.1. *Let the* PF-*operator* P *of the dynamical system* (T, X) *transform each bounded function on* X *to a bounded piecewise-continuous function, and the family of stochastic kernels* $\{q_\varepsilon(x, y)\}_{\varepsilon > 0}$ *be continuous with respect to the second variable. Then if* $\|P_\varepsilon - P\| \to 0$ *as* $\varepsilon \to 0$ *then all limit points (in the sense of weak convergence of measures) of smooth invariant measures of ε-perturbed systems are T-invariant measures.*

Proof. If the Markov chain corresponding to the ε-perturbed system satisfies the Feller condition (that an operator P transforms each continuous function to a continuous one) then an analogous statement has been proved in [4, 11]. The proof in given here is a simple generalization of the construction in [11].

Let a measure μ be a limit point for the sequence of smooth invariant measures of perturbed systems. Then there exists the subsequence $\mu_{\varepsilon_n} \to \mu$ as $n \to \infty$, where $\varepsilon_n \to 0$. Hence for all $f \in C(X) \int P_{\varepsilon_n} f \,\mathrm{d}\mu_{\varepsilon_n} = \int f \,\mathrm{d}\mu_{\varepsilon_n}$, as the measure μ_{ε_n} is T_{ε_n}-invariant. We shall demonstrate that in this case the limit measure μ is T-invariant. For this purpose we obtain the following estimate:

$$\left| \int Pf \,\mathrm{d}\mu - \int f \,\mathrm{d}\mu \right| \leqslant \left| \int Pf(\mathrm{d}\mu - \mathrm{d}\mu_{\varepsilon_n}) \right| + \left| \int (Pf - P_{\varepsilon_n} f) \,\mathrm{d}\mu_{\varepsilon_n} \right| +$$

$$+ \left| \int f(\mathrm{d}\mu_{\varepsilon_n} - \mathrm{d}\mu) \right|.$$

The first integral can be estimated by the sum of moduli of integrals over connected domains on each of which the function Pf is continuous and bounded (as $Pf \in PC(X)$). Then, because of the weak convergence of measures on such domains, the first integral tends to zero. The second integral tends to zero by the convergence of P_{ε_n} to P in $L_1(X)$. The third integral tends to zero by the weak convergence of the sequence of invariant measures.

Hence for each $f \in C(X)$ we have $|\int Pf\, d\mu - \int f\, d\mu| = 0$, from which, and from the definition of the operator P, it follows that the measure μ is T-invariant. □

It is interesting to remark that the condition of smallness of the ε-perturbation is not sufficient for $||P_\varepsilon - P|| \to 0$ as $\varepsilon \to 0$. In §5 we shall construct some counterexamples to show this.

Consider the family of stochastic kernels $(q_\varepsilon(x, y)\}_{\varepsilon > 0}$ satisfying the following conditions:

if $y \notin B_\varepsilon(x)$ then $q_\varepsilon(x, y) = 0$,

$$\int_{B_\varepsilon(y)} q_\varepsilon(x, y)\, dx \leqslant A < \infty \tag{3.2}$$

$$\text{there exists } \kappa > 0 \text{ such that } \lim_{\varepsilon \to 0} \inf \varepsilon^\kappa q_\varepsilon(x, x) > 0 \text{ for all } x \tag{3.3}$$

$$\text{there exists } \delta = \delta(\varepsilon) \text{ such that if } y \in B_\delta(x) \text{ then } q_\varepsilon(x, y) > 0. \tag{3.4}$$

A very wide class of stochastic perturbations satisfies these conditions. For instance, the conditions are satisfied for independent stochastic perturbations, Markov perturbations with stochastic kernels depending only on the norm of the difference of coordinates, and Markov perturbations with doubly stochastic kernels.

We now formulate the main theorem on the stability of intrinsically stable invariant measures of the dynamical system (T, X) with respect to the family of stochastic ε-perturbations with kernels satisfying conditions (3.2)–(3.4).

THEOREM 3.1. *Let the dynamical system* (T, X) *have a stochastic attractor* Λ, *and let one of the following conditions be valid:*

(a) $T \in PC^{1+\alpha}(X)$;
(b) $T \in PC^1(X)$ *and for all* $i = 1, \dots, w(T)$ $V(|\det(DT_i^{-1}x)|) < \infty$

and let stochastic perturbations satisfy conditions (3.2)–(3.4). *Then for all* $\varepsilon > 0$ *there exists only one invariant measure* μ_ε *of* (T_ε, X) *and* $\mu_\varepsilon \to \mu_*$ *weakly as* $\varepsilon \to 0$, *where* μ_* *is the intrinsically stable T-invariant measure on* Λ.

We first sketch the proof of the theorem. Let μ be a smooth measure with support in a small neighbourhood of the stochastic attractor. Examine iterations of this measure under the action of T^*: $\mu^{(n)} = T^{*n}\mu$ and T_ε^*: $\mu_\varepsilon^{(n)} = T_\varepsilon^{*n}\mu$. From the properties of Λ it follows that $\mu^{(n)} \to \mu_*$ as $n \to \infty$. We shall demonstrate that under the assumptions of Theorem 3.1, for all $\varepsilon > 0$ there exists only one T_ε-invariant measure μ_ε and $\mu_\varepsilon^{(n)} \to \mu_\varepsilon$ as $n \to \infty$. The following estimate is valid:

$$\rho(\mu_\varepsilon^{(n)}, \mu^{(n)}) \leqslant \varepsilon^\alpha \phi(\mu, n)$$

where $\alpha \in (0, 1)$ and the finite functional $\phi(.,.)$ depends only on the initial measure μ and the number of iterations n, but does not depend on ε for all $\varepsilon > 0$ small enough. The main statement of the theorem now follows from this consideration by Lemma 2.8.

To begin the proof formally we need some properties of systems with ε-perturbations satisfying conditions (3.2)–(3.4).

LEMMA 3.1. *Under the conditions of Theorem* 3.1, *for all small enough* $\varepsilon > 0$ *there exists only one* T_ε*-invariant measure* μ_ε.

Proof. From the definition of the stochastic attractor it follows that there exists a neighbourhood U such that $\overline{TU} \subset U$. Therefore, for sufficiently small $\varepsilon > 0$ the Markov chain defined by the ε-perturbed system never leaves U if $x_\varepsilon^{(0)} \in U$. Now this fact and the transitivity of Λ (which follows from part (c) of its definition), together with (3.2)–(3.4), imply the Doeblin condition (see Hypothesis D′ in [22]). From this condition the statement of Lemma 3.1 follows. □

REMARK 3.1. Under the conditions of Theorem 3.1 there exists an estimate of the rate of convergence of $\mu_\varepsilon^{(n)}$ to μ_ε which does not depend on ε for all $\varepsilon > 0$ small enough.

Proof. From the Doeblin condition one can obtain that the exponential rate of convergence to the invariant measure μ_ε exists for the sequence $\{\mu_\varepsilon^{(n)}\}_{n=1}$. In this estimate the index of the exponent is equal to $-(\text{const}\cdot n/n_0 - 1)$. The value n_0 in this expression is the 'mixing time' for the corresponding Markov chain. But from the definition of a stochastic attractor it follows that (T, Λ, μ_*) is a mixing system. The stochastic perturbations examined in this section can only increase the rate of mixing, because of the estimate above, so the statement of the remark is proved. □

LEMMA 3.2. *For all* $T \in PC^{1+\alpha}(X)$ *and stochastic perturbations satisfying conditions* (3.2)–(3.4) *there exists a finite functional* $F(f)$ *on* $PC^\alpha(X)$ *such that it depends only on* $\|f\|_\alpha$, $w(f)$ *and the total* $(v-1)$*-dimensional Lebesgue measure of the boundary of the Hölder partition elements* $X_i(f)$ $(l(\partial X_i))$ *and*

$$\|Q_\varepsilon f - f\| \leqslant A\varepsilon^\alpha F(f). \tag{3.5}$$

Proof.

$$\|Q_\varepsilon f - f\| = \int |Q_\varepsilon f(x) - f(x)|\,dx \leqslant \int_{B_\varepsilon(x)} q_\varepsilon(u, x)|f(u) - f(x)|\,du\,dx$$

$$\leqslant \int_{B_\varepsilon(x)} \operatorname{osc}(f) \int q_\varepsilon(u, x)\,du\,dx$$

where $\operatorname{osc}_E(f) = \sup_E(f(x)) - \inf_E(f(x))$. Hence it follows from the condition (3.2) that

$$\|Q_\varepsilon f - f\| \leqslant A \int_{B_\varepsilon(x)} \operatorname{osc}(f)\,dx = A \sum_{i=1}^{w(f)} \int_{X_i(f)\,B_\varepsilon(x)} \operatorname{osc}(f)\,dx$$

$$\leqslant A \sum_{i=1}^{w(f)} \left(\int_{B_\varepsilon(\partial X_i(f))\,B_\varepsilon(x)} \operatorname{osc}(f)\,dx + \int_{X_i(f)\setminus B_\varepsilon(X_i(f))\,B_\varepsilon(x)} \operatorname{osc}(f)\,dx \right)$$

$$\leqslant A \sum_{i=1}^{w(f)} \left(2\varepsilon l(\partial X_i(f)) \sup_X f(x) + \int_{X_i(f)} \varepsilon^\alpha(\|f\|_\alpha - \|f\|)\,dx\right) = \varepsilon^\alpha F(f)$$

where ∂X_i is the boundary of the set X_i. □

In this estimate we have essentially used properties of functions from $PC^{\alpha}(X)$ obtained in Lemma 2.5.

LEMMA 3.3. *Under the conditions of Lemma* 3.1, *let the condition* $T \in PC^{1+\alpha}(X)$ *be replaced by* $T \in PC^1(X)$ *and* $V(|\det(DT_i^{-1}x)|) < \infty$; *then for all* $f \in PC(X)$ *we have*

$$\|Q_\varepsilon f - f\| \leqslant 2\varepsilon AV(f). \tag{3.6}$$

Proof. Just as in the proof of Lemma 3.1 we can obtain that

$$\|Q_\varepsilon f - f\| \leqslant A \int_{B_\varepsilon(x)} \operatorname{osc}(f)\,dx \leqslant 2\varepsilon AV(f).$$

The last inequality follows from the properties of functions with bounded Tonelly variation [19] and from the total continuity of the space $L_1(X)$ [23]. □

Note that in the proofs above we have essentially used the condition (3.2) of smallness of ε-perturbations. In the case of Gaussian perturbations, for instance, such an approach is not valid.

LEMMA 3.4. *Let* $\mu^{(n)} = T^{*n}\mu$, $\mu_\varepsilon^{(n)} = T_\varepsilon^{*n}\mu$ *where* $\mu \in M(X)$, *be a smooth measure with support* U *in a small neighbourhood of the stochastic attractor and its density* $h = d\mu/dm = 1_U(x)/m(U)$. *Then there exists the finite function* ϕ *of a natural argument* $n \in Z_+$ *such that* $\rho(\mu_\varepsilon^{(n)}, \mu^{(n)}) \leqslant \varepsilon^\alpha \phi(n)$.

Proof. Denote $h_\varepsilon^{(n)} = d\mu_\varepsilon^{(n)}/dm$ and $h^{(n)} = d\mu^{(n)}/dm$. From Lemma 2.3 and the triangle inequality we obtain that

$$\begin{aligned}\rho(\mu_\varepsilon^{(n)}, \mu^{(n)}) &\leqslant \|h_\varepsilon^{(n)} - h^{(n)}\| = \|Q_\varepsilon P h_\varepsilon^{(n-1)} - Ph^{(n-1)}\| \\ &\leqslant \|Q_\varepsilon P h_\varepsilon^{(n-1)} - Q_\varepsilon P h^{(n-1)}\| + \|Q_\varepsilon P h^{(n-1)} - Ph^{(n-1)}\| \\ &= \|Q_\varepsilon P(h_\varepsilon^{(n-1)} - h^{(n-1)})\| + \|Q_\varepsilon(Ph^{(n-1)}) - Ph^{(n-1)}\|.\end{aligned}$$

From Lemmas 2.2 and 3.2 (or 3.3) we can derive

$$\begin{aligned}\rho(\mu_\varepsilon^{(n)}, \mu^{(n)}) &\leqslant \|h^{(n-2)} - h^{(n-2)}\| + \varepsilon^\alpha F(P^{n-2}h) + \varepsilon^\alpha F(P^{n-1}h) \\ &\leqslant \cdots \leqslant \varepsilon^\alpha \sum_{k=0}^{n-1} F(P^k h) = \varepsilon^\alpha \phi(n).\end{aligned}$$ □

Proof of Theorem 3.1. From Lemmas 3.1–3.4 it follows that the sequences $\{\mu_\varepsilon^{(n)}\}_{n=1}$ and $\{\mu^{(n)}\}_{n=1}$ satisfy the conditions of Lemma 2.8, and hence all the statements of the theorem can be derived from this lemma. Note that one of the conditions of Lemma 2.8 is the existence of the uniform (by ε) estimate of the rate of the convergence of $\mu_\varepsilon^{(n)}$ to μ_ε, a fact which follows from Remark 3.1. Theorem 3.1 is proved. □

Conditions (a) and (b) of this theorem are technical ones, and can be relaxed essentially in the same manner as in Theorem 2.2.

THEOREM 3.2. *Let the dynamical system* $(\hat{T}, X)$ *be non-singularly topologically conjugated with the system* (T, X) *with a stochastic attractor satisfying the conditions of Theorem* 3.1; *then all the statements of Theorem* 3.1 *remain valid for the system* $(\hat{T}, X)$.

For the proof of this theorem it is enough to check that after changing variables, by the conjugated homeomorphism, a new family of stochastic kernels satisfies the conditions (3.2)–(3.4). It is immediately obvious that these conditions really hold, possibly with other constants $\hat{A}$ and $\hat{\kappa}$. On the other hand, for each pair of perturbed maps T and $\hat{T}$ their invariant measures are connected with the same conjugated homeomorphism. Therefore the limit measures are connected in the same manner, which finishes the proof. □

REMARK 3.2. In the general case, when there exist several stochastic attractors in a dynamical system (T, X) and its spectrum has a discrete component, all the statements of Theorems 3.1 and 3.2 are also valid for the restriction of T to a small neighbourhood of each connected component of each attractor.

Proof. In this case iterations of a smooth measure μ with the density $h = 1_{U_i}(x)/m(U_i)$ (where U_i is a small neighbourhood of the ith stochastic attractor) do not converge to μ_{*_i}, but their Cesaro means do, $n^{-1}\Sigma_{k=0}^{n-1} T^{*k}\mu \to \mu_{*_i}$ as $n \to \infty$. Therefore we can replace the estimate from Lemma 3.4 by

$$\rho\left(n^{-1}\sum_{k=0}^{n-1}\mu_\varepsilon^{(k)}, n^{-1}\sum_{k=0}^{n-1}\mu^{(k)}\right) \leqslant n^{-1}\sum_{k=0}^{n-1}\rho(\mu_\varepsilon^{(n)}, \mu^{(n)}) \leqslant \varepsilon^\alpha\phi(n),$$

and hence we can again apply Lemma 2.8. □

Now we consider the smoothing effect of stochastic perturbations. Consider two dynamical systems (T, X) and (S, X) with small C-distance between maps T and S. We shall assume that there exists a stochastic attractor Λ_T of the system (T, X) and that there is no such attractor of (S, X) (for instance (S, X) has a stable cycle in the domain Λ_T). In this case invariant measures of these systems can differ widely in the weak convergence metric $\rho(.,.)$. Denote PF-operators of these systems by P_T and P_S.

THEOREM 3.3. *Let dynamical systems (T, X) and (S, X) satisfy conditions* (a) *or* (b) *of Theorem* 3.1, *and assume that there exists sufficiently small $\varepsilon > 0$ such that $|Tx - Sx| \leqslant \varepsilon$, $||P_T - P_S|| \leqslant \varepsilon$ for all $x \in X$. Then there exists $\delta = \delta(\varepsilon)$ such that ε-perturbed systems (T_δ, X), (S_δ, X) with stochastic perturbations satisfying conditions* (3.2)–(3.4) *have intrinsically stable invariant measures μ_δ and η_δ and $\rho(\mu_\delta, \eta_\delta) \leqslant \text{const}(\delta)$.*

Proof. The existence of such $\delta = \delta(\varepsilon)$ follows from the fact that for each $\delta > 0$ the system (T_δ, X) satisfies the Doeblin condition [22], and because of a small difference between T and S in the C-metric. Exactly as in the proof of Lemma 3.4 we obtain

$$\begin{aligned}\rho(\mu_{T,\delta}^{(n)}, \mu_{S,\delta}^{(n)}) &\leqslant ||Q_\delta P_T h_{T,\delta}^{(n-1)} - Q_\delta P_S h_{S,\delta}^{(n-1)}|| \leqslant ||P_T h_{T,\delta}^{(n-1)} - P_S h_{S,\delta}^{(n-1)}|| \\ &\leqslant ||P_T h_{T,\delta}^{(n-1)} - P_T h_{S,\delta}^{(n-1)}|| + ||P_T h_{S,\delta}^{(n-1)} - P_S h_{S,\delta}^{(n-1)}|| \\ &\leqslant ||h_{T,\delta}^{(n-1)} - h_{S,\delta}^{(n-1)}|| + ||(P_T - P_S)h_{S,\delta}^{(n-1)}|| \\ &\leqslant ||h_{T,\delta}^{(n-1)} - h_{S,\delta}^{(n-1)}|| + \varepsilon||h_{S,\delta}^{(n-1)}|| \leqslant \cdots \leqslant n\varepsilon||h||\end{aligned}$$

where $\mu_{T,\delta}^{(n)}$ and $\mu_{S,\delta}^{(n)}$ and iterations of a smooth measure μ with support in some neighbourhood of Λ_T and $h_{T,\delta}^{(n)}$, $h_{S,\delta}^{(n)}$ and h are their densities. This estimate allows us to apply Lemma 2.8, which completes the proof. The estimated rates of convergence here are exponential, as follows immediately from the Doeblin condition. □

We now fix some $\varepsilon > 0$, and consider what can be said about the T_ε-invariant measure μ_ε.

THEOREM 3.4. *Let (T, X) be a dynamical system with stochastic attractor Λ and stochastic ε-perturbations satisfying conditions* (3.3)–(3.4); *then for all $\varepsilon > 0$ there exists only one T_ε-invariant measure μ_ε and*

$$\mathrm{supp}(\mu_\varepsilon) \supset \mathrm{supp}(\mu_*) \tag{3.7}$$

Proof. The first part of this theorem follows from Theorem 3.1, therefore we need prove only (3.7). If the condition (3.4) is fulfilled then $\mathrm{supp}(G_\varepsilon^* \mu) \supset \mathrm{supp}(\mu)$ for all $\mu \in M(X)$. Denote $M = \mathrm{supp}(\mu_*)$, $M_\varepsilon = \mathrm{supp}(\mu_\varepsilon)$, $M_0 = M \setminus M_\varepsilon$, $M_1 = M \cap M_\varepsilon$. If $M_0 = \emptyset$ then (3.7) is fulfilled. We shall prove (3.7) by a contradiction. Let $M_0 \neq \emptyset$; then two situations are possible.

(a) $M_1 \neq \emptyset$; from the definition of a stochastic attractor it then follows that $TM_1 \cap M_0 \neq \emptyset$. As $M_\varepsilon = \mathrm{supp}(G_\varepsilon^* T^* \mu_\varepsilon)$ and $M_\varepsilon = M_1 \cup (M_\varepsilon \setminus M_1)$, hence $\mathrm{supp}(T^*\mu_\varepsilon) = TM_\varepsilon = TM_1 \cup T(M_\varepsilon \setminus M_1)$ and $\mathrm{supp}(T^*\mu_\varepsilon) \cap M_0 \supset TM_1 \cap M_0 \neq \emptyset$ (otherwise M_1 is a T-invariant set as $M_0 \cup M_1 = \Lambda$). From this fact and (3.4) it follows that $M_\varepsilon \cap M_0 \neq \emptyset$, that is a contradiction to the assumption $M_1 \neq \emptyset$.

(b) $M_1 = \emptyset$. Consider the set TM_ε. If $TM_\varepsilon \setminus M \neq \emptyset$ then $\mathrm{supp}(G_\varepsilon T^* \mu_\varepsilon) \setminus \mathrm{supp}(\mu_\varepsilon) \neq \emptyset$, but this is not valid. Hence only the case $TM_\varepsilon \subset M_\varepsilon$ remains. We define the sequence of compact sets $M^{(k)}$ by the formula $M_\varepsilon^{(k+1)} = TM_\varepsilon^{(k)}$. It is easy to show that for all $k \in Z_+$ $M_\varepsilon^{(k+1)} \subset M_\varepsilon^{(k)}$ (otherwise the previous inclusion is not fulfilled). Hence this sequence has a limit point – the set $M^{(\infty)}$ such that $TM^{(\infty)} = M^{(\infty)}$. Therefore the set $M^{(\infty)}$ is a T-invariant stable set which is not equal to Λ, which contradicts the definition of a stochastic attractor.

4. The Law of Exponential Decay and Small Stochastic Perturbations

So far in this paper we have dealt with maps transforming a compact set X into itself. It turns out that in some examples, especially in problems connected with pre-turbulent phenomena, this condition is restrictive. We therefore consider here dynamical systems (T, X) such that $X \subset R^\nu$ is an open bounded set and $T: X \to R^\nu$. The most interesting problems arising here are the investigation of conditionally invariant measures of (T, X) and the distribution of the exit time for points of the set X [24]. Here we shall study sufficient conditions by which smooth conditionally invariant measures are stable with respect to small stochastic perturbations, and the law of the exponential decay for ε-perturbed systems is valid. This is so because of the fact that the probability for a given point remaining in the set X after n iterations is an exponential function of x. In other words, we shall study in this section the regular case of such dynamical systems.

Let $X = \cup_{i=1}^{w} X_i$, where $X_i \subset R^\nu$ are disjoint open linear connected bounded sets. We shall assume that the endomorphism T satisfies the following properties:

(a) $TX \supset X$;
(b) $X \cap \partial TX = \emptyset$;
(c) there exists $N \in Z_+$ such that $T^N X_i \supset X$ for all $i = 1, \ldots, w$;
(d) $T \in C^2(X \cap T^{-1} X)$ and all partial derivatives are uniformly bounded;

(e) there exists $\lambda > 1$ such that the Jacobian of the map T is a λ-expanding matrix for all $x \in X$.
Remember that a matrix A is called λ-expanding if

$$\inf\{|Ax|: |x| = 1\} \geqslant \lambda.$$

The system (T, X) satisfying conditions (a)–(e) is called in [24] a Markov system. The normalized Borel measure $\mu \in M(X)$ is called conditionally T-invariant [24] if for each Borel subset $E \subset X$

$$\mu(E) = \mu(T^{-1}E)/\mu(T^{-1}X).$$

For the systems studied in this section one can define an operator describing the dynamics of conditional densities which is analogous to the PF-operator in the usual case (see §2). Let P be a linear operator defined on $L_1(R^\nu)$ by the formula (2.4), then for non-negative functions $f \in L_1(R^\nu)$ such that

$$||Pf||_X = \int_X Pf(x)dx > 0,$$

one can define the operator $\hat{P}: L_1(R^\nu) \to L_1(R^\nu)$ by the formula $\hat{P}f = Pf/||Pf||_X$. From [24] it follows that the properties of this operator with respect to densities of conditional measures are equal to the corresponding properties of PF-operators with respect to the usual densities of smooth measures.

By $C_+(X)$ we define a set of non-negative Lipschitz functions on $C(X)$. For functions $f \in C_+(X)$ one can define a functional $H(\cdot)$: $H(f) = \sup\{|f'(x)|/f(x): x \in X, f'(x)$ is defined and $f(x) > 0\}$. In [24] it was proved that if the dynamical system (T, X) satisfies the Markov conditions (a) – (e), then there exists a constant $\alpha < \infty$ such that

$$H(\hat{P}f) \leqslant \alpha + H(f)/\lambda. \tag{4.1}$$

In this section we shall assume that stochastic ε-perturbations satisfy conditions (3.2)–(3.4), and moreover that their transition probability densities depend only on the norm of differences of coordinates, $q_\varepsilon(x, y) = q_\varepsilon(|x - y|)$, and are piecewise continuous.

An operator $\hat{P}$ describing the dynamics of conditional probabilities densities of the ε-perturbed system (T_ε, X) is

$$\hat{P}_\varepsilon f = Q_\varepsilon Pf/||Q_\varepsilon Pf||_X. \tag{4.2}$$

We now obtain some properties of stochastic ε-perturbations examined in this section.

LEMMA 4.1. *For all* $f \in C_+(X)$, $H(Q_\varepsilon f) \leqslant H(f)$.

Proof. By definition

$$H(Q_\varepsilon f) = \sup_X |\mathrm{d}/\mathrm{d}x\, Q_\varepsilon f(x)|/Q_\varepsilon f(x) = \sup_X \left|\mathrm{d}/\mathrm{d}x \int q_\varepsilon(|u|)f(x-u)\mathrm{d}u\right|/Q_\varepsilon f(x)$$

$$= \sup_X \left|\int q_\varepsilon(u, x)f'(u)\mathrm{d}u\right|/Q_\varepsilon f(x).$$

But from the definition of the functional $H(f)$ one can obtain that for all $u \in X$ $|f'(u)| \leqslant H(f)f(u)$, therefore

$$H(Q_\varepsilon f) \leqslant \sup_X \left| \int q_\varepsilon(u,x) H(f) f(u) du \right| / Q_\varepsilon f(x) = H(f). \qquad \square$$

LEMMA 4.2. *There exists a finite functional* $\phi_1(.)$ *such that for all* $f \in C_+(X)$ *if* $\|f\|_X > 0$ *then*

$$\|f\|_X / \|Q_\varepsilon f\|_X \leqslant 1 - \varepsilon\phi_1(f).$$

Proof. In [24] it was proved that for a given system (T, X) there exists a constant $C > 0$ which depends only on the Markov partition $\{X_i\}_{i=1}^w$ such that for all $f \in C_+(X)$ $\sup_X |f(x)| \leqslant CH(f)\|f\|_X$. Therefore one can obtain the following inequalities:

$$\|Q_\varepsilon f\|_X = \int_X Qf(x)dx \geqslant \int_{X \setminus B_\varepsilon(\partial X)} f(x)dx \geqslant \int_X f(x)dx - \int_{B_\varepsilon(\partial X)} f(x)dx$$
$$\geqslant \|f\|_X - l(\partial X)\varepsilon \cdot \sup_X |f(x)| \geqslant \|f\|_X(1 - CH(f)\cdot l(\partial X)\varepsilon).$$

The second inequality follows from the condition (3.2), because for each Borel set $A \subset X$ $\text{supp}(Q_\varepsilon 1_A) \subset \text{supp}(1_A) \cup B_\varepsilon(\partial A)$. $\square$

We define the finite functional $\phi_2(f)$ by $\phi_2(f) = CH(f)l(\partial X)$.

$$\|f\|_X / \|Q_\varepsilon f\|_X \leqslant \|f\|_X / (\|f\|_X(1 - \varepsilon\phi_2(f))) = 1/(1 - \varepsilon\phi_2(f)) \leqslant 1 - \varepsilon\phi_1(f),$$

as for sufficiently small $\varepsilon > 0$ such an $\phi_1(.)$ exists.

We can now prove the main result of this section.

THEOREM 4.1. *If a dynamical system* (T, X) *is a Markov system and the stochastic* ε*-perturbations satisfy the conditions described above, then for all* $\varepsilon > 0$ *there exists only one smooth conditionally invariant measure* μ_ε *of the* ε*-perturbed system* (T_ε, X) *and* $\mu_\varepsilon \to \mu_*$ *weakly as* $\varepsilon \to 0$, *where* μ_* *is the only smooth conditionally T-invariant measure.*

Proof. Let $f \in C_+(X)$. We shall show that $H(\hat{P}_\varepsilon f) \leqslant H(\hat{P} f)$. In fact

$$H(\hat{P}_\varepsilon f) = H(Q_\varepsilon Pf / \|Q_\varepsilon Pf\|_X) = H(Q_\varepsilon Pf) \leqslant H(Pf) = H(\hat{P} f).$$

The second and last inequalities follow from the fact that for each constant $C > 0$ $H(Cf) = \sup|Cf'(x)|/(Cf(x)) = H(f)$. Therefore $H(\hat{P}_\varepsilon f) \leqslant H(\hat{P} f) \leqslant \alpha + H(f)/\lambda$.

Now, exactly as in [24] one can obtain that the equation $\hat{P}_\varepsilon f = f$ defining the density of the smooth conditionally invariant measure has only one normalized solution f_ε and for all $f \in C_+(X)$, $\|f\|_X = 1$ the sequence of iterations $\hat{P}_\varepsilon^n f \to f_\varepsilon$ in L_1-norm as $n \to 0$.

It remains only to prove that $f_\varepsilon \to f_*$ as $\varepsilon \to 0$, where f_* is the density of μ_*. Applying Lemma 4.2 one can derive

$$\|\hat{P}_\varepsilon f - \hat{P} f\|_X = \|Q_\varepsilon Pf / \|Q_\varepsilon Pf\|_X - Pf / \|Pf\|_X\|_X$$
$$= \|(Q_\varepsilon Pf \|Pf\|_X / \|Q_\varepsilon Pf\|_X - Pf) / \|Pf\|_X\|_X$$
$$\leqslant \|Q_\varepsilon Pf(1 - \varepsilon\phi_1(f)) - Pf\|_X / \|Pf\|_X \leqslant \|Q_\varepsilon Pf - Pf\|_X / \|Pf\|_X +$$
$$+ \varepsilon\phi_1(f) / \|Pf\|_X \leqslant \varepsilon A \phi_3(f).$$

The last inequality is a corollary of Lemma 3.2. A and $\phi_3(.)$ are an appropriative constant and some finite functional. Now just as in the proof of Theorem 3.1 we can apply Lemma 2.8.

The phenomenon of exponential decay is an easy consequence of this theorem and therefore we do not prove it here. □

5. Stochastic Perturbations (Singular Case)

In §§3 and 4 we mentioned that under given assumptions the measure μ_* which was described in the definition of the stochastic attractor is intrinsically stable (by means of the definition in §1). It is easy to demonstrate that smooth invariant measures of ε-perturbed systems in these cases are also intrinsically stable. Therefore statements proved in §§3 and 4 imply the convergence of intrinsically stable measures of ε-perturbed systems to the intrinsically stable measure of the initial system. This we call the *regular* case. The conditions obtained above are of a sufficiently general character to be applied to the broad class of dynamical systems and their stochastic perturbations. It therefore might seem that in the general case ε-perturbed systems are regular; but this is not true. Moreover, in this section we show that there always exists a sequence of stochastic perturbations tending to zero such that their intrinsically stable smooth invariant measures (conditionally invariant measures) converge to an intrinsically unstable invariant measure of the initial system.

Let us consider an endomorphism $T: X \to R^\nu$, where X is an open bounded subset of R^ν (as in §4).

THEOREM 5.1. *Let the dynamical system (T, X) have a periodic point $y \in X$ and in a small neighbourhood of its trajectory let the map T satisfy the Lipschitz condition with a constant λ. Then there exists a family $\{G_\varepsilon\}$ of stochastic ε-perturbations such that there exists a sequence of intrinsically stable $T_\varepsilon = G_\varepsilon \cdot T$-invariant measures converging as $\varepsilon \to 0$ to the T-invariant measure concentrated on the given periodic trajectory.*

Proof. Consider the case when y is a fixed point of (T, X), that is $Ty = y$. To prove this we shall construct the sequence of stochastic perturbations in an obvious form and find supports of their smooth invariant measures converging to the measure concentrated at the point y.

Fix numbers γ, η such that $1 < \gamma \geqslant \eta > 0$. Let us denote $\varepsilon = \gamma\eta$, $V_1^{\gamma,\eta} = B_\varepsilon(y) \setminus B_\eta(y)$, $V_2^\varepsilon = R^\nu \setminus B_\varepsilon(y)$. We are interested only in the set $B_\varepsilon(y)$, so outside this set stochastic perturbations can be defined in an arbitrary way; for instance one can fix the uniform distribution on balls with radius ε. If both points $x, u \in X$ belong to $B_\varepsilon(y)$, we define the kernel of the stochastic perturbation by

$$q^{\gamma,\eta}(x, u) = \begin{cases} 0, & \text{if } x \in B_\varepsilon(y),\ u \in V_1^{\gamma,\eta}, \\ 1/m(B_\eta(0)), & \text{if } x \in B_\varepsilon(y),\ u \in B_\eta(y). \end{cases}$$

Now if $\varepsilon > 0$ is small enough and $\gamma \geqslant \lambda$ then there exists a subset $V_0^{\gamma,\eta} \subset B(y)$ which is invariant and transitive with respect to the ε-perturbed dynamical system. Therefore from Lemma 3.1 it follows that there exists only one smooth T_ε-invariant measure μ_ε with the support on $V_0^{\gamma,\eta}$.

Fix $\gamma \geqslant \lambda$; then $\varepsilon = \gamma\eta$ tends to zero as $\eta \to 0$ and the sequence of densities of T_ε-

invariant measures converges to the delta-function at the point $y \in X$ (because their supports converge to this point). This completes the proof. □

In the general case of the periodic point $y \in X$ of an arbitrary period one can repeat the same constructions at the neighbourhood of each point of the trajectory of y. As these constructions are essentially similar to the above we do not describe them.

There remains the interesting question of which are the conditions among (3.2)–(3.4) that are violated in this case. Analysis shows that only the condition (3.2) is fulfilled, i.e. there exist points for which the probability of returning to their small neighbourhoods is equal to zero. This brings us to the situation when the Markov chain corresponding to the ε-perturbed system has a new ergodic component which was not there in the initial system. In other words, the singular phenomenon for stochastic perturbations arises.

One can remark that if trajectories of almost all points of the dynamical system (T, X) are dense in X, then the smooth measure constructed in the proof of Theorem 5.1 is the only invariant measure for the ε-perturbed system and hence this measure is clearly intrinsically stable.

Now let the point $y \in X$ be an unstable periodic point of the dynamical system (T, X). The results of Theorem 5.1 show that the unstable periodic trajectory of this point may be stochastically stabilized. Analogous results in the continuous-time case were previously obtained using absolutely different methods [25]. It is interesting to remark that in both cases, in spite of the difference between the methods, the structure of perturbations depends only on the Lipschitz constant of the map T in a small neighbourhood of the periodic trajectory.

6. Small Quasi-Stochastic Perturbations

In previous sections of this paper we have dealt with stochastic perturbations of dynamical systems with stochastic attractors. In §§6 and 7 we shall study two types of deterministic perturbations of such systems, which nevertheless have some stochastic properties.

In §6 we investigate only one-dimensional dynamical systems (T, X), where $X = [0, 1]$ and $T \in PC^1(X)$. This map is said to be *piecewise expanding* [6] if $\inf|T'x| \geqslant a > 1$ and $V(1/|T'x|) < \infty$. In [6, 8, 10], it was demonstrated that piecewise expanding maps satisfy the following condition: there exists $n_0 \in Z_+$, $\alpha \in (0, 1)$ and $\beta < \infty$ such that

$$||P^{n_0}f||_v \leqslant \alpha||f||_v + \beta||f|| \tag{6.1}$$

for all $f \in BV(X)$, where P is the PF-operator of the dynamical system (T, X). It is easily shown that if the statement (6.1) is fulfilled then there exists a constant $\gamma < \infty$ such that for all $f \in BV(X)$ and sufficiently large $n \in Z_+$, $||P^n f||_v \leqslant \gamma||f||$. All the results of Proposition 2.1 are therefore valid in this case and hence there exists a smooth intrinsically stable T-invariant measure μ_* (there may be several such measures). The support of each smooth T-invariant measure is a stochastic attractor for the dynamical system (T, X) (see also [6, 8, 10]).

The perturbations G_ε that we consider in this section are deterministic dynamical systems (G_ε, X). We shall assume that the map $G_\varepsilon \in PC^2(X)$ and $\inf|G'x| \geqslant \lambda_\varepsilon > 1$. The smallness condition of ε-perturbations, as in §1, means that for all $x \in X$

$$|G_\varepsilon x - x| \leqslant \varepsilon. \tag{6.2}$$

We denote points of discontinuities and changes of monotonic intervals of the map T by $a^{(1)} < \cdots < a^{(w)}$ and of the map G_ε by $a_\varepsilon^{(1)} < \cdots < a_\varepsilon^{(w_\varepsilon)}$; $d_\varepsilon = \min\{|x - y|: x, y \in \Omega_\varepsilon, x \neq y\}$ where

$$\Omega_\varepsilon = \bigcup_{i=1}^{w} a^{(i)} \cup T^{-1}\left(\bigcup_{i=1}^{w_\varepsilon} a_\varepsilon^{(i)}\right);$$

$$Y_\varepsilon(R) = \{x \in X: |G'_\varepsilon x| > R\}, \qquad \hat{Y}_\varepsilon(R) = X \setminus Y_\varepsilon(R),$$

$$X_\varepsilon = \{x \in X: \operatorname{card}(G_\varepsilon^{-1}x) > 1\}, \qquad \hat{X}_\varepsilon = \{x \in X: \operatorname{card}(G_\varepsilon^{-1}x) = 1\}.$$

THEOREM 6.1. *Let T be a piecewise expanding map on X and assume that there exists only one smooth T-invariant measure, $G_\varepsilon \in PC^2(X)$ and*

(a) $\operatorname{card}(G_\varepsilon^{-1}x) \leqslant K < \infty$;
(b) $d_\varepsilon \geqslant d > 0$;
(c) $m(Y_\varepsilon(1 + \beta\varepsilon)) \leqslant \alpha_\varepsilon(\beta) \to 0$ as $\varepsilon \to 0$ *for all* $\beta > 0$.

Then for all $f \in BV(X)$, $||f|| = 1, f \geqslant 0$ for all $\varepsilon > 0$ small enough

$$n^{-1} \sum_{k=0}^{n-1} P_\varepsilon^k f \to f_\varepsilon^* \quad as \quad n \to \infty$$

where $P_\varepsilon f_\varepsilon^* = f_\varepsilon^*$, $||f_\varepsilon^*|| = 1$, $||f_\varepsilon^*||_v \leqslant C$ (this constant does not depend on $\varepsilon > 0$).

To start the proof formally we need in some estimates for piecewise expanding ε-perturbations.

LEMMA 6.1. *For all $R > 1$, $m(X_\varepsilon \setminus Y_\varepsilon(R)) \leqslant R - 1$.*

Proof. Let us consider the set $A = X_\varepsilon \setminus Y_\varepsilon(R)$. For all $x \in A$ $\operatorname{card}(G_\varepsilon^{-1}x) > 1$ and $|G'_\varepsilon x| \leqslant R$. This set is divided by points $a_\varepsilon^{(i)}$ into sets A_i on each of which G_ε is a one-to-one map. Then

$$\sum_i m(G_\varepsilon A_i) \leqslant R \sum_i m(A_i) = Rm(A).$$

Therefore the summary measure of the set B, the set of overlappings of images of A_i under the action of G_ε, is smaller than or equal to

$$\sum_i m(GA_i) - 1 \leqslant Rm(A) - 1 \leqslant R - 1$$

as $m(A) \leqslant 1$. But then $m(A) \leqslant m(G_\varepsilon^{-1}B) \leqslant m(B) \leqslant R - 1$, where the first inequality follows from the fact that for all $x \in A$ $\operatorname{card}(G_\varepsilon^{-1}x) > 1$ and the second one from $|G'_\varepsilon x| > 1$. □

LEMMA 6.2. *Under the conditions of Theorem 6.1 for all $f \in BV(X)$*

$$||Q_\varepsilon f - f|| \leqslant \hat{\alpha}_\varepsilon ||f||_v \to 0 \quad as \quad \varepsilon \to 0$$

where Q_ε is the PF-*operator of the dynamical system (G_ε, X).*

Proof. Let us consider the set

$$\hat{Z}_\varepsilon(R) = \{x \in \hat{X}_\varepsilon : |G'_\varepsilon(G_\varepsilon^{-1}x)| > R\}.$$

For all $x \in \hat{Z}_\varepsilon(R)$ we have $G_\varepsilon^{-1}x \in Y_\varepsilon(R)$. The reverse map in this case is defined correctly as $x \in \hat{X}$, that is $G_\varepsilon^{-1}x$ is a one-to-one map. But G_ε is a piecewise expanding map, and therefore for each Borel set A, $m(G_\varepsilon^{-1}A) \leqslant Km(A)$, where K is a constant from Theorem 6.1 (a). Denote

$$\hat{A}_\varepsilon = \{x \in \hat{X} : |G'_\varepsilon(G_\varepsilon^{-1}x)| \leqslant 1 + \beta\varepsilon\},\ A_\varepsilon = X \setminus \hat{A}.$$

Then one can derive an estimate

$$m(A_\varepsilon) \leqslant m(Y_\varepsilon(R)) + m(X_\varepsilon) + m(\hat{Z}_\varepsilon(R)) \leqslant \alpha_\varepsilon + \beta\varepsilon + K\beta\varepsilon$$

where $R = 1 + \beta\varepsilon$. The second inequality follows from Lemma 6.1. On the other hand $||Q_\varepsilon f - f|| \leqslant ||(Q_\varepsilon f - f)|_{\hat{A}_\varepsilon}|| + ||(Q_\varepsilon f - f)|_{A_\varepsilon}||$.

$$||(Q_\varepsilon f - f)|_{\hat{A}_\varepsilon}|| = \int_{\hat{A}_\varepsilon} |Q_\varepsilon f(x) - f(x)|\mathrm{d}x = \int_{\hat{A}_\varepsilon} |f(G_\varepsilon^{-1}x)/|G'_\varepsilon(G_\varepsilon^{-1}x)| - f(x)|\mathrm{d}x$$

$$\leqslant \int_{\hat{A}_\varepsilon} |f(G_\varepsilon^{-1}x) - |G'_\varepsilon(G_\varepsilon^{-1}x)|f(x)|\mathrm{d}x/\inf|G'_\varepsilon(G_\varepsilon^{-1}x)|$$

$$\leqslant \int_{\hat{A}_\varepsilon} |f(G_\varepsilon^{-1}x) - f(x)|\mathrm{d}x + \beta\varepsilon||f||$$

by the definition of the set $\hat{A}_\varepsilon$.

As $f \in BV(X)$ then exist two monotone increasing functions f_1 and f_2 such that $f_1(x) = V(f_1; [0, x]), f_2(x) = f_1(x) - f(x)$ [20].

$$S = \int_{\hat{A}_\varepsilon} |f(G^{-1}x) - f(x)|\mathrm{d}x \leqslant \sum_{i=1}^{2} \int_{\hat{A}_\varepsilon} |f_i(G_\varepsilon^{-1}x) - f_i(x)|\mathrm{d}x$$

$$\leqslant \sum_{i=1}^{2} \int_X |f_i(x + \varepsilon) - f_i(x - \varepsilon)|\mathrm{d}x.$$

To obtain the last inequality we use the fact that the functions f_i are monotone increasing. To ensure the correctness of the above inequalities one can extend the functions f_i to the segment $[-\varepsilon, 1 + \varepsilon]$ in the following manner: $f_i(x) = f(0)$ for $x < 0$ and $f_i(x) = f_i(1)$ for $x > 1$.

$$S \leqslant \sum_{i=1}^{2} \left(\int_{1-\varepsilon}^{1+\varepsilon} f_i(x)\ \mathrm{d}x - \int_{-\varepsilon}^{\varepsilon} f_i(x)\ \mathrm{d}x \right) \leqslant 2\varepsilon \sum_i^2 (f_i(1) - f_i(0))$$

$$\leqslant 2\varepsilon(V(f) - 0 + V(f) - f(1) - f(0)) \leqslant 6\varepsilon V(f)$$

as $|f(1) - f(0)| \leqslant V(f)$.

So the first integral is estimated by

$$\int_{\hat{A}_\varepsilon} |Q_\varepsilon f(x) - f(x)|\mathrm{d}x \leqslant \alpha_\varepsilon^{(1)}||f||_v \to 0 \quad \text{as} \quad \varepsilon \to 0.$$

We now estimate the second integral:

$$\int_{A_\varepsilon} |Q_\varepsilon f(x) - f(x)|\,dx \leqslant \int_{A_\varepsilon} |Q_\varepsilon f(x)|\,dx + \int_{A_\varepsilon} |f(x)|\,dx$$
$$\leqslant m(A_\varepsilon)\Big(\operatorname*{esssup}_{A_\varepsilon} |Q_\varepsilon f(x)| + \operatorname*{esssup}_{A_\varepsilon} |f(x)|\Big).$$

But by the definition $\operatorname{card}(G_\varepsilon^{-1}x) \leqslant K$ and $|G'_\varepsilon x| > 1$, therefore

$$\operatorname{esssup}(|Q_\varepsilon f(x)|) \leqslant K \operatorname{esssup}(|f(x)|).$$

Hence

$$\int_{A_\varepsilon} |Q_\varepsilon f(x) - f(x)|\,dx \leqslant (K+1)m(A_\varepsilon)\operatorname{esssup}(|f(x)|)$$
$$\leqslant (K+1)m(A_\varepsilon)\|f\|_v = \alpha_\varepsilon^{(2)}\|f\|_v \to 0 \quad \text{as} \quad \varepsilon \to 0.$$

The estimates above show that

$$\|Q_\varepsilon f - f\| \leqslant \hat{\alpha}_\varepsilon \|f\|_v = \alpha_\varepsilon^{(1)}\|f\|_v + \alpha_\varepsilon^{(2)}\|f\|_v \to 0. \text{ as } \varepsilon \to 0.$$

□

Proof of Theorem 6.1. The partition of the segment X formed by points of the set Ω_ε is a C^1-partition for the map $T_\varepsilon = G_\varepsilon \cdot T$. Piecewise expanding conditions can be derived immediately from properties of maps G_ε and T. Therefore the PF-operator P_ε of the ε-perturbed dynamical system (T_ε, X) satisfies the condition (6.1) and hence the limit transition in (6.2) follows immediately from Proposition 2.1. To complete the proof it remains to obtain the estimate uniform in ε for $\|f_\varepsilon^*\|_v$. This estimate may be constructed in the same way as in [14], but note that the condition (b) of the theorem is necessary for the ε to be uniform by this estimate.

THEOREM 6.2. *Under the conditions of Theorem* 6.1 *for all* $f \in BV(X)$, $\|f\| = 1$, $f \geqslant 0$ *the following limit transitions are valid:*

$$f_\varepsilon^* \xrightarrow[\varepsilon \to 0]{} f^* \xleftarrow[\infty \leftarrow n]{} n^{-1} \sum_{k=0}^{n-1} P^k f. \tag{6.3}$$

Proof. The second limit transition is a corollary of Proposition 2.1, so it remains to prove the first one. This proof can be carried out in the same manner as in Theorem 3.1. By means of Lemma 6.2 and of estimates analogous to those in the proof of Lemma 3.4 we obtain

$$\left\| n^{-1} \sum_{k=0}^{n-1} P_\varepsilon^k f - n^{-1} \sum_{k=0}^{n-1} P^k f \right\| \leqslant \varepsilon\gamma(n)\|f\|_v$$

where $\gamma(n)$ depends only on the number of iterations and does not depend on ε for sufficiently small $\varepsilon > 0$. Hence we can apply Lemma 2.8 to obtain the first limit transition of (6.3). The estimates of rates of convergence of iterations here are exponential, as follows immediately from [8, 10]. □

Theorems 6.1 and 6.2 demonstrate the stability of the smooth invariant measure of the dynamical system (T, X) with respect to small piecewise expanding perturbations satisfying conditions (a)–(c) of Theorem 6.1. However, if these conditions are violated the singular phenomenon is possible.

THEOREM 6.3. *Let the dynamical system have a periodic point $y \in X$ and in a small neighbourhood of its trajectory let the map T be locally a C^2-map. Then there exists a sequence of piecewise expanding perturbations G_ε such that the corresponding sequence of smooth intrinsically stable measures of ε-perturbed systems converges to the singular invariant measure of the initial system concentrated on the given cycle.*

Proof. We shall consider only the case when $y \in X$ is a fixed point of the map T, as in the general case the repetition of analogous constructions around each point of the cycle gives the required result. Let $|T'y| = \lambda > 1$. We shall define the function $G_\varepsilon x$ only in a small neighbourhood of the point $y \in X$, as we want to construct the map transforming each small neighbourhood of the point y to its small internal part. Let us fix numbers $\gamma > 1 \geqslant \eta > 0$ and denote $\varepsilon = \gamma\eta$. For $x \in B_\varepsilon(y)$ we shall define the map G_ε by the following formula:

$$G_\varepsilon x = y + \eta \cdot \operatorname{sgn}(x - y)\{2|x - y|/\eta\},$$

where $\{\cdot\}$ is a fractional part of a number.

Then the piecewise expanding condition $|G'_\varepsilon x| > 1$ is equal to the inequality $\eta > 0$, as for almost all $x \in B(y)$ it holds that

$$|G'(x)| = 2\eta/\eta = 2.$$

The map G_ε transforms the set $B_\varepsilon(y)$ into $B_\eta(y)$ *and* $B_\eta(y)$ into itself. The diagram of the function $G_\varepsilon x$ at a small neighbourhood of the point $y \in X$ is given in Figure 1.

Let $\gamma \geqslant \lambda$, then the ε-perturbed map $T_\varepsilon = G_\varepsilon \cdot T$ transforms the set $B_\eta(y)$ into itself, as $TB_\eta(y) \subset B_\varepsilon(y)$ and $G_\varepsilon B_\varepsilon(y) \subset B_\eta(y)$. Hence the dynamical system $(T_\varepsilon, B_\eta(y))$ is piecewise expanding and therefore there exists the T_ε-invariant smooth intrinsically stable measure μ_ε with support in the segment $B_\eta(y)$. This fact follows from Proposition 2.1. But the length of this segment is equal to $2\eta < 2\varepsilon$, and hence measures μ_ε converge weakly to the measure concentrated at the point y as $\varepsilon \to 0$.

The diagram of the function $T_\varepsilon x$ at a small neighbourhood of the point $y \in X$ is given in Figure 2. The extension of the map G_ε to a piecewise expanding map for the full segment X is not difficult; for example, it can be done in the class of piecewise linear functions, so we do not describe it in detail.

If the dynamical system (T, X) has only one stochastic attractor Λ and a periodic point $y \in \Lambda$, then the smooth invariant measures μ_ε of ε-perturbed systems (T_ε, X), described above, are the only intrinsically stable measures of these systems.

It is interesting to remark that in the case of piecewise expanding (quasi-stochastic) perturbations the cause of the singular phenomenon is just the same as in the case of pure stochastic perturbations (see §5). □

7. Ergodic Properties of Dynamical System Discretizations

In this section we shall consider the influence of small deterministic perturbations of dynamical systems, arising as a result of space discretizations, on asymptotic properties of these systems, such as the existence and the stability of cycles, the ergodicity and so on. A typical example of such perturbations is round-off errors,

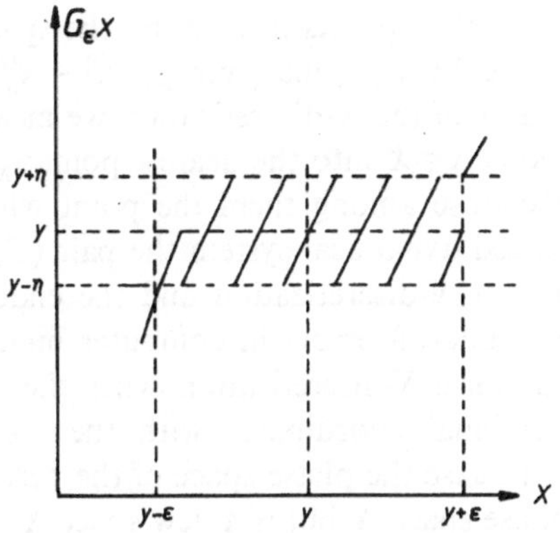

Fig. 1.

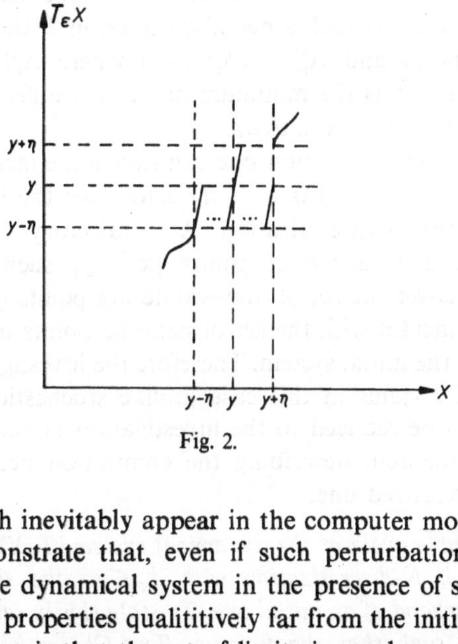

Fig. 2.

which inevitably appear in the computer modelling of dynamical systems. We shall demonstrate that, even if such perturbations are extremely small, a structurally stable dynamical system in the presence of space discretizations becomes a system with properties qualititively far from the initial state. One of the main results of this section is that the set of limits in weak-topology invariant measures of discretized systems coincides with the closure of the set of invariant measures of the initial system concentrated on its cycles. If an initial dynamical system satisfies the Bowen specification condition [2], then it follows that for each invariant measure of this system there exists a sequence of invariant measures of discretized systems converging to this measure, for instance to the measure of the maximal entropy [2] or to the intrinsically stable measure μ_* on the stochastic attractor (regular case). The other results are the phenomenon of 'periodic multiplication', and also the investigation of the statistical probability of the existence of the given unstable cycle of the initial system in the presence of space discretizations (stabilization of an unstable cycle) and their influence on the ergodicity property. Some of the results of this section were announced in [26].

Let us begin with some definitions. Just as in the previous sections of this paper X is a ν-dimensional unit cube (torus) on which there is a non-singular endomorphism

$T: X \to X$. We shall call each ordered subset of N points from X by the space N-discretization of $X \supset X_N = \{x_N^{(i)}\}_{i=1}^N$. Denote by $d_N = \max_i \min_{j \neq i} |x_N^{(i)} - x_N^{(j)}|$ the diameter of this discretization. By the operator of the N-discretization we mean the map $D_N: X \to X_N$ which transforms each point $x \in X$ into the nearest point $x_N \in X_N$ (if there are several such points then we choose among them the point with the minimal index). We shall call the N-discretized dynamical system the pair (T_N, X_N) where T_N is a composition of the operator of N-discretization and the endomorphism T, that is $T_N = D_N \cdot T$. The case of round-off errors in computer modelling with accuracy $\varepsilon_N = 1/(2N)$ corresponds to uniform N-discretization, when the set X_N is a union of all points of X having rational coordinates with the common denominator N. One may remark that in this case the phase space of the perturbed system does not coincide with the initial phase space X but is a new space $X_N \subset X$.

The set of different points $x^{(1)}, \ldots, x^{(n)}$ is called a *cycle* of the dynamical system (T, X) with period n if $Tx^{(i)} = x^{(i+1)}$ $i = 1, 2, \ldots, n-1$ and $Tx^{(n)} = x^{(1)}$. We shall say that this cycle is *stable* (*unstable*) if there exists such a neighbourhood in X that the function $Tx \in C^1$ on this neighbourhood and $\Lambda_n^{(1)} <$ ($\Lambda_n^{(1)} > 1$) where $\Lambda_k^{(i)}$ is a modulus of the maximal eigenvalue and $\lambda_k^{(i)}$ is the minimum one of the differential T^k at the point $x^{(i)}$ (if $k = 1$ this index may be omitted).

Note that from the definition of the N-discretization one can derive the fact that each trajectory of the dynamical system (T_N, X_N) is a d_N-trajectory for the initial system, where d_N is a diameter of the N-discretization. A δ-trajectory of the dynamical system (T, X) is an ordered sequence of points $\{x^{(i)}\}_{i=1}$ such that $|x^{(i+1)} - Tx^{(i)}| \leqslant \delta$ for all $i \in Z_+$. Moreover the set of non-wandering points [3] of the N-discretized dynamical system coincides with the set of periodic points of this system without any assumptions about the initial system. Therefore the investigation of asymptotic properties of perturbed systems in this case (unlike stochastic and piecewise expanding perturbations) can be reduced to the investigation of periodic points only. Now we formulate the theorem describing the connection between cycles of the initial system and the discretized one.

THEOREM 7.1. *If $x^{(1)}, \ldots, x^{(n)}$ is a stable cycle of the dynamical system (T, X) then there exists $\varepsilon_1 > 0$ such that for each N-discretization with $d_N \leqslant \varepsilon_1$ the system (T_N, X_N) has a cycle in a small neighbourhood of an initial one. To state this in reverse, if $x_N^{(1)}, \ldots, x_N^{(n)}$ is a cycle of (T_N, X_N) such that the function $Tx \in C^1$ on a small neighbourhood of this cycle in X and $\Pi_{i=1}^n \Lambda^{(i)} < 1$ or $\Pi_{i=1}^n \lambda^{(i)} > 1$ then there exists a cycle of the system (T, X) in a small neighbourhood of $x_N^{(1)}, \ldots, x_N^{(n)}$ in X.*

Proof. Let us first prove the direct statement. For this purpose we shall consider the sequence $\{x_N^{(i)}\}_{i=1}$, $x_N^{(i)} \in X_N$, $x_N^{(1)} = D_N x^{(1)}$ and $x_N^{(i+1)} = T_N x_N^{(i)}$ for all $i \in Z_+$. Hence $|x_N^{(1)} - x^{(1)}| \leqslant d_N/2 = \varepsilon$. Denote by T'_x the linear part of the increment of the map T at the point $x \in X$. If $|y| \leqslant \varepsilon$, which is small enough, then

$$T(x + y) = Tx + T'_x y + o(\varepsilon).$$

Denote $y^{(i)} = T_N x_N^{(i)} - Tx_N^{(i)}$, then $|y^{(i)}| \leqslant \varepsilon$ and

$$x_N^{(2)} = Tx_N^{(1)} + y^{(1)} = T(x^{(1)} + y^{(0)}) + y^{(1)} = Tx^{(1)} + T'_x(1) y^{(0)} + y^{(1)} + o(\varepsilon).$$

$$\cdots\cdots\cdots\cdots\cdots\cdots\cdots\cdots\cdots\cdots\cdots\cdots$$

$$x_N^{(n)} = T^{n-1} x^{(1)} + (T^{n-1})'_{x^{(1)}} y^{(0)} + (T^{n-2})'_{x^{(2)}} y^{(1)} + \cdots +$$

$$+ T'_{x^{(n-1)}}y^{(n-2)} + y^{(n-1)} + o(\varepsilon).$$

....

$$x_N^{(kn+1)} = T^{kn}x^{(1)} + (T^{kn})'_{x^{(1)}}y^{(0)} + (T^{kn-1})'_{x^{(2)}}y^{(1)} + \cdots +$$
$$+ T'_{x^{(kn)}}y^{(kn-1)} + y^{(kn)} + o(\varepsilon).$$

....

From the definition of the stability of the cycle $x^{(1)}, \ldots, x^{(n)}$ it follows that there exists such $L > 0$ and $q \in (0, 1)$ that for all $k \in Z_+$

$$|(T^k)'_{x^{(i)}}| \leqslant Lq^k \tag{7.1}$$

where $|\cdot|$ is the Euclidean norm of a matrix. Denote $z^{(k)} = x_N^{(kn+1)}$.

$$\varepsilon^{-1}|z^{(k+1)} - z^{(k)}| \leqslant |T^{(k+1)n}x^{(1)} - T^{kn}x^{(1)}| + |((T^{(k+1)n})'_{x^{(1)}} -$$
$$- (T^{kn})'_{x^{(1)}})y^{(0)}| + \cdots + o(\varepsilon) \leqslant 2L/(1-q) \tag{7.2}$$

for sufficiently small $\varepsilon > 0$. Analogously one can obtain that

$$|z^{(k)} - x^{(1)}| \leqslant \varepsilon 2L/(1-q).$$

The last inequality shows that there exists a ball with a centre at the point $x^{(1)}$ and a radius $\varepsilon L/(1-q)$ such that there exist k_0, $k_1 \in Z_+$ such that for all $k \geqslant 1$, $z^{(k_0 + k_1 k)} = z^{k_0}$. Therefore points $z^{(k_0)}$, $T_N z^{(k_0)}, \ldots,$ $T_N^{k_1 n - 1} z^{(k_0)}$ form the required cycle of the N-discretized system (T_N, X_N).

Now let us prove the reverse statement. First we consider the case of a stable cycle ($\Pi_{i=1}^n \Lambda^{(i)} < 1$). Introduce a map $F = T_N^n$; then $\Lambda^{(1)}(F) = \Pi_{i=1}^n \Lambda^{(i)}$. We consider the sequence $\{x^{(i)}\}_{i=1}$ with elements $x^{(i)} = Tx^{(i-1)}$ for $i \geqslant 2$ and $x^{(1)} = x_N^{(1)}$. Obtaining estimates in the same manner as in the previous part of the proof we get an estimate of the type (7.2) for $|x^{(nk+1)} - x_N^{(1)}|$ and as the map F is a contractive one (as $\Lambda^{(1)}(F) = \Pi_{i=1}^n \Lambda^{(i)} < 1$) then for sufficiently small diameters of discretizations the map T^n is also contractive. Therefore the sequence $\{x^{(nk+1)}\}_{k=1}$ has a limit point $x^* \in X$ and then points x^*, $Tx^*, \ldots, T^n x^*$ form a cycle of the dynamical system (T, X).

In the case of an unstable cycle ($\Pi_{i=1}^n \lambda^{(i)} > 1$) we consider the restriction of the map T to some small neighbourhoods of points $\{x_N^{(i)}\}_{i=1}^n$. By the condition of the Theorem this restriction of T is a diffeomorphism satisfying the condition

$$\prod_{i=1}^n \Lambda^{(i)}(T^{-1}) = \prod_{i=1}^n (\lambda^{(i)}(T))^{-1} < 1.$$

But the points $x_N^{(n)}$, $x_N^{(n-1)}, \ldots, x_N^{(1)}$ form a cycle for the dynamical system $(D_N T^{-1}, X_N)$, hence one can reduce this case to the one considered above. □

REMARK 7.1. The condition $\Pi_{i=1}^n \lambda^{(i)} > 1$ is a restrictive one, and can be replaced if we impose stronger assumptions on the structure of the dynamical system (T, X). For instance, it is enough to assume that the system (T, X) satisfies the hyperbolicity condition [3]. In this case one can use the Lemma about ε-trajectories instead of estimates (7.1) and (7.2).

THEOREM 7.2. *If $x^{(1)},\ldots,x^{(n)}$ is a cycle of (T, X) and the set*

$$C_\infty = \bigcup_{k\geq 0} \bigcup_{i=1}^{n} T^{-k}x^{(i)}$$

is dense in X, then there exists a sequence of discretizations such that their diameters tend to zero and for all $N \in Z_+$ the N-discretized dynamical system (T_N, X_N) has only one cycle coinciding with the initial one.

Proof. Denote

$$C_N = \bigcup_{k=0}^{N} \bigcup_{i=1}^{n} T^{-k}x^{(i)}.$$

By the condition of the theorem, distances between neighbouring points of the set C_N tend to zero as $N \to \infty$, because the set X is bounded. On the other hand, if we fix the set C_N as the set of N-discretization then the statement of the Theorem is fulfilled. □

The last theorem demonstrates the possibility of the global stabilization of each stable or unstable cycle of the dynamical system (T, X), as the construction ensures that each point of the set $X_N = C_N$ gets into the cycle $x^{(1)},\ldots,x^{(n)}$ up to $(N-1)$ iterations. Note that there are no smoothness conditions in this theorem.

THEOREM 7.3. *The set of weak limit points as $N \to \infty$ of sets of invariant measures of N-discretized systems (T_N, X_N) coincides with the closure of the set of invariant measures, concentrated on cycles of the initial system.*

This theorem is a reformulation in terms of invariant measures of the results of Theorems 7.1 and 7.2. On the other hand, this statement demonstrates the singular phenomenon in the case of space discretizations. Let us consider a simple example illustrating this result. Here X is the unit circle and T is given by the formula $Tx = 2x \pmod 1$. The dynamical system (T, X) has a dense family of cycles of all periods and for each point $x \in X$ the set of its prototypes is dense in X [3]. All the statements proved above remain valid for this system, even if we restrict ourselves only to uniform space discretizations (round-off errors).

The theorems above demonstrate only local connections between cycles of the initial system and the perturbed one, but do not give any information about their periods.

THEOREM 7.4. *If $x^{(1)},\ldots,x^{(n)}$ is a cycle of a dynamical system (T, X) and in its small neighbourhood U, consisting of disjoint neighbourhoods U_i of points of this cycle, the map T is a local homeomorphism and there exists a cycle $x_N^{(1)},\ldots,x_N^{(k)}$ of the dynamical system (T_N, X_N) then*

(a) if $\nu = 1$ *then* $k/n \in \{1, 2\}$;
(b) if $\nu > 1$ *then* $k/n \in \{1, 2, \ldots, \mathrm{const}(T, d_N)\}$.

Proof. We consider first the one-dimensional case ($\nu = 1$). As T is a local homeomorphism it follows that the function Tx is monotone on the set U and hence $T_N x$ is also a monotone function, being a composition of two monotone functions T and D_N.

Let $T_N^{(*)}$ be a derivative map [3] constructed for the map T_N for the set of points of the set X_N belonging to the given neighbourhood $U_1 \subset U$ of the point $x^{(1)}$. Denote by $y_N^{(1)}, \ldots, y_N^{(m)}$ the points of the cycle $x_N^{(1)}, \ldots, x_N^{(k)}$ lying in U_1. As T is a local homeomorphism then the number m_i of points of the cycle $x_N^{(1)}, \ldots, x_N^{(k)}$, belonging to the corresponding neighbourhood U_i of the point $x^{(i)}$, does not depend on i and is equal to some constant $m \in Z_+$. This number defines the *multiplicity* $k/n = m$ of a cycle in (T_N, X_N). We shall assume that points $y_N^{(1)}, \ldots, y_N^{(m)}$ are numbered so that they form the cycle of (T_N, X_N) and the point $y_N^{(1)}$ is minimal among them (in any case, one can renumber these points). The following cases may arise:

(a) $T_N^{(*)} y_N^{(1)} = y_N^{(1)}$; this means that $m = 1$ and $k = n$, that is the cycle is of the same period as the initial one.

(b) $T_N^{(*)} y_N^{(1)} = y_N^{(1)}$ but $T_N^{(*)2} y_N^{(1)} = y_N^{(1)}$. Then $m = 2$ and $k = 2n$. In this case the N-, discretized system has a cycle with a double period. In Figure 3 the general construction of periodic doubling for a cycle with an arbitrary period $n \in Z_+$ is demonstrated. The solid lines in this diagram show the trajectory of the cycle of (T, X) and the dotted lines the trajectory of the cycle of (T_N, X_N).

Remark that the necessary condition for doubling is the property that the function Tx decreases monotonically, i.e. it changes its order. In the case of uniform discretization with a round-off error $\varepsilon = 1/(2N)$, the necessary and sufficient condition for the existence of a cycle with the double period $2n$ described in Figure 3 is defined by the following system of inequalities:

$$\begin{cases} \lambda^{(1)} z_1 > \varepsilon - z_2 \\ \lambda^{(1)}(2\varepsilon - z_1) < \varepsilon + z_2 \\ \lambda^{(i)}(2\varepsilon - z_i) > \varepsilon - z_i, \quad i = 2, 3, \ldots, n-1 \\ \lambda^{(i)} z_i < \varepsilon + z_{i+1} \\ \lambda^{(n)}(2\varepsilon - z_n) > \varepsilon - z_1 \\ \lambda^{(n)} z_n < \varepsilon + z_1 \end{cases} \tag{7.3}$$

where $z_i = |x^{(i)} - D_N x^{(i)}|$.

The necessary condition for doubling in the one-dimensional system for the given cycle is $\Pi_{i=1}^n T' x^{(i)} \leqslant 0$.

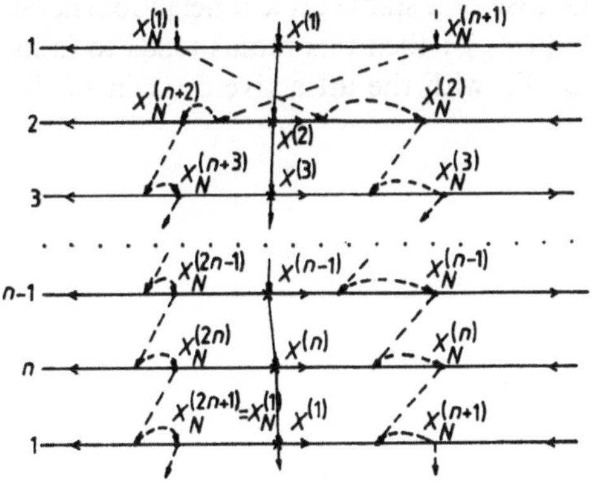

Fig. 3.

(c) $T_N^{(*)}y_N^{(1)} = y_N^{(1)}$, $T_N^{(*)2}y_N^{(1)} = y_N^{(1)}$. We shall demonstrate that this case is not valid. We first assume that the number $m = 2l$ is even. But as $T_N^{(*)}$ is a monotone function, hence $T_N^{(*)2}$ is monotone non-decreasing. Therefore

$$y_N^{(1)} < T_N^{(*)2}y_N^{(1)} \leqslant T_N^{(*)4}y_N^{(1)} \leqslant \cdots \leqslant T_N^{(*)2l}y_N^{(1)} = y_N^{(1)}$$

This is a contradiction, as the first inequality is a strict one. Therefore the cycle $y_N^{(1)}, \ldots, y_N^{(2l)}$ with an even period greater than 2 does not exist, because $y_N^{(1)} < y_N^{(2l)}$.

It remains to prove that if $l \in Z_+$ then there is no cycle with period $m = 2l + 1$. If $T_N^{(*)}$ is monotone non-decreasing then the orientations of pairs $(y_N^{(i)}, y_N^{(i+1)})$ are equal for all $i = 1, \ldots, n-1$. This means that $y_N^{(i)} < y_N^{(i+1)}$, and the strict inequality here follows from the fact that these points are points of one cycle. Hence the points considered do not form a cycle. We now consider the last case when $T_N^{(*)}$ is monotone non-increasing. In this case the orientation of the pair $(y_N^{(i)}, y_N^{(i+1)})$ is opposite to the orientation of the next pair, as $T_N^{(*)}y_N^{(i)} = y_N^{(i+1)}$ by the definition of the map $T_N^{(*)}$. Hence the pairs $(y_N^{(1)}, y_N^{(2)})$ and $(T_N^{(*)}y_N^{(2l+1)}, T_N^{(*)}y_N^{(2l+2)})$ are oriented oppositely, but $T_N^{(*)}y_N^{(2l+1)} = y_N^{(1)}$, so again we come to a contradiction.

Now let us consider the multidimensional case ($v > 1$). In this case even for $v = 2$ a cycle with the period of any multiplicity with respect to the initial one may arise. If $v > 1$ the case of a cycle with an arbitrary period $n > 1$ does not differ from a fixed point from our point of view. We therefore construct an example demonstrating the phenomenon of the *period multiplication* of the fixed point in two-dimensional space. Let the map T be defined by the formula $T = SU_\alpha$, where the operator S is a linear contractive one and U_α is an operator for a linear rotation around the fixed point $x_0 \in X$ of the map T. We choose the N-discretization set such that the point $x_0 \in X_N$. Then $T_N x_0 = x_0$ and for a suitable α and S there exists a cycle with the period $]2\pi/\alpha[$ of (T_N, X_N). (Here $]\cdot[$ represents the integer part of a number.) The trajectory of this cycle is demonstrated in Figure 4 for the case of the uniform discretization.

The effect that in the one-dimensional case only the doubling of the initial period is possible is explained by the fact that on R^1 there exists only a rotation through the angle $\alpha = \pi$ or 2π. □

Points of the set X_N belonging to the neighbourhood U given in Theorem 7.4 divide into sets such that in each of them there exists one or several cycles of an equal period divisible by the initial one. The other points of X_N hit these sets or leave U under the action of T_N. In the case of a stable cycle a neighbourhood arises around it with the radius d_N const(T) (if $\Lambda_n \uparrow 1$ then this radius tends to infinity), in which all points of the intersection of X_N with the attractive domain of the given

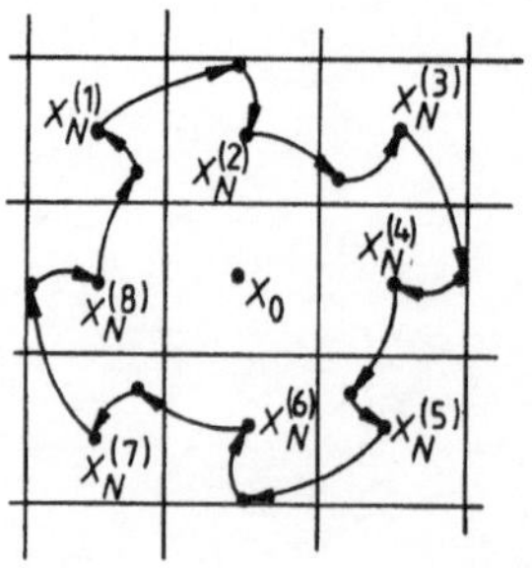

Fig. 4.

cycle hit. The partition described above for a stable case occurs only in this neighbourhood.

The phenomenon described in Theorem 7.4 is called the *period multiplication.* Note that the smoothness of a map T is not important for this phenomenon, but on the other hand there exists an effective condition for the map T renders the period multiplication phenomenon invalid all discretizations. If $\nu = 1$ the condition is equal to the fact that the map T^n is monotone increasing.

The question remains, when in a small neighbourhood of an unstable cycle of the initial system does there exist a cycle of a discretized system? This situation does not satisfy the conditions of Theorem 7.1. For different discretizations a cycle of the discretized system can appear or disappear (even if the diameters of the discretizations are extremely small). To give an answer to this question it is therefore necessary to make some additional assumptions concerning the construction of the N-discretization operator. In the present paper we shall investigate the case of uniform discretizations, which is the most interesting for applications. Even in this case the presence or absence of a cycle of the N-discretized system is connected to the number $N \in Z_+$ in an extremely complicated way. It is therefore most natural here, from our point of view, to use a statistical approach; in other words, a description of the portion of that uniform N-discretization for which such a cycle exists, with respect to the number of all uniform discretizations. We denote by $M(N; x^{(1)}, \ldots, x^{(n)})$ the number of uniform k-discretizations, $k \leqslant N$, for which in a small neighbourhood of a cycle $x^{(1)}, \ldots, x^{(n)}$ there exists a cycle of (T_k, X_k). The limit of the ratio $M(N; x^{(1)}, \ldots, x^{(n)})/N$ as $N \to \infty$ (if it exists) we denote by $p(x^{(1)}, \ldots, x^{(n)})$.

In computer modelling of dynamical systems, the fact that the discretized system has a cycle in a small neighbourhood of the given unstable cycle of the initial system means that we really observe an unstable trajectory. In this situation it is thus natural to call the stabilization and the value $p(x^{(1)}, \ldots, x^{(n)})$ *the stabilization probability* of the unstable cycle. Numerical experiments in [27] demonstrate that in the computer modelling of unstable dynamical systems (i.e. systems with only unstable cycles), certain unstable cycles are observed significantly more often than others. The following theorem gives an explanation of this phenomenon.

THEOREM 7.5. *Let* $\nu = 1$ *and* $x^{(1)}, \ldots, x^{(n)}$ *be an unstable cycle of the dynamical system* (T, X), $\lambda = \min_i\{\lambda^{(i)}\} \geqslant 2$ *and the points of this cycle are rationally independent. Then the value* $p(x^{(1)}, \ldots, x^{(n)})$ *is correctly defined and coincides with the volume of the domain defined in coordinates* $(z_1, \ldots, z_n)$ *by the system of semi-linear restrictions:*

$$\begin{cases} \lambda^{(1)}|z_1| \leqslant 1/2 + \operatorname{sgn}(z_1 \cdot T'x^{(1)})z_2 \\ \lambda^{(2)}|z_2| \leqslant 1/2 + \operatorname{sgn}(z_2 \cdot T'x^{(2)})z_3 \\ \cdots\cdots\cdots\cdots\cdots\cdots \\ \lambda^{(n)}|z_n| \leqslant 1/2 + \operatorname{sgn}(z_n \cdot T'x^{(n)})z_1 \end{cases} \tag{7.4}$$

To begin the proof of this statement formally we prove the following number-theory proposition which is also of independent interest.

LEMMA 7.1. *For all $\delta \in (0, 1)$ and almost all $x \in [0, 1]$, the density of $N \in Z_+$ such that x is approximated by a rational number with the denominator N with accuracy $\delta/(2N)$ is equal to δ.*

Proof. Let us consider the number $M(x; N) \in Z_+$ (if it exists) such that $|x - M(x; N)/N| \leqslant \delta/(2N)$ or $|Nx - M(x; N)| \leqslant \delta/2$. The last inequality is equal to

$$\{Nx\} \in [0, \delta/2] \cup [1 - \delta/2, 1) \tag{7.5}$$

where $\{x\}$ is a fractional part of a number x, $\{x\} = x -]x[$. We consider a dynamical system (F, Y) on the unit circle Y, where F is a rotation through the angle x. The trajectory of this system beginning at the point $x \in Y$ is by definition the sequence of points $\{x^{(k)}\}_{k=1}$ such that $x^{(k)} = \{kx\}$. Therefore the density of $N \in Z_+$ for which this trajectory is found in any Borel subset of Y in the ergodic case coincides with the F-invariant measure of this set. But if the number x is irrational then the dynamical system (F, Y) is ergodic and there exists only one F-invariant measure coinciding with Lebesgue measure on the circle. The Lebesgue measure of the set defined by the inequality (7.5) is equal to δ. □

Proof of Theorem 7.5. If $x^{(1)}, \dots, x^{(n)}$ is a cycle of the initial system and in a small neighbourhood of it there exists a cycle of the N-discretized system $x_N^{(1)}, \dots, x_N^{(k)}$, then from the proof of Theorem 7.1 one can derive an estimate of the distance between them. In the one-dimensional case this estimate satisfies the inequality $\delta < \varepsilon/(\lambda - 1)$, where $\varepsilon = d_N/2 = 1/(2N)$. Hence if $\lambda \geqslant 2$, then $\delta < \varepsilon$ and therefore for all $i = 1, \dots, n$ $x_N^{(i)}$ is of the form $x_N^{(i)} = D_N x^{(i)}$ and $n = k$. Note that the condition $\lambda \geqslant 2$ is sufficient to reject the 'period multiplication' phenomenon in the one-dimensional case.

Fix a natural number N and $\varepsilon = 1/(2N)$. If $\lambda \geqslant 2$, necessary and sufficient conditions for the fact that points $\{D_N x^{(i)}\}_{i=1}^n$ form a cycle of (T_N, X_N) are

$$x_N^{(i-1)} \in T^{-1}(D_N x^{(i)} - \varepsilon, D_N x^{(i)} + \varepsilon)), \quad x_N^{(0)} = x_N^{(n)}.$$

For $\varepsilon > 0$ small enough (i.e., large N) these inclusions are equal to the following system of inequalities:

$$\begin{aligned} &\lambda^{(1)}|y_1| \leqslant \varepsilon + \operatorname{sgn}(y_1 \cdot T'x^{(1)})y_2 \\ &\lambda^{(2)}|y_2| \leqslant \varepsilon + \operatorname{sgn}(y_2 \cdot T'x^{(2)})y_3 \\ &\dots\dots\dots\dots\dots\dots\dots\dots \\ &\lambda^{(n)}|y_n| \leqslant \varepsilon + \operatorname{sgn}(y_n \cdot T'x^{(n)})y_1 \end{aligned} \tag{7.4'}$$

where $y_i = x^{(i)} - D_N x^{(i)}$. Note that for a given fixed N this system may be not combined.

The idea of the proof in this case is the same as in Lemma 7.1. Let us consider the dynamical system (F, Y) given on an n-dimensional unit torus Y, where F is a rotation through the angle $\alpha = (x^{(1)}, \dots, x^{(n)})$. If we consider the trajectory of this system beginning at the point α, then after N iterations the ith coordinate of the point of this trajectory is $(F^N\alpha)^{(i)} = \{Nx^{(i)}\}$. But $\{Nx^{(i)}\} = \{Ny_i - ND_N x^{(i)}\}$ and $ND_N x^{(i)} \in Z$ by the definition of uniform discretization. Therefore, just as in the proof of Lemma 7.1, the density of numbers N such that (7.4′) is a combined system coincides with the F-invariant measure of the domain defined by the system of inequalities (7.4) and this measure is the only invariant measure of (F, Y) and coincides with the Lebesgue measure on the torus Y [3]. □

The stabilization probability obtained in Theorem 7.5 depends only on derivatives along a cycle and its period, but it is fulfilled only for 'typical' cycles. There are exclusive cycles (even exclusive dynamical systems, all cycles of which are exclusive) for which stabilization probabilities depend only on coordinates of points of these cycles. Among them there exist cycles with extremely large stabilization probabilities and cycles with smaller ones (with respect to the 'typical' situation).

In the general multidimensional case we can prove statements analogous to Theorem 7.5, but in this case the system of restrictions replacing (7.4) is essentially non-linear and extremely complicated and so we do not prove it here.

Note that from the system of inequalities (7.3) one can derive, in the same manner as in the proof of Theorem 7.5, the probabilities of all possible partitions arising in the presence of the period multiplication phenomenon around cycles.

We now consider how space discretizations influence dynamical systems whose trajectories are neither stable nor unstable. Let X be a ν-dimensional unit torus and T_α be its notation through the angle $\alpha = (\alpha_1, \ldots, \alpha_\nu)$. The dynamical system (T_α, X) is ergodic iff values $\alpha_1, \ldots, \alpha_\nu$ are rationally independent [3]. The investigation of a separate trajectory makes no sense in this case, and we shall study the influence of discretizations on the ergodicity property itself. For different natural N perturbed systems $(T_{\alpha,N}, X_N)$ may be either ergodic or non-ergodic, with no evident connections with the ergodicity of the initial system. Therefore here, as in Theorem 7.5, it is natural to use a statistical approach. We shall study here only uniform discretizations. By $p(\alpha)$ we denote the density of natural N such that $(T_{\alpha,N}, X_N)$ is ergodic.

THEOREM 7.6. *If values $\alpha_1, \ldots, \alpha_\nu$ are rationally independent then the value $p(\alpha)$ is correctly defined and $p(\alpha) = 0$.*

Proof. We consider first the one-dimensional case ($\nu = 1$). In this case $T_\alpha x = x + \alpha$ (mod 1), where $\alpha \in (0, 1)$ is irrational. Then

$$p(\alpha) = \lim_{N \to \infty} N^{-1} \operatorname{card}(n \leqslant N :]n\alpha[/n \text{ is irreducible}) \tag{7.6}$$

where $]x[$ is the whole number closest to $x \in R^1$. The dynamical system $(T_{\alpha,N}, X_N)$ is ergodic iff $]N\alpha[$ has no common divisors with $N \in Z_+$.

We now sketch the proof of this theorem. Consider the partition of the set of natural numbers to disjoint series $Q_1, \ldots, Q_n, \ldots$. Each series is defined by the fact that all its members have some given common divisor and have no common divisors with some other given number. In each of these series it is easy to calculate the conditional density of the set of $N \in Z_+$ such that the N-discretized system $(T_{\alpha,N}, X_N)$ is ergodic. Then

$$p(\alpha) = 1 - \sum_i p_i \sum_j p_{ij}(\alpha),$$

where p_i is the density of the set Q_i with respect to all natural numbers and $p_{ij}(\alpha)$ is the conditional density of such numbers $n \in Q_i$, that $]n\alpha[\in Q_j$ and the common divisor of the series Q_j is also the divisor of the number n.

Denumerate by increasing all prime numbers $2 = r_1 < r_2 < \cdots$, denote $r_0 = 1$ and $r_k? = \Pi_{i=1}^{k} r_i$. Now we can define the series Q_i as the set of natural numbers having the common divisor r_i and having no common divisors with $r_{i-1}?$ Obviously each natural number belongs to one and only one of these series.

For each $k \in Z_+$ the following representation of the series Q_k is valid:

$$Q_k = \{m \in Z_+ : m = r_k(r_{k-1}?n + q), n \in Z_+, q \in H_k\} \tag{7.7}$$

where H_k is the set of all prime numbers from r_k to $r_{k-1}?$. This representation enables us to calculate the density of each series:

$$p_k = \text{card}(H_k)/r_k?.$$

Let us study conditions under which $N \in Q_k$, r_l is a divisor of N and $]N\alpha[\in Q_l$ for arbitrary natural k, l. From the representation (7.7) it follows that these conditions are equal to the following inclusion:

$$\begin{aligned} r_l r_k(r_{k-1}?n + q_k)\alpha \in (r_l(r_{l-1}?m + q_l) - 1/2, \\ r_l(r_{l-1}?m + q_l) + 1/2), \end{aligned} \tag{7.8}$$

where $q_k \in H_k$, $q_l \in H_l$, $m \in Z_+$.

We rewrite this inclusion as

$$\begin{aligned} r_k?n\alpha/r_{l-1}? \in (m + q_l/r_{l-1}? - r_k q_k \alpha/r_{l-1}? - 1/(2r_l?), \\ m + q_l/r_{l-1}? - r_k q_k \alpha/r_{l-1}? + 1/(2r_l?)). \end{aligned}$$

Denote $R_{kl} = r_k?/r_{l-1}?$ and $L_{kl} = (q_l - r_k q_k \alpha)/r_{l-1}?$. Then in these terms the last inclusion is equal to

$$R_{kl} n\alpha \in (m + L_{kl} - 1/(2r_l?), m + L_{kl} + 1/(2r_l?)). \tag{7.9}$$

The natural number m on the right-hand side of the inclusion (7.9) may be chosen in an arbitrary way, so (7.9) is equivalent to the statement that the point $\{R_{kl}n\alpha\}$ lies on the smaller segment of the unit circle between points $(L_{kl} \pm 1/(2r_l?))$ (mod 1). But the length of this segment is equal to $1/r_l?$.

Let us consider again, as in the proof of Lemma 7.1, the dynamical system (F, Y) on the unit circle Y, where F is a rotation through the angle $\{R_{kl}\alpha\}$. The trajectory of this system beginning at the point $y = \{R_{kl}\alpha\} \in Y$ after n iterations hits the point $y_n = \{yn\} = \{R_{kl}n\alpha\}$. Therefore the density of $n \in Z_+$ such that this trajectory is found in arbitrary Borel set in Y in the ergodic case coincides with the F-invariant measure of this set. But the number $\{R_{kl}\alpha\}$ is irrational by the definition of the angle α, therefore the dynamical system (F, Y) has only one F-invariant measure, coinciding with Lebesgue measure on the circle.

From this consideration one can obtain the conditional density

$$p_{ij}(\alpha) = \text{card}(H_j)/r_j?.$$

From this equality it follows that the value $p_{ij}(\alpha)$ does not depend on the angle α and the number of the series Q_i. Hence the value $p(\alpha)$, which is equal to

$$p(\alpha) = 1 - \sum_i p_i \sum_j p_{ij}(\alpha)$$

$$= 1 - \sum_i \operatorname{card}(H_i)/r_i? \sum_j \operatorname{card}(H_j)/r_j?$$

$$= 1 - \left(\sum_i \operatorname{card}(H_i)/r_i?\right)^2 = 1 - \left(\sum_i p_i\right)^2 = 0,$$

as the total density of all series Q_i is equal to 1, so we prove that the value $p(\alpha)$ in the one-dimensional case does not depend on α for irrational angles and is equal to 0.

Now let us study the multidimensional case ($\nu > 1$). Fix $N \in Z_+$. The N-discretized dynamical system $(T_{\alpha,N}, X_N)$ is ergodic iff the following conditions are satisfied:

(a) $]N\alpha[$ and N are prime one to another;
(b) numbers $]N\alpha_i[$ are prime to one another.

In the first part of the proof we demonstrated that the density of the set of natural numbers N, satisfying the condition (a), is equal to 0. It remains to show that the density of the set of natural N satisfying the condition (b) is correctly defined, and this fact can be proved in the same manner as the first part of the proof. □

In the case of rational angles α, the values $p(\alpha)$ in effect chaotically fill in the unit segment. For example, if $\nu = 1$, then $p(1/2) = 1/2$, $p(1/3) = 2/3$.

In the case of smooth perturbations, estimates analogous to ones in Theorem 7.6 were studied in [28], where it was shown that if the initial system is ergodic, then, on average, perturbed systems are also ergodic for small perturbations. This result demonstrates the qualitative difference between the influence of space discretizations (round-off errors) and smooth perturbations, as from Theorem 7.6 it follows that for large dimensions ($\nu \to \infty$) discretized systems are non-ergodic as a rule. It is interesting to note the connection between chaotic births and disappearances of cycles in systems $(T_{\alpha,N}, X_N)$ as $N \to \infty$ and the subfurcation phenomenon which arises in the case of smooth perturbations.

The results of this section demonstrate that even highly accurate calculations change the asymptotic properties of dynamical systems in computer modelling. Thus Theorem 7.4 shows that the method of calculating the histogram of a numerically obtained trajectory of the dynamical system, which is often used to estimate the density of the smooth intrinsically stable invariant measure of the system considered, is subject to large errors even in L_1-norm. The assumption often used in the literature, that round-off errors can be replaced in the investigation by uniform distributed independent random values, brings us (because of the results of §3 above) of the weak convergence of invariant measures of perturbed systems to an intrinsically stable measure of the initial system (regular case). This essentially differs from the results of §7.

The phenomenon of period multiplication described in Theorem 7.4 causes great difficulties in numerical investigations of bifurcations of dynamical systems, such as the birth of a cycle, doubling of the period (the Feigenbaum construction) and so on. Moreover, in some numerical investigations the period multiplication phenomenon was assumed to be a bifurcation of the initial system. We now formulate a rule to distinguish these effects.

PROPOSITION 7.1. *Let the discretized system* (T_N, X_N) *have a cycle* $x_N^{(1)}, \ldots, x_N^{(n)}$, *in a small neighbourhood* $Tx \in C^1$, $\max \Lambda^{(i)} = \Lambda < 1$ and $s = \min\{|x^{(i)} - x^{(j)}|: i \neq j\}$. *The sufficient condition for the absence of period multiplication is equivalent to the satisfaction of one of the following conditions:*

(a) N *is a prime number (if* $v = 1$ *then it is sufficient that* N *is odd);*
(b) $Ns > 1/(1 - \Lambda)$;
(c) *for* $k \in Z_+$, *provided* $k > 2/(s(1 - \Lambda))$ *the sequence* $\{T_k^i D_k x_N^{(1)}\}_{i=1}$ *converges to a cycle of the period n, lying in the s/4-neighbourhood of the cycle considered.*

Analogous conditions also exist if the cycle considered is unstable, but their checking them numerically is extremely difficult (unlike the case above), so we do not describe them here.

Acknowledgement

The author wishes to express his gratitude to Ya. G. Sinai for numerous useful discussions.

References

1. Sinai, Ya. G.: 'Stochasticity in Dynamical Systems', in *Nonlinear Waves*, Nauka, Moscow, 1979 (in Russian).
2. Bowen, R.: *Methods of Symbolic Dynamics*, Mir, Moscow, 1979.
3. Nitecki, Z.: *Differentiable Dynamics*, MIT Press, Cambridge, 1971.
4. Sinai, Ya. G.: 'Gibbs Measures in Ergodic Theory', *Russian Math. Surveys* **27** (1972), 21–70.
5. Lasota, A. and Yorke, J. A.: 'On the Existence of Invariant Measures for Piecewise Monotonic Transformations', *Trans. Amer. Math. Soc.* **186** (1973), 481–488.
6. Wong, S.: 'Some Metric Properties of Piecewise Monotonic Mappings of the Unit Interval', *Trans. Amer. Math. Soc.* **246** (1978), 493–500.
7. Bunimovich, L. A. and Sinai, Ya. G.: 'Stochasticity of an Attractor in the Lorenz Model', in *Nonlinear Waves*, Nauka, Moscow, 1979 (in Russian).
8. Blank, M. L.: 'Small perturbations of Quasistochastic Dynamical Systems', *Trudy Sem. L. G. Petrovsky* **11** (1986), 166–189 (in Russian).
9. Golosov, A. O.: 'Small Stochastic Perturbations of Dynamical Systems', *Trudy Mosk. Mat. Obsch.* **46** (1983), 243–261.
10. Keller, G.: 'Stochastic Stability in Some Chaotic Dynamical Systems', Heidelberg, Preprint, 1980.
11. Khasminskii, R. Z.: 'The Averaging Principle for Parabolic and Elliptic Differential Equations and Markov Processes with Small Diffusions', *Theory Prob. Appl.* **8** (1963), 1–21.
12. Kifer, Yu.: 'General Random Perturbations of Hyperbolic and Expanding Transformations', Maryland, Preprint, 1983.
13. Kifer, Yu.: 'On Small Random Perturbations of Some Smooth Dynamical Systems', *Izv. Acad. Nauk SSSR, Ser. Mat.* **8** (1974), 1083–1107 (in Russian).
14. Ventcel, A. D. and Freidlin, M. I.: 'On Small Random Perturbations of Dynamical Systems', *Russian Math. Surveys* **25** (1970), 1–56.

15. Hamming, R. W.: *Numerical Methods for Scientists and Engineers*, McGraw-Hill, New York, 1962.
16. Blank, M. L.: 'An Estimate of the Rate of Correlation Decay in One-Dimensional Dynamical Systems', *Funct. Anal. Appl.* **18** (1984), 61–62.
17. Hofbauer, F. and Keller, G.: 'Ergodic Properties of Invariant Measures for Piecewise Monotonic Transformations', *Math. Z.* **180** (1982), 119–140.
18. Blank, M. L.: 'A Conjunction of Some Class of One-Dimensional Mappings with a Class of Piecewise Expanding Mappings', *Uspehi Mat. Nauk.* **40** (1985), 187–188 (in Russian).
19. Clarcson, J. A. and Adams, C. R.: 'Properties of Functions of Bounded Variation', *Trans. Amer. Math. Soc.* **36** (1934), 711–760.
20. Natanson, I. P.: *Theory of Functions of Real Variables*. Nauka, Moscow, 1974.
21. Yosida, K.: *Functional Analysis*, Mir, Moscow, 1967.
22. Doob, J. L.: *Stochastic Processes*, Wiley, New York, 1953.
23. Besov, O. B. and Iljin, P. V.: *Integral Function Representations and Embedding Theorems*, Nauka, Moscow 1975 (in Russian).
24. Pianigiani, G. and Yorke, J. A.: 'Expanding Maps on Sets which are Almost Invariant: Decay and Chaos', *Trans. Amer. Math. Soc.* **252** (1979), 351–366.
24. Blank, M. L.: 'Stochastic Stabilization of Unstable Dynamical Systems', *Russ. Math. Surveys* **36** (1981), 165–166.
26. Blank, M. L.: 'Ergodic Properties of Discretizations of Dynamical Systems', *Dokl. Akad. Nauk SSSR* **278** (1984), 779–782 (in Russian).
27. Matsumoto, K. and Ishida, I.: 'Noise-Induced Order', *J. Statist. Phys.* **31** (1983), 87–106.
28. Arnold, V. I.: 'Small Denominators', *Izv. Akad. Nauk SSSR, Ser. Mat.* **25** (1961), 21–86 (in Russian).

N. N. ČENCOVA

Statistical Properties of Smooth Smale Horseshoes

Introduction

For a broad class of dynamical systems in continuous time, chaotic behaviour of trajectories is caused [1–4] by the fact that their Poincaré transformation T contains a Smale horseshoe configuration [5–7], see also [8–11], Ya. G. Sinai [12] posed the problem of defining statistic characteristics of trajectories inherent in the horseshoe configuration and of describing the evolution of smooth conditional probability measures on the horseshoe induced by the Smale mechanism. In the present paper this problem is solved for smooth horseshoes (see Definition 1.7). To formulate our results we first introduce some notation and definitions.

In modern analysis a recursive process for constructing a fixed point for a contraction operator is widely employed. Let a diffeomorphism $T: \Delta \to T\Delta \subset \Delta$, where Δ is a convex compact set, be a contraction on Δ. Then the sequence of embedded sets $T^n\Delta$ is contracted to a fixed point, unique in Δ. A more complicated prototype of the Smale mechanism is the Hadamard–Perron technique of constructing a fixed point of a pre-hyperbolic diffeomorphism which is an expansion along some directions and a contraction along others. Suppose the following assumptions are satisfied:

(a) The set $\mathscr{D} = U \times V$ is a compact convex 'rectangle' in a finite-dimensional space $\mathbb{Z} = \mathbb{X} \times \mathbb{Y}$;

(b) The trivaial projection of $\mathscr{D}$ onto U generates for a closed domain $S^{(-1)} \subset \mathscr{D}$ a fibration with the base U onto fibres $\Phi(u) = S^{(-1)} \cap (\{u\} \times V)$ homeomorphic to V;

(c) T is defined in a neighbourhood of $S^{(-1)}$ and is uniform pre-hyperbolic with expansion along $\mathbb{Y}$ on $S^{(-1)}$, see Definition 1.3 (it suffices that the cells of the matrix dT satisfy inequalities (1.20));

(d) $TS^{(-1)} = S^{(1)} \subset \mathscr{D}$ and $T(\partial\Phi(u)) \subset U \times \partial V$ for all $u \in U$.

Then in $S^{(-1)}$ there exists a unique T-fixed point z_0. The recursive sequences of embedded sets $S^{(n)} = T(S^{(n-1)} \cap S^{(-1)})$ and $S^{(-m)} = T^{-1}(S^{(-m+1)} \cap S^{(1)})$ are contracted as $n \to \infty$ and $m \to \infty$ to a locally unstable (or stable, respectively) sets $W^{(u)}(z_0)$ and $W^{(s)}(z_0)$ of the point z_0 and $\{z_0\} = W^{(u)}(z_0) \cap W^{(s)}(z_0)$. To prove this we establish that the $S^{(n)}$ are $\mathbb{Y}$-sets in $\mathscr{D}$, i.e. are fibrated onto T-expanding $\mathbb{Y}$-leaves with a boundary belonging to $U \times \partial V$, see [1]. Corresponding fibrations are constructed recursively starting from the trivial fibration $S^{(-1)}$, since $S^{(1)} = \cup \mathfrak{F}(u)$, where $u \in U$ and $\mathfrak{F}(u) = T\Phi(u)$, etc. We similarly verify that $S^{(-m)}$ is an $\mathbb{X}$-set in $\mathscr{D}$. The corresponding fibrations are constructed recursively starting from the trivial fibration $S^{(1)}$ over the base V onto leaves $\Gamma(v) = S^{(1)} \cap (U \times \{v\})$. Finally we show that the recursive passage from leaves of one fibration to leaves of the next can be described in terms of a contraction, see [11].

R. L. Dobrushin (ed.), Mathematical Problems of Statistical Mechanics and Dynamics, 199–256.

This construction is applicable in particular to the case when T is defined everywhere on $\mathscr{D}$ and is uniform hyperbolic on $\mathscr{D}(-1) = T^{-1}(\mathscr{D} \cap T\mathscr{D})$ with expansion along $\mathbb{Y}$, and the image $T\mathscr{D}$ stabs, roughly speaking, the whole rectangle $\mathscr{D}$ in the $\mathbb{Y}$-direction, i.e. $\mathscr{D} \cap T\mathscr{D}$ is a $\mathbb{Y}$-set in $\mathscr{D}$ and $\partial(\mathscr{D} \cap T\mathscr{D}) \subset T(\partial U \times V) \cap (U \times \partial V)$ (see Fig. 1) where the boundary of $T\mathscr{D}$ is depicted by dotted lines. Smale [5] drew attention to the situation which arises when the T-image of $\mathscr{D}$ is bent and pokes $\mathscr{D}$ along the $\mathbb{Y}$-direction $k \geqslant 2$ times. For $k = 2$ this image resembles a horseshoe, hence its name; in Fig. 2 this image is depicted by dotted lines. The Smale construction of an invariant hyperbolic set is analogous to the Hadamard–Perron process. We shall sketch his scheme without assuming that T is defined on the whole of $\mathscr{D}$. The latter is important for applications to Poincaré transformations. Assume that: (i)

(i) $\mathscr{D}$ satisfies the assumption (a) above;

(ii) the closed domain $\mathscr{D}(-1) \subset \mathscr{D}$ consists of k components $S^{(-1)}(j)$;

(iii) each component $S^{(-1)}(j)$ and the action of T on it satisfy assumptions (b)-(d).

Let us construct the set $\mathscr{D}(1) = T\mathscr{D}(-1)$ with components $S^{(1)}(i) = TS^{(-1)}(i)$, where $1 \leqslant i \leqslant k$, and construct their fibrations onto T-expanding $\mathbb{Y}$-leaves $\mathfrak{F}(u, i)$ and fibrations of components $S^{(-1)}(j)$ onto T^{-1}-expanding $\mathbb{X}$-leaves $\mathfrak{G}(v, i)$. Then let us construct recursively sequences of embedded fibrations $\mathscr{D}(n) = T(\mathscr{D}(n-1) \cap \mathscr{D}(-1))$ and $\mathscr{D}(-m) = T^{-1}(\mathscr{D}(-m+1) \cap \mathscr{D}(1))$. Each $\mathscr{D}(n)$ consists of k^n components which are $\mathbb{Y}$-sets in $\mathscr{D}$ with k components in each component of the previous rank. The structure of $\mathscr{D}(-m)$ is similar. Each T-expanding $\mathbb{Y}$-leaf of $\mathscr{D}(n)$ intersects in the unique point with the T^{-1}-expanding $\mathbb{X}$-leaf of $\mathscr{D}(-m)$. Therefore each connected component of $\mathscr{D}(n)$ intersects with each component of $\mathscr{D}(-m)$ and all of them constitute together a kind of a lattice in $\mathscr{D}$.

Now let us construct limit sets $\mathscr{D}(\infty), \mathscr{D}(-\infty)$ and $\Omega = \mathscr{D}(\infty) \cap \mathscr{D}(-\infty)$. The set Ω of the horseshoe is the Cantor discontinuum of a 'Scotch plaid' type. It is T- and T^{-1}-invariant and contains exactly k fixed points: one in each $S^{(-1)}(j)$. In Ω an everywhere countable dense set of periodic points is contained, exactly k^n points fixed with respect to T^n. The action of T on Ω is described by the symbolic dynamics. Limit sets $\mathscr{D}(\infty)$ and $\mathscr{D}(-\infty)$ consist of a Cantor discontinuum of smooth $\mathbb{Y}$-leaves $\mathfrak{F}(I)$ and $\mathbb{X}$-leaves $\mathfrak{G}(J)$, respectively, which are local expanding (contracting) manifolds for points of Ω that belong to them.

The above method of constructing Ω was axiomatized by Smale in [6]. His axioms define a broad class of topological Smale horseshoes. We restrict ourselves to smooth Smale horseshoes only, and are interested in the asymptotics of approximating trajectories of T which do not leave $\mathscr{D}(-1)$ by trajectories of T on Ω. Our Definition 1.7, of the situation when T possesses a smooth horseshoe in $\mathscr{D}$, is

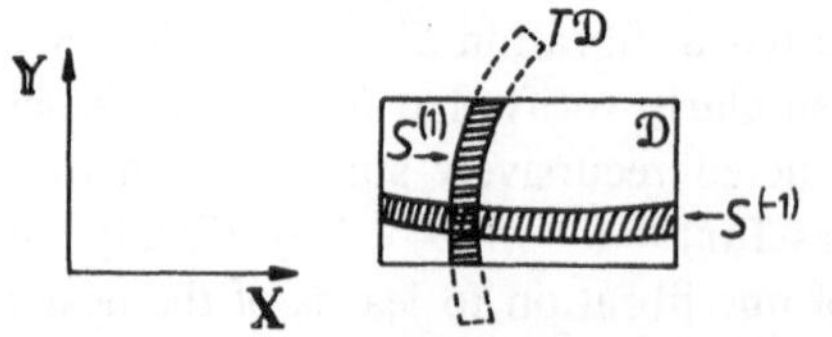

Fig. 1.

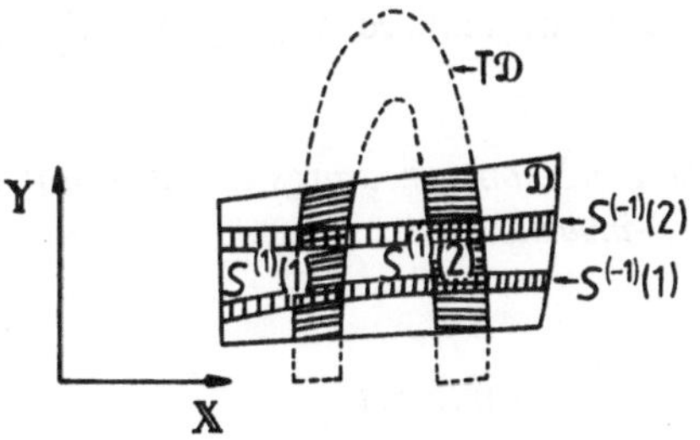

Fig. 2.

somewhat wider than the above description: we allow $\mathcal{D}$ for dim $\mathbb{X} = 1$ or dim $\mathbb{Y} = 1$ to be a diffeomorphic image of a rectangle with some $\mathbb{X}$, $\mathbb{Y}$-convexity. All mappings participating in the definition of a horseshoe are assumed to be C^{1+1}-smooth. We also define a uniform pre-hyperbolic diffeomorphism according to Alekseev [1] and Sinai [12].

Let the C^{1+1}-diffeomorphism T possess in $\mathcal{D} \subset \mathbb{Z} = \mathbb{X} \times \mathbb{Y}$ a smooth Smale horseshoe (with expansion along $\mathbb{Y}$). The main results of our paper can then be briefly formulated as follows.

THEOREM (3.7). *On fibres $\mathfrak{F}(I)$ of the bundle $\mathcal{D}(\infty)$ there is a unique family of conditional probability distributions $P\{\cdot \mid I\}$ with the following properties*:

(a) *each $P\{\cdot \mid I\}$ has a density $p(y \mid I)$ relative to the $\mathbb{Y}$-volume on a chart of the fibre $\mathfrak{F}(I)$*;
(b) *$\ln p(y \mid I)$ is Lipschitzizable in y*;
(c) *the family is T-invariant, i.e. for $(x^{(1)}, y^{(1)}) = T(x, y)$ and $\mathfrak{F}(Ij) = T(\mathfrak{F}(I) \cap S^{(-1)}(j))$ we have $p(y^{(1)} \mid I_j) \cdot P\{\mathfrak{F}(I) \cap S^{(-1)}(j)\} = p(y) . \mid Jac(dy \mid dy^{(1)}) \mid$*.

The density function $p(y \mid I)$, where y is a $\mathbb{Y}$-coordinate on a chart of $\mathfrak{F}(I)$, is uniformly continuous on $\mathcal{D}(\infty)$ in both arguments.

Let us denote by $\mathfrak{S}$ the class of measures on $\mathcal{D}(\infty)$ inducing on the fibre $\mathfrak{F}(I)$ the conditional probability distribution $P\{\cdot \mid I\}$ introduced in Theorem 3.7.

THEOREM. *There exists a unique probability measure $\mu \in \mathfrak{S}$ such that $\mu\{T^{-1}H\} = \lambda\mu\{H\}$ for some $\lambda > 0$ for every measurable subset $H \subset \mathcal{D}(\infty)$.*

This is why we call λ an *eigenvalue* of the horseshoe and μ an *eigenmeasure* of the horseshoe. The measure μ is given for every measurable $H \subset \mathcal{D}(\infty)$ by the integral

$$\mu\{H\} = \int \hat{\mu}(dI) P\{H \mid I\}.$$

THEOREM. *There is a unique (up to multiplication by a constant) continuous function $e(I)$ satisfying the system of equations*

$$e(I) = \sum_{1 < j < k} P\{S^{(-1)}(j) \mid I\} e(Ij), \qquad ((6.31))$$

where $P\{\cdot \mid I\}$ is introduced in Theorem 3.7 and λ is an eigenvalue of the horseshoe.

The function $e(I)$ is of fixed sign and is normalized throughout the paper: $\int e(I)\hat{\mu}(dI) = 1$.

THEOREM. *The weak $*$-limit $\mu_0\{\}$ of the conditional measures $\mu\{\cdot\,|\,\mathcal{D}(-m)\}$ exists. The measure μ_0 is supported in Ω and is T-invariant, i.e.*

$$\mu_0\{T^{-1}H\} = \mu_0\{TH\} = \mu_0\{H\} \tag{(5.50)}$$

for each measurable $H \subset \Omega$. It can be given by the integral

$$\mu_0\{H\} = \int \hat{\mu}(dI)e(I)P_0\{H\,|\,I\}. \tag{6.10}$$

LIMIT THEOREM. *For measures ν having Lipshitzizable densities relative to Lebesgue measure in $\mathcal{D}$ and any continuous function $h(z)$ on $\mathcal{D}$ we have*

$$\lim_{n\to\infty} \lambda^{-n} \int_{\mathcal{D}(-n)} h(T^n z)\nu\{dz\} = c[\nu] \int_{\mathcal{D}(\infty)} h(z)\mu\{dz\} \tag{7.30}$$

$$\lim_{m,n\to\infty} \lambda^{-m-n} \int_{\mathcal{D}(-n-m)} h(T^n z)\nu\{dz\} = c[\nu] \int_{\Omega} h(z)\mu_0\{dz\} \tag{7.32}$$

where $c[\nu]$ is a linear functional in ν. Moreover, if the density of the measure ν is strictly positive in the closure of $\mathcal{D}$, then $c[\nu] > 0$.

The dynamical system (Ω, T, μ_0) induced by the horseshoe is a mixing one and possesses a weak Bernoulli partition $\Omega = \cup_{1 \leqslant j \leqslant k} (\Omega \cap S^{(-1)}(j))$. Let ν, h_1, h_2 satisfy the restrictions of the Limit Theorem, then

$$\lim_{m,n,r\to\infty} \lambda^{-m-n-r} \int_{\mathcal{D}(-m-n-r)} h_1(T^n z)h_2(T^{n+m}z)\nu\{dz\}$$

$$= c[\nu] \int_{\Omega} h_1(z)\mu_0\{dz\} \int_{\Omega} h_2(z)\mu_0\{dz\}. \tag{7.36}$$

$T^{-1}: \mathcal{D}(1) \to \mathcal{D}(-1)$, the inverse of T, also possesses a smooth Smale horseshoe in $\mathcal{D}$, with the same invariant set Ω but with expansion relative to $\mathbb{X}$. We show that the invariant measure μ'_0 for the inverse horseshoe is not generally speaking equal to μ_0, and $\lambda' \neq \lambda$.

Constructing a fixed point of an operator by successive approximations, we can vary the initial contracting neighbourhood within in reasonable limits. Similarly, constructing Ω we can accordingly vary somewhat the configuration of $\mathcal{D}$ and $\mathcal{D}(-1)$. Then Ω and the probability measure μ_0 on it do not vary. Functions whose graphs define leaves $\mathfrak{F}(I) \subset \mathcal{D}(\infty)$ and $\mathfrak{G}(J) \subset \mathcal{D}(-\infty)$ of fibrations coincide up to a domain of definition. The eigenvalue $\lambda = \mu\{\mathcal{D}(\infty) \cap \mathcal{D}(-1)\}$ does not, generally speaking, vary. To make the Limit Theorem invariant we have to pass from the smooth measure ν on $\mathcal{D}$ to conditional probability measures $\nu\{\cdot\,|\,\mathcal{D}(N)\} = \nu\{\cdot\}/\nu\{\mathcal{D}(N)\}$ as in the ergodic theorem, e.g. instead of (7.32) we get

$$\lim_{m,n\to\infty} \int_{\mathcal{D}} h(T^n z)\nu\{dz\,|\,\mathcal{D}(-n-m)\} = \int_{\Omega} h(z)\mu_0\{dz\}$$

ERGODIC THEOREM FOR CONDITIONAL DISTRIBUTIONS

For any Borel probability distribution Q on $\mathscr{D}$ with the Lipschitzizable logarithm of density relative to $\mathbb{Z}$-volume, arbitrary continuous function $h(z)$ in $\mathscr{D}$ and any $\varepsilon > 0$ we have

$$\lim_{\substack{M+N\to\infty\\ M,N\geqslant 0}} Q_{M,N}\Big\{z: |(N+M+1)^{-1}\sum_{-M\leqslant n\leqslant N} h(T^n z) - - N(N+M)^{-1}\int_\Omega h(w)\mu_0\{dw\} - M(N+M)^{-1}\int_\Omega h(w)\mu_0'\{dw\}| > \varepsilon\Big\} = 0 \tag{7.50}$$

where $Q_{M,N}\{\cdot\} = Q\{\cdot\,|D(-N)\cap D(M)\}$ is a conditional distribution.

In our constructions of the invariant measure μ_0 and investigation of its properties we generally follow the Sinai scheme [12–15] with modifications caused by non-invariance of the domain $\mathscr{D}(-1)$ of definition of T. In particular, we have to make use of ergodic properties of positive substochastic matrices (not stochastic matrices as, in Sinai's work); a list of these properties is given in the Appendix. To prove the conditional ergodic theorem we need the limit relations (7.32) and (7.36) with uniform estimate of remainders. We therefore make a practice of writing with accuracy the constants in all inequalities in terms of the norm and Lipschitz constants of the functions encountered and characteristics of the hyperbolicity of T. For lack of space we omit most of the proofs. In the construction of §1 we have used approaches developed by V. M. Aleseev [1] and in §2 we have applied some ideas of Anosov [16, 17] and of Bowen and Ruelle [18, 19]. The results of this work have been used in [4, 20] in the investigation of the behaviour of trajectories of a system describing properties of an autogenerator on a tunnel diode proposed in [21].

1. General Background

The clearest description of Smale's example consists of a description of its three ingredients: (1) a rectangular domain $\mathscr{D}$, (2) a transformation $T:\mathscr{D}\to T\mathscr{D}$ which is hyperbolic in the domain $T^{-1}(\mathscr{D}\cap T\mathscr{D}) = \mathscr{D}(-1)$, (3) a horseshoe $T\mathscr{D}$ which is geometrically superimposed over $\mathscr{D}$ in such a way that $\mathscr{D}\cap T\mathscr{D}$ consists of two components. The main geometrical property of Smale's construction is the lattice formed by components of the closed domain $\mathscr{D}(-1)$ and of components of $T\mathscr{D}(-1) = \mathscr{D}(1)$ which intersect them. This lattice is the object of our investigation.

1.1. STRUCTURES IN THE PRODUCT $\mathbb{Z} = \mathbb{X}\times\mathbb{Y}$

Let $\mathbb{X}$ and $\mathbb{Y}$ be normed vector spaces and $\mathbb{Z} = \mathbb{X}\times\mathbb{Y}$. If x is an $\mathbb{X}$-component of $z\in\mathbb{Z}$ and y is a $\mathbb{Y}$-component of z, then put $z = (x, y)$ or $z = x\oplus y$. For $z = (x, y)\in\mathbb{Z}$ define the $\mathbb{X}$-length and the $\mathbb{Y}$-length of z to be

$$|z|_{\mathbb{X}} = |x|, \qquad |z|_{\mathbb{Y}} = |y|. \tag{1.1}$$

Define vector cones putting

$$\mathscr{K}_{\mathbb{X}}^{\alpha} = \{z\in\mathbb{Z}: |z|_{\mathbb{Y}} \leqslant \alpha|z|_{\mathbb{X}}\}, \qquad \mathscr{K}_{\mathbb{Y}}^{\beta} = \{z\in\mathbb{Z}: |z|_{\mathbb{X}} \leqslant \beta|z|_{\mathbb{Y}}\}, \tag{1.2}$$

where $0 < \alpha, \beta < \infty$. It is evident that $\mathscr{K}_{\mathbb{X}}^{\alpha} \cap \mathscr{K}_{\mathbb{Y}}^{\beta} = \{(0,0)\}$ if and only if $\alpha\beta < 1$. Further, fix α and β such that

$$0 < \alpha, \beta < \infty; \qquad \alpha\beta < 1.$$

In $\mathrm{Lin}(\mathbb{X}, \mathbb{Y})$, the linear normed space of linear operators from $\mathbb{X}$ to $\mathbb{Y}$, define the ball

$$N_{\mathbb{X}}^{\alpha} = \{A \in \mathrm{Lin}(\mathbb{X}, \mathbb{Y}) : \|A\| \leqslant \alpha\}. \tag{1.3}$$

For each $A \in N_{\mathbb{X}}^{\alpha}$ the graph $\{x \oplus Ax, x \in \mathbb{X}\}$ is the subspace which belongs to $\mathscr{K}_{\mathbb{X}}^{\alpha}$. The set of these subspaces

$$\mathscr{N}_{\mathbb{X}}^{\alpha} = \{\{x \oplus Ax, x \in \mathbb{X}\}, A \in \mathrm{Lin}(\mathbb{X}, \mathbb{Y}), \|A\| \leqslant \alpha\} \tag{1.4}$$

is the closed region in the Grassmann manifold $G_{\dim \mathbb{X}}(\mathbb{Z})$. The subset $\{x \oplus (Ax + v), x \in \mathbb{X}\}$, where $v \in \mathbb{Y}$ and $A \in N_{\mathbb{X}}$, is called an $\mathbb{X}$-*plane in* $\mathbb{Z}$.

DEFINITION 1.1. The $\mathbb{Y}$-distance from the $\mathbb{X}$-plane $\{x \oplus (Ax + v), x \in \mathbb{X}\}$ to 0 is

$$\rho_{\beta}(A, v) = \max |z|_{\mathbb{Y}}, \ z \in \mathscr{K}_{\mathbb{Y}}^{\beta} \cap \{x \oplus (Ax + v), x \in \mathbb{X}\}. \tag{1.5}$$

Definitions of the *ball of operators* $N_{\mathbb{Y}}^{\beta}$, the *graph of the operator* $B \in N_{\mathbb{Y}}^{\beta}$, the *set of subspaces* $\mathscr{N}_{\mathbb{Y}}^{\beta}$, a $\mathbb{Y}$-*plane*, the $\mathbb{X}$-*distance from a* $\mathbb{Y}$-*plane to* 0 are analogous. The $\mathbb{X}$-plane $\{x \oplus (Ax + v), x \in \mathbb{X}\}$ for $A \in N_{\mathbb{X}}^{\alpha}$ and $v \in \mathbb{Y}$ intersects the subspace $\{By \oplus y, y \in \mathbb{Y}\}$, where $B \in N_{\mathbb{Y}}^{\beta}$, by the unique vector $z(A, v, B)$ because $\alpha\beta < 1$.

DEFINITION 1.2. The $\alpha\beta$-distance between subspaces $\{B_i y \oplus y, y \in \mathbb{Y}\}$, where $B_i \in N^{\beta}{}_{\mathbb{Y}}$, is

$$\rho_{\alpha\beta}(B_1, B_2) = \max_{A \in N_{\mathbb{X}}^{\alpha}, v \in \mathbb{Y} \setminus \{0\}} \{|z(A, v, B_1) - z(A, v, B_2)|_{\mathbb{X}} [\rho_{\beta}(A, v)]^{-1}\}. \tag{1.6}$$

LEMMA 1.1. *The* $\alpha\beta$-*distance is a metric on* $N_{\mathbb{Y}}^{\beta}$ *and* $\mathscr{N}_{\mathbb{Y}}^{\beta}$ *equivalent to* $\rho_{0\beta}(B_1, B_2) = \|B_1 - B_2\|$, *i.e.*

$$\rho_{0\beta} \leqslant \rho_{\alpha\beta} \leqslant (1 - \alpha\beta)^{-1} \cdot \rho_{0\beta}. \tag{1.7}$$

The tangent space at each point of $\mathbb{Z}$ is isomorphic to $\mathbb{Z}$ and inherits its linear structure. Denote points of $\mathbb{Z}$ by p and q, i.e. $p = x + y$, where x and y are coordinates of p, and denote vectors of the tangent space by Greek letters, e.g. $\zeta = \xi \oplus \eta$. Vector cones $\mathscr{K}_{\mathbb{X}}^{\alpha}(p)$ and $\mathscr{K}_{\mathbb{Y}}^{\beta}(p)$, sets $\mathscr{N}_{\mathbb{X}}^{\alpha}(p)$ and $\mathscr{N}_{\mathbb{Y}}^{\beta}(p)$, etc. in the tangent space $\mathbb{Z}(p)$ are similarly introduced.

1.2. UNIFORMLY HYPERBOLIC TRANSFORMATIONS OF $\mathbb{Z} = \mathbb{X} \times \mathbb{Y}$

Let $T: \mathscr{D} \to \mathbb{Z}$ be a C^1-transformation of a closed region $\mathscr{D} \subset \mathbb{Z}$. In coordinates, T is defined by the formula $T(x \oplus y) = f(x, y) \oplus g(x, y)$. The differential $\mathrm{d}T$ of T is given by blocks $d_x f$, $d_y f$, $d_x g$, $d_y g$ of the matrix $(d_x f / d_x g)(d_y f / d_y g)$. Each block as an element of the corresponding operator space depends continuously on p as parameter. We say that a C^1-transformation T belongs to the class C^{1+1} if each of the blocks satisfies the Lipschitz condition, e.g.

$$|d_x f(p_1) - d_x f(p_2)| \leqslant \mathrm{Lip}_{\mathbb{X}}(d_x f) |p_1 - p_2|_{\mathbb{X}} + \mathrm{Lip}_{\mathbb{Y}}(d_x f) |p_1 - p_2|_{\mathbb{Y}}.$$

DEFINITION 1.3. The C^{1+1}-diffeomorphism T of a compact domain $\mathcal{D} \subset \mathbb{Z}$ is uniformly hyperbolic with respect to the splitting $\mathbb{Z} = \mathbb{X} \times \mathbb{Y}$ with expansion along $\mathbb{Y}$ (and with parameters α and β) if there are constants λ_s, $\lambda_u > 1$ such that

$$dT(p)\cdot\zeta \in \mathcal{K}^{\beta}_{\mathbb{Y}}(T_p), \quad |dT(p)\cdot\zeta|_{\mathbb{Y}} \geqslant \lambda_u\cdot|\zeta|_{\mathbb{Y}} \tag{1.8}$$

holds at $p \in \mathcal{D}$ for any vector $\zeta \in \mathcal{K}^{\beta}_{\mathbb{Y}}(p)$ and

$$(dT(q))^{-1}\cdot\zeta \in \mathcal{K}^{\alpha}_{\mathbb{X}}(T^{-1}q), \quad |(dT(q))^{-1}\cdot\zeta|_{\mathbb{X}} \geqslant \lambda_s\cdot|\zeta|_{\mathbb{X}} \tag{1.9}$$

holds at any $q \in T\mathcal{D}$ for any $\zeta \in \mathcal{K}^{\alpha}_{\mathbb{X}}$(q).

We also say that systems of vector cones $\mathcal{K}^{\alpha}_{\mathbb{X}}(p)$, $\mathcal{K}^{\beta}_{\mathbb{Y}}(p)$ *define the hyperbolic prestructure in* $\mathcal{D}$ according to Alekseev [1] and Sinai [12, 14].

REMARK. We can also consider a smaller set of uniform hyperbolic diffeomorphisms $T: \mathcal{D} \to \mathbb{Z}$ satisfying

$$dT(p)\zeta \in \operatorname{int}\mathcal{K}^{\beta}_{\mathbb{Y}}(Tp), \quad |dT(p)\zeta|_{\mathbb{Y}} > \lambda_u\cdot|\zeta|_{\mathbb{Y}}$$

at each $p \in \mathcal{D}$ and $\zeta \in \mathcal{K}^{\beta}_{\mathbb{Y}}(p)$ and

$$(dT(q))^{-1}\zeta \in \operatorname{int}\mathcal{K}^{\alpha}_{\mathbb{X}}(T^{-1}q), \quad |(dT(q))^{-1}\zeta|_{\mathbb{X}} > \lambda_s\cdot|\zeta|_{\mathbb{X}}$$

at each $q \in \mathbb{T}\mathcal{D}$ and $\zeta \in \mathcal{K}^{\alpha}_{\mathbb{X}}(q)$. This set is open in the C^{1+1}-topology.

Let us investigate the action of the hyperbolic diffeomorphism T on subspaces of the tangent space. From Definition 1.3 we deduce that

$$dT(p)\mathcal{N}^{\beta}_{\mathbb{Y}}(p) \subset \mathcal{N}^{\beta}_{\mathbb{Y}}(Tp); \tag{1.10}$$

$$(dT(q))^{-1}\mathcal{N}^{\alpha}_{\mathbb{X}}(q) \subset \mathcal{N}^{\alpha}_{\mathbb{X}}(T^{-1}q) \tag{1.11}$$

for any $p \in \mathcal{D}$ and for any $q \in T\mathcal{D}$ respectively.

LEMMA 1.2. *The uniform hyperbolic transformation T generates a strict contraction of sets* $\mathcal{N}^{\beta}_{\mathbb{Y}}(p)$, *i.e.*

$$\rho_{\alpha\beta}(\mathcal{B}_1, \mathcal{B}_2) \geqslant \lambda_u\cdot\lambda_s\cdot\rho_{\alpha\beta}(dT(p)\mathcal{B}_1, dT(p)\mathcal{B}_2) \tag{1.12}$$

for any $p \in \mathcal{D}$ *and any two subspaces* $\mathcal{B}_i \in \mathcal{N}^{\beta}_{\mathbb{Y}}(p)$, *where* $i = 1, 2$.

The diffeomorphism T^{-1} *generates a strict contraction of sets* $\mathcal{N}^{\alpha}_{\mathbb{X}}(q)$,

$$\rho_{\beta\alpha}(\mathcal{A}_1, \mathcal{A}_2) \geqslant \lambda_u\cdot\lambda_s\cdot\rho_{\beta\alpha}((dT(q))^{-1}\mathcal{A}_1, (dT(q))^{-1}\mathcal{A}_2) \tag{1.13}$$

for any $q \in T\mathcal{D}$ *and any two subspaces* $\mathcal{A}_i \in \mathcal{N}^{\alpha}_{\mathbb{X}}(q)$, *where* $i = 1, 2$.

REMARK. It is not necessary for T to be a contraction with respect to the functional norm $\rho_{0\beta}$, see [22].

The contraction (1.12) of sets $\mathcal{N}^{\beta}_{\mathbb{Y}}(p)$ generates the contraction of balls $N^{\beta}_{\mathbb{Y}}(p)$. Let $B \in N_{\mathbb{Y}}(p)$. Denote by $\mathcal{T}^{+}[p](B)$ a linear operator which belongs to $N^{\beta}_{\mathbb{Y}}(Tp)$ if its graph in $\mathbb{Z}(Tp)$ is the image of the graph of B *in* $\mathbb{Z}(p)$ under $dT(p)$. Define $\mathcal{T}^{-}[q](A)$ in a similar way; see [1, 17].

STATEMENT 1.3. Let T be a uniform hyperbolic diffeomorphism of the domain D. Then

$$\mathcal{T}^+[p](B) = [d_x f(p)\cdot B + d_y f(p)]\cdot[d_x g(p)\cdot B + d_y g(p)]^{-1}, \tag{1.14}$$

$$d_B\mathcal{T}^+[p]\cdot\Delta = [d_x f(p) - \mathcal{T}^+[p](B)\cdot d_x g(p)]\cdot\Delta\cdot[d_x g(p)\cdot B + d_y g(p)]^{-1} \tag{1.15}$$

hold for $p\in\mathcal{D}$ and any $B\in N^\beta_{\mathbb{Y}}(p)$ and

$$\mathcal{T}^-[q](A) = [d_y g(p) - A\cdot d_y f(p)]^{-1}\cdot[A\cdot d_x f(p) - d_x g(p)], \tag{1.16}$$

$$d_A\mathcal{T}^-[q]\cdot\Delta = [d_y g(p) - A\cdot d_y f(p)]^{-1}\cdot\Delta\cdot[d_y f(p)\mathcal{T}^-[q](A) + d_x f(p)] \tag{1.17}$$

hold for any $q\in T\mathcal{D}$, and any $A\in N^\alpha_{\mathbb{X}}$, where $Tp = q$,

$$d_B\mathcal{T}^+[p]\Delta = \frac{d}{dt}\mathcal{T}^+[p](B + t\cdot\Delta)|_{t=0}.$$

1.3. A SUFFICIENT CONDITION FOR UNIFORM HYPERBOLICITY IN $\mathbb{Z} = \mathbb{X}\times\mathbb{Y}$

Define the functional norms of the fields of operators $d_x f, d_y f, (d_y g)^{-1}, (d_y g)^{-1}\cdot d_x g$ to be the least upper bounds of the norms at $p\in\mathcal{D}$. Let us recover from this data the quadratic polynomials

$$r(\theta) = \|(d_y g)^{-1} d_x g\|\cdot\theta^2 - [1 - \|d_x f\|\cdot\|(d_y g)^{-1}\|]\cdot\theta + \|d_y f\|\cdot\|(d_y g)^{-1}\|, \tag{1.18}$$

$$s(\lambda) = \|(d_y g)^{-1}\|\cdot\lambda^2 - [1 + \|d_x f\|\cdot\|(d_y g)^{-1}\|]\cdot\lambda + \|(d_y g)^{-1} d_x g\|\cdot\|d_y f\| + \|d_x f\| \tag{1.19}$$

and let $\theta^\pm, \lambda^\pm$ be the roots of these polynomials, respectively.

STATEMENT 1.4. If

$$\|d_x f\| < 1, \qquad \|(d_y g)^{-1}\| < 1, \tag{1.20}$$
$$\|(d_y g)^{-1}\cdot d_x g\|\cdot\|d_y f\| < [1 - \|d_x f\|]\cdot[1 - \|(d_y g)^{-1}\|],$$

then $\theta^\pm$ are real and positive, $0 < \theta^- < \theta^+$ and $\lambda^\pm$ are real and positive, $\lambda^+ > 1 > \lambda^- > 0$, cf. [23, 24].

STATEMENT 1.5. If a C^{1+1}-diffeomorphism $T:\mathcal{D}\to\mathbb{Z}$ of the compact domain $\mathcal{D}\subset\mathbb{Z}$ satisfies (1.20) then T is uniformly hyperbolic with expansion along $\mathbb{Y}$ and parameters of cones

$$\alpha = (\theta^+)^{-1}, \qquad \beta = \theta^- \tag{1.21}$$

with constants

$$\lambda_s = (\lambda^-)^{-1}, \qquad \lambda_u = \lambda^+. \tag{1.22}$$

Besides,

$$\|d_B\mathcal{T}^+\| \leqslant (\lambda_s\cdot\lambda_u)^{-1}, \|d_A\mathcal{T}^-\| \leqslant (\lambda_s\cdot\lambda_u)^{-1}. \tag{1.23}$$

REMARKS

(a) Inequalities (1.20) also yield that

$$\lambda_s\cdot|dT(p)\cdot\zeta|_{\mathbb{X}} \leqslant |\zeta|_{\mathbb{X}}$$

at any $p \in \mathscr{D}$ and any $\zeta \in \mathscr{K}^{\alpha}_{\mathbb{X}}$

(b) We can choose α and β from the sets

$$\theta^{-} \leqslant \beta < \|(d_y g)^{-1} d_x g\|^{-1} \cdot [1 - \|(d_y g)^{-1}\|],$$

$$[\theta^{+}]^{-1} \leqslant \alpha < \|d_y f\|^{-1}[1 - \|d_x f\|]$$

but such a choice might given smaller $\lambda_u(\beta)$ and $\lambda_s(\alpha)$ than (1.22).

1.4. LEAVES AND RECTANGLES

DEFINITION 1.4. The graph $\{x \oplus G(x), x \in U\}$ of the C^1-transformation $G: U \to \mathbb{Y}$ is a smooth $\mathbb{X}$-leaf over the domain $U \subset \mathbb{X}$, if $\sup_U \|dG\| \leqslant \alpha$. A connected smooth manifold in $\mathbb{Z}$ is a smooth $\mathbb{X}$-surface, if locally it is a smooth $\mathbb{X}$-leaf. A smooth $\mathbb{Y}$-leaf and a smooth $\mathbb{Y}$-surface are similarly defined, [1].

Hereafter T will be a uniform hyperbolic diffeomorphism (with expansion along $\mathbb{Y}$) of compact domain $\mathscr{D} \subset \mathbb{Z}$ into $\mathbb{Z}$.

STATEMENT 1.6. Let $\mathscr{F}$ be a smooth $\mathbb{Y}$-surface, $\mathscr{F} \subset \mathscr{D}$. Then $T\mathscr{F}$ is also a smooth $\mathbb{Y}$-surface. Similarly, if $\mathfrak{G}$ is a smooth $\mathbb{X}$-surface, $\mathfrak{G} \subset T\mathscr{D}$, then $T^{-1}\mathfrak{G}$ is also a smooth $\mathbb{X}$-surface.

REMARK. If $\dim \mathbb{Y} = 1$, then the image of a $\mathbb{Y}$-leaf under T is a $\mathbb{Y}$-leaf; when $\dim \mathbb{Y} \geqslant 2$ it is not.

STATEMENT 1.7. Let $\mathfrak{F}$ be a smooth $\mathbb{Y}$-leaf, $\mathfrak{F} \subset \mathscr{D}$ and $T\mathfrak{F}$ a smooth $\mathbb{Y}$-leaf over the convex domain $V \subset \mathbb{Y}$. Then $|Tp - Tq|_{\mathbb{Y}} \geqslant \lambda_u \cdot |p - q|_{\mathbb{Y}}$ for any two points $p, q \in \mathfrak{F}$.

Similarly, if $\mathfrak{G}$ is a smooth $\mathbb{X}$-leaf, $\mathfrak{G} \subset T\mathscr{D}$, and $T^{-1}\mathfrak{G}$ is a smooth $\mathbb{X}$-leaf over the convex domain $\mathscr{U} \subset \mathbb{X}$, then $|p - q|_{\mathbb{X}} \geqslant \lambda_s |T_p - T_q|_{\mathbb{X}}$ for any $Tp, Tq \in \mathfrak{S}$.

Therefore we sometimes call $\mathbb{Y}$-leaves expanding ones (with respect to the T-action), and $\mathbb{X}$-leaves contracting ones.

STATEMENT 1.8. Let $W = U \times V = \{x \oplus y, x \in U, y \in V\}$, U be a convex compact domain in $\mathbb{X}, V$ a convex compact domain in $\mathbb{Y}$. Let $\mathfrak{G}$ be a compact smooth $\mathbb{X}$-surface such that

$$\mathfrak{G} \cap (\text{int } W) \neq \emptyset, \; \partial\mathfrak{G} \cap (\text{int } W) = \emptyset, \qquad \mathfrak{G} \cap \{(\text{int } U) \times \partial V\} = \emptyset.$$

Then each component of $\mathfrak{G} \cap W$ is a smooth $\mathbb{X}$-leaf over U. The analogous statement holds for a $\mathbb{Y}$-leaf.

DEFINITION 1.5. The compact domain $\mathscr{D}$ is a φ-rectangle (or curvilinear rectangle when φ is not defined explicitly) if φ is a local C^{1+1}-diffeomorphism $\varphi: \mathbb{Z} \to \mathbb{Z}$ such that

(a) $\mathscr{D} = \varphi(U \times V)$, where U is a convex compact domain in $\mathbb{X}, V$ is a convex compact domain in $\mathbb{Y}$.

(b) For any $v \in V$ the fibre $\mathfrak{G}(v) = \varphi(U \times \{v\})$ is a smooth $\mathbb{X}$-leaf and for any $u \in U$ the fibre $\mathscr{F}(u) = \varphi(\{u\} \times V)$ is a smooth $\mathbb{Y}$-leaf.

Such a φ-rectangle is said to be convex with respect to $\mathbb{X}$ if either $\dim \mathbb{X} = 1$ or

$\varphi(\partial U \times V) \subset U' \times \mathbb{Y}$, where U' is a convex compact set in $\mathbb{X}$, and convex with respect to $\mathbb{Y}$ if either $\dim \mathbb{Y} = 1$ or $\varphi(U \times \partial V) \subset \mathbb{X} \times V'$, where V' is a convex compact set in $\mathbb{Y}$.

DEFINITION 1.6. Let $\mathscr{D}$ be a φ-rectangle, $\mathfrak{G}$ a smooth $\mathbb{X}$-leaf, such that

$$\mathfrak{G} \cap \mathscr{D} \neq \emptyset, \qquad \partial\mathfrak{G} \cap (\operatorname{int}\mathscr{D}) = \emptyset, \qquad \mathfrak{G} \cap \varphi([\operatorname{int} U] \times \partial V) = \emptyset.$$

Then $\mathfrak{G}$ is a smooth $\mathbb{X}$-leaf in $\mathscr{D}$. Similarly, if $\mathfrak{F}$ is a smooth $\mathbb{Y}$-leaf such that

$$\mathfrak{F} \cap \mathscr{D} \neq \emptyset, \qquad \partial\mathfrak{F} \cap (\operatorname{int}\mathscr{D}) = \emptyset, \qquad \mathfrak{F} \cap \varphi(\partial U \times [\operatorname{int} V]) = \emptyset,$$

then $\mathfrak{F}$ is a smooth $\mathbb{Y}$-leaf in $\mathscr{D}$.

STATEMENT 1.9. Assume that $\mathscr{D}$ is an $\mathbb{X}, \mathbb{Y}$-convex φ-rectangle, Γ is a smooth $\mathbb{X}$-surface such that

$$\Gamma \cap (\operatorname{int}\mathscr{D}) \neq \emptyset, \qquad \partial\Gamma \cap (\operatorname{int}\mathscr{D}) = \emptyset, \qquad \Gamma \cap \varphi([\operatorname{int} U] \times \partial V) = \emptyset. \tag{1.24}$$

Then each component of $\mathfrak{G} = \Gamma \cap \mathscr{D}$ is a smooth $\mathbb{X}$-leaf in $\mathscr{D}$, over a convex domain in $\mathbb{X}$. A similar statement holds on $\mathbb{Y}$-surfaces.

REMARK. The property formulated in Statement 1.9 is a characteristic one for $\mathbb{X}, \mathbb{Y}$-convex rectangles.

STATEMENT 1.10. In the $\mathbb{X}, \mathbb{Y}$-convex φ-rectangle $\mathscr{D}$ each fibre $\mathfrak{S}(v_0), v_0 \in V$ is a graph of the C^{1+1}-smooth function

$$\varphi(u, v_0) = \bar{x}(u, v_0) \oplus \bar{y}(u, v_0) = \bar{x}(u, v_0) \oplus G(\bar{x}(u, v_0), v_0)$$

defined on the convex set $\{x = \bar{x}(u, v_0), u \in U]$. Similarly, each fibre $\mathfrak{F}(u_0)$, where $u_0 \in U$, is a graph of the C^{1+1}-smooth function $x = F(y, u_0)$. Partial differentials of G and F with respect to the parameter are Lipschitzizable, i.e. satisfy the Lipschitz condition with respect to x and y respectively, so that $\det d_v G \neq 0$ and $det\, d_u F \neq 0$. The partial differentials of six functions $\bar{x}, \bar{y}, G, F, \bar{u}, \bar{v}$, where $\bar{u}(x,y) \oplus \bar{v}(x,y) = \varphi^{-1}(x,y)$, are related by a number of identities, e.g.

$$(d_v \bar{y})^{-1} = d_y \bar{v} + d_x \bar{v} \cdot d_y F, \qquad d_u F = [\mathbb{1}_{\mathbb{X}} - d_y F \cdot d_x G] \cdot d_u \bar{x},$$
$$(d_u \bar{x})^{-1} = d_x \bar{u} + d_y \bar{u} \cdot d_x G, \qquad d_v G = [\mathbb{1}_{\mathbb{Y}} - d_x G \cdot d_y F] \cdot d_v \bar{y}$$

and $\|\mathrm{d}_x G\| \cdot \|\mathrm{d}_y F\| \leqslant \alpha \cdot \beta < 1$.

STATEMENT 1.11. Let $\mathscr{D}$ be a φ-rectangle, $\mathfrak{G}$ a smooth $\mathbb{X}$-leaf in $\mathscr{D}$ and $\mathfrak{F}$ a smooth $\mathbb{Y}$-leaf in $\mathscr{D}$. Then $\mathfrak{S} \cap \mathfrak{F} \neq \emptyset$. If the φ-rectangle $\mathscr{D}$ is $\mathbb{X}, \mathbb{Y}$-convex, then $\mathfrak{G} \cap \mathscr{F}$ consists of the unique point.

STATEMENT 1.12. Suppose $\{\hat{\Gamma}(v), v \in V\}$ is a smooth family of (mutually non-intersecting smooth $\mathbb{Y}$-leaves in the $\mathbb{X}, \mathbb{Y}$-convex rectangle $\mathscr{D}$, $\hat{\Gamma}(v) = \{(x,y)\colon x = \Gamma(y,v)\}$ and $d\Gamma(y,v) = d_y \Gamma + d_v \Gamma$, where partial differentials are Lipschitzizable with respect to $\mathbb{Y}$ and $\det d_v \Gamma \neq 0$ everywhere. Suppose $\{\Phi(u), u \in U\}$ is an analogous smooth family of smooth $\mathbb{X}$-leaves in $\mathscr{D}$. Then points of mutual intersections of both fibrations fill in a curvilinear rectangle $\mathscr{D}_0$ diffeomorphic to $U \times V$, i.e. $\{\Psi(u,v)\} = \Phi(u) \cap \hat{\Gamma}(v)$ for any $u \in U, v \in V$.

1.5. THE SMALE HORSESHOE

DEFINITION 1.7. Let $\mathscr{D} = \varphi(U \times V)$ be an $\mathbb{X}$, $\mathbb{Y}$-convex curvilinear rectangle, $\mathscr{D}' \subset \mathbb{Z}$. A C^{1+1}-diffeomorphism $T: \mathscr{D}' \to \mathbb{Z}$ possesses in $\mathscr{D}$ a smooth Smale horseshoe (with expansion along $\mathbb{Y}$) if there is a compact domain $\mathscr{D}(-1) \subset \mathscr{D} \cap \mathscr{D}'$ such that

(a) $D(-1)$ consists of a finite number of components $S^{(-1)}(1), \ldots, S^{(-1)}(k)$.
(b) Each component $S^{(-1)}(j)$ is smoothly fibrated onto fibres $\Phi(u,j) = \mathfrak{F}(u) \cap S^{(-1)}(j)$ with base space U.
(c) The diffeomorphism T is uniform hyperbolic on $\mathscr{D}(-1)$ with expansion along $\mathbb{Y}$ and $T\mathscr{D}(-1) \subset \mathscr{D}$. Each $\mathfrak{F}(u,j) = T\Phi(u,j)$ $\mathbb{Y}$-leaf in $\mathscr{D}$, in particular $\partial\mathfrak{F}(u,j) \subset \varphi(U \times \partial V)$ for any $u \in U$ and $j = 1, \ldots, k$.

REMARK. The smaller set of diffeomorphisms T, which possesses on $\mathscr{D}$ a smooth Smale horseshoe with expansion along $\mathbb{Y}$, uniform hyperbolic in the sense of the remark to Definition 1.3 in a domain $\mathscr{D}' \supset \mathscr{D}(-1)$, and such that in condition (b) $\Phi(u,j) = \varphi(\{u\} \times \operatorname{int} V) \cap S^{(-1)}(j)$ and in condition (c) $\varphi(U \times \partial V)$ is replaced by $\varphi(\operatorname{int} U \times \partial V)$, is open in the C^{1+1}-topology.

LEMMA 1.13. *Each component $S^{(-1)}(j)$ is a curvilinear rectangle $S^{(-1)}(j) = \varphi_j^{(-1)}(U \times V)$ and its structure is generated by fibration by $\mathbb{X}$-leaves $\mathfrak{G}(v,j) = T^{-1}[\mathfrak{G}(v) \cap S^{(-1)}(j)]$ and $\mathbb{Y}$-leaves $\mathfrak{F}(u)$. The image $S^{(1)}(i) = TS^{(-1)}(i)$ of each component $S^{(-1)}(i)$ is also a curvilinear rectangle generated by fibration by $\mathbb{X}$-leaves $\mathfrak{G}(v)$ and $\mathbb{Y}$-leaves $\mathfrak{F}(u,i)$.*

COROLLARY: *If the diffeomorphism T possesses in an $\mathbb{X}, \mathbb{Y}$-convex curvilinear rectangle $\mathscr{D}$ a Smale horseshoe (with expansion along $\mathbb{Y}$) then the inverse diffeomorphism T^{-1} also possesses in $\mathscr{D}$ a Smale horseshoe (with expansion along $\mathbb{X}$). Its configuration is generated by the compact domain $\mathscr{D}(1) = T\mathscr{D}(-1)$ with components $S^{(1)}(i) = TS^{(-1)}(i)$, where $i = 1, \ldots, k$.*

LEMMA 1.14. *Let $\mathscr{D}$* be an $\mathbb{X}$, $\mathbb{Y}$-*convex φ-rectangle and T: $\mathscr{D} \to \mathbb{Z}$* a diffeomorphism. If

(a) *$\mathscr{D} \cap T\mathscr{D} \neq \emptyset$ and contains a finite number of components;*
(b) $\varphi([\partial U] \times V) \cap (\operatorname{int} T\mathscr{D}) = \emptyset$, $(\operatorname{int} \mathscr{D}) \cap T\varphi(U \times [\partial V]) = \emptyset$;
(c) *T is a uniform hyperbolic (with expansion along $\mathbb{Y}$) in the domain $T^{-1}(\mathscr{D} \cap T\mathscr{D})$*

then T possesses in $\mathscr{D}$ the Smale horseshoe with $\mathscr{D}(-1) = T^{-1}(\mathscr{D} \cap T\mathscr{D})$.

DEFINITION 1.8. For a smooth Smale horseshoe we set

$$\mathscr{D}(-n) = \{z \in \mathscr{D}(-1), \ldots, T^{n-1} z \in \mathscr{D}(-1)\}, \qquad \mathscr{D}(n) = T^n \mathscr{D}(-n)$$

or recursively for $r > 0$ set

$$\mathscr{D}(r+1) = T(\mathscr{D}(-1) \cap \mathscr{D}(r)), \qquad \mathscr{D}(-r) = T^{-r}(\mathscr{D}(r)). \tag{1.25}$$

STATEMENT 1.15. For any positive integer n the set $\mathscr{D}(n)$ consists of k^n components and in each component of the set $\mathscr{D}(n-m)$, where $n > m$, there are k^{n-m} components of $\mathscr{D}(n)$. Each component of $\mathscr{D}(n)$ intersects with each component of $\mathscr{D}(-1)$ and the T-image of this intersection is a component of $\mathscr{D}(n+1)$.

Let us describe the natural k-adic numeration of components proposed by Smale, see [6, 7]. Let all components of $\mathscr{D}(-1)$ be enumerated in a certain order. Let us enumerate components of $\mathscr{D}(1)$ so that $S^{(1)}(j) = TS^{(-1)}(j)$. Further, to the component of $\mathscr{D}(n)$ which is the T-image of $S^{(n-1)}(I_{n-1}) \cap S^{(-1)}(j)$ assign the number $I_n = I_{n-1}j$, i.e. $S^{(n)}(I_{n-1}j) = T(S^{(n-1)}(I_{n-1}) \cap S^{(-1)}(j))$. Similarly, to the component of $\mathscr{D}(-m)$ which is the T^{-1}-image of $S^{(-m+1)}(J_{m-1}) \cap S^{(1)}(i)$ assign the number $J_m = iJ_{m-1}$, i.e. $S^{(-m)}(iJ_{m-1}) = T^{-1}(S^{(-m+1)}(J_{m-1}) \cap S^{(1)}(i))$. Later we shall assign to the intersection of a component of $\mathscr{D}(n)$ with a component of $\mathscr{D}(-m)$ a two-part number, i.e. $S(I_n, J_m) = S^{(n)}(I_n) \cap S^{(-m)}(J_m)$.

STATEMENT 1.16. With respect to the above numbers we have

$$S^{(n)}(I_n) \subset S^{(n-1)}(I_{n-1}), \qquad S^{(-m)}(J_m) \subset S^{(-m+1)}(J_{m-1}), \tag{1.26}$$

$$TS(I_n, J_m) = S(I_n j_1, j_2 \ldots j_m), \qquad T^{-1}S(I_n, J_m) = S(i_n \ldots i_2, i_1 J_m), \tag{1.27}$$

i.e. the T-action is described by a shift.

STATEMENT 1.17. The set $S^{(n)}(I_n)$ is fibrated by $\mathbb{Y}$-leaves $\mathfrak{F}(u, I_n) = T^n(\mathfrak{F}(u) \cap S^{(-n)}(I_n))$ and by $\mathbb{X}$-leaves $\mathfrak{G}(v) \cap S^{(n)}(I_n)$. The set $S^{(-m)}(J_m)$ is fibrated by $\mathbb{Y}$-leaves $\mathfrak{F}(u) \cap S^{(-m)}(J_m)$ and by $\mathbb{X}$-leaves $\mathfrak{G}(v, J_m) = T^{-m}(\mathfrak{G}(v) \cap S^{(m)}(J_m))$. The set $S(I_n, J_m)$ is fibrated by $\mathbb{X}$-leaves $\mathfrak{G}(v, J_m)$ and by $\mathbb{Y}$-leaves $\mathfrak{F}(u; I_n)$. Therefore $S^{(n)}(I_n)$, $S^{(-m)}(J_m)$, $S(I_n, J_m)$ are topological rectangles homeomorphic to $U \times V$.

The fibration of $\mathscr{D}(n)$ by $\mathbb{Y}$-leaves $\mathfrak{F}(u, I_n)$ described in Statement 1.16 is referred to as the *canonical expanding fibration of rank n* and the fibration of $\mathscr{D}(-m)$ by $\mathbb{X}$-leaves $\mathfrak{G}(v, J_m)$ as the *canonical contracting fibration of rank m.* In §2 we shall verify in detail (see corollary of Theorem 2.7) that homeomorphism defined by these fibrations is a diffeomorphism, i.e. that all the above-mentioned sets are curvilinear rectangles.

Write $\sigma^-(I_n, J_m)$ for the set of $\mathbb{X}$-leaves in $\mathscr{D}$ such that their intersection with $S(I_n, J_m)$ is a $\mathbb{X}$-leaf in $S(I_n, J_m)$ and $\sigma^+(I_n, J_m)$ for the similar set of $\mathbb{Y}$-leaves.

DEFINITION 1.9. The $\mathbb{X}$-*distance* between $\mathbb{Y}$-leaves $\mathfrak{F}_1, \mathfrak{F}_2 \in \sigma^+(I_n, J_m)$ is

$$\rho_{\mathbb{X}}^{I_p, J_m}(\mathfrak{F}_1, \mathfrak{F}_2) = \sup |z(\mathfrak{F}_1, \mathfrak{G}) - z(\mathfrak{F}_2, \mathfrak{G})|_{\mathbb{X}}, \qquad \mathfrak{G} \in \sigma^-(I_n, J_m), \tag{1.28}$$

where $z(\mathfrak{F}_l, \mathfrak{G})$ is the unique point of intersection of $\mathfrak{G}$ and $\mathfrak{F}_l$. The $\mathbb{X}$-*diameter* of the set $S(I_n, J_m)$ is

$$\rho_{\mathbb{X}}(S(I_n, J_m)) = \sup \rho_{\mathbb{X}}^{I_p, J_m}(\mathfrak{F}(u_1, I_n), \mathfrak{F}(u_2, I_n)), u_1, u_2 \in U. \tag{1.29}$$

The $\mathbb{Y}$-*diameter* is similarly defined.

STATEMENT 1.18. The following properties of diameters hold:

(a) $\rho_{\mathbb{X}}^{I_p, J_m}(\mathfrak{F}_1, \mathfrak{F}_2) \leqslant \rho_{\mathbb{X}}(S(I_n, J_m))$ holds for $\mathfrak{F}_1, \mathfrak{F}_2 \in \sigma^+(I_n, J_m)$;

(b) $\rho_{\mathbb{Y}}^{I_p, J_m}(\mathfrak{G}_1, \mathfrak{G}_2) \leqslant \rho_{\mathbb{Y}}(S(I_n, J_m))$ holds for $\mathfrak{G}_1, \mathfrak{G}_2 \in \sigma^-(I_n, J_m)$;

(c) $\rho_{\mathbb{X}}(S(I, J)) \leqslant \rho_{\mathbb{X}}(S(I, J'))$ holds for $S(I, J) \subset S(I, J')$;

(d) $\rho_{\mathbb{Y}}(S(I, J)) \leqslant \rho_{\mathbb{Y}}(S(I', J))$ holds for $S(I, J) \subset S(I', J)$.

The distance ρ is a generalization of the usual functional distance between graphs of functions and is equivalent to it for $\alpha\beta < 1$ and necessary reservations concerning domains of definition of functions. The metric ρ is the Lyapunov metric with respect to T. It is known that the functional metric is not generally speaking the Lyapunov metric, cf. [9].

STATEMENT 1.19. For any set $S^{(-m)}(J_m)$ and any $S^{(n)}(I_n)$ we have

$$\rho_{\mathbb{X}}(S^{(n)}(I_n)) \leqslant (\lambda_s)^{-n}\cdot\rho_{\mathbb{X}}(\mathcal{D}), \qquad \rho_{\mathbb{Y}}(S^{(-m)}(J_m)) \leqslant (\lambda_u)^{-m}\cdot\rho_{\mathbb{Y}}(\mathcal{D}) \tag{1.30}$$

DEFINITION 1.10. Put

$$\mathcal{D}(\infty) = \bigcap_{n>0} \mathcal{D}(n), \qquad \mathcal{D}(-\infty) = \bigcap_{m>0} \mathcal{D}(-m), \qquad \Omega = \mathcal{D}(\infty)\cap\mathcal{D}(-\infty). \tag{1.31}$$

STATEMENT 1.20. The closed set $\mathcal{D}(\infty)$ is fibrated by components which are fibres $\mathfrak{F}(I)$, where $(I) = (\dots i_n \dots i_1)$, numbered by all possible infinite-to-the-left k-adic sequences such that $\mathfrak{F}(I) \subset S^{(n)}(I_n)$ for any $n \in \mathbb{N}$. Each fibre is the graph of the corresponding Lipschitzizable function $x = F(y, I)$ with the convex domain of definitions and the constant $\mathrm{Lip}_{\mathbb{Y}} F \leqslant \beta$. The intersection of the fibre $\mathfrak{F}(I)$ with any $\mathbb{X}$-leaf in $\mathcal{D}$ consists of the unique point. Similarly, the closed set $\mathcal{D}(-\infty)$ is fibrated by non-intersecting graphs $\mathfrak{G}(J)$ of Lipschitzizable functions $y = G(x, J)$ with the constants $\mathrm{Lip}_{\mathbb{X}} G \leqslant \alpha$, and to any infinite-to-the-right k-adic sequence $(J) = (j_1 \cdots j_m j_{m+1} \cdots)$ the fibre $\mathfrak{G}(J) \subset S^{(-m)}(J_m)$ for any $m \in \mathbb{N}$ corresponds. The intersection of the fibre $\mathfrak{G}(J)$ with any $\mathbb{Y}$-leaf in $\mathcal{D}$ consists of the unique point.

STATEMENT 1.21. Any pair of components $\mathfrak{F}(I)$ and $\mathfrak{G}(J)$ of limit sets $\mathcal{D}(\infty)$ and $\mathcal{D}(-\infty)$ intersects, the intersection being the unique point $z(I, J)$. The Cantor compact set Ω of all these point is T- and T^{-1}-invariant and the T-action on it is described by the shift to the left of the numerating sequence

$$(\cdots i_n \cdots i_1, j_1 \cdots j_m \cdots) \to (\cdots i_n \cdots i_1 j_1, j_2 \cdots j_m \cdots).$$

The hyperbolic pre-structure for T in $\mathcal{D}$ defines the hyperbolic structure T on Ω.

2. Expanding and Contracting Fibrations of a Smale Horseshoe

2.1. THE SMOOTHNESS OF EXPANDING AND CONTRACTING FIBRES

We shall repeatedly make use of the following well-known statement, see e.g. [17].

STATEMENT 2.1. Suppose the sequence $\{a_n\}$ of positive real numbers satisfies the recursive inequality

$$a_{n+1} \leqslant \mu a_n + b,$$

where $0 < \mu < 1$, $0 < b$ and $a_0 \leqslant c$. Then $a_n \leqslant \max\{c, (1-\mu)^{-1}\cdot b\}$ for any $n \in \mathbb{N}$ and if $c > (1-\mu)^{-1}\cdot b$ then for any $\Delta > 1$ there is N such that $a_n \leqslant \Delta\cdot(1-\mu)^{-1} b$ for any $n \geqslant N$.

Let a $\mathbb{Y}$-leaf in $\mathcal{D}$ be defined by the graph $\{F(y) \oplus y\}$, where $d_y F(y) = B(y)$. Put

$$\mathrm{Lip}_{\mathbb{Y}}^{\alpha\beta}(d_y F) = \sup_{y_1 \neq y_2} \{\rho_{\alpha\beta}(B(y_1), B(y_2))\cdot|y_1 - y_2|^{-1}\}.$$

Let us prove that leaves of canonical fibrations of sets $\mathscr{D}(n)$ and $\mathscr{D}(-m)$ are graphs of C^{1+1}-maps.

LEMMA 2.2. *Let a* $\mathbb{Y}$*-leaf in* $\mathscr{D}$ *be defined by the graph* $\mathfrak{F} = \{F(y) \oplus y\}$. *Then* $T(\mathfrak{F} \cap S^{(-1)}(j))$ *is the* $\mathbb{Y}$*-leaf* Φ *in* $\mathscr{D}$ *which belongs to* $S^{(1)}(j)$ *with the graph* $\Phi(y^{(1)}) \oplus y^{(1)} = f(F(y), y) \oplus g(F(y), y)$ *and*

$$\mathrm{Lip}_{\mathbb{Y}}^{\alpha\beta}(d_y\Phi) \leqslant (\lambda_s\lambda_u)^{-1}\cdot \mathrm{Lip}_{\mathbb{Y}}^{\alpha\beta}(d_yF) + c_{21}(1-\alpha\beta)^{-1}, \tag{2.1}$$

$$\begin{aligned} c_{21} = (\lambda_u)^{-2}[\beta\,\mathrm{Lip}_{\mathbb{Y}}(d_xf) + \beta^2\,\mathrm{Lip}_{\mathbb{X}}(d_xf) + \mathrm{Lip}_{\mathbb{Y}}(d_yf) + \beta\,\mathrm{Lip}_{\mathbb{X}}(d_yf)] +\\ + (\lambda_u)^{-3}[\beta\cdot\|d_xf\| + \|d_yf\|][\beta\,\mathrm{Lip}_{\mathbb{Y}}(d_xg) +\\ + \beta^2\,\mathrm{Lip}_{\mathbb{X}}(d_xg) + \mathrm{Lip}_{\mathbb{Y}}(d_yg) + \beta\,\mathrm{Lip}_{\mathbb{X}}(d_yg)] \end{aligned} \tag{2.2}$$

COROLLARY. *For any* $\Delta > 1$ *there is an* N *such that for any* $n > N$ *the Lipschitz constant of the canonical fibration of the curvilinear rectangle* $S^{(n)}(I_n)$ *onto* $\mathbb{Y}$*-leaves* $\mathfrak{F}(u, I_n)$ *satisfies*

$$\mathrm{Lip}[d_yF(\cdot, u, I_n)] \leqslant \Delta\cdot c_{21}', \tag{2.3}$$

$$c_{21}' = c_{21}\cdot[1-(\lambda_s\lambda_u^2)^{-1}]^{-1}\cdot(1-\alpha\beta)^{-1}. \tag{2.4}$$

Besides, for all n *we have*

$$\mathrm{Lip}_{\mathbb{Y}}[d_yF(\cdot, u, I_n)] \leqslant \max\{\mathrm{Lip}_{\mathbb{Y}}[d_yF(\cdot, u)]\cdot(1-\alpha\beta)^{-1}, c_{21}'\}.$$

Similar statements are valid for the fibration by $\mathbb{X}$*-leaves with special constants* c_{22} *and* c_{22}'.

Proof. As follows from Statements 1.6 and 1.8, $\Phi = T(\mathfrak{F} \cap S^{(-1)}(j))$ is a $\mathbb{Y}$-leaf in $\mathscr{D}$.

Put $B_l = d_yF(p_l)$, $\mathscr{F}_l^+ = \mathscr{T}^+[p_l]$. Making use of (1.12) and (1.7) we get

$$\begin{aligned} \rho_{\alpha\beta}(\mathscr{T}_1^+(B_1), \ \mathscr{T}_0^+(B_0)) &\leqslant \rho_{\alpha\beta}(\mathscr{T}_1^+(B_1), \ \mathscr{T}_1^+(B_0)) + \rho_{\alpha\beta}(\mathscr{T}_1^+(B_0), \ \mathscr{T}_0^+(B_0))\\ &\leqslant (\lambda_s\lambda_u)^{-1}\,\rho_{\alpha\beta}(B_1, B_0) + (1-\alpha\beta)^{-1}\|\mathscr{T}_1^+(B_0) - \mathscr{T}_0^+(B_0)\|. \end{aligned}$$

In expression (1.14) for $\mathscr{T}^+[p](B_0)$ the operator $d_xg\cdot d_yF + d_yg$ occurs which defines the transformation of the $\mathbb{Y}$-coordinate of a vector tangent to $\mathfrak{F}$ under $\mathrm{d}T$. From the hyperbolicity of T this implies $\|(d_xg\cdot d_yF + d_yg)^{-1}\| \leqslant \lambda_u^{-1}$. Besides, $\|p_1 - p_0\|_{\mathbb{X}} \leqslant \beta\|p_1 - p_0\|_{\mathbb{Y}}$ due to the convexity of the domain of definition of $\mathfrak{F}$. Estimating the upper bound of $\|\mathscr{F}_1^+(B) - \mathscr{F}_0^+(B)\|$ with the help of these considerations we get (2.3).

The estimation of $\mathrm{Lip}_{\mathbb{X}}^{\beta\alpha}\mathscr{T}^-$ for $\mathbb{X}$-leaves is carried out similarly, except that the operator $d_yg - d_xG\cdot d_yf$ possesses no direct geometric meaning. However, since $d_yf)\eta \oplus (d_yg)\eta \in \mathscr{K}_{\mathbb{Y}}^{\alpha}(Tp)$ and $\|(d_yf)(d_yg)^{-1}\| \leqslant \beta$ for any $\eta \in \mathbb{Y}(p)$, then

$$\sup_{p\in\mathscr{D}(-1)}\ \sup_{A\in N^{\alpha}} \|[d_yg(p) - Ad_yf(p)]^{-1}\| = c_{23} \leqslant [\lambda_s\cdot(1-\alpha\beta)]^{-1}. \tag{2.5}$$

Under the conditions of Statement 1.4 we have $c_{23} < (\lambda_u)^{-1}$.

It follows from Statement 1.10 that fibres of the canonical fibration of the φ-rectangle $\mathscr{D}(-1)$ are leaves of class C^{1+1} satisfing the conditions of Lemma 2.2, and therefore so are the fibres of $\mathscr{D}(n)$ and $\mathscr{D}(-m)$. Statement 2.1 applied to the inductive inequality (2.1) gives the estimate of the corollary for $\mathrm{Lip}_{\mathbb{Y}}^{\alpha\beta}$. We can similarly estimate $\mathrm{Lip}_{\mathbb{X}}^{\beta\alpha}$, but $\mathrm{Lip}_{\mathbb{Y}}$ and $\mathrm{Lip}_{\mathbb{X}}$ are only known to be smaller.

Let us investigate components of limit spaces $\mathscr{D}(\infty)$ and $\mathscr{D}(-\infty)$.

THEOREM 2.3. *Each component* $\mathfrak{F}(I)$, *where* $(I) = (\cdots i_n \cdots i_1)$, *of* $\mathscr{D}(\infty)$ *is a* C^{1+1}-*smooth* $\mathbb{Y}$-*leaf in* $\mathscr{D}$, *i.e. is the graph of a function with the convex domain of definition, i.e.*

$$x = F(y, I), \qquad \|d_y F\| \leqslant \beta, \qquad \mathrm{Lip}_{\mathbb{Y}}(d_y F) \leqslant c'_{21}.$$

A similar statement holds for the fibration of $\mathscr{D}(-\infty)$.

COROLLARY. *For any point* $z(\cdots i_n \cdots i_1, j_1 \cdots j_m \cdots) \in \Omega$ *the smooth* $\mathbb{Y}$-*leaf* $\mathfrak{F}(\cdots i_n \cdots i_1)$ *is an expandable local fibre and the smooth* $\mathbb{X}$-*leaf* $\mathfrak{G}(j_1 \cdots j_m \cdots)$ *is a contractable local fibre.*

2.2. EXPANDING AND CONTRACTING FIBRATIONS ARE HÖLDERIAN

Let us estimate how close the fibres of fibrations are in the C^1-metric.

LEMMA 2.4. *Let* $\mathfrak{F}_l = \{F_l(y) \oplus y\}$, *where* $l = 1, 2$, *are two* $\mathbb{Y}$-*leaves in* $\mathscr{D}$. *Each* $\mathbb{X}$-*leaf* $\mathfrak{G} = \{x \oplus G(x)\}$ *in* $S^{(1)}(j)$ *intersects* $\Phi_l = T(\mathfrak{F}_l \cap S^{(-1)}(j))$ *at the unique point* $T(p_l)$. *Then setting* $B_l = d_y F_l(p_l)$ *we have*

$$\rho_{\alpha\beta}(\mathscr{T}^+[p_1](B_1), \mathscr{T}^+[p_2](B_2)) \leqslant (\lambda_s \lambda_u)^{-1} \rho_{\alpha\beta}(B_1, B_2) + c_{24}(1 - \alpha\beta)^{-1} |p_1 - p_2|_{\mathbb{X}} \tag{2.6}$$

$$\begin{aligned} c_{24} = {} & (\lambda_u)^{-1}[\beta\, \mathrm{Lip}_{\mathbb{X}}(d_x f) + \alpha\beta\, \mathrm{Lip}_{\mathbb{Y}}(d_x f) + \mathrm{Lip}_{\mathbb{X}}(d_y f) + \alpha\, \mathrm{Lip}_{\mathbb{Y}}(d_y f)] + \\ & + (\dot{\lambda}_u)^{-2}[\beta \|d_x f\| + \|d_y f\|] \cdot [\beta\, \mathrm{Lip}_{\mathbb{X}}(d_x g) + \alpha\beta\, \mathrm{Lip}_{\mathbb{Y}}(d_x g) + \\ & + \mathrm{Lip}_{\mathbb{X}}(d_y g) + \alpha\, \mathrm{Lip}_{\mathbb{Y}}(d_y g)]. \end{aligned} \tag{2.7}$$

THEOREM 2.5. *Let* $\Phi_1^{(n)}$ *and* $\Phi_2^{(n)}$ *be two* $\mathbb{Y}$-*leaves in* $S^{(n)}(J_n)$ *and suppose* $\Phi_1^{(0)} = T^{-n}\Phi_1^{(n)}$ *and* $\Phi_2^{(0)} = T^{-n}\Phi_2^{(n)}$ *are* $\mathbb{Y}$-*leaves in* $S^{(-n)}(J_n)$. *Let* $\mathfrak{G}$ *be an arbitrary* $\mathbb{X}$-*leaf in* $\mathscr{D}$ *intersecting with* $\Phi_1^{(n)}$ *and* $\Phi_2^{(n)}$ *at* p_1 *and* p_2. *Then*

$$\|d_y \Phi_1^{(n)}(p_1) - d_y \Phi_2^{(n)}(p_2)\| \leqslant (\lambda_s)^{-n} \cdot c_{25}, \tag{2.8}$$

$$c_{25} = (1 - \alpha\beta)^{-1} \cdot 2\beta + c_{24}(1 - \alpha\beta)^{-1} \rho_{\mathbb{X}}(\mathscr{D}) \cdot \lambda_s \cdot [1 - (\lambda_u)^{-1}]^{-1}. \tag{2.9}$$

COROLLARY. *The fibration of the limit space* $\mathscr{D}(\infty)$ *by smooth fibres* $\mathfrak{F}(I)$ *satisfies the Hölder condition with the exponent* $\gamma = \lambda_s(\hat{\lambda}_s)^{-1}$, *where*

$$\hat{\lambda}_s = \sup_{p \in \mathscr{D}(1)} \; \sup_{\zeta \in \mathscr{K}_{\mathbb{X}}^{+}(p), |\zeta|_{\mathbb{X}} = 1} |(dT(p))^{-1}\zeta|_{\mathbb{X}}.$$

2.3. THE LOCAL SMOOTHNESS OF EXPANDING AND CONTRACTING FIBRATIONS

The canonical fibration of each component $S^{(-m)}(J_m)$ of $\mathscr{D}(-m)$ by $\mathbb{X}$-leaves $\mathfrak{G}(v, J_m)$ is smooth with respect to v. Analogously, the canonical fibration of the set $S^{(n)}(I_n)$ by $\mathbb{Y}$-leaves $F(u, I_n)$ is smooth with respect to u as was found in [17]. Some characteristics of smoothness are independent of m or n, respectively.

We shall make use of the following simple inequality for $B \in \mathrm{Lin}(\mathbb{Y}, \mathbb{Y})$:

$$\ln|\det(E + B)| \leqslant \dim \mathbb{Y}\cdot||B||. \tag{2.10}$$

LEMMA 2.6 [17]. *Let* $\mathfrak{G}(v, J_m) \cap S^{(1)}(i) = T\mathfrak{G}(v, iJ_m)$. *Then*

$$\mathrm{Lip}_{\mathbb{X}} \ln|\det d_v G(\cdot, v, iJ_m)| < (\lambda_s)^{-1} \mathrm{Lip}_{\mathbb{X}} \ln|\det d_v G(\cdot, v, J_m)| + c_{27}, \tag{2.11}$$

where

$$c_{27} = \dim \mathbb{Y}\cdot c_{23}\cdot[\mathrm{Lip}_{\mathbb{X}}(d_y g) + \alpha\, \mathrm{Lip}_{\mathbb{Y}}(d_y g) + (1 - \alpha\beta)^{-1}||d_y f||\cdot\lambda_s^{-1} \times$$
$$\times \max\{c_{22}\cdot[1 - (\lambda_s^2\lambda_u)^{-1}]^{-1}, \quad \mathrm{Lip}_{\mathbb{X}} d_v G^{(0)}\} + \alpha\, \mathrm{Lip}_{\mathbb{X}}(d_y f) + \alpha^2\, \mathrm{Lip}_{\mathbb{Y}}(d_y f)].$$

Proof. Set $x^{(1)} = f(x, G^{(m+1)}(x, v))$ and

$$y^{(1)} = g(x, G^{(m+1)}(x, v)) = G^{(m)}(x^{(1)}(x, v), v).$$

Let us differentiate both parts of this equality with respect to v for a fixed x:

$$(d_y g) d_v G^{(m+1)} = d_x G^{(m)}(d_y f) d_v G^{(m+1)} + d_v G^{(m+1)},$$
$$d_v G^{(m+1)} = [d_y g - (d_x G^{(m)})(d_y f)]^{-1}\cdot(d_v G^{(m)}). \tag{2.12}$$

As previously mentioned in the proof of Lemma 2.2, see (2.5), the operator $\Gamma^{(m)} = d_y g - (d_x G^{(m)})(d_y f)$ is invertible, therefore $\det \Gamma^{(m)} \neq 0$. By statement 1.10, for the initial fibration by $\mathbb{X}$-leaves we have $\det(d_v G^{(0)}) \neq 0$. The recursion (2.12) implies that $\det(d_v G^{(m)}) \neq 0$ and the sign of $\det(d_v G^{(m)})$ does not depend on x for a fixed v, hence

$$\mathrm{Lip}_{\mathbb{X}} \ln|\det(d_v G^{(m+1)})| \leqslant (\lambda_s)^{-1} \mathrm{Lip}_{\mathbb{X}} \ln|\det(d_v G^{(m)})| + \mathrm{Lip}_{\mathbb{X}} \ln|\det \Gamma^{(m)}|.$$

From (2.10) we have an estimate

$$\ln|\det \Gamma(x_1, v)| - \ln|\det \Gamma(x_2, v)| = \ln \det[E + (\Gamma_1 - \Gamma_2)\Gamma_2^{-1}]$$
$$\leqslant \dim \mathbb{Y}||\Gamma(x_1, v) - \Gamma(x_2, v)||\cdot||\Gamma^{-1}(x, v)||.$$

Interchanging x_1 and x_2 we obtain the estimate for the module of difference of logarithms. □

For $m > M$ the constant C_{27} in (2.11) can be replaced by

$$c'_{27} = \dim \mathbb{Y}\cdot c_{23}\cdot[\mathrm{Lip}_{\mathbb{X}}(d_y g) + \alpha\, \mathrm{Lip}_{\mathbb{Y}}(d_y g) + 2c'_{22}||d_y f||\lambda_s^{-1} +$$
$$= \alpha^2\, \mathrm{Lip}_{\mathbb{Y}}(d_y f)][1 - \lambda_s^{-1}]^{-1}.$$

THEOREM 2.7. *Let* $\mathfrak{G}(v, J_m)$, *where* $v \in V$, *is the canonical fibration of the component* $S^{(-m)}(J_m)$ *of* $\mathcal{D}(-m)$ *onto* $\mathbb{X}$*-leaves,* $y = G(x, v, J_m)$ *and* $d_v G \in \mathrm{Lin}(\mathbb{Y}, \mathbb{Y})$ *the partial differential of the vector function* G *with respect to the parameter* $v \in V$ *at* $p \in \mathcal{D}(-m)$. *Then*

$$\mathrm{Lip}_{\mathbb{X}} \ln|\det(d_v G)| < c_{28}. \tag{2.13}$$

For any $\Delta > 1$ *there is an* M *such that*

$$\mathrm{Lip}_{\mathbb{X}} \ln|\det(d_v G)| < \Delta\cdot c'_{27}.$$

for any $m > M$.

REMARK. The number M depends on constants $\mathrm{Lip}_{\mathbb{X}} \ln|\det d_v G^{(0)}|$, $\mathrm{Lip}_{\mathbb{X}} d_v G^{(0)}$, on openings of cones α and β, on Lipschitz constants and norms of blocks of the operator dT, while c'_{27} does not depend on smoothness of the initial boundary.

Hence $c_{28} = \max\{c'_{27}, \mathrm{Lip}_{\mathbb{X}} \ln|\det d_v G(\cdot, v)|\}$

COROLLARY. *Components* $S^{(-m)}(J_m)$ *and* $S^{(n)}(I_n)$ *are curvilinear rectangles in the sense of Definition* 1.5, cf. [1].

2.4. THE HÖLDER PROPERTY OF THE CANONICAL ISOMORPHISM DEFINED BY A FIBRATION

In the construction of the smooth invariant conditional measure for an Anosov system the canonical isomorphism of expanding fibers generated by contracting fibers is used (see [13]). In the present paper, within each $\mathscr{D}(-m)$ we make use of the mth rank correspondence between expanding fibres defined by the canonical fibration of $\mathscr{D}(-m)$ by $\mathbb{X}$-leaves. It defines the product bundle on $\mathscr{D}(-m)$.

DEFINITION 2.1. For any $\mathbb{Y}$-leaf $\{\mathfrak{F} = F(y) \oplus y\}$ in $\mathscr{D}$ the intersection $\mathfrak{F} \cap \mathscr{D}(-m)$ consists of k^m-components, i.e. $\mathbb{Y}$-leaves $\mathfrak{F} \cap S^{(-m)}(J_m)$. For any two $\mathbb{Y}$-leaves $\mathfrak{F}_1$ and $\mathfrak{F}_2$ the canonical fibration of $S^{(-m)}(J_m)$ by $\mathbb{X}$-leaves $\mathfrak{G}(v, J_m) = \{x \oplus G^m(x, v, J_m)\}$ defines the correspondence between points of $\mathfrak{F}_1 \cap S^{(-m)}(J_m)$ and $\mathfrak{F}_2 \cap S^{(-m)}(J_m)$ so that

$$\mathfrak{F}_2 \cap \mathfrak{G} = p_2(\mathfrak{F}_2, v, J_m) \to p_1(\mathfrak{F}_1, v, J_m) = \mathfrak{F}_1 \cap \mathfrak{G} \tag{2.14}$$

which we may refer to as the *canonical isomorphism of rank m*.

The $\mathbb{Y}$-coordinate y_l of the point of intersection is the function in v implicitly defined by the equation

$$y_l = G^{(m)}(F_l(y_l), v, J_m), \qquad y_2(v) \to y_1(v). \tag{2.15}$$

It warrants a one-to-one correspondence. Put $\mathrm{Jac}(dy_1 | dy_2)$ for the Jacobian of the transformation. Then we get

$$\begin{aligned}\mathrm{Jac}(dy_1 | dy_2) = {} & \det[E - d_x G^{(m)}(x_1, v) d_y F_1(y_1)]^{-1} \times \\ & \times \det[E - d_x G^{(m)}(x_2, v) d_y F_2(y_2)] \times \\ & \times \det[d_v G^{(m)}(x_1, v)(d_v G^{(m)}(x_2, v))^{-1}].\end{aligned} \tag{2.16}$$

Now let us check the uniform (in m) absolute continuity of the canonical isomorphism.

THEOREM 2.8. *Suppose that the pre-images* $T^{-n}\mathfrak{F}_l^{(n)}$ *of* $\mathbb{Y}$*-leaves* $\mathfrak{F}_l^{(n)}$*, where* $l = 1, 2$*, in* $S^{(n)}(I_n)$ *are* $\mathbb{Y}$*-leaves in* $S^{(-n)}(I_n)$*. Then the canonical isomorphism of rank m between* $\mathfrak{F}_1^{(n)}$ *and* $\mathfrak{F}_2^{(n)}$ *within* $S^{(-m)}(J_m)$ *almost preserves the* $\mathbb{Y}$*-volume. More precisely, the Jacobian* (2.16) *of the correspondence* (2.15) *is positive,* $\mathrm{Jac}(dy_1 | dy_2) > 0$, *and*

$$\exp\{-c_{29}(\lambda_s)^{-n}\} \leqslant \mathrm{Jac}(dy_1 | dy_2) \leqslant \exp\{c_{29}(\lambda_s)^{-n}\}, \tag{2.17}$$

$$\exp\{-c_{29}(\lambda_s)^{-n}\} \leqslant \frac{\mathrm{vol}_{\mathbb{Y}}[\mathfrak{F}_1^{(n)} \cap S^{(-m)}(J_m)]}{\mathrm{vol}_{\mathbb{Y}}[\mathfrak{F}_2^{(n)} \cap S^{(-m)}(J_m)]} \leqslant \exp\{c_{29}(\lambda_s)^{-n}\}, \tag{2.18}$$

where

$c_{29} = c_{28}\rho_{\mathrm{X}}(\mathscr{D}) + \dim \mathrm{Y}(1-\alpha\beta)^{-1}[\alpha c_{25} + \beta\rho_{\mathrm{X}}(\mathscr{D})\max\{c_{22}', (1-\alpha\beta)^{-1}\,\mathrm{Lip}_{\mathrm{X}}(\mathrm{d}_u G)\}]$.

REMARK. There is an M such that for $m > M$ the constant c_{29} can be replaced by c_{29}' which is expressed in terms of the norm and Lipschitz constants of blocks of dT.

Let $\mathfrak{F}_l$ be a Y-leaf in $S^{(-1)}(j)$ and $\mathfrak{F}_l = \{F_l(y) \oplus y\}$. Its T-image $\mathfrak{F}_l^{(1)} = T\mathfrak{F}_l$ is a Y-leaf in $S^{(1)}(j)$ and $\mathfrak{F}_l^{(1)} = \{F_l^{(1)}(y^{(1)}) \oplus y^{(1)}\}$, where

$$y^{(1)} = g(F_l(y), y) \tag{2.19}$$

is the natural parameter on the surface. Put $\mathrm{Jac}(dy^{(1)}|dy)$ for the determinant of the operator $(d_x g)(d_y F) + d_y g$. Let us show that for leaves of expanding fibrations this Jacobian is continuous along a contracting fibre.

Note that the map T transforms for $m > 1$ the canonical isomorphism of rank m between Y-leaves $\mathfrak{F}_2 \cap S^{(-m)}(j_1 J_{m-1}) \to \mathfrak{F}_1 \cap S^{(-m)}(j_1 J_{m-1})$ into the canonical isomorphism of rank $m-1$ between Y-leaves $T\mathfrak{F}_2 \cap S^{(-m+1)}(J_{m-1}) \to T\mathfrak{F}_1 \cap S^{(-m+1)}(J_{m-1})$.

THEOREM 2.9. *Suppose that pre-images $T^{-n}\mathfrak{F}^{(n)}$ of Y-leaves $\mathfrak{F}_l^{(n)}$, where $l = 1, 2$, in $S^{(n)}(I_n)$ are Y-leaves in $S^{(-n)}(I_n)$. Let $y_2 = y_2(v) \to y_1 = y_1(v)$ be the correspondence (2.15) of Y-coordinates of points under the canonical isomorphism of rank m, where $m > 1$, between Y-leaves $\mathfrak{F}_2^{(n)} \cap S^{(-m)}(J_m)$ and $\mathfrak{F}_1^{(n)} \cap S^{(-m)}(J_m)$. Put $y_l^{(1)}$ for Y-coordinates of their T-images $y_l^{(1)} = g(F_l^{(n)}(y), y)$. Then*

$$\exp\{-2c_{29}(\lambda_s)^{-n}\} \leqslant \frac{\mathrm{Jac}(dy_1, dy_1^{(1)})}{\mathrm{Jac}(dy_2, dy_2^{(1)})} \leqslant \exp\{2c_{29}(\lambda_s)^{-n}\}. \tag{2.20}$$

3. Smooth Invariant Conditional Probability Distributions on Fibrations

In this section we establish the existence on expanding fibres $\mathfrak{F}(I)$ of a Smale horseshoe of the unique invariant family of smooth conditional probability distributions, and trace how smooth conditional probability distributions on $\mathscr{D}(n)$ tend to distributions of the family as $n \to \infty$. It turns out that it is more convenient to consider these two problems together, so all auxiliary statements (as well as the statements from the end of the preceding section) are formulated in terms equally applicable to components of $\mathscr{D}(\infty)$ and to fibres from $\mathscr{D}(n)$ when $n < \infty$.

By *smooth probability measures* on fibres we understand measures with Lipschitzizable densities. Recall that a continuous density of a probability distribution (if it exists) can be uniquely recovered from the Baire measure. Since all the fibres considered are Y-leaves and their presentation in the form of a graph provides a convenient chart of the fibre, we consider all densities with respect to a Y-volume, i.e. the Lebesgue measure on the chart. Note that this approach is compatible with the method introduced in §1 of measurement of distances on a chart, i.e. with the Y-length.

3.1. THE EVOLUTION OF DENSITIES OF CONDITIONAL PROBABILITY DISTRIBUTIONS ON FIBRES INDUCED BY $\mathscr{T}$

STATEMENT 3.1. Let the measure μ be defined on the Y-leaf

$\mathfrak{F} \cap S^{(-1)}(j) = \{F(y) \oplus y\}$ in $S^{(-1)}(j)$ with a density continuous in $\mathbb{Y}$, $\rho(y)$, with respect to the $\mathbb{Y}$-volume. Then the measure $\mu^{(1)}\{\cdot\} = \mu\{T^{-1}\cdot\}$ on $T(\mathfrak{F} \cap S^{(-1)}(j))$ is defined by the density $\rho^{(1)}(y^{(1)}) = \rho(y)|\mathrm{Jac}(dy|dy^{(1)})|$, where the Jacobian is defined on (2.20) and $y^{(1)} = g(F(y), y)$. If on the $\mathbb{Y}$-leaf $\mathfrak{F}$ in $\mathscr{D}$ we define the probability distribution ν by continuous density ω with respect to the $\mathbb{Y}$-volume, then on each of $\mathbb{Y}$-leaves $\mathfrak{F}^{(1)}(j) = T(\mathfrak{F} \cap S^{(-1)}(j))$ the probability distribution with the continuous density

$$\omega^{(1)}(y^{(1)}) = \omega(y)\cdot|\mathrm{Jac}(dy|dy^{(1)})|\cdot[\nu\{T^{-1}\mathfrak{F}^{(1)}(j)\}]^{-1}. \tag{3.1}$$

is induced.

LEMMA 3.2. *If the probability distribution on the* $\mathbb{Y}$*-leaf* $\mathfrak{F} = \{F(y) \oplus y\}$ *in* $\mathscr{D}$ *possesses a density with respect to the* $\mathbb{Y}$*-volume with the logarithm Lipschitzizable in* y*, then the logarithm of the density* (3.1) *of the induced probability distribution is also Lipschitzizable and their Lipschitz constants are related by the inequality*

$$\mathrm{Lip}_{\mathbb{Y}}(\ln \omega^{(1)}(y^{(1)})) \leqslant (\lambda_u)^{-1}\,\mathrm{Lip}_{\mathbb{Y}}(\ln \omega(y)) + c_{31} + c_{32}\,\mathrm{Lip}_{\mathbb{Y}}\,(d_y F), \tag{3.2}$$

where

$$c_{31} = \dim \mathbb{Y}\cdot(\lambda_u)^{-2}[\mathrm{Lip}_{\mathbb{Y}}(d_y g) + \beta\,\mathrm{Lip}_{\mathbb{X}}(d_y g) + \beta\,\mathrm{Lip}_{\mathbb{Y}}(d_x g) + \beta^2\,\mathrm{Lip}_{\mathbb{X}}\,(d_x g)] \tag{3.3}$$

$$c_{32} = \dim \mathbb{Y}\cdot||d_x g||\cdot(\lambda_u)^{-2}. \tag{3.4}$$

STATEMENT 3.3. Let $\mathfrak{F}$ be a $\mathbb{Y}$-leaf in $\mathscr{D}$ and $\omega(y)$ a Lipschitzizable positive density with respect to the $\mathbb{Y}$-volume of a probability distribution ν on it. Let $\{j_l\}_{l=1}^{\infty}$ be the sequence of indices and $\mathfrak{F}^{(0)} = \mathfrak{F}$, $\mathfrak{F}^{(l)} = T(\mathfrak{F}^{(l-1)} \cap S^{(-1)}(j_l))$. Let $\{\omega^{(l)}\}_{l=1}^{\infty}$ be the recursive sequence induced by the action of T, i.e.

$$\omega^{(l+1)}(y^{(l+1)}) = \omega^{(l)}(y^{(l)})\cdot|\mathrm{Jac}(dy^{(l)}|dy^{(l+1)})|\cdot[\nu_l\{T^{-1}\mathfrak{F}^{(l+1)}(J_{l+1})\}]^{-1}.$$

Then the nth term of the sequence is expressed in terms of the density of the initial distribution via the formula

$$\omega^{(n)}(y^{(n)}) = \omega(y)\cdot|\prod \mathrm{Jac}(dy^{(k-1)}|dy^{(k)})\;[\nu\{T^{-n}\mathfrak{F}^{(n)}(J_n)\}]^{-1}. \tag{3.5}$$

LEMMA 3.4. *For the recursive sequence of probability densities* (3.5) *the estimation*

$$\mathrm{Lip}_{\mathbb{Y}}\,\omega^{(n)} < \max\{\mathrm{Lip}_{\mathbb{Y}}\,\omega, c_{33}\} \tag{3.6}$$

holds, where

$$c_{33} = [c_{31} + c_{32}\cdot\max\{(1-\alpha\beta)^{-1}\,\mathrm{Lip}_{\mathbb{Y}}\,(d_y F), c'_{21}\}]\cdot[1 - (\lambda_u)^{-1}]^{-1},$$
$$c_{34} = [c_{31} + 2c_{32}\cdot c'_{21}]\cdot[1 - (\lambda_u)^{-1}]^{-1},$$

and for any $\Delta > 1$ *there is* N *such that* $\mathrm{Lip}_{\mathbb{Y}}\,\omega^{(n)} < \Delta\cdot c_{34}$ *for any* $n > N$.

The proof is based on Statement 2.1 and Lemma 3.2.

3.2. THE EXISTENCE OF A T-INVARIANT SMOOTH FAMILY OF PROBABILITY DISTRIBUTIONS ON FIBRES OF $\mathscr{D}(\infty)$

LEMMA. 3.5. *Suppose on the* $\mathbb{Y}$*-leaf* $\mathfrak{F}$ *the probability distribution* $\mathscr{P}$ *is defined with the density* p *with respect to the* $\mathbb{Y}$*-volume such that* $\mathrm{Lip}_{\mathbb{Y}} \ln p < c$. *Then for any*

$J_m \in (\mathbb{N}_k)_1^m$ *and* y_0 *such that* $F(y_0) \oplus y_0 \in \mathfrak{F} \cap S^{(-m)}(J_m)$ *we have*

$$\exp\{-c(\lambda_u)^{-m}\cdot\rho_{\mathbb{Y}}(\mathscr{D})\} \leqslant \frac{\mathscr{P}\{\mathfrak{F} \cap S^{(-m)}(J_m)\}}{p(y_0)\,\mathrm{vol}_{\mathbb{Y}}\{\mathfrak{F} \cap S^{(-m)}(J_m)\}} \leqslant \exp\{c(\lambda_u)^{-m}\cdot\rho_{\mathbb{Y}}(\mathscr{D})\}, \quad (3.7)$$

$$p(y_0)\cdot\exp\{-c(\lambda_u)^{-m}\cdot\rho_{\mathbb{Y}}(\mathscr{D})\} \leqslant p(y) \leqslant p(y_0)\cdot\exp\{c(\lambda_u)^{-m}\cdot\rho_{\mathbb{Y}}(\mathscr{D})\}. \quad (3.8)$$

LEMMA 3.6. *Suppose on the* $\mathbb{Y}$*-leaf* $\mathfrak{F}$ *two probability distributions are defined with densities p and q with respect to the* $\mathbb{Y}$*-volume and* $\mathrm{Lip}_{\mathbb{Y}}(\ln p) < c$, $\mathrm{Lip}_{\mathbb{Y}}(\ln q) < c$. *Then for densities* $p^{(n)}$ *and* $q^{(n)}$ *on the* $\mathbb{Y}$*-leaf* $\mathfrak{F}^{(n)}(I_n)$ *induced via* (3.5) *at any point* $F^{(n)}(y_0^{(n)}) \oplus y_0^{(n)}$ *of this* $\mathbb{Y}$*-leaf the inequality*

$$\exp\{-2c(\lambda_u)^{-n}\cdot\rho_{\mathbb{Y}}(\mathscr{D})\} \leqslant q^{(n)}(y_0^{(n)})\cdot[p^{(n)}(y_0^{(n)})]^{-1} \leqslant \exp\{2c(\lambda_u)^{-n}\cdot\rho_{\mathbb{Y}}(\mathscr{D})\} \quad (3.9)$$

holds.

DEFINITION 3.1. The family of smooth probability distributions on fibres of $\mathscr{D}(\infty)$ is *T-invariant* if for any fibre $\mathfrak{F}(Ij)$ the density $\omega(y, Ij)$ is recovered from the density $\omega(y, I)$ on the fibre $\mathfrak{F}(I)$ via (3.1).

THEOREM 3.7. *On fibres* $\mathfrak{F}(I)$ *of the limit set* $\mathscr{D}(\infty)$ *there is the T-invariant family of conditional probability distributions* $P\{\cdot|I\}$ *with density* $p(y|I)$ *relative to the* $\mathbb{Y}$*-volume on the chart of* $\mathfrak{F}(I)$, $\ln p(y|I)$ *being Lipschitzizable in y. This family is unique.*

The proof is close to that in [13].

3.3. COMPARISON OF DENSITIES OF CONDITIONAL PROBABILITY DISTRIBUTIONS ON DIFFERENT FIBRES

THEOREM 3.8. *Suppose on expanding leaves* $\mathfrak{F}_1^{(0)}$ *and* $\mathfrak{F}_2^{(0)}$ *in* $\mathscr{D}$, *where* $\mathfrak{F}_l^{(0)} = \{F_l^{(0)}(y) \oplus y\}$ *for* $l = 1, 2$, *measures are defined such that* $\ln \omega_l^{(0)}(y)$ *are Lipschitzizable. Further, let expanding leaves* $\mathfrak{F}_1^{(n)}$ *and* $\mathfrak{F}_2^{(n)}$ *in* $\mathscr{D}$ *be such that* $T^{(-n)}\mathfrak{F}_l^{(n)} \subset \mathfrak{F}_l^{(0)}$ *for* $l = 1, 2$. *Then there is a constant* R' *and a number* $N_{31}(\upsilon)$ *such that for* $n \geqslant N_{31}(\upsilon)$ *and any m and the canonical correspondence* $\pi^{(m)}$ *between* $\mathfrak{F}_1^{(n)}$ *and* $\mathfrak{F}_2^{(n)}$ *defined by the canonical fibration* $\mathscr{D}(-m)$ *the inequality*

$$(R')^{-1}\cdot\omega_1^{(n)}(y) \leqslant \omega_2^{(n)}(\pi^{(m)}(y))\cdot\mathrm{Jac}(\mathrm{d}\pi^{(m)}(y)|\mathrm{d}y) \leqslant R'\cdot\omega_1^{(n)}(y) \quad (3.10)$$

holds, where

$$R' = \exp\{2c_{29} + 4c_{34}\cdot\rho_{\mathbb{Y}}(\mathscr{D})\} \quad (3.11)$$

and the constant c_{29} *depends on the initial horizontal fibration.*

In what follows $P\{\cdot|I\}$ stands for the family constructed in Theorem 3.7 of *T*-invariant conditional probabilities with density logarithm $p(y|I)$ Lipschitzizable in *y*. A parameter is the number $I \in (\mathbb{N}_k)_{k=-\infty}^{-1}$, where $\mathbb{N}_k = \{1, \ldots, k\}$ and $(\mathbb{N}_k)_{-\infty}^{-1}$ is endowed with the topology of the Cantor discontinuum.

THEOREM 3.9. *For a Smale horeshoe there is a sequence* $R_n \downarrow 1$ *with the following property. For any two expanding smooth leaves* $\mathfrak{F}_l^{(0)} = \{F_l^{(0)}(y) \oplus y\}$, *where* $l = 1, 2$, *in* $\mathscr{D}$ *with defined probability distributions on them with Lipschitzizable* $\ln(\omega_l^{(0)}(y))$ *there is* $N_{32}(\upsilon) > N_{31}$ *such that for any* $n \geqslant N_{32}$ *and* $S^{(-n)}(I_n)$ *for the canonical correspondence* $y_1 \to y_2$ *of rank m between expanding leaves* $\mathfrak{F}_l^{(n)}(I_n) = T^n(\mathfrak{F}_l^{(0)} \cap$

$S^{(-n)}(I_n))$ *the inequality*

$$(R_n)^{-1}\cdot\omega_1^{(n)}(y_1)\leqslant\omega_2^{(n)}(y_2)\cdot\mathrm{Jac}(dy_2|dy_1)\leqslant R_n\cdot\omega_1^{(n)}(y_1) \tag{3.12}$$

holds.

REMARK. There is an M such that for $m\geqslant M$ the inequality (3.12) holds with better constants R_n which do not depend on the Lipschitz constants of the initial contracting fibration.

Using a less accurate estimate we may write

$$\ln R_n\leqslant c_{35}(\lambda_u)^{-n/2}+c_{36}\cdot(\lambda_s)^{-n/2}, \tag{3.13}$$

where

$$c_{35}=4c_{34}\cdot\rho_{\mathbb{Y}}(\mathscr{D}),\qquad c_{36}=c_{29}\cdot(4\lambda_s^2-\lambda_s-1)\cdot(\lambda_s-1)^{-1}$$

The smoothness of the initial horizontal fibration affects only c_{29}. For $m>M$ it is replaced by c'_{29}.

3.4. THE DEPENDENCE OF T-INVARIANT CONDITIONAL DENSITIES ON THE NUMBER OF THE FIBRE

The inequality (3.12) implies in particular the Hölderian nature of the dependence of the density of invariant conditional probability distributions with the Lipschitzizable logarithm with respect to y on fibres $\mathfrak{F}(I)$ of $\mathscr{D}(\infty)$ on the parameter I. We need it in terms of the topology of the space of numbers only.

COROLLARY 3.10. *Estimate* (3.11) *and* (3.12) *of Theorems* 3.8 *and* 3.9 *are applicable to the density* $p(y|I)$. *The function* $p(y|I)=p(\pi_I^0 y|I)$ *is uniform on* $\mathscr{D}(\infty)$ *both in the number* I *and the coordinate* y *along the fixed fibre* $\mathfrak{F}(E)$, *where* $(E)=(\cdots1\cdots1)$.

For the expanding fibre $\mathfrak{F}(I)$ of $\mathscr{D}(\infty)$, where $(I)=(\cdots i_n\cdots i_1)$, and $S^{(-m)}(J_m)$, where $(J_m)=(j_1\cdots j_m)$, set $\mathfrak{F}(I,J_m)=\mathfrak{F}(I)\cap S^{(-m)}(J_m)$.

LEMMA 3.11. *On any two fibres* $\mathfrak{F}(I)$ *and* $\mathfrak{F}(L)$ *of the fibration* $\mathscr{D}(\infty)$ *and any component* $S^{(-m)}(J_m)$ *of the family constructed in Theorem* 3.7 *of T-invariant conditional probabilities we have*

$$(R')^{-1}\cdot P\{\mathfrak{F}(L,J_m)|L\}\leqslant P\{\mathfrak{F}(I,J_m)|I\}\leqslant R'\cdot P\{\mathfrak{F}(L,J_m)|L\}. \tag{3.14}$$

If, moreover, $\mathfrak{F}(I)$ *and* $\mathfrak{F}(L)$ *belong to the interior of a component of rank n, i.e.* $i_n=l_n,\ldots,i_1=l_1$, *then*

$$(R_n)^{-1}\cdot P\{\mathfrak{F}(L,J_m)|L\}\leqslant P\{\mathfrak{F}(I,J_m)|I\}\leqslant R_n\cdot P\{\mathfrak{F}(L,J_m)|L\}, \tag{3.15}$$

where R' *and* R_n *are introduced in Theorems* 3.8 *and* 3.9.

THEOREM 3.12. *For the T-invariant family of conditional probabilities constructed in Theorem* 3.7 *we have*

$$P\{\mathfrak{F}(I,J_m)|I\}\geqslant c_{37}\cdot\exp\{-m\cdot\dim\mathbb{Y}\cdot\ln\hat{\lambda}_u\}, \tag{3.16}$$

for all expanding fibres $\mathfrak{F}(I)$ *of the fibration* $\mathscr{D}(\infty)$ *and any* $S^{(-m)}(J_m)$, *where* $\hat{\lambda}_u$ *is defined like* $\hat{\lambda}_s$, *i.e.*

$$\hat{\lambda}_u = \sup_{p\in\mathscr{D}(-1)} \sup_{\zeta\in\mathscr{K}_{\xi}^{\mathfrak{c}}(p),|\zeta|_Y=1} |dT(p)\cdot\zeta|_Y, \tag{3.17}$$

$$\hat{\lambda}_u \leqslant ||d_y g|| + \beta||d_x g||, \qquad c_{37} = \exp\{-c_{29} - c_{34}\rho_Y(\mathscr{D})\}. \tag{3.18}$$

THEOREM 3.13. *For the T-invariant family of conditional probabilities constructed in Theorem* 3.7 *on each fibre* $\mathfrak{F}(I)$ *of the fibration* $\mathscr{D}(\infty)$ *and any* $S^{(-r)}(J_r)$ *and* $S^{(-m)}(L_m)$ *we have*

$$P\{\mathfrak{F}(I, J_r)|I\}\cdot P\{\mathfrak{F}(IJ_r, L_m)|IJ_r\} = P\{\mathfrak{F}(I, J_m L)|I\}. \tag{3.19}$$

4. Smooth Non-Singular Probability Distributions on a Smale Horseshoe

4.1. DEFINING MEASURES ON MEASURABLE RECTANGLES IN TERMS OF CONDITIONAL PROBABILITY DISTRIBUTIONS ON FIBRES

The initial domain is $\mathscr{D}$; sets $\mathscr{D}(n)$ and $\mathscr{D}(-m)$, their pairwise intersections $\mathscr{D}(n)\cap\mathscr{D}(-m)$, limit sets $\mathscr{D}(\infty)$ and $\mathscr{D}(-\infty)$, and also the hyperbolic set $\Omega = \mathscr{D}(\infty)\cap\mathscr{D}(-\infty)$ possess the structure of a measurable rectangle. This structure is defined by the pairwise intersections of fibres of canonical expanding and contracting fibrations described in Statements 1.17, 1.20, 1.21. In the study of smooth probability measures on these sets we shall present these measures by integrals of conditional distributions on expanding fibres. For measurable rectangles the theory of conditional distributions is notably simpler.

STATEMENT 4.1. Let the measurable space $(W, \mathfrak{C})$ be the product of two Lebesgue-measurable spaces $(U, \mathfrak{A})$ and $(V, \mathfrak{B})$, i.e. $W = U \times V$, $\mathfrak{C} = \mathfrak{A} \otimes \mathfrak{B}$. Then for any measure μ on $\mathfrak{C}$ there is a transition probability distribution $P\{u, B\}$, where $u\in U, B\in\mathfrak{B}$, from $(U, \mathfrak{A})$ to $(V, \mathfrak{B})$ such that

$$\mu\{H\} = \int_U \hat{\mu}(du)\cdot P\{u, H_u\} \tag{4.1}$$

for any $H\in\mathfrak{C}$, where H_u is the section of H at $u\in U$, i.e. $H_u = \{v\in V:(u,v)\in H\}$, and $\hat{\mu}$ is the restriction of μ onto $\mathfrak{A}$ such that $\hat{\mu}\{A\} = \mu\{A \times V\}$ for any $A\in\mathfrak{A}$. This transition probability distribution $P\{u, B\}$ defines the conditional probability distribution with respect to the σ-algebra $\mathfrak{A} \times V$ via the formula $P\{H|\mathfrak{A} \times V\}|_{(u,v)} = P\{u, H_u\}$ for any $H\in\mathfrak{C}$.

REMARK. For any measure on the measurable space $(W, \mathfrak{C}) = (U \times V, \mathfrak{A} \otimes \mathfrak{B})$ defined by the formula

$$\mu\{H\} = \int_U \hat{\mu}(du) P\{u, H_u\},$$

where $\hat{\mu}(\cdot)$ is an arbitrary measure on the measurable space $(U, \mathfrak{A})$ transition probabilities $P\{u, H_u\}$ constitute the canonical system of measures [25,26] with respect to the splitting into atoms $u \times V$ for $u\in U$.

Let $Q(\cdot)$ be a fixed probability measure on $(V, \mathfrak{B})$. If the measure $\mu(\cdot|u)$ is absolutely continuous with respect to Q for a fixed $u\in U$, then there is a probability density $p(v|u)$ such that $\mu\{H_u|u\} = \int_{H_u} p(v|u)Q(dv)$ for any $H_u\in\mathfrak{B}$.

STATEMENT 4.2. Let U and V be compact normal topological spaces, $\mathfrak{A}$ and $\mathfrak{B}$ algebras of their Borel sets. Suppose that for measures μ on $\mathfrak{C}$ and Q on $\mathfrak{B}$ there is a variant of conditional densities $p(v|u)$ continuous in $U \times V$ in all variables. If measures μ and Q are strictly positive (on any non-empty open set), then this variant is uniquely defined by the formula

$$p(v|u) = \lim_{K \times N} \{\mu(K \times N) \cdot [\mu(K \times V) \cdot Q(N)]^{-1}\},$$

where the limit is taken with respect to the system of contracting neighbourhoods of the point (u,v) of the form $K \times N$, cf. [27].

The action T on coordinates of expanding fibrations has a special form. Suppose the expanding fibre $\mathfrak{F}(u,I_n)$ is the graph $\{F(u,I_n,y) \oplus y\}$, then

$$T\{\mathfrak{F}(u,I_n) \cap S^{(-1)}(j)\} = \mathfrak{F}(u,I_n j). \tag{4.2}$$

We now show how to describe the evolution of the corresponding probability distributions.

STATEMENT 4.3. Let measurable spaces $(W,\mathfrak{C})$ and $(W',\mathfrak{C}')$ satisfy the conditions of Statement 4.1 and assume that the bijective measurable map T whose structure is described by (4.2) maps the domain $S \subset W$ onto W'. Then the conditional probability distributions $P(\cdot|u)$ and $P_1(\cdot|u')$ for measures μ on W and μT^{-1} on W' are related by the formula

$$P_1(N'|u') = P(N|u)[P(S_u|u)]^{-1}, \qquad p(v'|u') = p(v|u) \cdot [P(S_u|u)]^{-1}, \tag{4.3}$$

where

$$(u',I') = (u,Ij), \qquad N = \{v : v' \in N'\}, \qquad S_u = \{v \in V : (u,v) \in S\}$$

STATEMENT 4.4. Let the Borel measure μ with strictly positive density $\rho(x,y)$ with respect to the $\mathbb{Z}$-volume continuous in variables (x,y) be defined on $\mathscr{D}(n)$. Then it can be defined by the restriction onto the algebra $\mathfrak{A}$ of numbers (u,I_n) of fibres $\mathfrak{F}(u,I_n)$ of the canonical rank n fibration and the family of densities $p^{\mathrm{Y}}(y_o|u,I_n)$ of conditional probability distributions on $\mathfrak{F}(u,I_n)$.

These densities are defined via the formula

$$p^{\mathrm{Y}}(y_0|u,I_n) = \frac{\rho(F(u,I_n,y_0),y_0) \cdot |\det d_u F|}{\int \rho(F(u,I_n,y),y) \cdot |\det d_u F| \, dy_1 \wedge \cdots \wedge dy_s}. \tag{4.4}$$

REMARK. The T-invariant family of conditional densities $p(\cdot|I)$ is uniformly continuous with respect to the index $(I) = (\cdots i_n \cdots i_1)$ in the topology of the Cantor discontinuum $(\mathbb{N}_k)_{-\infty}^{-1}$.

DEFINITION 4.1. The class of all measures on $\mathscr{D}(\infty)$ with the T-invariant smooth conditional probability distributions $P\{\cdot|I\}$ on fibres $\mathfrak{F}(I)$ is referred to as the class $\mathfrak{S}$.

REMARK. Any measure $\nu \in \mathfrak{S}$ is completely defined according to the formula (4.1) by its restriction $\hat{\nu}$ onto the algebra $\mathfrak{A}$ of numbers of fibres, i.e. sequences $(I) = (\cdots i_n \cdots i_1)$. Thus, to define a measure $\nu \in \mathfrak{S}$ it suffices to define values of ν on basic sets $\Gamma(I_n) = \mathscr{D}(\infty) \cap S^{(n)}(I_n)$ for any $n \in \mathbb{N}$ and $I_n \in (\mathbb{N}_k)_{-n}^{-1}$.

Set

$$\Gamma(I_n, J_m) = \mathscr{D}(\infty) \cap S^{(n)}(I_n) \cap S^{(-m)}(J_m). \tag{4.5}$$

Define the action of T^* on measures induced by the map T. Assign to each Borel measure ν defined on $\mathscr{D}(\infty)$ the restriction onto $\mathscr{D}(\infty)$ of $\nu\{T^{-1}\}$, i.e. for any measurable subset $H \subset \mathscr{D}(\infty)$ put

$$(T^*\nu)\{H\} = \nu^{(1)}\{H\} = \nu\{T^{-1}H\}. \tag{4.6}$$

Powers of T^* are similarly defined. If the measure ν is normed, then it is not necessarily so for $\nu^{(1)}$. If

$$\nu\{\mathscr{D}(\infty) \cap \mathscr{D}(-n)\} \neq 0 \quad \text{put}$$
$$\nu_n\{\mathrm{H}\} = [\nu\{\mathscr{D}(\infty) \cap \mathscr{D}(-n)\}]^{-1} \cdot \nu\{T^{-n}H\} \tag{4.7}$$

for any measurable $H \subset \mathscr{D}(\infty)$.

STATEMENT 4.5. The class $\mathfrak{S}$ is T^*-invariant, i.e. if $\nu \in \mathfrak{S}$ then $T^*\nu \in \mathfrak{S}$ and

$$\nu^{(1)}\{\Gamma(I_n j)\} = \nu\{\Gamma(I_n, j)\}. \tag{4.8}$$

If ν is not the zero measure, then $\nu\{\mathscr{D}(\infty) \cap \mathscr{D}(-1)\} \neq 0$.

DEFINITION 4.2. The measure $\mu \in \mathfrak{S}$ is the eigenmeasure of the Smale horseshoe if $T^*\mu\{\cdot\} = \lambda \cdot \mu\{\cdot\}$.

REMARK. If P is an eigenprobability measure then $P_1\{\cdot\} = P\{\cdot\}$. Conversely, if $P \in \mathfrak{S}$ and $P\{\cdot\} = P_1\{\cdot\}$ then P is the eigen probability measure corresponding to $\lambda = P\{\mathscr{D}(\infty) \cap \mathscr{D}(-1)\}$.

4.2. AN AVERAGE DESCRIPTION OF THE EVOLUTION OF MEASURES FROM THE CLASS $\mathfrak{S}$

To each measure $\nu \in \mathfrak{S}$ assign the sequence of k^n-dimensional row-vectors with components

$$q_{I_n}^{(n)}(\nu) = \nu\{\Gamma(I_n)\}, \qquad \vec{q}^{(n)}(\nu) = \pi_n(\hat{\nu}), \tag{4.9}$$

which are measures of basic sets $\Gamma(I_n)$. For $m < n$ put

$$(\vec{q}^{(n)}(\nu)\pi_{\overline{n,m}})_{I_m} = \sum_{i_n \cdots i_{m+1}} q_{I_n}^{(n)}(\nu). \tag{4.10}$$

The distance, i.e. the variation of the difference, for $n > m$ satisfies

$$|\vec{q}^{(m)}(\nu) - \vec{q}^{(m)}(\mu)|_m \leqslant |\vec{q}^{(n)}(\nu) - \vec{q}^{(n)}(\mu)|_n \leqslant |\hat{\nu} - \hat{\mu}|. \tag{4.11}$$

Note also that for the sequence ν_l the weak convergence to μ is equivalent to the convergence as $l \to \infty$ (non-uniform, generally speaking) of $\vec{q}^{(m)}(\nu_l)$ to $\vec{q}^{(m)}(\mu)$ for all m.

From the fibration $\Gamma(I_n)$ choose the standard representative, i.e. the fibre $\mathfrak{F}(EI_n)$, where E is the infinite-to-the-left sequence of units.

DEFINITION 4.3. Put $A^{(n)}$ for the $k^n \times k^n$ matrix with entries

$$a_{I_n, L_n}^{(n)} = P\{\Gamma(, L_n) | EI_n\}, \tag{4.12}$$

where $P\{\cdot|EI_n\}$ is a T-invariant family of smooth conditional probability distributions on fibres $\mathfrak{F}(EI_n)$ and $I_n \in (\mathbb{N}_k)_{-n}^{-1}$, $L_n \in (\mathbb{N}_k)_1^n$.

DEFINITION 4.4. Denote by $B^{(n)}(v)$ the $k^n \times k^n$ matrix with entries $b_{I_n,L_n}^{(n)}(v)$ equal to

$$[v\{\Gamma(I_n,)\}]^{-1}\cdot v\{\Gamma(I_n,L_n)\} \quad \text{for} \quad v\{\Gamma(I_n,)\} \neq 0 \tag{4.13}$$

$$a_{I_n,L_n}^{(n)} \quad \text{for} \quad v\{\Gamma(I_n,)\} = 0. \tag{4.14}$$

REMARK. $A^{(n)} = B^{(n)}(\mu)$, where $\mu \in \mathfrak{S}$ is supported on fibres $\mathfrak{F}(EI_n)$ and the measure of each of the fibres is positive.

STATEMENT 4.6. The matrix $B^{(n)}(v)$ is a strictly positive substochastic matrix, i.e.

$$b_{I_n,L_n}^{(n)}(v) > 0, \qquad \sum_{L_n} b_{I_n,L_n}^{(n)}(v) \leqslant 1. \tag{4.15}$$

$$v^{(n)}\{\Gamma(L_n,)\} = \sum_{I_n} b_{I_n,L_n}^{(n)}(v)\cdot v\{\Gamma(I_n)\}. \tag{4.16}$$

for any $I_n \in (\mathbb{N}_k)_{-n}^{-1}$, $L_n \in (\mathbb{N}_k)_1^n$.

COROLLARY. If $v \in \mathfrak{S}$ then

$$q_{L_n}^{(n)}(v^{(n)}) = \sum_{I_n} q_{I_n}^{(n)}(v)\cdot b_{I_n,L_n}^{(n)}(v). \tag{4.17}$$

The formula (4.17) is a purely formal corollary from the definition of $B^{(n)}(v)$. However, it turns out that for any $v \in \mathfrak{S}$ these matrices are close to each other and to $A^{(n)}$ and possess nice ergodic properties.

LEMMA 4.7. *For any* $v \in \mathfrak{S}$ *and* $I_n, H_n \in (\mathbb{N}_k)_{-n}^{-1}$, $L_n \in (\mathbb{N}_k)_1^n$ we have

$$(R')^{-1}\cdot b_{H_n,L_n}^{(n)}(v) \leqslant b_{I_n,L_n}^{(n)}(v) \leqslant R'\cdot b_{H_n,L_n}^{(n)}(v), \tag{4.18}$$

where R' *is defined in the formula* (3.11) *of Theorem* 3.8. *In particular, this inequality holds for entries of* $A^{(n)}$.

Proof. For the T-invariant families of conditional probabilities on the intersection of fibres $\mathfrak{F}(M_1)$ and $\mathfrak{F}(M_2)$ of the fibration $\mathscr{D}(\infty)$ with $S^{(-n)}(L_n)$ the inequality (3.14) of Lemma 3.11 is applicable:

$$P\{\mathfrak{F}(M_1)\cap S^{(-n)}(L_n)|M_1\} \leqslant R'\cdot P\{\mathfrak{F}(M_2)\cap S^{(-n)}(L_n)|M_2\}.$$

The desired relation (4.18) is obtained by integration of this inequality with respect to the product-measure $\hat{v}\{dM_1\} \oplus \hat{v}\{dM_2\}$ along the set

$$\{(M_1,M_2): \mathfrak{F}(M_1) \subset S^{(n)}(I_n), \mathfrak{F}(M_2) \subset S^{(n)}(H_n)\}. \qquad \square$$

LEMMA 4.8. *For any measures* $\mu, v \in \mathfrak{S}$ *and* $I_n \in (\mathbb{N}_k)_{-n}^{-1}$, $L_n \in (\mathbb{N}_k)_1^n$ *we have*

$$(R_n)^{-1}\cdot b_{I_n,L_n}^{(n)}(\mu) \leqslant b_{I_n,L_n}^{(n)}(v) \leqslant R_n\cdot b_{I_n,L_n}^{(n)}(\mu). \tag{4.19}$$

In particular, we may take $B^{(n)}(\mu)$ *instead of* $A^{(n)}$.

REMARK. For any $v \in \mathfrak{S}$, if I_n and H_n are such that $i_1 = h_1, \ldots, i_m = h_m$, then

$$(R_m)^{-1}\cdot b_{H_n,L_n}^{(n)}(v) \leqslant b_{I_n,L_n}^{(n)}(v) \leqslant R_m\cdot b_{H_n,L_n}^{(n)}(v). \tag{4.20}$$

LEMMA 4.9. *For any* $\nu \in \mathfrak{S}$ *and* $I_n \in (\mathbb{N}_k)_{-n}^{-1}, L_n \in (\mathbb{N}_k)_1^n$ *we have*

$$b_{I_n, L_n}^{(n)}(\nu) \geqslant c_{37} \cdot \exp\{-n \cdot \dim \mathbb{Y} \cdot \ln \hat{\lambda}_u\}, \tag{4.21}$$

$$\sum_{L_n} b_{I_n, L_n}^{(n)}(\nu) \geqslant c_{37} \cdot k^n \cdot \exp\{-n \cdot \dim \mathbb{Y} \cdot \ln \hat{\lambda}_u\}. \tag{4.22}$$

COROLLARY. *If* $\nu \in \mathfrak{S}$ *then*

$$\nu^{(1)}\{\mathscr{D}(\infty)\} \geqslant c_{41} \cdot \nu\{\mathscr{D}(\infty)\} \tag{4.23}$$

$$c_{41} = c_{37} \cdot k \cdot \exp(-\dim \mathbb{Y} \cdot \ln \hat{\lambda}_u)$$

For proof we use inequality (3.16) of Theorem 3.12. □

4.3. THE CONSTRUCTION OF AN EIGENMEASURE FOR A SMALE HORSESHOE

The proof of existence and uniqueness of an eigenmeasure may be reduced to the classical Perron–Frobenius theorem for matrices with positive entries. A similar reduction for the case of a positive Markov kernel was applied by Sinai [13] in the theory of Anosov systems, and later by Bunimovich and Sinai [24] for the Lorentz system. The construction of the eigenmeasure of a sub-Markov operator Π is complicated by the fact that the variation of the difference of two measures is not a Lyapunov metric for Π. Such a metric exists, but is expressed in terms of a previously unknown eigenfunction of the dual operator Π^*. Estimations of norms with respect to the variation necessary in this and subsequent sections are presented in the Appendix.

$$\vec{q}^{(n)}(\nu_s) = |\vec{q}^{(n)}(\nu^{(s)})|^{-1} \cdot \vec{q}^{(n)}(\nu^{(s)}). \tag{4.24}$$

Any strictly positive substochastic matrix $A^{(n)}$ possesses, from the Frobenius–Perron theorem, a normed eigenvector $\vec{p}^{(n)}$ with positive coordinates $p_{I_n}^{(n)} > 0$ corresponding to the positive eigenvalue λ_n. Let us estimate the proximity of the eigenvector $\vec{p}^{(n)}$ of the matrix $A^{(n)}$ introduced in (4.12) to $\vec{q}^{(n)}(\nu_s)$.

LEMMA 4.10. *If n is so large that* $R_n \theta < 1$, *where*

$\theta = 1 - (R')^{-1}$, *then*

$$|\vec{q}^{(n)}(\nu_s) - \vec{p}^{(n)}| \leqslant 4 \cdot R' \cdot \{(R_n \theta)^{E[s/n]} + R_n \cdot (R_n - 1)[1 - R_n \theta]^{-1}\} \tag{4.25}$$

for any $\nu \in \mathfrak{S}$, *where* $E[s/n]$ *means the integer part of* $s \cdot n^{-1}$.

Proof. For a given n, the matrices $B^{(n)}$ make it possible to trace the evolution of subsequences $\nu^{(t)}, \nu^{(t+n)}, \nu^{(t+2n)}, \ldots$ where t is an integer such that $0 \leqslant t < n$. By Lemmas 4.7 and 4.8, the estimate of Theorem 4.10 is applicable to each of recursive subsequences $\vec{q}^{(n)}(\nu_{rn+t})$ yielding (4.25).

THEOREM 4.11. *Let* $\vec{p}^{(n)}$ *be a positive normed eigenvector of* $A^{(n)}$ *defined in* (4.12). *Then for any positive integer l there exists the limit*

$$\vec{r}^{(l)} = \lim_{m \to \infty} \vec{p}^{(m)} \pi_{\overrightarrow{m,l}}. \tag{4.26}$$

The family of vectors $\vec{r}^{(l)}$ *is such that*

$$\bar{r}^{(l+1)}\pi_{l+\overrightarrow{1,l}} = \bar{r}^{(l)}. \tag{4.27}$$

It defines the normed measure $\mu \in \mathfrak{S}$ via the formula

$$\mu\{\Gamma(I_l,)\} = r^{(l)}_{I_l}. \tag{4.28}$$

For any probability measure $v \in \mathfrak{S}$ the sequence of probability measures $\hat{y}_s\{\cdot\}$ converges weakly to $\hat{\mu}$, so that

$$|\bar{q}^{(l)}(v_s) - \bar{q}^{(l)}(\mu)| < 4\cdot R'\{(R_l\cdot\theta)^{E[s/l]} + R_l\cdot(R_l - 1)\cdot[1 - R_l\cdot\theta]^{-1}\}. \tag{4.29}$$

Proof. Take any measure $v \in \mathfrak{S}$.

By (4.25) for fixed $l \leqslant m < n$ and any $\varepsilon > 0$ for a sufficiently large s we have

$$|\bar{p}^{(m)} - \bar{q}^{(m)}(v_s)|_m < 4\cdot R'\cdot[\varepsilon + R_m\cdot(R_m - 1)\cdot(1 - R_m\theta)^{-1}],$$

$$|\bar{p}^{(n)} - \bar{q}^{(n)}(v_s)|_n < 4\cdot R'\cdot[\varepsilon + R_n\cdot(R_n - 1)\cdot(1 - R_n\cdot\theta)^{-1}].$$

Hence the projection onto vectors of the rank l yields

$$|\bar{p}^{(m)}\pi_{\overrightarrow{m,l}} - \bar{p}^{(n)}\pi_{\overrightarrow{n,l}}|_l \leqslant 8\cdot R'\cdot R_m\cdot(R_m - 1)\cdot(1 - R_m\cdot\theta)^{-1}.$$

Thus for a fixed l the sequence $\bar{p}^{(m)}\pi_{\overrightarrow{m,l}}$ is a fundamental one and

$$|\bar{p}^{(m)}\pi_{\overrightarrow{m,l}} - \bar{r}^{(l)}|_l \leqslant 8\cdot R'\cdot R_m\cdot(R_m - 1)\cdot(1 - R_m\cdot\theta)^{-1}. \tag{4.30}$$

These sequences converge in agreement because from Definition 4.10 we have $\pi_{\overrightarrow{m,l}} = \pi_{m,\overrightarrow{l+1}}\cdot\pi_{l+\overrightarrow{1,l}}$ for $m > l + 1$, which proves (4.26) and (4.27). Since the finite-additive measure on various rectangles $\Gamma(I_n,)$ is extendable up to a countably-additive measure $\hat{\mu}$ on $\mathfrak{A}$, then (4.28) also holds. Finally, comparing (4.30) and (4.25) we get (4.29).

LEMMA 4.12. *Let $\mu, v \in \mathfrak{S}, \mu\{\mathscr{D}(\infty)\} = v\{\mathscr{D}(\infty)\} = 1$. Then*

$$|\bar{q}^{(n)}(\mu^{(1)}) - \bar{q}^{(n)}(v^{(1)})| < |\bar{q}^{(n)}(\mu) - \bar{q}^{(n)}(v)| + 2\cdot(R_n - 1), \tag{4.31}$$

$$|\bar{q}^{(n)}(\mu_1) - \bar{q}^{(n)}(v_1)| < 2\cdot(c_{41})^{-1}\cdot\{|\bar{q}^{(n)}(\mu) - \bar{q}^{(n)}(v)| + 2(R_n - 1)\}. \tag{4.32}$$

THEOREM 4.13.

(a) *The measure μ constructed in Theorem* 4.11 *is an eigenmeasure corresponding to the eigenvalue*

$$\lambda = |\mu^{(1)}| \geqslant k\cdot\exp\{-\dim\mathbb{Y}\cdot\ln\hat{\lambda}_u\}, \tag{4.33}$$

(b) *The measure μ is strictly positive, in particular*

$$\mu\{\Gamma(I_n,)\} > c_{35}\cdot(k)^n\cdot\exp\{-n\cdot\dim\mathbb{Y}\cdot\ln\hat{\lambda}_u\}, \tag{4.34}$$

(c) *The class $\mathfrak{S}$ contains no other eigenmeasures.*

Proof. Let $v \in \mathfrak{S}$, then by Theorem 4.11

$$|\bar{q}^{(l)}(v_s) - \bar{q}^{(l)}(\mu)|_l \leqslant \varepsilon(N,m) + \varepsilon(m),$$

for $l < m$ and any $s > N$, where $\varepsilon(m) \to 0_{m\to\infty}$, and $\varepsilon(N,m) \to 0$ when $N \to \infty$ and m is fixed implying

$$|\bar{q}^{(l)}(\mu) - \bar{q}^{(l)}(\mu_1)| \leqslant |\bar{q}^{(l)}(\mu) - \bar{q}^{(l)}(v_{s+1})| + |\bar{q}^{(l)}(v_{s+1}) - \bar{q}^{(l)}(\mu_1)|$$

$$\leqslant \varepsilon(N,m) + \varepsilon(m) + [\varepsilon(N,m) + \varepsilon(m) + 2(R_m - 1)]\cdot 2\cdot (c_{35}\cdot k)^{-1}\cdot \exp\{\dim \Upsilon \cdot \ln \hat{\lambda}_u\},$$

where for the estimation of $|\bar{q}(v_{s+1}) - \bar{q}(\mu_1)|$ we have made use of (4.32). Letting N tend to infinity and then passing to the limit as $m \to \infty$ we get $\mu\{\Gamma(I_l,)\} = \mu_1\{\Gamma(I_l,)\}$ for all sets of indices I_l. Hence, $\hat{\mu} = \hat{\mu}_1$ and $\mu = \mu_1$.

Since $\lambda^n = \mu^{(n)}\{\mathscr{D}(\infty)\} = |\bar{q}^{(n)}(v^{(n)})|$, then λ^n is an eigenvalue of the matrix $B^{(n)}(\mu)$. Therefore due to (4.21) we have

$$\lambda^n = \sum_{L_n} \lambda^n \cdot q_{L_n}^{(n)}(\mu) = \sum_{L_n}\sum_{I_n} q_{I_n}^{(n)}(\mu)\cdot b_{I_n,L_n}^{(n)}(\mu) \geqslant c_{35}\cdot k^n \cdot \exp\{-n \dim \Upsilon \cdot \ln \hat{\lambda}_u\}.$$

Extracting the nth root from both sides of the inequality and passing to the limit as $n \to \infty$, we get (4.33). Finally, since $v_n = v$ for an eigen probability measure $v \in \mathfrak{S}$ and any n, the inequality (4.29) yields $\bar{q}^{(l)}(v) = \bar{q}^{(l)}(\mu)$ for all l, i.e. $v = \mu$. Similarly, due to (4.21) we have

$$q_{I_n}^{(n)}(\mu) = \lambda^n \cdot \sum_{L_n} b_{I_n,L_n}^{(n)}(\mu)\cdot q_{L_n}^{(n)}(\mu) \geqslant c_{35}\cdot k^n \cdot \exp\{-n\cdot \dim \Upsilon \cdot \ln \hat{\lambda}_u\} \sum_{L_n} q_{L_n}^{(n)}(\mu)$$

which implies (4.34) since $|\bar{q}| = |\bar{\mu}| = 1$. □

5. A Natural Invariant Probability Distribution on the Hyperbolic Set of a Smale Horseshoe

5.1. THE SEQUENCE OF PROBABILITY DISTRIBUTIONS $\hat{\mu}_{(m)}$

The structure of the measurable rectangle for the set $\mathscr{D}(\infty, -m) = \mathscr{D}(\infty) \cap \mathscr{D}(-m)$ is defined by pairwise intersections of expanding fibres $\mathfrak{F}(I)$, where $I \in \mathfrak{A}$, with contracting fibres $\mathfrak{G}(v, J_m)$ of the canonical fibration of rank m. Any measure $v \in \mathfrak{S}$ is defined on Borel subsets of $\mathscr{D}(\infty)$. It is in one-to-one correspondence with the measure which coincides with v on Borel subsets of $\mathscr{D}(\infty)$ and vanishes on subsets from $\mathscr{D} \setminus \mathscr{D}(\infty)$. We shall not distinguish these two measures. Thus, considering $v \in \mathfrak{S}$ we have

$$v_{(m)}\{\cdot\} = v\{\cdot | \mathscr{D}(\infty, -m)\} = v\{\cdot | \mathscr{D}(-m)\}. \tag{5.1}$$

Similarly, we sometimes make use of the notation $v\{S^{(n)}(I_n)\} = v\{\Gamma(I_n,)\}$. Also, in denoting values of T-invariant conditional probabilities $P\{\cdot | I\}$ we sometimes omit the indication of the intersection with $\mathfrak{F}(I)$, i.e. we write $P\{H|I\} = P\{H \cap \mathfrak{F}(I)|I\}$.

On leaves $\mathfrak{F}(I)$ of the bundle $\mathscr{D}(\infty)$, introduce the following 'doubly conditional' probability distributions

$$P_m\{H|I\} = P\{H|I|\mathscr{D}(-m)\} = [P\{\mathscr{D}(-m)|I\}]^{-1}\cdot P\{H \cap \mathscr{D}(-m)|I\}, \tag{5.2}$$

where $P\{\cdot|I\}$ is the family of T-invariant conditional probability distributions constructed in §3. These families are also defined by transition probabilities.

THEOREM 5.1. *Let μ be a normed eigenmeasure on a horseshoe, λ an eigenvalue, $\mu_{(m)}\{\cdot\} = \mu\{\cdot|\mathscr{D}(-m)\}$ conditional probability distribution recovered from μ. Then*

$$\mu_{(m)}\{H\} = \lambda^{-m} \cdot \mu\{H \cap \mathscr{D}(-m)\} \tag{5.3}$$

for any measurable subset $H \subset \mathscr{D}(\infty)$. *The integral presentation*

$$\mu_{(m)}\{\cdot\} = \int \hat{\mu}_{(m)}\{\mathrm{d}I\} \cdot P_m\{\cdot | I\}, \tag{5.4}$$

is also valid where the restriction $\hat{\mu}_{(m)}$ *of* $\mu_{(m)}$ *onto* $\mathfrak{A}$ *is absolutely continuous with respect to* $\hat{\mu}$ *with the Radon–Nikodym derivative*

$$\Psi_m(I) = \frac{\mathrm{d}\hat{\mu}_{(m)}}{\mathrm{d}\hat{\mu}}(I) = \lambda^{-m} \cdot P\{\mathscr{D}(-m)|I\}. \tag{5.5}$$

Proof. First let us prove the second equality in

$$\int \hat{\mu}\{\mathrm{d}I\} P\{\mathscr{D}(-m)|I\} = \mu\{\mathscr{D}(-m)\} = \lambda^m. \tag{5.6}$$

The property of T-eigenmeasure for $\mathscr{D}(-m) = T^{-m}\mathscr{D}(m)$ implies

$$\mu\{\mathscr{D}(-m)\} = \mu\{\mathscr{D}(\infty) \cap \mathscr{D}(-m)\} = \lambda^m \cdot \mu\{\mathscr{D}(\infty) \cap T^m \mathscr{D}(\infty)\} = \lambda^m \cdot \mu\{\mathscr{D}(\infty)\} = \lambda^m.$$

The first equality in (5.6) is based on the presentation (4.1). The definition of a conditional probability implies

$$\mu_{(m)}\{H\} = [\mu\{\mathscr{D}(-m)\}]^{-1} \cdot \mu\{H \cap \mathscr{D}(-m)\} = \lambda^{-m} \cdot \mu\{H \cap \mathscr{D}(-m)\}.$$

Hence the presentation (4.1) implies

$$\mu_{(m)}\{H\} = \lambda^{-m} \cdot \int \mu\{\mathrm{d}I\} \cdot P\{H \cap \mathscr{D}(-m)|I\}.$$

Dividing and multiplying the integrand by $P\{\mathscr{D}(-m)|I\}$ and making use of Definition 5.2 we get (5.4), where the Radon–Nikodym derivative is expressed via (5.5).

LEMMA 5.2. *For* $m > 0$, *the probabilities* $P\{\mathscr{D}(-m)|I\}$ *are strictly positive and equicontinuous with respect to* I:

$$(R')^{-1} \cdot P\{\mathscr{D}(-m)|I\} \leqslant P\{\mathscr{D}(-m)|L\} \leqslant R' \cdot P\{\mathscr{D}(-m)|I\} \tag{5.7}$$

for any $I, L \in \mathfrak{A}$ *and for* $I, I' \in \mathfrak{A}$ *such that* $i_n = i'_n, \ldots, i_1 = i'_1$ *we have*

$$(R_n)^{-1} \cdot P\{\mathscr{D}(-m)|I\} \leqslant P\{\mathscr{D}(-m)|I'\} \leqslant R_n \cdot P\{\mathscr{D}(-m)|I\}. \tag{5.8}$$

COROLLARY. *The Radon–Nikodym derivative* (5.5) *is positive,*

$$(R')^{-1} \leqslant \Psi_m(I) \leqslant R' \tag{5.9}$$

and equicontinuous with respect to I (*with the same constants* R_n).

To compute $P\{\mathscr{D}(-m)|I\}$ approximately, let us make use of the matrix technique developed in the preceding section.

LEMMA 5.3. *Let* $v = v_I \in \mathfrak{S}$ *be a singular probability measure* $\hat{v}_I\{I\} = 1$. *Then*

$$P\{\mathscr{D}(-m)|I\} = v_I^{(m)}\{\mathscr{D}(\infty)\} \tag{5.10}$$

and for $l^2 \leqslant m \leqslant l\cdot(l+1)$, $m = l^2 + r$ *the presentation*

$$P\{\mathscr{D}(-m)|I\} = \left\langle \overleftarrow{q}^{(l+1)}(v_l) \prod_{s=0}^{r-1} B^{(l+1)}(v_l^{s(l+1)}) \pi_{\overrightarrow{l+1},l} \prod_{s=r}^{l-1} B^{(l)}(v_l^{(sl+r)}), \vec{E}_l \right\rangle \quad (5.11)$$

holds, where $\vec{E}_l$ *is the column vector of* k^l *units, the row vector* $\overleftarrow{q}^{(l+1)}(v_l)$ *is multiplied by substochastic matrices from the right, the embedding* $\pi_{l+1,l}$ *is defined by* (4.10) *and the scalar product of row vectors by the column vectors is defined by the formula*

$$\langle \overleftarrow{q}^{(l)}, f_{(l)} \rangle = \sum_{I_l} q_{I_l}^{(l)} \cdot f_{(l)}^{(I_l)}. \quad (5.12)$$

For $l\cdot(l+1) \leqslant m \leqslant (l+1)^2$, $m = l\cdot(l+1) + r$ *we can then write*

$$P\{\mathscr{D}(-m)|I\} = \left\langle \overleftarrow{q}^{(l+1)}(v_l) \prod_{s=0}^{r-1} B^{(l+1)}(v_l^{s(l+1)}) \pi_{\overrightarrow{l+1,l}} \prod_{s=0}^{l-r} B^{(l)}(v_l^{(rl+sl+r)}), \vec{E}_l \right\rangle \quad (5.13)$$

Now let us estimate the expressions obtained. First, let us make use of the fact that for any measures v of class $\mathfrak{S}$ the matrices $B(v)$ recovered from them are closed.

LEMMA 5.4. *Under the conditions of Lemma* 5.3 *put*

$$\lambda^{-l^2-r} \langle \overleftarrow{q}^{(l+1)}(v_l)[B^{(l+1)}(\mu)]^r \pi_{\overrightarrow{l+1,l}}[B^{(l)}(\mu)]^{l-r}, \vec{E}_l \rangle = \Psi(I, l, r), \quad (5.14)$$

$$\lambda^{-l\cdot(l+1)-r} \langle \overleftarrow{q}^{(l+1)}(v_l)[B^{(+1)}(\mu)]^r \pi_{\overrightarrow{l+1,l}}[B^{(l)}(\mu)]^{l+1-r}, \vec{E}_l)\rangle = \Psi(I, l+1, -l-1+r), \quad (5.15)$$

where μ *is the eigenmeasure of the horseshoe. Then for* $m = l^2 + r$ *and* $m = l(l+1) + r$ *the following estimations of the Radon–Nikodym derivative* (5.9) *are valid, respectively:*

$$(R_{l+1})^{-r} \cdot (R_l)^{-l+r} \leqslant [\Psi(I, l, r)]^{-1} \cdot \Psi_m(I) \leqslant (R_{l+1})^r (R_l)^{l-r}, \quad (5.16)$$

$$(R_{l+1})^{-r} \cdot (R_l)^{-l-1+r} \leqslant [\Psi(I, l+1, -l-1+r)]^{-1} \cdot \Psi_m(I)$$
$$\leqslant (R_{l+1})^r (R_l)^{l+1-r}. \quad (5.17)$$

The constants R_n constructed in Theorem 3.9 are such that $\ln R_n = O(\lambda_u^{-n/2}) + O(\lambda_s^{-n/2})$, where $\lambda_s, \lambda_u > 1$. Hence, $\ln (R_n)^n = n \ln R_n$ and $\ln R_n$ tend to 0 when n tends to ∞.

This means that $\Psi(I, l, \pm r)$ are the asymptotics of $\Psi_m(I)$. However, to find the indicated limit we should make use of the ergodicity properties of $B^{(n)}(\mu)$.

5.2. THE COMPUTATION OF THE ASYMPTOTICS OF $\hat{\mu}_{(m)}$ VIA THE MATRIX TECHNIQUE

The probability distribution μ constructed in §4 is the eigendistribution on the Smale horseshoe. Since $\mu^{(n)}\{\cdot\} = \lambda^n \cdot \mu\{\cdot\}$, then $\overleftarrow{q}^{(n)}(\mu^{(n)})$ might be presented as $\lambda^n \cdot \overleftarrow{q}^{(n)}(\mu) = \overleftarrow{q}^{(n)}(\mu) \cdot B^{(n)}(\mu)$, i.e. the vector $\overleftarrow{q}^{(n)}(\mu)$ is a normed strictly positive eigenvector of the square matrix $B^{(n)}(\mu)$ corresponding to its positive eigenvalue. By the Frobenius–Perron theorem $B^{(n)}(\mu)$ also possesses an eigen strictly positive column vector $\vec{e}_n$ corresponding to the same eigenvalue λ^n.

DEFINITION 5.1. Components of the normed positive eigen column vector $\vec{e}_n$ of $B^{(n)}(\mu)$ are denoted by $e_n(I_n)$, where $(I_n) = (i_n \cdots i_1) \in (\mathbb{N}_k)^{-1}_{-n}$, and normed so that

$$\langle \overleftarrow{q}^{(n)}(\mu), \vec{e}_n \rangle = 1. \tag{5.18}$$

We shall also sometimes set $e_n(I) = e_n(I_n)$, $(I) = (\cdots i_{n+1} I_n)$.

Since matrices $B^{(n)}(\mu)$ satisfy statements from Statement A.4, then components of the normed eigenvector $\vec{e}_n$ satisfy

$$(R')^{-1} \leqslant e_n(I) \leqslant R', \tag{5.19}$$

$$(R_l)^{-1} \cdot e_n(I) \leqslant e_n(J) \leqslant R_l \cdot e_n(I) \tag{5.20}$$

for all sets $(I) = (\cdots i_{l+1} i_l \cdots i_1)$ and J such that $j_1 = i_1, \ldots, j_l = i_l$.

The whole space of column vectors splits into the direct sum of the $B^{(n)}(\mu)$-invariant space $\mathscr{E}_n$ spanned by the eigenvector $\vec{e}_n$ and the $B^{(n)}(\mu)$ – invariant subspace $\mathscr{H}_n = \{\vec{h}\}$ of column vectors orthogonal to the eigenrow vector $\overleftarrow{q}^{(n)}(\mu)$;

$$\langle \overleftarrow{q}^{(n)}(\mu), \vec{h} \rangle = 0. \tag{5.21}$$

Each column vector $\vec{f}$ is decomposable so that

$$\vec{f} = \langle \overleftarrow{q}^{(n)}(\mu), \vec{f} \rangle \vec{e}_n + \vec{h}(\vec{f}), \qquad \vec{h}(\vec{f}) = \vec{f} - \langle \overleftarrow{q}^{(n)}(\mu), \vec{f} \rangle \vec{e}_n. \tag{5.22}$$

The space of column vectors is normed by the formula

$$\|\vec{f}_{(n)}\|_{\rightarrow} = \max_{I_n} \{ f_{(n)}(I_n) \cdot [e_n(I_n)]^{-1} \}. \tag{5.23}$$

The dual norm on the space of row vectors vectors is defined by the formula

$$\|\overleftarrow{q}^{(n)}\|_{\leftarrow} = \sum_{I_n} |q^{(n)}_{I_n}| \cdot e_n(I_n). \tag{5.24}$$

On $\mathscr{E}_n$ the matrix $B^{(n)}(\mu)$ acts by multiplying a vector by λ and on $\mathscr{H}_n$ we have

$$\|B^{(n)}(\mu)\vec{h}\|_{\rightarrow} \leqslant [1 - (R')^{-2}] \cdot \lambda \cdot \|\vec{h}\|_{\rightarrow}. \tag{5.25}$$

Note that $B^{(n)}(\mu)$ acts similarly on row vectors, i.e. the operator norm of $B^{(n)}(\mu)$ is a Liapunov norm with respect to (5.23) and (5.24).

If for row vectors that describe measures in mean there exists a 'projective' embedding $\pi_{\overleftarrow{n,l}}$, see (4.10), of the space of rows of rank n into the space of rows of rank l, where $l < n$, so that the diagram

$$\overleftarrow{q}^{(n)}(\mu) \xrightarrow[\pi_{\overleftarrow{n,l}}]{\mu} \overleftarrow{q}^{(l)}(\mu)$$

commutes, then for the column vectors there is a natural (inductive) embedding $\sigma_{\overrightarrow{n,l}}$ of the space of columns of rank l into the space of columns of rank n, where $l < n$, by the formula

$$(\sigma_{\overrightarrow{n,l}} \vec{f}_{(l)})(i_n \cdots i_l \cdots i_1) = f_{(l)}(i_l \cdots i_1). \tag{5.26}$$

The generalization of (5.21) to the case of infinite sequences $(I) = (\cdots i_n \cdots i_1)$ gives the embedding $\sigma_{\overrightarrow{\infty,l}}$ of column vectors into the limit space of (continuous) functions in $I \in (\mathbb{N}_k)^{-1}_{-\infty}$. Our notation is compatible with the possibility of such an embedding. Note also that $\sigma_{\overleftarrow{r,l}} = \sigma_{\overleftarrow{r,m}} \cdot \sigma_{\overleftarrow{m,l}}$ for any $l < m < r \leqslant \infty$.

The embedding $\sigma_{\overleftarrow{n,l}}$ is dual to $\pi_{\overrightarrow{n,l}}$ in the sense that

$$\langle \overleftarrow{q}^{(n)} \pi_{\overrightarrow{n,l}}, \vec{f}_{(l)} \rangle = \langle \overleftarrow{q}^{(n)}, \sigma_{\overleftarrow{n,l}} \vec{f}_{(l)} \rangle. \tag{5.27}$$

However, the norm of $\sigma_{\overleftarrow{n,l}}$ with respect to (5.23) is greater than 1 and is defined by the proximity of components of $\sigma_{\overleftarrow{n,l}}\vec{e}_l$ to the corresponding components of $\vec{e}_n$. A similar statement is valid for $\pi_{\overrightarrow{n,l}}$. It goes without saying that these embeddings have the unit norm with respect to natural metrics, i.e. the maximum of the module of components of the column vector and the sum of absolute values of components of the row vectors, but these standard metrics are inconvenient in the estimation of $B^{(n)}(\mu)$-action. Note that since $\overleftarrow{q}^{(n)}(\mu) \cdot \pi_{\overrightarrow{n,l}} = \overleftarrow{q}^{(l)}(\mu)$ then

$$\sigma_{\overleftarrow{n,l}} \mathscr{H}_l = \mathscr{H}_n. \tag{5.28}$$

Generally speaking $\sigma_{\overleftarrow{n,l}}\vec{e}_l \neq \vec{e}_n$, but these two vectors are close to each other. To prove this we make use of an artificial trick comparing two estimates of the Radon–Nikodym derivative $\Psi_{l(l+1)}$.

THEOREM 5.5. *For eigen column vectors $\vec{e}_l$ of matrices $B^{(l)}(\mu)$ for l so large that $2\theta^l < (R')^{-3}$ we have*

$$(\mathscr{K}_l)^{-1} \leqslant \min_I \{e_{l+1}(I) \cdot [e_l(I)]^{-1}\} \leqslant \max_I \{e_{l+1}(I) \cdot [e_l(I)]^{-1}\} \leqslant \mathscr{K}_l \tag{5.29}$$

$$\mathscr{K}_l = (R_l)^{l+1} \cdot (R_{l+1})^l \cdot (1 + c_{51} \cdot \theta^l) \cdot (1 - c_{51} \cdot \theta^l)^{-1}, \tag{5.30}$$

$$c_{51} = 2 \cdot (R')^3, \qquad \theta = 1 - (R')^{-1} < 1. \tag{5.31}$$

For small l the trivial estimate (5.29) holds, where $\mathscr{K}_l = (R')^2$.

COROLLARY. *The sequence of equicontinuous normed positive functions $e_n(I) = (\sigma_{\overleftarrow{\infty,n}}\vec{e}_n)(I)$ and $\ln e_n(I)$ are fundamental ones. The limit $e(I)$ is normed positive and continuous, i.e.*

$$\int e(I)\hat{\mu}\{dI\} = 1, \qquad (R')^{-1} \leqslant e(I) \leqslant R',$$

$$(R_l)^{-1} \cdot e(I) \leqslant e(J) \leqslant R_l \cdot e(I) \quad \textit{if} \quad j_l = i_l, \ldots, j_1 = i_1.$$

The function $\ln e(I)$ is a uniform limit of $\ln e_n(I)$ and

$$\max_I |\ln e(I) - \ln e_n(I)| \leqslant \mathscr{L}_n, \tag{5.32}$$

where

$$\mathscr{L}_n = \min\left\{(R')^2, \sum_{l=n}^{\infty} \ln \mathscr{K}_l\right\} \xrightarrow[u \to \infty]{} 0. \tag{5.33}$$

Proof. For $m = l \cdot (l+1)$ formulas (5.16) and (5.17) give two estimations for $\Psi_m(I)$: one in terms of $\Psi(I, l, l+1)$, the other one in terms of $\Psi(I, l+1, -l-1)$. Since $\sigma_{\overleftarrow{l+1,l}}\vec{E}_l = \vec{E}_{l+1}$, $\overleftarrow{q}^{(l+1)}(\nu)\pi_{\overrightarrow{l+1,l}} = \overleftarrow{q}^{(l)}(\nu)$, then the values mentioned converge to

$$\lambda^{-l(l+1)} \langle \overleftarrow{q}^{(l)}(\nu_I)[B^{(l)}(\mu)]^{l+1}, \vec{E}_l \rangle \quad \text{and} \quad \lambda^{-l(l+1)} \langle \overleftarrow{q}^{(l+1)}(\nu_I)[B^{(l+1)}(\mu)]^l, \vec{E}_l \rangle,$$

respectively. For a singular measure ν_l we have $\langle \overleftarrow{q}^{(n)}(\nu_l), \vec{e}_n\rangle = e_n(I)$. Therefore

$$\overleftarrow{q}^{(n)}(\nu_l) = e_n(I)\cdot\overleftarrow{q}^{(n)}(\mu) + \overleftarrow{h}^{(n)}, \qquad \langle \overleftarrow{h}^{(n)}, \vec{e}_n\rangle = 0$$

for $n = l$, $n = l+1$. Multiplying from the right both parts of the last equation by $B^{(n)}$ as often as necessary and taking the convolution of the result with $\vec{E}_n$ we get

$$\lambda^{-n(n\pm 1)}\langle \overleftarrow{q}^{(n)}(\nu_l)[B^{(n)}(\mu)]^{n\pm 1}, \vec{E}_n\rangle = e_n(I) + \langle \overleftarrow{h}^{(n)}[\lambda^{-n}\cdot B^{(n)}(\mu)]^{n\pm 1}, \vec{E}_n\rangle,$$

where from the estimate (A.18) we have

$$\begin{aligned}|\langle \overleftarrow{h}^{(n)}[\lambda^{-n}\cdot B^{(n)}(\mu)]^{n\pm 1}, \vec{E}_n\rangle| &\leqslant \|\overleftarrow{h}^{(n)}\|_{\leftarrow}\cdot[1-(R')^{-1}]^{n\pm 1}\|\vec{E}_n\|_{\rightarrow}\\ &\leqslant 2\cdot(R')^2[1-(R')^{-1}]^{n\pm 1} \leqslant e_n(I)\cdot 2\cdot(R')^3[1-(R')^{-1}]^{n\pm 1},\end{aligned}$$

since $\|\overleftarrow{h}^{(n)}\|_{\leftarrow} \leqslant \|\overleftarrow{q}^{(n)}(\nu)\|_{\leftarrow} + \|e_n(I)\cdot\overleftarrow{q}^{(n)}(\mu)\|_{\leftarrow} \leqslant 2\cdot R'$, $\|\vec{E}_n\|_{\rightarrow} \leqslant R'$, $e_n(I), \geqslant (R')^{-1}$. This implies the following estimates for Ψ:

$$\begin{aligned}e_l(I)\cdot(1-c_{51}\cdot\theta^{l+1}) &\leqslant \Psi(I,l,l+1) \leqslant e_l(I)\cdot(1+c_{51}\cdot\theta^{l+1}),\\ e_{l+1}(I)\cdot(1-c_{51}\cdot\theta^{l}) &\leqslant \Psi(I,l+1,-l-1) \leqslant e_{l+1}(I)\cdot(1+c_{51}\cdot\theta^{l}),\end{aligned}$$

where for brevity we have made use of (5.31). Let us now sharpen these estimates, replacing θ^{l+1} by $\theta^l > \theta^{l+1}$ and substituting in (5.16) and (5.17):

$$\begin{aligned}(R_{l+1})^{-l}\cdot(1-c_{51}\cdot\theta^l)\cdot e_l(I) &\leqslant \Psi_{l(l+1)}(I) \leqslant (R_{l+1})^{l}\cdot(1+c_{51}\cdot\theta^l)\cdot e_l(I),\\ (R_l)^{-l-1}\cdot(1-c_{51}\cdot\theta^l)\cdot e_{l+1}(I) &\leqslant \Psi_{l(l+1)}(I) \leqslant (R_l)^{l+1}\cdot(1+c_{51}\cdot\theta^l)\cdot e_{l+1}(I).\end{aligned}$$

Hence we deduce that the right-hand side of the first inequality is larger than the left-hand side of the second one, while the right-hand side of the second one is larger than the left-hand side of the first. Suppose that l is such that $c_{51}\cdot\theta^l < 1$. Then the inequalities above are equivalent to (5.29) with the constant (5.30).

To prove the corollary it suffices to verify if $\Sigma \ln \mathscr{K}_l$ converges. Due to (3.13) $\ln R_l \leqslant c_{35}(\lambda_u)^{-l/2} + c_{36}(\lambda_s)^{-l/2}$. Further, $\ln(1-c_{51}\cdot\theta^l)^{-1} < 2\cdot c_{51}\cdot\theta^l$, $\ln(1+c_{51}\cdot\theta^l) < c_{51}\cdot\theta^l$ for l such that $c_{51}\cdot\theta^l < 0.5$ and the last inequality always holds implying

$$\ln\mathscr{K}_l \leqslant c_{52}(l+1)\cdot(\theta_2)^l, \qquad \theta_2 = \max\{\theta, (\lambda_u)^{-1/2}, (\lambda_s)^{-1/2}\}.$$

The majorizing series for $\Sigma \ln \mathscr{K}_l$ converges yielding

$$\mathscr{L}_n \leqslant c_{52}(\theta_2)^n\cdot(n\cdot(1-\theta_2)^{-1} + (1-\theta_2)^{-2}) \tag{5.34}$$

for $n \geqslant -(\ln\theta)^{-1}\cdot\ln(2\cdot c_{51})$. □

THEOREM 5.6. *The sequence* $\Psi_m(I)$ *of Radon–Nikodym derivatives* (5.9) *converges uniformly to* $e(I)$ *(described in the corollary to Theorem* 5.5*). There are two sequences* $\mathscr{K}'_m \to 1$, $m \to \infty$ *and* $\mathscr{K}''_m \to 1$, $m\to\infty$ *expressed only in terms of characteristics of smoothness; each term of these sequences is bigger than* 1 *and*

$$(\mathscr{K}'_m)^{-1}\cdot e(I) \leqslant \Psi_m(I) \leqslant \mathscr{K}''_m\cdot e(I). \tag{5.35}$$

COROLLARY. *Restrictions* $\hat{\mu}_{(m)}$ *onto* $\mathfrak{A}$, *the algebra of conditional probability distributions* $\mu_{(m)}$, *converge to the limit probability measure* $\hat{\mu}_0$ *on* $\mathfrak{A}$ *with Radon–*

Nikodym derivative $e(I)$ with respect to $\hat{\mu}$, i.e.

$$\frac{\mathrm{d}\hat{\mu}_0}{\mathrm{d}\hat{\mu}}(I) = e(I). \tag{5.36}$$

where the function $e(I)$ is described in the corollary to Theorem 5.5.

Proof. Suppose that $l^2 \leqslant m \leqslant l\cdot(l+1)$, and try to estimate $\Psi(I, l, r)$ which is asymptotically equivalent to $\Psi_n(I)$

$$\begin{aligned}&\lambda^{-m}\langle \overleftarrow{q}^{(l+1)}(v_I)[B^{(l+1)}(\mu)]^r \pi_{\overrightarrow{l+1,l}}[B^{(l)}(\mu)]^{l-r}, \vec{E}_l\rangle \\ &\quad = \lambda^{-m}\langle \overleftarrow{q}^{(l+1)}(v_I), [B^{(l+1)}(\mu)]^r \sigma_{\overleftarrow{l+1,l}}[B^{(l)}(\mu)]^{l-r}\vec{E}_l\rangle.\end{aligned}$$

We shall multiply column vectors by matrices B. We have

$$\begin{aligned}&\vec{E}_l = \vec{e}_l + \vec{\varepsilon}_l, \qquad \|\vec{\varepsilon}_l\|_{\rightarrow} \leqslant R' - 1, \qquad \vec{\varepsilon}_l \in \mathscr{H}_l, \\ &\exp\{l\cdot(l-r)\ln\lambda\}[B^{(l)}(\mu)]^{l-r}(\vec{e}_l + \vec{\varepsilon}_l) = \vec{e}_l + \vec{h}_l, \\ &\|\vec{h}_l\|_{\rightarrow} \leqslant [1-(R')^{-2}]^{l-r}\cdot(R'-1), \qquad \vec{h}_l \in \mathscr{H}_l.\end{aligned}$$

Therefore from (5.28), (5.25) and (5.29) we have

$$\begin{aligned}&\sigma_{\overleftarrow{l+1,l}}\vec{h}_l \in \mathscr{H}_{l+1}, \qquad \|\sigma_{\overleftarrow{l+1,l}}\vec{h}_l\|_{\rightarrow} \leqslant \mathscr{K}_l\cdot(R'-1)\cdot[1-(R')^{-2}]^{l-r}, \\ &(\mathscr{K}_l)^{-1}\cdot\vec{e}_{l+1} \leqslant \sigma_{\overleftarrow{l+1,l}}\vec{e}_l \leqslant \mathscr{K}_l\cdot\vec{e}_{l+1}.\end{aligned}$$

Further, multiplying the vector $\sigma_{\overleftarrow{l+1,l}}\vec{h}_l$ and the inequality for $\sigma_{\overleftarrow{l+1,l}}\cdot\vec{e}_l$ by $\lambda^{(l+1)r}[B^{(l+1)}(\mu)]^r$ and convoluting with the row vector $\overleftarrow{q}^{(l+1)}(v_I)$ we get

$$\begin{aligned}&\Psi(I,l,r) \leqslant \mathscr{K}_l\cdot e_{l+1}(I) + \mathscr{K}_l\cdot(R'-1)\cdot(\theta_3)^l\cdot R', \\ &(\mathscr{K}_l)^{-1}\cdot e_{l+1}(I) - \mathscr{K}_l\cdot(R'-1)\cdot(\theta_3)^l\cdot R' \leqslant \Psi(I,l,r),\end{aligned}$$

where $\theta_3 = 1-(R')^{-2}$. Since $[e_{l+1}(I)]^{-1} \leqslant R'$, the last inequality may be written as

$$(\mathscr{K}_l)^{-1}\cdot e_{l+1}(I)\cdot(1-(\mathscr{K}_l)^2\cdot(R'-1)\cdot(R')^2\cdot(\theta_3)^l) \leqslant \Psi(I,l,r).$$

For sufficiently large l the expression in brackets is positive. Then on the left-hand side of the inequality we can estimate $e_{l+1}(I)$ from below due to (5.32) in terms of $e(I)$ and substitute the result into the estimation from below (5.16) for $\Psi_n(I)$. We have

$$(R_{l+1})^{-r}\cdot(R_l)^{-l+r}\cdot(\mathscr{K}_l)^{-1}\cdot\exp\{-\mathscr{L}_{l+1}\}\cdot(1-c_{53}(\mathscr{K}_l)^2\cdot(\theta_3)^l)\cdot e(I) \leqslant \Psi_m(I).$$

Hence $l = E[\sqrt{m}]$, and all factors of $e(I)$ tend to 1 as $m \to \infty$. Similarly,

$$\Psi_m(I) \leqslant (R_{l+1})^r(R_l)^{l-r}\cdot\mathscr{K}_l\cdot\exp\{\mathscr{L}_{l+1}\}\cdot(1+c_{53}(\theta_3)^l)\cdot e(I).$$

Thus for $m \leqslant (E[\sqrt{m}])^2 + E[\sqrt{m}]$ we can put

$$\begin{aligned}&\mathscr{K}'_m = (R_{l+1})^r\cdot(R_l)^{l-r}\cdot\mathscr{K}_l\cdot\exp\{\mathscr{L}_{l+1}\}\cdot(1-c_{53}(\mathscr{K}_l)^2\cdot(\theta_3)^l), \\ &l = E[\sqrt{m}], \qquad r = m - l^2,\end{aligned} \tag{5.37}$$

only if the expression in brackets is positive. Otherwise we can take $\mathscr{K}'_m = (R')^2$. Similarly

$$\mathscr{K}''_m = (R_{l+1})^r\cdot(R_l)^{l-r}\cdot\mathscr{K}_l\cdot\exp\{\mathscr{L}_{l+1}\}\cdot(1+c_{53}(\theta_3)^l)$$

The case when $l\cdot(l+1) \leqslant m \leqslant (l+1)^2$ is worked out on exactly the same lines.

$$\mathscr{K}'_m = (R_{l+1})^r (R_l)^{l+1-r}\cdot\mathscr{K}_l\cdot\exp\{\mathscr{L}_{l+1}\}(1 - c_{53}(\mathscr{K}_l)^2(\theta_3)^{l+1}), \qquad l = E[\sqrt{m}],$$

$$\mathscr{K}''_m = (R_{l+1})^r\cdot(R_l)^{l+1-r}\cdot\mathscr{K}_l\cdot\exp\{\mathscr{L}_{l+1}\}(1 + c_{53}(\theta_3)^{l+1}), \qquad r = m - l\cdot(l+1).$$

The statement of the Corollary follows automatically from that of the theorem. The proven convergence is the convergence of the sequence of measures in the strongest sense.

5.3. THE WEAK LIMIT $\mu_0\{\cdot\}$ OF THE SEQUENCE OF MEASURES $\mu_{(m)}\{\cdot\}$

Let us pass on to computation of the asymptotics of values

$$\mu_{(m)}\{\Gamma(I_n, J_r)\} = \mu_{(m)}\{\mathscr{D}(\infty)\cap\Gamma(I_n, J_r)\} \tag{5.38}$$

of measures $\mu_{(m)}$ supported on $\mathscr{D}(\infty)\cap\mathscr{D}(-m) = \mathscr{D}(\infty, -m) \subset \mathscr{D}(\infty)$.

LEMMA 5.7. *For $r > 0$ and any rectangle $\Gamma(I_n, J_r)$ we have*

$$\mu_{(m)}\{\Gamma(I_n, j_1 j_2\cdots j_r)\} = \mu_{(m-1)}\{\Gamma(I_n j_1, j_2\cdots j_r)\} \textit{ for } m \geqslant 1 \tag{5.39}$$

$$\mu_{(m)}\{\Gamma(I_n, J_r)\} = \mu_{(m-r)}\{\Gamma(I_n J_r,)\} \textit{ for } m \geqslant r. \tag{5.40}$$

Proof. The formula (5.39) follows from the relation (5.3) of $\mu_{(m)}$ and μ, the main property $\mu = \lambda^{-1}\cdot\mu^{(1)} = \lambda^{-1}\cdot\mu T^{-1}$ and (5.38).

THEOREM 5.8. *For any rectangle $\Gamma = \Gamma(I_n, J_r)$ there is a limit of values $\mu_{(m)}\{\Gamma\}$ which will be denoted by $\mu_0\{\Gamma\}$. For a fixed r this convergence is uniform; for $m > r$ we have*

$$(\mathscr{K}'_{m-r})^{-1}\cdot\mu_0\{\Gamma\} \leqslant \mu_{(m)}\{\Gamma\} \leqslant \mathscr{K}''_{m-r}\cdot\mu_0\{\Gamma\}. \tag{5.41}$$

The limit $\mu_0\{\Gamma\}$ satisfies

$$\mu_0\{\Gamma(I_n, J_r)\} = \mu_0\{\Gamma(I_n j_1, j_2\cdots j_r)\} = \mu_0\{\Gamma(I_n J_r,)\}. \tag{5.42}$$

Proof. Let $r = 0$, $\Delta(\Gamma) = \{L\in(\mathbb{N}_k)_{-\infty}^{-1} : l_s = i_s, 1 \leqslant s \leqslant n\}$. Then

$$\mu_{(m)}\{\Gamma\} = \int_{\Delta(\Gamma)} \hat{\mu}\{\mathrm{d}L\}\cdot\Psi_m\{L\}$$

and inequalities (5.41) with $r = 0$ are obtained by integrating inequalities (5.35) where

$$\mu_0\{\Gamma\} = \int_{\Delta(\Gamma)} \hat{\mu}\{\mathrm{d}L\}\cdot e(L). \tag{5.43}$$

From (5.41) it follows that $\mu_{(m)}\{\Gamma\} \to \mu_0\{\Gamma\}$. Now let $r \geqslant 1$. Since the right-hand side of (5.40) tends to a limit as $m \to \infty$, then so does the left-hand side and deviations from the limit are estimated by constants $\mathscr{K}'_{m-r}$ and $\mathscr{K}''_{m-r}$.

LEMMA 5.9. *We have*

$$\mu_{(m)}\{\Gamma(I_n, J_{r-1})\} = \sum_{1\leqslant j\leqslant k} \mu_{(m)}\{\Gamma(I_n, J_{r-1} j)\} \tag{5.44}$$

for $r \geqslant 1$ and $m \geqslant r$ and

$$\mu_{(m)}\{\Gamma(I_{n-1}, J_r)\} = \sum_{1 \leqslant i \leqslant k} \mu_{(m)}\{\Gamma(iI_{n-1}, J_r)\} \tag{5.45}$$

for $n \geqslant 1$ and any $m > 0$. In particular

$$\sum_{1 \leqslant i \leqslant k} \mu_{(m)}\{\Gamma(i,)\} = \sum_{1 \leqslant j \leqslant k} \mu_{(m)}\{\Gamma(,j)\} = 1 \tag{5.46}$$

for $n = 0$ or $r = 0$ and $m \geqslant r$.

COROLLARY. *For limit values of $\mu_0\{\Gamma\}$ equations* (5.44), (5.45) *and* (5.46) *also hold.*

The hyperbolic set $\Omega = \mathscr{D}(\infty) \cap \mathscr{D}(-\infty)$ is the one-to-one and continuous (even homeomorphic) image of the compact space $(\mathbb{N}_k)_{-\infty}^{\infty}$ of left-and-right infinite k-valued sequences L. The cylindric sets of the form

$$\Delta(I_n, J_m) = \{L \in (\mathbb{N}_k)_{-\infty}^{\infty} : l_{-s} = i_s, 1 \leqslant s \leqslant n, l_s = j_s, 1 \leqslant s \leqslant m\} \tag{5.47}$$

are open and at the same time closed sets in the Tikhonov topology of this space and constitute its open pre-base and generate the whole σ-algebra of its Borel subsets. They also generate the algebra of cylindric sets. The simplest cylinders Δ of the form (5.47) (for brevity referred to as Π-*cylinders* or Π_m^n-*cylinders*) are in one-to-one correspondence with measurable subsets of $\mathscr{D}$ of the form

$$\Delta(I_n, J_m) = \Omega \cap S^{(n)}(I_n) \cap S^{(-m)}(J_m) = \Gamma(I_n, J_m) \cap \mathscr{D}(-\infty). \tag{5.48}$$

Therefore the correspondence between $(\mathbb{N}_k)_{-\infty}^{\infty}$ and Ω is measurable both ways.

STATEMENT 5.10. Any finite-additive normed measure on the algebra of cylindric subsets of $(\mathbb{N}_k)_{-\infty}^{\infty}$ is extendable up to a countably additive measure on the σ-algebra of Borel sets.

THEOREM 5.11. *On $\mathscr{D}$, there exists a unique normed countably additive measure μ_0 which takes on rectangles Γ limit values $\mu_0\{\Gamma\}$ constructed in Theorem 5.8. Its support is Ω and*

$$\mu_0\{\Gamma(I_n, J_m)\} = \mu_0\{\mathscr{D}(-\infty) \cap \Gamma(I_n, J_m)\}. \tag{5.49}$$

The measure $\mu_0\{\cdot\}$ as a measure on the compact $\Omega = T\Omega$ is T-invariant and

$$\mu_0\{H\} = \mu_0\{T^{-1}H\} = \mu_0\{TH\} \tag{5.50}$$

for any measurable set $H \subset \Omega$.

Let a normed eigenmeasure μ on the horseshoe and conditional probability measures $\mu_{(m)}$ be considered as measures on $\mathscr{D}$ with the support in $\mathscr{D}(\infty)$. Then μ_0 is the $*$-weak limit of measures $\mu_{(m)}$, i.e. for any continuous function $f(z)$ on $\mathscr{D}$

$$\lim_{m \to \infty} \int_{\mathscr{D}} f(z)\mu_{(m)}\{dz\} = \int_{\mathscr{D}} f(z)\mu_0\{dz\} = \int_{\Omega} f(z)\mu_0\{dz\}. \tag{5.51}$$

We omit the proof since it is obvious.

6. Some Properties of the Constructed Limit Probability Distributions on a Smale Horseshoe

6.1. THE T-INVARIANT CONDITIONAL PROBABILITY DISTRIBUTIONS $P\{\cdot|I\}$ ON EXPANDING FIBRES OF THE HYPERBOLIC SET Ω

Let us define properties of the natural T-invariant measure μ_0 constructed in §5 from conditional probability distributions $\mu_{(m)}$ by passage to the limit.

LEMMA 6.1. *For any component $S^{(-r)}(J_r)$ of $\mathscr{D}(-r)$ the conditional probabilities $P_m\{S^{(-r)}(J_r)|I\} = P_m\{\Gamma(,J_r)|I\}$ defined by the formula* (5.2) *satisfy*

$$(R')^{-2}\cdot P_m\{S^{(-r)}(J_r)|I\} \leqslant P_m\{S^{(-r)}(J_r)|L\} \leqslant (R')^2\cdot P_m\{S^{(-r)}(J_r)|I\} \tag{6.1}$$

for any I, $L\in\mathfrak{A}$ and

$$(R_n)^{-2}\cdot P_m\{S^{(-r)}(J_r)|I\} \leqslant P_m\{S^{(-r)}(J_r)|L\} \leqslant (R_n)^2\cdot P_m\{S^{(-r)}(J_r)|I\} \tag{6.2}$$

for I and L such that $i_n = l_n, \ldots, i_1 = l_1$.

LEMMA 6.2. *For any component $S^{(-r)}(J_r)$ and any number $I\in\mathfrak{A}$ the following presentation of the conditional probability holds for $m \geqslant r$:*

$$P_m\{S^{(-r)}(J_r)|I\} = \lambda^{-r}\cdot P\{S^{(-r)}(J_r)|I\}\cdot\Psi_{m-r}(IJ_r)\cdot[\Psi_m(I)]^{-1}, \tag{6.3}$$

where the Radon–Nikodym densities Ψ are defined by the formula (5.5).

LEMMA 6.3. *For the fixed component $S^{(-r)}(J_r)$ the sequence of conditional probabilities converges uniformly to the positive limit*

$$P_0\{S^{(-r)}(J_r)|I\} = \lambda^{-r}P\{S^{(-r)}(J_r)|I\}\cdot e(IJ_r)\cdot[e(I)]^{-1}, \tag{6.4}$$

where the Radon–Nikodym derivative $e(I)$ of $\hat{\mu}_0$ with respect to μ is described in the Corollary to Theorem 5.5. Limit values $P_0\{S^{(-r)}(J_r)|I\}$ as functions in I satisfy (6.1) *and* (6.2) *and $P_0\{\mathscr{D}|I\} = 1$. For different r they are compatible, i.e.*

$$P_0\{S^{(-r)}(J_r)|I\} = \sum_{1\leqslant j\leqslant k} P_0\{S^{(-r-1)}(J_rj)|I\}. \tag{6.5}$$

COROLLARY. *The function $e(I)$ satisfies the family of identities*

$$\lambda\cdot P\{S^{(-r)}(J_r)|I\}\cdot e(IJ_r) = \sum_{1\leqslant j\leqslant k} P\{S^{(-r-1)}(J_rj)|I\}\cdot e(IJ_rj). \tag{6.6}$$

THEOREM 6.4. *For a T-invariant measure μ_0 with support in Ω there is a transition probability distribution $P_0(I, \mathrm{d}J)$ from the space $(\mathbb{N}_k)^{-1}_{-\infty}$ of sequences $(I) = (\cdots i_n\cdots i_1)$ which denumerate leaves $\mathfrak{F}(I)$ with the algebra $\mathfrak{A}$ of Borel sets into the algebra $\mathfrak{B}$ of Borel sets of the space $(\mathbb{N}_k)^{\infty}_1$ of sequences $(J) = (j_1\cdots j_r\cdots)$ denumerating leaves $\mathfrak{G}(J)$ of the fibration $\mathscr{D}(-\infty)$. These transition probabilities, $P_0(I,\cdot)$, are uniform continuous in I, i.e.*

$$(R')^{-2}\cdot P_0(I, B) \leqslant P_0(L, B) \leqslant (R')^2\cdot P_0(I, B) \tag{6.7}$$

for all $B\in\mathfrak{B}$ and I, $L\in\mathfrak{A}$,

$$(R_n)^{-2}\cdot P_0(I, B) \leqslant P_0(L, B) \leqslant (R_n)^2 P_0(I, B) \tag{6.8}$$

for all $B \in \mathfrak{B}$ *and* $I, L \in \mathfrak{A}$ *such that* $i_s = l_s$ *for* $1 \leqslant s \leqslant n$. *The transition distribution* $P_0(I, \mathrm{d}J)$ *defines for* μ_0 *the family of conditional probability distributions* $P_0\{\cdot|I\}$ *according to the formula*

$$P_0\{H|I\} = P_0(I, H_I), \qquad H_I = \{J \in (\mathbb{N}_k)_1^\infty : \mathfrak{F}(I) \cap \mathfrak{G}(J) \in H\} \tag{6.9}$$

so that the integral presentation

$$\mu_0\{H\} = \int \hat{\mu}\{\mathrm{d}I\} \cdot e(I) \cdot P_0(I, H_I) \tag{6.10}$$

holds, where $e(I)$ *is the Radon–Nikodym derivative* (5.36) *of* $\hat{\mu}_0$ *with respect to* $\hat{\mu}$. *The family of conditional distributions* $P_0\{\cdot|I\}$ *with respect to the fibration of* Ω *by fibres* $\mathfrak{F}(I) \cap \mathscr{D}(-\infty)$ *is T-invariant, meaning that*

$$P_0\{H|Ij\} = [P_0\{S^{(-1)}(j)|I\}]^{-1} \cdot P_0\{T^{-1}H|I\} \tag{6.11}$$

for any $H \subset \mathfrak{F}(Ij)$, *where* $T^{-1}[\mathfrak{F}(Ij) \cap \Omega] = \mathfrak{F}(I) \cap S^{(-1)}(j) \cap \Omega$.

REMARK. We can show that with respect to the restriction $\hat{\mu}_0$ of μ_0 onto $\mathfrak{B}$ transition distributions $P_0(I, \cdot)$ have densities $p_0(J|I)$ uniformly continuous in I together with $\ln p_0(J|I)$. They are T-invariant in the sense of Definition 3.1.

6.2. WEAK BERNOULLI PARTITION FOR THE T-INVARIANT MEASURE μ_0

Our proof of weak Bernoulli partition of the hyperbolic set Ω onto $\Omega \cap S^{(-1)}(j)$, where $1 \leqslant j \leqslant k$ and $S^{(-1)}(j)$ are components of $\mathscr{D}(-1)$, makes use of a matrix technique similar to that used in the previous sections except that we take $P_0\{\cdot|I\}$ instead of the family of conditional distributions $P\{\cdot|I\}$ on the expanding fibration of Ω.

DEFINITION 6.1. The class of all measures on Ω with T-invariant conditional probability distributions $P_0\{\cdot|I\}$ on fibres $\mathfrak{F}(I) \cap \mathscr{D}(-\infty)$ is referred to as the *class* $\mathfrak{S}_0$.

Any measure $\nu \in \mathfrak{S}_0$ is completely defined by the formula

$$\nu\{\cdot\} = \int \hat{\nu}\{\mathrm{d}I\} \cdot P_0\{\cdot|I\} \tag{6.12}$$

by its restriction $\hat{\nu}$ on the algebra $\mathfrak{A}$ of subsets of the space $(\mathbb{N}_k)_{-\infty}^{-1}$ of numbers of fibres $\mathfrak{F}(I) \cap \mathscr{D}(-\infty)$.

STATEMENT 6.5. The class $\mathfrak{S}_0$ is T-invariant, i.e. $\nu \in \mathfrak{S}_0$ implies $\nu T^{-1} = \nu^{(1)} \in \mathfrak{S}_0$.

For brevity, set

$$\Omega^{(n)}(I_n) = \Omega \cap S^{(n)}(I_n), \qquad \Omega^{(-m)}(J_m) = \Omega \cap S^{(-m)}(J_m),$$
$$\Omega(I_n, J_m) = \Omega^{(n)}(I_n) \cap \Omega^{(-m)}(J_m). \tag{6.13}$$

Any measure ν of class $\mathfrak{S}_0$ is completely defined by its values on basic sets $\Omega^{(n)}(I_n)$. To each measure $\nu \in \mathfrak{S}_0$ assign the sequence of k^n-dimensional row vectors with components

$$q_{I_n}^{(n)}(\nu) = \nu\{\Omega^{(n)}(I_n)\} = \nu\{S^{(n)}(I_n)\}, \qquad \bar{q}^{(n)}(\nu) = \pi_n(\hat{\nu}), \tag{6.14}$$

i.e. measures of basic sets.

DEFINITION 6.2. The rectangular matrix indexed by sets $(I_n) = (i_n \cdots i_1)$ and $(L_m) = (l_m \cdots l_1)$ with elements such that

$$b_{I_n,L_m}^{(n,m)}(\nu) = [\nu\{\Omega^{(n)}(I_n)\}]^{-1} \cdot \nu\{\Omega(I_n, L_m)\} \quad \text{for} \quad \nu\{\Omega^{(n)}(I_n)\} \neq 0 \tag{6.15}$$

$$b_{I_n,L_m}^{(n,m)}(\nu) = b_{I_n,L_m}^{(n,m)}(\mu_0) \quad \text{for} \quad \nu\{\Omega^{(n)}(I_n)\} = 0 \tag{6.16}$$

is referred to as the *matrix* $B^{(n,m)}(\nu)$ *of ranks* n and m.

STATEMENT 6.6. The matrix $B^{(n)}(\nu)$ for $\nu \in \mathfrak{S}_0$ is a strictly positive stochastic matrix, i.e.

$$b_{I_n,L_n}^{(n)}(\nu) > 0, \qquad \sum_{L_n} b_{I_n,L_n}^{(n)}(\nu) = 1 \tag{6.17}$$

for any I_n, L_n. If $\nu^{(n)} = \nu T^{-n}$ then

$$\nu^{(n)}\{\Omega^{(n)}(L_n)\} = \sum_{I_n} b_{I_n,L_n}^{(n)}(\nu) \cdot \nu\{\Omega(I_n, L_n)\}. \tag{6.18}$$

COROLLARY. *If* $\nu^{(n)}\{\cdot\} = \nu\{T^{-n}\cdot\}$ *then*

$$q_{L_n}^{(n)}(\nu^{(n)}) = \sum_{I_n} q_{I_n}^{(n)}(\nu) \cdot b_{I_n,L_n}^{(n)}(\nu). \tag{6.19}$$

As in §4, the matrices $B(\nu)$ are close to $B(\mu_0)$ and square matrices $B^{(n)}$ possess almost the same nice ergodic properties. Besides, they are stochastic, which enables us to apply stronger estimates.

LEMMA 6.7. *For any measure* $\nu \in \mathfrak{S}_0$ *and any* I_n, L_m, H_m *we have*

$$(R')^{-2} \cdot b_{I_n,L_m}^{(n,m)}(\nu) \leqslant b_{I_n,H_m}^{(n,m)}(\nu) \leqslant (R')^2 \cdot b_{I_n,L_m}^{(n,m)}(\nu), \tag{6.20}$$

where R' *is defined by* (3.11). *For any* $\nu \in \mathfrak{S}_0$ *and* μ_0 *we have*

$$(R_n)^{-2} \cdot b_{I_n,L_m}^{(n,m)}(\mu_0) \leqslant b_{I_n,L_m}^{(n,m)}(\nu) \leqslant (R_n)^2 \cdot b_{I_n,L_m}^{(n,m)}(\mu_0). \tag{6.21}$$

LEMMA 6.8. *The estimate*

$$\begin{aligned} |\nu\{\Omega(I_n, J_m)\} - \mu_0\{\Omega(I_n, J_m)\}| &\leqslant (R_n^2 - 1) \cdot \mu_0\{\Omega(I_n, J_m)\} + \\ &+ (R_n)^2 \cdot \mu_0\{\Omega^{(-m)}(J_m) | \Omega^{(n)}(I_n)\} \cdot |\nu\{\Omega^{(n)}(I_n)\} - \mu_0\{\Omega^{(n)}(I_n)\}| \end{aligned} \tag{6.22}$$

holds for any measure $\nu \in \mathfrak{S}_0$.

Let us recall two definitions.

The partition $\mathcal{R}$ of the T-invariant set Ω is a weak Bernoulli partition with respect to the measure μ_0, $\mu_0\{\Omega\} = 1$, see [18, 19], if for any $\varepsilon > 0$ there exists $N(\varepsilon)$ such that the induced partitions

$$\mathcal{D} = \mathcal{R} \vee T^{-1}\mathcal{R} \vee \cdots \vee T^{-s}\mathcal{R}, \qquad \mathfrak{H} = T^{-t}\mathcal{R} \vee \cdots \vee T^{-t-r}\mathcal{R} \tag{6.23}$$

are ε-independent for all $s \geqslant 0$, $r \geqslant 0$, $t \geqslant s + N(\varepsilon)$.

Partitions $\mathfrak{D} = \{\mathfrak{d}(I)\}$ and $\mathfrak{H} = \{\mathfrak{h}(J)\}$ of the set Ω, $\mu_0\{\Omega\} = 1$, are called ε-independent relative to μ_0 if

$$\sum_I \sum_J |\mu_0\{\mathfrak{h}(J) \cap \mathfrak{d}(I)\} - \mu_0\{\mathfrak{h}(J)\} \cdot \mu_0\{\mathfrak{d}(I)\}| < \varepsilon. \tag{6.24}$$

THEOREM 6.9. *The partition $\mathcal{R}$ of the hyperbolic set Ω into sets $\Omega^{(-1)}(j) = \Omega \cap S^{(-1)}(j)$, where $1 \leqslant j \leqslant k$ and each $S^{(-1)}(j)$ is a component of $\mathcal{D}(-1)$, is a weak Bernoulli partition with respect to T and μ_0.*

Proof. Since μ_0 is T-invariant then the partitions $T^t\mathfrak{D}$ and $T^t\mathfrak{H}$ are ε-independent with respect to μ_0 if $\mathfrak{D}$ and $\mathfrak{H}$ are, and vice versa. We therefore check that the ε-independence of partitions

$$\mathfrak{D} = T^{N+p}\mathcal{R} \vee \cdots \vee T^{N+p+s}\mathcal{R}, \qquad \mathfrak{H} = \mathcal{R} \vee \cdots \vee T^{-r}\mathcal{R}, \tag{6.25}$$

where $r \geqslant 0$, $s \geqslant 0$, $p = t - s - N \geqslant 0$, obtained under the T^t-action from partitions (6.23) and $N = n^2 + 1$ is such that $R_n \cdot \theta < 1$ and

$$(R_n^2 - 1) + 2R_n^2 \cdot (R_n \cdot \theta)^n + 2R_n^3 \cdot (R_n - 1) \cdot (1 - R_n \theta)^{-1} < \varepsilon. \tag{6.26}$$

The atoms of the partition $\mathfrak{H}$ are $\mathfrak{h}(J_{r+1}) = \Omega^{(-r-1)}(J_{r+1})$. The atoms of $\mathfrak{D}$ are $\mathfrak{d}(I_{s+1}) = T^{n^2+p}[\Omega^{(s+1)}(I_{s+1})]$ and

$$T^p\Omega^{(s+1)}(I_{s+1}) = \bigcup_{J_p} \Omega^{(p+s+1)}(I_{s+1}J_p). \tag{6.27}$$

For any $(I_{s+1}) = (i_{s+1} \cdots i_1)$ we construct the measure

$$\nu_{I_{s+1}}\{\cdot\} = \mu_0\{\cdot \mid T^p\Omega^{(s+1)}(I_{s+1})\}. \tag{6.28}$$

As follows from (6.27) the measure $\nu_{I_{s+1}}$ belongs to the class $\mathfrak{S}_0$, since $T^p\Omega^{(s+1)}(I_{s+1})$ is the union of expanding fibres $\mathfrak{F}(L)$:

$$\{L \in (\mathbb{N}_k)_{-\infty}^{-1} : l_1 = j_1, \ldots, l_p = j_p, l_{p+1} = i_1, \ldots, l_{p+s+1} = i_{s+1}\}.$$

Therefore, $\nu_{I_{s+1}}^{(mn)} \in \mathfrak{S}_0$ and

$$\nu_{I_{s+1}}^{(n^2)}\{H\} = \frac{\mu_0\{T^{-n^2}H \cap T^p\Omega^{(s+1)}(I_{s+1})\}}{\mu_0\{T^p\Omega^{(s+1)}(I_{s+1})\}} = \frac{\mu_0\{H \cap T^{n^2+p}\Omega^{(s+1)}(I_{s+1})\}}{\mu_0\{T^{n^2+p}\Omega^{(s+1)}(I_{s+1})\}}$$
$$= \mu_0\{H \mid T^{n^2+p}\Omega^{(s+1)}(I_{s+1})\} = \mu_0\{H \mid \mathfrak{d}(I_{s+1})\} \tag{6.29}$$

due to the T-invariance of μ_0 for any $H \in \mathfrak{C}$. By the corollary (6.29), from Statement 6.6 we have

$$\vec{q}^{(n)}(\nu_{I_{s+1}}^{(n^2)}) = \vec{q}^{(n)}(\nu_{I_{s+1}}) \cdot \prod_{m=0}^{n-1} B^{(n)}(\nu_{I_{s+1}}^{(mn)}).$$

Matrices $B^{(n)}(\nu_{I_{s+1}}^{(nm)})$ are in proximity with $B^{(n)}(\mu_0)$. By Theorem A.10 for the stochastic matrices we have

$$|\vec{q}^{(n)}(\nu_{I_{s+1}}^{(n^2)}) - \vec{q}^{(n)}(\mu_0)| = \|\vec{q}^{(n)}\nu_{I_{s+1}}^{(n^2)} - \vec{q}^{(n)}(\mu_0)\|_{-}$$
$$\leqslant 2(R_n \cdot \theta)^n + 2 \cdot R_n \cdot (R_n - 1)(1 - R_n \cdot \theta)^{-1}. \tag{6.30}$$

We now proceed to the first step of the estimation for the fixed $\mathfrak{d}(I_{s+1})$. We have

$$\sum_{J_{r+1}} |\mu_0\{\mathfrak{h}(J_{r+1})|\mathfrak{d}(I_{s+1})\} - \mu_0\{\mathfrak{h}(J_{r+1})\}|$$

$$= \sum_{J_{r+1}} |\nu^{(n2)}_{I_{s+1}}\{\Omega^{(-r-1)}(J_{r+1})\} - \mu_0\{\Omega^{(-r-1)}(J_{r+1})\}|$$

$$\leqslant \sum_{J_{r+1}, L_n} |\nu^{(n2)}\{\Omega(L_n, J_{r+1})\} - \mu_0\{\Omega(L_n, J_{r+1})\}|$$

$$\leqslant \sum_{J_{r+1}, L_n} [(R_n^2 - 1)\mu_0\{\Omega(L_n, J_{r+1})\} +$$

$$+ R_n^2 \cdot \mu_0\{\Omega^{(-r-1)}(J_{r+1})|\Omega^{(n)}(L_n)\}|\nu^{(n2)}_{I_{s+1}}\{\Omega^{(n)}(L_n)\} - \mu_0\{\Omega^{(n)}(L_n)\}|]$$

$$\leqslant (R_n^2 - 1) + R_n^2 \cdot \sum_{L_n} |\nu^{(n2)}_{I_{s+1}}\{\Omega^{(n)}(L_n)\} - \mu_0\{\Omega^{(n)}(L_n)\}|$$

$$\leqslant R_n^2 - 1 + R_n^2 \cdot \{2(R_n \cdot \theta)^n + 2 \cdot R_n \cdot (R_n - 1) \cdot (1 - R_n \cdot \theta)^{-1}\} \leqslant \varepsilon$$

where we have consequently made use of the presentation (6.29) for the conditional measure, estimated differences along stripes in terms of differences along rectangles, made use of the estimation (6.22) of Lemma 6.8 with the subsequent summation of these estimates and estimated the sum of modules of differences of matrix stripes with respect to (6.30). The obtained expression is less than ε by the choice of n, see (6.26). The second part of the estimate is trivial, since

$$\sum_{J_{r+1}} |\mu_0\{\mathfrak{h}(J_{r+1})|\mathfrak{d}(I_{s+1})\} - \mu_0\{\mathfrak{h}(J_{r+1})\}| \leqslant \varepsilon,$$

$$\sum_{J_{r+1}} |\mu_0\{\mathfrak{h}(J_{r+1}) \cap \mathfrak{d}(I_{s+1})\} - \mu_0\{\mathfrak{h}(J_{r+1})\} \cdot \mu_0\{\mathfrak{d}(I_{s+1})\}| \leqslant \varepsilon \cdot \mu_0\{\mathfrak{d}(I_{s+1})\}.$$

Summing the last inequality with respect to all I_{s+1} we get the required relation (6.24).

6.3. THE EIGENFUNCTION $e(I)$ OF A SMALE HORSESHOE

In the constructions of this and preceding sections we have essentially made use of the function $e(I)$ which first appeared in the Corollary to Theorem 5.5. In the Corollary to Lemma 6.3 we established the system of identities (6.6) for $e(I)$. We now show that these properties are characteristic.

DEFINITION 6.3. A non-negative function $e(I)$ continuous (in the Tikhonov topology) and defined on the space $(\mathbb{N}_k)^{-1}_{-\infty}$ of indices of expanding leaves of the horseshoe is the *eigenfunction of the fibration of the limit set of the Smale horseshoe* if it satisfies

$$\lambda \cdot e(I) = \sum_{1 \leqslant j \leqslant k} P\{S^{(-1)}(j)|I\} \cdot e(Ij) \tag{6.31}$$

with $\lambda = \lambda_e > 0$, where $P\{\cdot|I\}$ is a T-invariant family of conditional probability distributions on the expanding leaf of the horseshoe constructed in Theorem 3.7.

REMARK. The function $e(I)$ constructed in the Corollary to Theorem 5.5 is an eigenfunction since it is positive, continuous, and due to Lemma 6.3 satisfies (6.31) which is the special case of (6.6) for $r = 0$.

The system of coefficients in (6.31) is a continual analogue of the matrix with non-negative elements, and as we shall show, the Frobenius–Perron theory holds for it.

LEMMA 6.10. *For any eigenfunction* $e(I)$ *of a horseshoe the system of identities*

$$\lambda^m \cdot e(I) = \sum_{J_m} P\{S^{(-m)}(J_m)|I\} \cdot e(IJ_m) \tag{6.32}$$

and also the system of identities (6.6) *hold for* $\lambda = \lambda_e$.

COROLLARY. *Each eigenfunction* $e(I)$ *is strictly positive and* $\min e(I) > 0$.

LEMMA 6.11. *There is the unique up to a constant factor eigenfunction* $e(I)$.

THEOREM 6.12. *The function* $e(I)$ *constructed in the Corollary to Theorem* 5.5 *satisfies* (6.31) *and is the unique eigenfunction of the horseshoe up to a constant factor. The corresponding eigenfunction coincides with that of the horseshoe.*

The proof follows from the remark to Definition 6.3 and Lemma 6.11.

7. Evolution of Probability Distributions on a Smale Horseshoe

7.1. ASYMPTOTIC INEQUALITIES FOR MEASURES AND INTEGRALS

In §1 we described the decreasing sequence of domains $\mathcal{D}(n)$, on each of which the action of powers $T^{-1}, \ldots, T^{-n}$ of the inverse mapping T^{-1} and also the sequence of corresponding pre-images $\mathcal{D}(-n) = T^{-n}\mathcal{D}(n)$ are defined.

To any probability distribution ν on the measurable space $(\mathcal{D}, \mathfrak{C})$ the sequence of measures

$$\nu^{(n)}\{H\} = \nu\{T^{-n}[H \cap \mathcal{D}(n)]\} \tag{7.1}$$

is assigned, each of them with support in its domain $\mathcal{D}(n)$. Since $\mathcal{D}$ is not T-invariant, this sequence will not be normed (but only in the case where ν is not supported on $\mathcal{D}(-\infty)$). Moreover, starting from some $n = N$ all subsequent measures might become zero. The last possibility being excluded, it is convenient to consider the sequence of normed measures $\nu_n\{\cdot\}$, i.e. induced probability distributions

$$\nu_n\{H\} = [\nu\{\mathcal{D}(-n)\}]^{-1} \cdot \nu\{T^{-n}[H \cap \mathcal{D}(n)]\} \tag{7.2}$$

and the sequence of corresponding norming divisors $\nu^{(n)}\{\mathcal{D}(n)\} = \nu\{\mathcal{D}(-n)\}$. Thus the problem of the study of the evolution of the probability distributions on the Smale horseshoe differs essentially from similar problem for Anosov systems and attractors, cf. [28]. As we shall prove, $\nu_n \Rightarrow \mu$, where μ is the eigen probability distribution on the horseshoe from §4, which makes it possible to compute the asymptotics of a number of simple integrals. Another class of integrals requires to estimate the evolution of normed measures $\nu_{n,m}\{\cdot\}$, where

$$\nu_{n,m}\{H\} = [\nu\{\mathscr{D}(-n-m)\}]^{-1}\cdot\nu\{T^{-n}[H\cap\mathscr{D}(n)\cap\mathscr{D}(-m)]\} = \nu_n\{H|\mathscr{D}(-m)\} \quad (7.3)$$

for $n, m \to \infty$. We shall see that $\nu_{n,m} \Rightarrow \mu_0$, where μ_0 is the natural T-invariant measure on the hyperbolic set Ω of the horseshoe constructed in §5.

DEFINITION 7.1. The measurable subset $\Delta \subset \mathscr{D}$ is an *expanding* one if it is a continuous union of $\mathbb{Y}$-leaves in $\mathscr{D}$, i.e. $\Delta = \cup_a \Phi(a)$, and $\partial\Phi(a) \subset \varphi(U \times \partial V)$ for any a so that the set A of values of the parameter a is a compact topological space with the algebra $\mathfrak{A}$ of Borel sets and for any $m \geqslant 0$ the mapping $A \times [V \times (\mathbb{N}_k)_1^m] \to \mathscr{D}$ defined by the intersection of leaves

$$(a, v, L_m) \to \Phi(a) \cap \mathfrak{G}(v, L_m) = x(a, v, L_m) \oplus y(a, v, L_m)$$

is continuous with respect to all arguments.

DEFINITION 7.2. Let Δ be an expanding subset in $\mathscr{D}$. Write $\mathfrak{M}(\Delta)$ for the convex cone of measures ν on the algebra $\mathfrak{C}$ of Borel subsets of the initial domain $\mathscr{D}$ such that their support belongs to Δ and on leaves $\Phi(a)$ of the fibration of the set Δ they have conditional distributions with a variant of the density $\omega_\nu(y|a)$ y-Lipschitzizable, with respect to the $\mathbb{Y}$-volume on the $\mathbb{Y}$-projection of the leaf which depends continuously on values of the function $(a, v) \to \omega_\nu(y(a, v)|a)$. The union of these cones $\mathfrak{M}(\Delta)$ is referred to as the *cone* $\mathfrak{N}$. Similarly, the subcone of measures ν for which the above-mentioned continuous variant of the density is strictly positive is denoted $\mathfrak{M}^+(\Delta)$, and $\mathfrak{N}^+$ denotes their union. If $\nu\in\mathfrak{N}$ and $\nu\in\mathfrak{N}^+$, we write $\nu^{(n)}\in\mathfrak{N}(n)$ and $\nu^{(n)}\in\mathfrak{N}^+(n)$, respectively.

Since after n iterations of T we obtain k^n leaves $\Phi(a, I_n) = T^n[\Phi(a)\cap S^{(-n)}(I_n)]$ in $\mathscr{D}(n)$ from the leaf $\Phi(a)$, then $\Delta(n) = T^n[\Delta\cap\mathscr{D}(-n)]$ is also an expanding subset and by Lemma 4.3 $\nu^{(n)}\in\mathfrak{M}(\Delta(n))\subset\mathfrak{N}$ if $\nu\in\mathfrak{M}(\Delta)$. The evolution of conditional distributions $\nu^{(n)}\{\cdot|a, I_n\}$ on fibres and their densities is completely defined from formulas (3.5) and (3.9) by the evolution of the corresponding fibre $\Phi(u) \to \Phi(u, i_1) \to \cdots \to \Phi(u, I_n)$. This follows from Lemma 4.4, which implies in particular $\mathfrak{N}^+(n)\in\mathfrak{N}^+$. Note also that to the class $\mathfrak{M}^+(\mathscr{D})\subset\mathfrak{N}^+$ belongs any measure on $\mathfrak{C}$ which possesses a strictly positive variant of the density $\rho(x, y)$ with respect to the $\mathbb{Z}$-volume on $\mathscr{D}$ Lipschitzizable in variables (x, y), i.e.

$$|\rho(x_2, y_2) - \rho(x_1, y_1)| \leqslant \mathrm{Lip}_{\mathbb{X}}\,\rho\cdot|x_2 - x_1| + \mathrm{Lip}_{\mathbb{Y}}\,\rho\cdot|y_2 - y_1|.$$

This follows from (4.4), and also $\mathfrak{S} \subseteq \mathfrak{M}^+(\mathscr{D}(\infty)) \subset \mathfrak{N}^+$. The class $\mathfrak{N}^+$ is therefore sufficiently broad.

As was shown in §3 the positive densities of conditional distributions on expanding leaves tend exponentially to Lipschitzizable T-invariant densities on the fibration $\mathscr{D}(\infty)$. It remains to trace the evolution of measures $\hat{\nu}^{(n)}$ on algebras $\mathfrak{A}(n)$ of Borel sets of leaves. Since the support of $\nu^{(n)}$ is in $\mathscr{D}(n)$, then the asymptotics of values $\nu^{(n)}\{S^{(m)}(L_m)\}$ when $n\to\infty$ are of great interest, starting from $n = m$. We now show how this information enables us to judge the asymptotics of certain integrals of continuous functions.

DEFINITION 7.3. Let $h(z)$ be a continuous function on the compact set $\mathscr{D}(-s)$ where $s = s(h)$ such that

$$1 \leqslant h(z) \leqslant M(h) \tag{7.4}$$

for any $z \in \mathscr{D}(-s)$. Put

$$R_m(h) = \sup_{r \geqslant s} \max_{L_m} \max_{v, J_r} \sup_{z', z'' \in S^{(m)}(L_m) \cap \mathfrak{G}(v, J_r)} \{h(z') \cdot [h(z'')]^{-1}\}, \tag{7.5}$$

where $(L_m) = (l_m \cdots l_1) \in (\mathbb{N}_k)_{-m}^{-1}$ is the number of the component $S^{(m)}(L_m)$ and $\mathfrak{G}(v, J_r)$ is the leaf of the foliation of the component $S^{(-r)}(J_r)$, where $(J_r) = (j_1 \cdots j_r) \in (\mathbb{N}_k)_1^r$. Also let $\mathscr{H}_m(s, M, R)$ for the class of continuous functions $h(\cdot)$ satisfying (7.4) with $s(h) \leqslant s$, $M(h) \leqslant M$, $R_m(h) \leqslant R$.

It is evident that any continuous function $h(z)$ can be presented as the difference of functions satisfying (7.4) and a constant. Further, since $|z'' - z'|_{\mathbb{X}} \leqslant \rho_m$, $|z'' - z'|_{\mathbb{Y}} \leqslant \alpha \cdot \rho_m$ in (7.5), where ρ_m is the maximal $\mathbb{X}$-diameter of components $S^{(m)}(L_m)$, and $\rho_m \downarrow 0$ due to (1.30) and a function continuous on the compact set $\mathscr{D}(-s)$ is uniformly continuous, then $R_m(h) \downarrow 1$ when $m \to \infty$ assuming (7.4). The following properties of classes $\mathscr{H}_m$ are also valid:

$$\begin{aligned} &\mathscr{H}_m(s, M, R) \subset \mathscr{H}_{m'}(s', M', R'), \\ &\mathscr{H}_m(s, M, R) \mathscr{H}_m(s', M', R') \subset \mathscr{H}_m(\max\{s, s'\}, M \cdot M', R \cdot R') \end{aligned} \tag{7.6}$$

for any (m, s, R) and (m', s', R') such that $m < m'$, $s \leqslant s'$, $R \leqslant R'$, and $h \in \mathscr{H}_m(s, M, R)$ implies $hT^n \in \mathscr{H}_t(s + n, M, R)$ for any $t \geqslant \max\{m - n, 0\}$.

LEMMA 7.1. *Let $v \in \mathfrak{N}^+$, $h \in \mathscr{H}_m(s, M, R)$ and μ be an probability eigenmeasure on the expanding fibration $\mathscr{D}(\infty)$ of the horseshoe. Then for any pair of components $S^{(m)}(L_m)$ and $S^{(-r)}(J_r)$ we have*

$$\mu\{S^{(m)}(L_m)\} \cdot \int_{S(L_m, J_r)} h(z) \cdot v_n\{dz\} \lesseqgtr [R_m \cdot R_m(h)]^{\pm 1} \cdot v\{S^{(m)}(L_m)\} \cdot \int_{S(L_m, J_r)} h(w) \cdot \mu\{dw\} \tag{7.7}$$

for $r \geqslant s$, $n \geqslant N_{32}$ *and* $n \geqslant m$ *and* $<$ *corresponds to the choice of exponent* $+1$.

Let us connect with each measure $v \in \mathfrak{N}^+(n + r)$, where $0 \leqslant r \leqslant \infty$, the row vector $\vec{q}^{(n)}(v)$ with k^n components

$$q_{I_n}^{(n)}(v) = v\{S^{(n)}(I_n)\}, \qquad (I_n) = (i_n \cdots i_1) \in (\mathbb{N}_k)_{-n}^{-1}. \tag{7.8}$$

It is evident that the vector functional $\vec{q}^{(n)}(v)$ is a linear functional in v on each cone $\mathfrak{M}^+$-normed, i.e. $|\vec{q}^{(n)}(v)| = 1$, if v is so, i.e. $v\{\mathscr{D}\} = v\{\mathscr{D}(n + r)\} = 1$. In the general case we have

$$v\{\mathscr{D}\} = v\{\mathscr{D}(n)\} = v\{\mathscr{D}(n + r)\} = |\vec{q}^{(n)}(v)| = \sum_{I_n} q_{I_n}^{(n)}(v) \tag{7.9}$$

for $v \in \mathfrak{N}(n + r)$.

DEFINITION 7.4. Unlike Definition 4.3, here the $k^n \times k^n$ matrix with entries

$$a_{I_n, J_n}^{(n)} = [\mu\{S^{(n)}(I_n)\}]^{-1} \cdot \mu\{S(I_n, J_n)\},$$

where μ is an eigenmeasure of the horseshoe and $\mu \in \mathfrak{N}^+(\infty)$ is referred to as the *matrix* $A^{(n)}$. Generalizing Definition 4.4, the $k^n \times k^n$-matrix with entries

$$[v\{S^{(n)}(I_n)\}]^{-1} \cdot v\{S^{(n)}(I_n) \cap S^{(-n)}(J_n)\} \quad \text{for} \quad v\{S^{(n)}(I_n)\} \neq 0; \; a^{(n)}_{I_n, J_n} \text{ otherwise} \tag{7.10}$$

will be referred to as the *matrix* $B^{(n)}(v)$ *of rank n* generated by values of the measure $v \in \mathfrak{N}^+(n)$.

Note that $A^{(n)} = B^{(n)}(\mu)$.

LEMMA 7.2. *If two measures* $v', v'' \in \mathfrak{N}^+$ *differ by a numerical multiple, in particular* $v' = v^{(m)}$, $v'' = v_m$, *then* $B^{(n)}(v') = B^{(n)}(v'')$. *For* $r \geqslant 0$ *we have*

$$\overleftarrow{q}^{(n)}(v^{(2n+r)}) = \overleftarrow{q}^{(n)}(v^{(n+r)}) \cdot B^{(n)}(v^{(n+r)}), \tag{7.11}$$

where the matrix $B^{(n)}$ *is multiplied from the left by the row vector. For* $r \geqslant 0$ *all components of the vector* $\overleftarrow{q}^{(n)}(v^{(n+r)})$ *and all entries of* $B^{(n)}(v^{(n+r)})$ *are strictly positive.*

LEMMA 7.3. Let $v \in \mathfrak{N}^+$. *Then*

$$(R')^{-1} \cdot b^{(n)}_{I_n, J_n}(v^{(n+r)}) \leqslant b^{(n)}_{I'_n, J_n}(v^{(n+r)}) \leqslant R' \cdot b^{(n)}_{I_n, J_n}(v^{(n+r)}), \tag{7.12}$$

$$(R_n)^{-1} \cdot a^{(n)}_{I_n, J_n} \leqslant b^{(n)}_{I_n, J_n}(v^{(n+r)}) \leqslant R_n \cdot a^{(n)}_{I_n, J_n} \tag{7.13}$$

for $r \geqslant 0$, $n \geqslant N_{31}(v)$ *and all* $I_n, I'_n \in (\mathbb{N}_k)^{-1}_{-n}$, $J_n \in (\mathbb{N}_k)^n_1$.

THEOREM 7.4. *For any* $v \in \mathfrak{N}^+$ *there is an* $N_{71}(v)$ *such that*

$$|\overleftarrow{q}^{(m)}(v_n) - \overleftarrow{p}^{(m)}| < c_{71} \cdot (\theta_2)^{m-1} \tag{7.14}$$

for $n > N_{71}(v)$ *and* $m = E[\sqrt{n}] \geqslant \sqrt{n} - 1$, *where* $\overleftarrow{p}^{(m)}_n = \overleftarrow{q}^{(m)}(\mu)$, v_n *is the normalized* $v^{(n)}$, *while*

$$\theta_2 = \max\{1 - (R')^{-1}, (\lambda_u)^{-1/2}, (\lambda_s)^{-1/2}\}, \tag{7.15}$$

c_{71} *and* $N_{71}(v)$ *are defined below by formulas* (7.16) *and* (7.17).

Proof. Let $n = m^2 + r < (m+1)^2$. For an normalized measure $v_{m+r} \in \mathfrak{N}^+(m)$ let us construct the vector $\overleftarrow{q}^{(m)}(v_{m+r})$ and trace its evolution under the repeated action of matrices

$$B^{(m)}(v^{(sm+r)}) = B^{(m)}(v_{sm+r}), \text{ where } 1 \leqslant s \leqslant m-1.$$

Since

$$\overleftarrow{q}^{(m)}(v_n) = |\overleftarrow{q}^{(m)}((v_{m+r})^{(m \cdot (m-1))})|^{-1} \cdot \overleftarrow{q}^{(m)}((v_{m+r})^{(m \cdot (m-1))})$$

and $\overleftarrow{p}^{(m)}$ is an eigenvector of $A^{(m)}$, then the inequality (A.26) of Lemma A.10 implies

$$|\overleftarrow{q}^{(m)}(v_n) - \overleftarrow{p}^{(m)}| \leqslant 4 \cdot R' \cdot (R_m \theta)^{m-1} + 4 \cdot R' \cdot R_m \cdot (R_m - 1) \cdot (1 - R_m \theta)^{-1}.$$

This inequality is only valid for sufficiently large m. First we should have $m > N_{32}(v)$, so that all entries of $B^{(m)}(v^{(sm+r)})$ satisfy (7.12) and (7.13). Secondly, by the hypothesis of Theorem A.10 we should have

$$R_m \cdot \theta < 1, \qquad \theta = 1 - (R')^{-1}.$$

To simplify estimations we will require a stronger inequality

$2\cdot(1-R_m\cdot\theta)\geqslant 1-\theta>0$ to be satisfied for $m>N_{72}$, where

$$N_{72}=[\ln(c_{35}+c_{36})-\ln\ln(1+(2R'-2)^{-1})]\cdot(\ln\theta_2)^{-1}.$$

We may also require, in order to simplify the notation, that $(R_m)^m<2$. It is not difficult to show that this inequality holds for $m>N_{73}$, where

$$N_{73}=2(c_{35}+c_{36})\cdot(\ln\theta_2)^{-2}\cdot(\ln 2)^{-1}.$$

Under these conditions we have $1\leqslant R_m<2$ and $R_m-1\leqslant R_m\cdot\ln R_m\leqslant 2\cdot\ln R_m$. All this yields the estimate (7.14), where

$$c_{71}=4\cdot(R')^2\cdot(1+(R')^2)\cdot(1+4\cdot(c_{35}+c_{36})\cdot(1-\theta)^{-1}) \tag{7.16}$$

$$N_{71}(\nu)=(\max\{N_{32}(\nu),N_{72},N_{73}\})^2 \tag{7.17}$$

and N_{72} and N_{73} were defined in the above proof. □

LEMMA 7.5. *Let* $\nu\in\mathfrak{N}^+$, $h\in\mathscr{H}_m(s,M,R)$, $n\geqslant N_{71}(\nu)$, $n\geqslant m$. *Then*

$$\left|\int_{\mathscr{D}}h(z)\cdot\nu_{n,r}\{dz\}-\int_{\mathscr{D}(\infty)}h(w)\cdot\mu_r\{dw\}\right|\leqslant M(h)\cdot(R_m^2\cdot R_m(h)-$$
$$-1)+M(h)\cdot R'\cdot|\overleftarrow{q}^{(m)}(\nu_n)-\overleftarrow{p}^{(m)}|+$$
$$+M(h)\cdot R'\cdot[(R_m-1)\cdot(c_{n/r}[\nu])^{-1}+(c_{n/r}[\nu])^{-1}-1] \tag{7.18}$$

for any $r\geqslant s$, *where*

$$c_{n/r}[\nu]=\lambda^{-r}\cdot\nu_n\{\mathscr{D}(-r)\}. \tag{7.19}$$

Proof. Let us make use of the inequalities

$$\mu\{S^{(m)}(L_m)\}\cdot\nu_n\{\mathscr{D}(-r)\}\cdot\nu_{n,r}\{S^{(m)}(L_m)\}\lesseqgtr(R_m)^{\pm 1}\cdot\nu\{S^{(m)}(L_m)\}\cdot\mu\{\mathscr{D}(-r)\}\cdot\mu_r\{S^{(m)}(L_m)\},$$

$$\left|\int_{S^{(m)}(L_m)}h(z)\nu_{n,r}\{dz\}-\int_{S^{(m)}(L_m)}h(w)\mu_r\{dw\}\right| \tag{7.20}$$

$$\leqslant M(h)\cdot(R_m^2\cdot R_m(h)-1)\cdot\nu_{n,r}\{S^{(m)}(L_m)\}+M(h)\cdot|\nu_{n,r}\{S^{(m)}(L_m)\}-\mu_r\{S^{(m)}(L_m)\}| \tag{7.21}$$

7.2. THE ASYMPTOTICS OF INTEGRALS

The measure ν_n differs from $\nu^{(n)}$ by the normalizing factor $\nu\{\mathscr{D}(-n)\}$. Let us study its asymptotics. Put

$$c_n[\nu]=\lambda^{-n}\cdot\nu\{\mathscr{D}(-n)\}=\lambda^{-n}\cdot\langle\overleftarrow{q}^{(m)}(\nu^{(n)}),\vec{E}_m\rangle,\qquad m\leqslant n. \tag{7.22}$$

THEOREM 7.6. *For any measure* $\nu\in\mathfrak{N}$ *there is a finite limit*

$$c[\nu]=\lim_{n\to\infty}c_n[\nu], \tag{7.23}$$

which is a linear functional on each cone $\mathfrak{M}(\Delta)$, *and* $c[\nu]>0$ *if* $\nu\in\mathfrak{N}^+$.

COROLLARY. *If* $\nu\in\mathfrak{N}^+$ *then*

$$c_{n/r}[\nu]=\lambda^{-r}\cdot\nu_n\{\mathscr{D}(-r)\}\xrightarrow[n\to\infty]{}1. \tag{7.24}$$

uniform in r.

Proof. First let us compare $c_{n(n-1)}[v]$ and $c_{n(n+1)}[v]$. Making use of the matrix technique we may write

$$c_{mn}[v] = \lambda^{-n\cdot m}\left\langle \overleftarrow{q}^{(n)}(v^{(n)}) \prod_{s=1}^{m-1} B^{(n)}(v^{(sn)}), \vec{E}_n \right\rangle$$

starting from $v^{(n)} \in \mathfrak{N}^+(n)$. If $n \geqslant N_{32}(v)$, then the inequalities (7.10) fo Lemma 7.3 hold for entries of all matrices $B^{(n)}$ yielding

$$(R_n)^{-m+1}\cdot c_{nm}[v] \leqslant \lambda^{-n}\langle \overleftarrow{q}^{(n)}(v^{(n)})[\lambda^{-n}A^{(n)}]^{m-1}, \vec{E}_n\rangle \leqslant (R_n)^{m-1}\cdot c_{nm}[v], \quad (7.25)$$

where $A^{(n)} = B^{(n)}(\mu)$, μ is the eigen probability distribution on the horseshoe and λ^n an eigenvalue of $A^{(n)}$. The normed vector $\overleftarrow{p}^{(n)} = \overleftarrow{q}^{(n)}(\mu)$ is a positive eigen row vector of $A^{(n)}$. Let $\vec{e}_n$ be a positive eigen column vector of $A^{(n)}$ normed by the condition $\langle \overleftarrow{p}^{(n)}, \vec{e}_n\rangle = 1$ (see (5.18)). To estimate the scalar product in (7.25) according to the general theory let us distinguish from $\overleftarrow{q}^{(n)}(v^{(n)})$ the eigencomponent

$$\overleftarrow{q}^{(n)}(v^{(n)}) = \langle \overleftarrow{q}^{(n)}(v^{(n)}), \vec{e}_n\rangle\cdot \overleftarrow{p}^{(n)} + \overleftarrow{h}^{(n)}(v^n), \text{ where } \langle \overleftarrow{h}^{(n)}, \vec{e}_n\rangle = 0.$$

Inequality (A.18) yields

$$\langle \overleftarrow{q}^{(n)}[\lambda^{-n}A^{(n)}]^{m-1}, \vec{E}_n\rangle \lesseqgtr \langle \overleftarrow{q}^{(n)}, \vec{e}_n\rangle \pm \|\overleftarrow{h}^{(n)}\|_{\leftarrow}\cdot\theta^{m-1}\cdot\|\vec{E}_n\|_{\rightarrow},$$

where we have made use of Liapunov with respect to $A^{(n)}$ norms of vectors. By the definition (A.5) and (A.6) of these norms we have

$$\langle \overleftarrow{q}^{(n)}, \vec{e}_n\rangle = \|\overleftarrow{q}^{(n)}\|_{\leftarrow}, \; \|\overleftarrow{p}^{(n)}\|_{\leftarrow} = 1, \; \|\overleftarrow{h}^{(n)}\|_{\leftarrow} \leqslant 2, \; \|\overleftarrow{q}^{(n)}\|_{\leftarrow}, \; \|\vec{E}_n\|_{\rightarrow} \leqslant R'.$$

Thus we have

$$\langle \overleftarrow{q}^{(n)}[\lambda^{-n}\cdot A^{(n)}]^{m-1}, \vec{E}_n\rangle \lesseqgtr \|\overleftarrow{q}^{(n)}\|_{\leftarrow}\cdot(1 \pm 2\cdot R'\cdot\theta^{m-1}), \quad (7.26)$$

where we have put $\theta = 1 - (R')^{-1} < 1$ as in the formula (5.31). From (7.25) and (7.26) for $m = n-1$, $m = n+1$ we have

$$\lambda^{-n}\|\overleftarrow{q}^{(n)}\|_{\leftarrow}\cdot(R_n)^{-m+1}\cdot(1 - 2\cdot R'\cdot\theta^{m-1}) \leqslant c_{nm}[v]$$
$$\leqslant \lambda^{-n}\|\overleftarrow{q}^{(n)}\|_{\leftarrow}\cdot(R_n)^{m-1}\cdot(1 + 2\cdot R'\cdot\theta^{m-1}).$$

Suppose that $n > N'$ is such that $\theta^{n-2} < (2R')^{-1}$ and $1 - 2\cdot R'\cdot\theta^{n-2} > 0$. Then dividing inequalities with $m = n+1$ by inequalities with $m = n-1$ we get

$$(R_n)^{-2n+2}\cdot(1 - 2\cdot R'\cdot\theta^n)\cdot(1 + 2\cdot R'\cdot\theta^{n-2})^{-1} \leqslant c_{n(n+1)}[v]\cdot(c_{n(n-1)}[v])^{-1}$$
$$\leqslant (R_n)^{2n-2}\cdot(1 + 2\cdot R'\cdot\theta^n)\cdot(1 - 2\cdot R'\cdot\theta^{n-2})^{-1}.$$

A rougher estimate gives

$$|\ln c_{n(n+1)}[v] - \ln c_{(n-1)n}[v]| \leqslant \ln M_n,$$
$$\ln M_n = 2\cdot(n-1)\cdot\ln R_n + \ln(1 + 2\cdot R'\cdot\theta^n) - \ln(1 - 2\cdot R'\cdot\theta^{n-2}).$$

To prove the convergence of the subsequence $c_{n(n+1)}[v]$ it suffices to prove that of $\Sigma \ln M_n$. Arguments similar to the proof of corollary to Theorem 5.5 and (3.13) yield

$$\ln M_n \leqslant 2\cdot(n-1)\cdot c_{35}(\lambda_u)^{-n/2} + 2\cdot(n-1)\cdot c_{36}(\lambda_s)^{-n/2} + 2\cdot R'\cdot\theta^n + 4\cdot R'\cdot\theta^{n-2}$$

if n is such that $2\cdot R'\cdot\theta^{n-2} < 0.5$, and then $-\ln(1 - 2\cdot R'\cdot\theta^{n-2}) < 4\cdot R'\cdot\theta^{n-2}$.

Hence

$$\ln M_n \leqslant c_{72}\cdot n\cdot(\theta_2)^n, \qquad \theta_2 = \max\{\theta, (\lambda_u)^{-1/2}, (\lambda_s)^{-1/2}\}. \tag{7.27}$$

Since $\theta_2 < 1$, the infinite sum $\Sigma \ln M_n$ exists and is finite and

$$|\ln c[v] - \ln c_{n(n-1)}[v]| \leqslant c_{72}(\theta_2)^n\cdot(n(1-\theta_2)^{-1} + (1-\theta_2)^{-2}). \tag{7.28}$$

Meanwhile we have defined $c[v]$ as the limit of the subsequence $c_m[v]$ for $m = (n-1)\cdot n$. Consider the general case. Let $n(n-1) < m \leqslant n(n+1)$. Two subcases are possible: $n(n-1) < m \leqslant n^2$, $m = n(n-1) + r$ and $n^2 < m \leqslant n(n+1)$, $m = n^2 + r$. In the first subcase we make use of the presentation

$$c_m[v] = \left\langle \overleftarrow{q}^{(n)}(v^{(n)})\cdot \prod_{s=1}^{r-1} B^{(n)}(v^{(sn)})\pi_{\overrightarrow{n,n-1}} \prod_{s=0}^{n-r-1} B^{(n-1)}(v^{(rn+s(n-1))}), \vec{E}_{n-1}\right\rangle$$

and compare it with $c_{n(n-1)}[v]$, and in the second one we make use of

$$c_m[v] = \left\langle \overleftarrow{q}^{(n+1)}(v^{(n+1)}) \prod_{s=1}^{r-1} B^{(n+1)}(v^{(s(n+1))})\pi_{\overrightarrow{n+1,n}} \prod_{s=0}^{n-r-1} B^{(n)}(v^{(r(n+1)+sn)}), \vec{E}_n\right\rangle$$

and compare it to $c_{n(n+1)}[v]$. We omit the computations.

Now let us remove the restriction $v \in \mathfrak{N}^+$. Let $v \in \mathfrak{M}(\Delta) \subset \mathfrak{N}$. Take an arbitrary $v' \in \mathfrak{M}^+(\Delta) \subset \mathfrak{N}^+$ and form the measure $v'' = v + v'$, where $v'' \subset \mathfrak{N}^+$. From what we have already proved, $c_n[v'] \to c[v']$ as $n \to \infty$ and $c_n[v''] \to c[v'']$ as $n \to \infty$. Since

$$c_n[v] = \lambda^{-n}\cdot v\{\mathscr{D}(-n)\} = \lambda^{-n}\cdot(v''\{\mathscr{D}(-n)\} - v'\{\mathscr{D}(-n)\}) = c_n[v''] - c_n[v'],$$

then the limit of $c_n[v]$ exists and $c[v] = c[v''] - c[v']$. The homogeneity of $c[v]$ is evident. To prove the corollary, note that

$$c_{n/r}[v] = (c_n[v])^{-1}\cdot c_{n+r}[v], \tag{7.29}$$

where both the numerator and the denominator tend to the same limit $c[v]$.

THEOREM 7.7. *For any measure $v \in \mathfrak{N}$ and any continuous function $h(z)$ on $\mathscr{D}$ we have*

$$\lim_{n\to\infty} \lambda^{-n}\int_{\mathscr{D}(-n)} h(T^n z)v\{dz\} = c[v]\cdot\int_{\mathscr{D}(\infty)} h(w)\mu\{dw\}, \tag{7.30}$$

where μ is the normed eigenmeasure on the horseshoe. If also $v \in \mathfrak{N}^+$, then

$$\lim_{n\to\infty}\int_{\mathscr{D}} h(z)\cdot v_n\{dz\} = \int_{\mathscr{D}(\infty)} h(w)\mu\{dw\}. \tag{7.31}$$

COROLLARY. *If $v \in \mathfrak{N}^+$ then $v_n \Rightarrow \mu$.*

Proof. We begin with the second statement. For constants it is evident. Let $v \in \mathfrak{N}$, $n \geqslant N_{71}(v)$, $m = E[\sqrt{n}]$, $h \in \mathscr{H}_m(0, M, R)$. Then by Lemma 7.5 (7.19) holds with $r = 0$, i.e. $c_{n/0}[v] = 1$. Estimating $|\overleftarrow{q}^{(m)}(v_n) - \overleftarrow{p}^{(m)}|$ by making use of Theorem 7.4 we come to the conclusion that

$$\left| \int_{\mathscr{D}} h(z)\nu_n\{dz\} - \int_{\mathscr{D}(\infty)} h(w)\mu\{dw\} \right| \leqslant M(h)\cdot(R_m^2\cdot R_m(h) - 1) + M(h)\cdot R'\cdot(R_m - 1) + \\ + M(h)\cdot R'\cdot c_{71}\cdot(\theta_2)^{m-1} \to 0$$

for $m = E[\sqrt{n}] \to \infty$ since $R_m \downarrow 1$, $R_m(h) \downarrow 1$ as $m \to \infty$ and $\theta_2 < 1$. Since any continuous function $h(\cdot)$ can be decomposed into the difference of $h' \in \mathscr{H}_m$ and a constant and both parts of (7.31) are linear in h, then (7.31) holds for any continuous function $h(z)$.

Let us prove (7.30). Take $T^n z$ for the new variable of integration. Then

$$\lambda^{-n}\cdot \int_{\mathscr{D}(-n)} h(T^n z)\nu\{dz\} = \lambda^{-n} \int_{\mathscr{D}(n)} h(z)\nu^{(n)}\{dz\} = c_n[\nu]\cdot \int_{\mathscr{D}} h(z)\nu_n\{dz\}.$$

By Theorem 7.6 and the already proved formula (7.31) this implies (7.30) for any measure $\nu \in \mathfrak{N}^+$. In the general case we present $\nu \in \mathfrak{N}$ in the form $\nu = \nu'' - \nu'$, where $\nu', \nu'' \in \mathfrak{N}^+$, and make use of the linearity of (7.30) in ν.

THEOREM 7.8. *For any measure $\nu \in \mathfrak{N}$ and any continuous function $h(z)$ on $\mathscr{D}$ we have*

$$\lim_{n,r\to\infty} \lambda^{-n-r} \int_{\mathscr{D}(-n-r)} h(T^n z)\nu\{dz\} = c[\nu]\cdot \int_{\Omega} h(w)\mu_0\{dw\}, \tag{7.32}$$

where μ_0 is the invariant measure on the hyperbolic set Ω of the Smale horseshoe. If in addition $\nu \in \mathfrak{N}^+$, then

$$\lim_{n,r\to\infty} \int_{\mathscr{D}} h(z)\nu_{n,r}\{dz\} = \int_{\Omega} h(w)\mu_0\{dw\}. \tag{7.33}$$

Proof. For $\nu \in \mathfrak{N}^+$, $n \geqslant N_{71}(\nu)$, $m = E[\sqrt{n}]$, $h \in \mathscr{H}_m(0, M, R)$ and any $r \geqslant 0$ the estimation (7.19) of Lemma 7.5 and (7.14) of Theorem 7.4 hold, implying

$$\sup_r \left| \int_{\mathscr{D}} h(z)\nu_{n,r}\{dz\} - \int_{\mathscr{D}(\infty)} h(w)\mu_r\{dw\} \right| \xrightarrow[n\to\infty]{} 0.$$

Since by Theorem 5.13 we have $\mu_r \Rightarrow \mu_0$ and

$$\left| \int_{\mathscr{D}(\infty)} h(w)\mu_r\{dw\} - \int_{\Omega} h(w)\mu_0\{dw\} \right| \xrightarrow[r\to\infty]{} 0,$$

see (5.51), then (7.33) holds for any h satisfying (7.4). Since for constants (7.33) is trivial, then by linearity (7.33) holds for all continuous functions $h(z)$ on $\mathscr{D}$. The formula (7.33) is derived from (7.32) as (7.30) is derived from (7.31).

Before we begin to study intermixing properties of T of the Smale horseshoe, we consider intermixing properties of T on $(\mathscr{D}(\infty), \mu)$ and make an auxiliary estimation.

LEMMA 7.9. *Suppose the continuous function $h(z)$ defined on $\mathscr{D}(\infty) \cap \mathscr{D}(-s)$ satisfies*

(7.7), *i.e.* $1 \leqslant h(z) \leqslant M(h)$. *Put*

$$R'_r(h) = \max_I \max_{J_r} \max_{z,z' \in \mathfrak{F}(I) \cap S^{(-r)}(J_r)} \{(h(z))^{-1} \cdot h(z')\} \tag{7.34}$$

for $r \geqslant s$, *where* $(I) = (\cdots i_n \cdots i_1)$ *is the number of the fibre* $\mathfrak{F}(I)$ *of the expanding fibration* $\mathcal{D}(\infty)$ *and* $(J_r) = (j_1 \cdots j_r)$ *is the number of the component* $S^{(-r)}(J_r)$ *of* $\mathcal{D}(-r)$. *Then for* $s \leqslant r \leqslant m$ *the estimation*

$$\int_{\mathcal{D}(\infty)} h(z)\mu_m\{dz\} \lesseqgtr (R'_r \cdot \mathcal{K}^{\pm}_{m-r})^{\pm 1} \cdot \int_{\Omega} h(w)\mu_0\{dw\}, \tag{7.35}$$

holds, where $<$ *corresponds to the exponent* $+1$ *and the constant* $\mathcal{K}^{+}_{m-r}$, *constants* $\mathcal{K}^{\pm}_m$ *being defined in Theorem* 5.6.

LEMMA 7.10. *For any* $\nu \in \mathfrak{N}$ *and any two continuous functions* $h_1(z)$ *and* $h_2(z)$ *on* $\mathcal{D}$ *we have*

$$\lim_{n,s,l \to \infty} \lambda^{-n-s-l} \int_{\mathcal{D}(-n-s-l)} h_1(T^n z) h_2(T^{n+s} z)\nu\{dz\}$$

$$= c[\nu] \cdot \int_{\Omega} h_1(w)\mu_0\{dw\} \cdot \int_{\Omega} h_2(w)\mu_0\{dw\}. \tag{7.36}$$

If in addition $\nu \in \mathfrak{N}^+$ *then*

$$\lim_{n,s,l \to \infty} \int_{\mathcal{D}} h_1(z) \cdot h_2(T^s z) \cdot \nu_{n,s+l}\{dz\}$$

$$= \int_{\Omega} h_1(w)\mu_0\{dw\} \cdot \int_{\Omega} h_2(w)\mu_0\{dw\}. \tag{7.37}$$

REMARK. In the conditions of the lemma, analogues of the interchanging property of any multiplicity are valid.

To prove the Lemma it suffices to consider the case when h_1 and h_2 satisfy inequalities (7.4), to check the uniform convergence

$$\sup_{s,l} \left| \int_{\mathcal{D}} h_1(z)h_2(T^s z)\nu_{n,s+l}\{dz\} - \int_{\mathcal{D}(\infty)} h_1(w)h_2(T^s w)\mu_{s+l}\{dw\} \right| \xrightarrow[n \to \infty]{} 0 \tag{7.38}$$

$$\sup_{s} \left| \int_{\mathcal{D}(\infty)} h_1(w)h_2(T^s w)\mu_{s+l}\{dw\} - \int_{\Omega} h_1(w)h_2(T^s w)\mu_0\{dw\} \right| \xrightarrow[l \to \infty]{} 0 \tag{7.39}$$

and to make use of the fact that T is an intermixing transformation, i.e.

$$\left| \int_{\Omega} h_1(w)h_2(T^s w)\mu_0\{dw\} - \int_{\Omega} h_1(w)\mu_0\{dw\} \cdot \int_{\Omega} h_2(w)\mu_0\{dw\} \right| \xrightarrow[s \to \infty]{} 0 \tag{7.40}$$

on (Ω, μ_0) which follows from Theorem 6.9. □

THEOREM 7.11. *For any* $\nu \in \mathfrak{N}^+$, *any continuous function* $h(z)$ *on* $\mathcal{D}$ *and any* $\varepsilon > 0$

we have

$$\nu_{0,N}\left\{z:\left|N^{-1}\cdot\sum_{1\leqslant n\leqslant N}h(T^n z)-\int_\Omega h(w)\mu_0\{\mathrm{d}w\}\right|>\varepsilon\right\}\underset{N\to\infty}{\to}0, \tag{7.41}$$

where $\nu_{0,N}\{\cdot\}=\nu\{\cdot|\mathscr{D}(-N)\}$.

Proof. Let $M=M(h)=\max|h(z)|$. If we show that

$$\mathscr{I}(N)=\int_{\mathscr{D}(-n)}\left(\sum_{1\leqslant n\leqslant N}h(T^n z)-N\int_\Omega h(w)\mu_0\{\mathrm{d}w\}\right)^2\nu_{0,N}\{\mathrm{d}z\}=\bar{o}(N^2) \tag{7.42}$$

then the required asymptotics of the measure will follow from the Chebyshev inequality. For brevity put $h(T^n z)=h_n(z)$, $\int h(w)\mu_0\{\mathrm{d}w\}=h_*$. Let us integrate the identity

$$\left(\sum_{1\leqslant n\leqslant N}h_n(z)-Nh_*\right)^2=\sum_{1\leqslant m,n\leqslant N}(h_m(z)\cdot h_n(z)-h_*^2)-2\cdot N\cdot h_*\sum_{1\leqslant n\leqslant N}(h_n(z)-h_*)$$

with respect to $\nu_{0,N}$ and estimate the module of the result in terms of modules of integrals of each bracket. We then obtain N^2 summands of one type

$$\mathscr{I}_{m,n}=\left|\int(h(T^m z)h(T^n z)-h_*^2)\nu_{0,N}\{\mathrm{d}z\}\right|, 1\leqslant m,n\leqslant N$$

and N summands of the second type

$$\mathscr{I}_n=\left|\int(h(T^n z)-h_*)\nu_{0,N}\{\mathrm{d}z\}\right|,\ 1\leqslant n\leqslant N.$$

Put

$$\mathscr{I}(N)\leqslant\sum_{1\leqslant m,n\leqslant N}\mathscr{I}_{m,n}+2\cdot N\cdot|h_*|\cdot\sum_{1\leqslant n\leqslant N}\mathscr{I}_n. \tag{7.43}$$

For further estimations let us split summands in (7.43) into 'good' and 'bad'. Let $A^{(1)}(N)=\{n: 1\leqslant n\leqslant\sqrt{N}\}$,

$$A^{(2)}(N)=\{n:\sqrt{N}<n<N-\sqrt{N}\},\ A^{(3)}(N)=\{n: N-\sqrt{N}\leqslant n<N\}.$$

For 'bad' n let us estimate $\mathscr{I}_n$ trivially in terms of the maximum M of the module of $h(z)$:

$$2\cdot N\cdot|h_*|\cdot\mathscr{I}_n\leqslant 8\cdot N\cdot M^2, n\in A^{(1)}\cup A^{(3)},\ \operatorname{card}(A^{(1)}\cup A^{(3)})\leqslant 2\cdot\sqrt{N}. \tag{7.44}$$

To estimate the integral of $\mathscr{I}_n$ for 'good' n let us first make a variable transform in the integral. For any $1\leqslant n\leqslant N$ we have

$$\int_{\mathscr{D}}h(T^n z)\nu_{0,N}\{\mathrm{d}z\}=\int_{\mathscr{D}}h(z)\nu_{n,N-n}\{\mathrm{d}z\}, \tag{7.45}$$

where $\nu_{0,N}$ has support in $\mathscr{D}(-N)$ and $\nu_{n,N-n}$ in $\mathscr{D}(n)\cap\mathscr{D}(-N+n)$, as follows from the definition of conditional measures $\nu_{n,m}$. Therefore

$$\mathscr{I}_n=\left|\int_{\mathscr{D}}h(z)\nu_{n,N-n}\{\mathrm{d}z\}-\int_\Omega h(w)\mu_0(\mathrm{d}w)\right|.$$

From the second statement of Theorem 7.8, see formula (7.33) for any $\delta > 0$, there is $N_{74}(v, h, \delta)$ such that for $n \geqslant N_{74}$ and $N - n > N_{74}$ we have $\mathcal{I}_n < \delta(2\cdot M)^{-1}$, i.e.

$$2\cdot N\cdot|h_*|\cdot\mathcal{I}_n \leqslant N\cdot\delta, \qquad n\in A^{(2)}(N), \quad N \geqslant [N_{74}(v,h,\delta)]^2, \quad \text{Card}\, A^{(2)}(N) < N. \tag{7.46}$$

Let us similarly distinguish the sets of 'bad' pairs (m, n):

$$B^{(1)}(N) = \{(m,n): 1 \leqslant m \leqslant \sqrt{N}\}, \qquad B^{(2)}(N) = \{(m,n): 1 \leqslant n \leqslant \sqrt{N}\},$$

$$B^{(3)}(N) = \{(m,n): N - \sqrt{N} \leqslant m \leqslant N\}, \quad B^{(4)}(N) = \{(m,n): N - \sqrt{N} \leqslant m \leqslant N\}$$

$$B^{(5)}(N) = \{(m,n): |m - n| \leqslant \sqrt{N}\}.$$

The complement to their union is the set of 'good' pairs $B^{(6)}(N)$. For 'bad' pairs let us estimate $\mathcal{I}_{m,n}$ trivially in terms of $M = M(h)$:

$$\mathcal{I}_{m,n} < 2\cdot M^2, \qquad (m,n)\in \bigcup_{1\leqslant i\leqslant 5} B^{(i)}, \quad \text{Card} \bigcup_{1\leqslant i\leqslant 5} B^{(i)} \leqslant 6\cdot N\cdot\sqrt{N}. \tag{7.47}$$

For 'good' pairs let us first make a variable transform in the integral. For $n < m$ take $m - n = s$, $N - m = r$. Then by (7.45) we have

$$\mathcal{I}_{m,n} = \left|\int_{\mathcal{D}} h(z)h(T^s z)v_{n,s+r}\{dz\} - \left(\int h(w)\mu_0\{dw\}\right)^2\right|.$$

Due to the second statement of Lemma 7.10, see formula (7.37), for any $\delta > 0$ there is $N_{75}(v, h, \delta)$ such that for $n \geqslant N_{75}$, $s \geqslant N_{75}$, $r \geqslant N_{75}$ we have $\mathcal{I}_{m,n} < \delta$. For $m < n$ put $n - m = s, N - n = r$, then $\mathcal{I}_{m,n} < \delta$ for $m \geqslant N_{75}, s \geqslant N_{75}, r \geqslant N_{75}$. Thus

$$\mathcal{I}_{m,n} < \delta, (m,n)\in B^{(6)}(N), N \geqslant (N_{75}(v,h,\delta))^2, \text{Card}\, B^{(6)}(N) < N^2. \tag{7.48}$$

Substituting estimations (7.44), (7.46), (7.47), and (7.48) in (7.43) we get $\mathcal{I}(N) \leqslant 28\cdot M^2\cdot N\cdot\sqrt{N} + 2\cdot N^2\cdot\delta$ for any $N \geqslant (\max\{N_{74}(\delta), N_{75}(\delta)\})^2$. Since $\delta > 0$ is arbitrary, then (7.42) holds. □

7.3. MAPPINGS WHICH POSSESS A SMALE HORSESHOE

It is clear that our definition of a smooth Smale horseshoe is stable with respect to small adjustments of the domain of definition: we may cut the protruding ends of components a little. In the two-dimensional case such cutting may be carried out up to separators of fixed points of the horseshoe. When we cut the domain of definition, fibres of the set $\mathcal{D}(\infty)$ are also cut; however, the T-invariant conditional density $p(y|I)$ on them is preserved up to a multiple. The set Ω of non-wandering points and the invariant measure μ_0 on it are stable under the cutting, whereas the eigennumber λ of the horseshoe and the eigenfunction $e(I)$ on the expanding foliation are not generally speaking stable.

Our definition of the Smale horseshoe is invariant with respect to the 'time inversion', i.e. symmetric with respect to T and T^{-1}. The hyperbolic set Ω is common for them. The question immediately arises whether T and T^{-1} generate the same limit dynamical system: the following example of the Smale horseshoe shows that this is not generally speaking the case.

EXAMPLE. Put $\mathscr{D} = \{(x, y): 0 \leqslant x \leqslant 1, 0 \leqslant y \leqslant 1\} \subset \mathbb{R}^2 = \mathbb{Z}$,

$$\mathscr{D}(-1) = S^{(-1)}(1) \cup S^{(-1)}(2)$$

and $T(x \oplus y) = f(x) \oplus g(y)$ on $\mathscr{D}(-1)$, where

$$f(x) = \tfrac{1}{3}x, g(y) = 2y \quad \text{for} \quad (x, y) \in S^{(-1)}(1) = \{(x, y): 0 \leqslant x \leqslant 1, 0 \leqslant y \leqslant \tfrac{1}{2}\}$$

and

$$f(x) = \tfrac{1}{3}(2 + x), g(y) = 4(y - \tfrac{3}{4}) \quad \text{for} \quad (x, y) \in S^{(-1)}(2) = \{(x, y): 0 \leqslant x \leqslant 1, \tfrac{3}{4} \leqslant y \leqslant 1\}.$$

THEOREM 7.12. *In the example considered the invariant measure μ_0 on Ω generated naturally by T as the weak limit of smooth measures $\nu_{n,m}$ differs from the measure μ_0' on Ω generated likewise by T^{-1}. In particular*

$$\mu_0\{S^{(1)}(1)\} = \mu_0\{S^{(-1)}(1)\} = \tfrac{2}{3}, \qquad \mu_0'\{S^{(1)}(1)\} = \mu_0'\{S^{(-1)}(1)\} = \tfrac{1}{2}.$$

Proof. The definition of T implies

$$S^{(1)}(1) = \{(x, y): 0 \leqslant x \leqslant \tfrac{1}{3}, 0 \leqslant y \leqslant 1\},$$
$$S^{(1)}(2) = \{(x, y): \tfrac{2}{3} \leqslant x \leqslant 1, 0 \leqslant y \leqslant 1\}.$$

For an initial fibration take the fibration into segments $x = \text{const.}$ and $y = \text{const.}$ and for an initial smooth measure take the flat Lebesgue measure $\nu\{\mathrm{d}x\, \mathrm{d}y\} = \mathrm{d}x \wedge \mathrm{d}y$ with density $\rho(x, y) \equiv 1$. The map T is linear on $S^{(-1)}(1)$ and $S^{(-1)}(2)$ with eigendirections parallel to coordinate axes. Therefore all subsequent fibrations of $\mathscr{D}(n)$ consist of segments $x = \text{const.}$ and carry a uniform distribution with respect to y as a conditional one. Hence, so does the limit foliation $\mathscr{D}(\infty)$ with the invariant uniform conditional density. Let $\hat{\mu}\{\mathrm{d}I\}$ be a restriction of the eigenmeasure onto the algebra of numbers of leaves. The definition of the eigenmeasure implies that

$$\lambda \cdot \mu\{S^{(1)}(1)\} = \int P\{S^{(-1)}(1) \mid I\}\, \hat{\mu}\{\mathrm{d}I\} = \tfrac{1}{2}, \qquad \lambda \cdot \mu\{S^{(1)}(2)\} = \tfrac{1}{4},$$

$$\lambda = \lambda \cdot (\mu\{S^{(1)}(1)\} + \mu\{S^{(1)}(2)\}) = \tfrac{3}{4}, \qquad \mu\{S^{(1)}(1)\} = \tfrac{2}{3}.$$

Further, $\Psi_m(I) = \lambda^{-m} \cdot P\{\mathscr{D}(-m) | I\} = \lambda^{-m} \cdot P\{\mathscr{D}(-m)\} = 1$

since $P\{\mathscr{D}(-m)|I\}$ does not depend on I. Hence $e(I) \equiv 1$, $\hat{\mu}_0\{\cdot\} \equiv \hat{\mu}\{\cdot\}$ identically. Therefore $\mu_0\{S^{(1)}(1)\} = \mu\{S^{(1)}(1)\} = \tfrac{2}{3}$ and, due to the T-invariance, $\mu_0\{S^{(-1)}(1)\} = \tfrac{2}{3}$. Similar arguments for T^{-1}, where x and y are interchanged, make it possible to conclude that the fibration $\mathscr{D}(-\infty)$ consists of segments $y = \text{const.}$ $0 \leqslant x \leqslant 1$ and the invariant conditional distribution on them is also uniform. Further by definition of the T^{-1}-eigenmeasure μ' we have

$$\lambda' \cdot \mu'\{S^{(-1)}(1)\} = \int P\{S^{(1)}(1) \mid J\}\, \hat{\mu}'\{\mathrm{d}J\} = \tfrac{1}{3}, \qquad \lambda' \cdot \mu'\{S^{(-1)}(2)\} = \tfrac{1}{3}$$

$$\lambda' = \lambda' \cdot (\mu'\{S^{(-1)}(1)\} + \mu'\{S^{(-1)}(2)\}) = \tfrac{2}{3}, \qquad \mu'\{S^{(-1)}(1)\} = \tfrac{1}{2}.$$

Quite similarly $e'(J) \equiv 1, \hat{\mu}_0'\{\cdot\} \equiv \hat{\mu}'\{\cdot\}$

implying $\mu_0'\{S^{-1}(1)\} = \mu'\{S^{-1}(1)\} = \frac{1}{2}$ and $\mu_0'\{S^{(1)}(1)\} = \frac{1}{2}$ due to T^{-1}-invariance.

THEOREM 7.13. *For any Borel probability distribution $\mathbb{Q}$ on $\mathcal{D}$ with strictly positive Lipschitzizable density with respect to the $\mathbb{Z}$-volume, arbitrary continuous function $h(z)$ on $\mathcal{D}$, any $M \geqslant 0$ and $\varepsilon > 0$ we have*

$$\lim_{N\to\infty} \mathbb{Q}\{z: |N^{-1}\cdot \sum_{1\leqslant n\leqslant N} h(T^n z) - \int_\Omega h(w)\mu_0\{dw\}| > \varepsilon || \mathcal{D}(M)\cap\mathcal{D}(-N)\} = 0 \tag{7.49}$$

uniform in M.

Proof. Let $\rho(x, y)$ be the density of $\mathbb{Q}$. Then for the measure $\nu\{\cdot\} = \mathbb{Q}\{\cdot|\mathcal{D}(M)\}$ the logarithm of the density of the conditional distribution on each fibre of the canonical fibration $\mathcal{D}(M)$ equals

$$\ln\omega(y\,|\,u, I_M) = \ln\rho(F^{(M)}(u, I_M, y), y) + \ln|\det d_u F^{(M)}| +$$

$$+ \ln\int \rho(F^{(M)}(u, I_M, \hat{y}), \hat{y})\,|\det d_u F^{(M)}|\,d\hat{y}_1\wedge\cdots\wedge d\hat{y}_{\dim\mathbb{Y}}$$

due to Lemma 4.4. The Lipschitz constant of the first summand is bounded by $\mathrm{Lip}_{\mathbb{Y}}(\ln\rho) + \beta\,\mathrm{Lip}_{\mathbb{X}}(\ln\rho)$, by hypothesis. The constant for the second summond is bounded by c_{28}' for any M, from an analogue of Theorem 2.7. Therefore the constant $N_{32}(\nu)$ of Theorem 3.9 which occurs in estimations of the asymptotics may also be taken common for all M. Since $\nu\in\mathfrak{M}^+(D(M))\subset\mathfrak{N}^+$ and no individual numerical characteristics of ν except $N_{32}(\nu)$ are used in the subsequent constructions, cf. (7.17), then in Theorem 7.11 the convergence (7.41) for $\nu\{\cdot\} = \mathbb{Q}\{\cdot|\mathcal{D}(M)\}$ is uniform in M.

THEOREM 7.14. *Suppose the diffeomorphism $T\in C^{1+1}(\mathcal{D})$ possesses in the $\mathbb{X}$, $\mathbb{Y}$-convex curvilinear rectangle $\mathcal{D}$ a smooth Smale horseshoe. Let $\mu_0\{\cdot\}$ be an invariant measure in Ω constructed with respect to T, while $\mu_0'\{\cdot\}$ is an invariant measure on Ω constructed with respect to T^{-1} and p, q are non-negative numbers such that $p + q = 1$. Then for any Borel probability distribution $\mathbb{Q}$ on $\mathcal{D}$ with strictly positive Lipschitzizable density with respect to the $\mathbb{Z}$-volume, arbitrary continuous function $h(z)$ on $\mathcal{D}$ and any $\varepsilon > 0$ the conditional probability*

$$\mathbb{Q}\Big\{z: |(N + M + 1)^{-1}. \sum_{-M\leqslant n\leqslant -N} h(T_z^n) -$$

$$- \int_\Omega h(w)(p\mu_0\{dw\} + q\mu_0'\{dw\})| > \varepsilon || \mathcal{D}(M, -N)\Big\}$$

tends to zero when M and N tend to infinity and

$$\lim_{N\to\infty, M\to\infty} [N\cdot(M + N)^{-1}] = p, \tag{7.51}$$

$$\mathcal{D}(M, -N) = \mathcal{D}(M)\cap\mathcal{D}(-N).$$

Appendix: Ergodic Properties of Positive Matrices with Bounded Ratio of Rows

Put $\mathfrak{F}$ for the real space of column vectors $\vec{f}$ with n rows and $\mathfrak{G}$ for the dual space of row vectors $\overleftarrow{g}$ with norms

$$|\vec{f}| = \max_{1 \leqslant i \leqslant n} |f_i|, \quad |\overleftarrow{g}| = \sum_{j=1}^{n} |g_j| \tag{A.1}$$

and convolution (scalar product)

$$(\overleftarrow{g},\vec{f}) = \sum_{j=1}^{n} f_j \cdot g_j, \qquad |(\overleftarrow{g},\vec{f})| \leqslant |\overleftarrow{g}|\cdot|\vec{f}|. \tag{A.2}$$

Let $A = (a_{ij})$ be an $n \times n$-matrix with positive entries, in particular a substochastic matrix, i.e.

$$a_{ij} > 0 \quad \text{for any} \quad 1 \leqslant i, \quad j \leqslant n. \tag{A.3}$$

We consider simultaneously its action from the left on column vectors and from the right on row vectors.

STATEMENT A.1 (Perron's theorem) [29]. The positive matrix A possesses eigen column vectors and row vectors with positive coordinates (*Perron vectors*). They are unique up to a multiple and correspond to the eigenvalue of A with the maximal module. This eigenvalue is positive and is a simple root of the characteristic polynomial of A.

Thus Perron vectors are defined up to a multiple. Put $\overleftarrow{p}$ and $\vec{e}$ for Perron vectors normed so that

$$\overleftarrow{p}A = \lambda\overleftarrow{p}, \qquad \sum_{1 \leqslant i \leqslant n} p_i = |\overleftarrow{p}| = 1, \tag{A.4}$$

$$A\vec{e} = \lambda\vec{e}, \qquad \sum_{1 \leqslant i \leqslant n} p_i e_i = (\overleftarrow{p},\vec{e}) = 1. \tag{A.5}$$

Our next goal is to describe ergodic properties of A. For this let us introduce (cf. [15]) special norms on $\mathfrak{F}$ and $\mathfrak{G}$

$$\|\vec{f}\|_{\rightarrow} = \max_{1 \leqslant j \leqslant n} \{(e_j)^{-1}|f_j|\}, \qquad \|\overleftarrow{g}\|_{\leftarrow} = \sum_{1 \leqslant i \leqslant n} |g_i| e_i. \tag{A.6}$$

For the row vector $\overleftarrow{q}$ with positive coordinates we have

$$\|\overleftarrow{q}\|_{\leftarrow} = (\overleftarrow{q},\vec{e}), \qquad \|\overleftarrow{q}A\|_{\leftarrow} = \lambda\cdot\|\overleftarrow{q}\|_{\leftarrow}. \tag{A.7}$$

Note that when A is a stochastic matrix, then $\vec{e} = \vec{E}$ and norms (A.6) coincides with (A.1).

STATEMENT A.2. The norms (A.6) are dual, therefore

$$|(\overleftarrow{g},\vec{f})| \leqslant \|\overleftarrow{g}\|_{\leftarrow}\cdot\|\vec{f}\|_{\rightarrow}. \tag{A.8}$$

The operator norm of A with respect to (A.6) equals λ.

Let A be the matrix with positive entries. We say that $R' \geqslant 1$ is an estimate of the ratio of rows of A if

$$(R')^{-1}\cdot a_{kj} \leqslant a_{ij} \leqslant R'\cdot a_{kj} \tag{A.9}$$

for all $1 \leqslant i,j,k \leqslant n$. For matrices satisfying (A.9), which is called the *Doeblin condition*, we may estimate ergodicity only in terms of R', see [15].

STATEMENT A.3. If A satisfies (A.3) and (A.9) then

$$(R')^{-1}\cdot e_k \leqslant e_i \leqslant R'\cdot e_k \quad \textit{for all} \quad 1 \leqslant i,k \leqslant n. \tag{A.10}$$

STATEMENT A.4. For normed Perron vectors of the matrix A with positive entries and the estimation $R' \geqslant 1$ of the ratio of rows (A.9) we have

$$(R')^{-1}\cdot\lambda\cdot p_j \leqslant a_{ij} \leqslant R'\cdot\lambda\cdot p_j, \tag{A.11}$$

$$(R')^{-1} \leqslant e_i \leqslant R' \tag{A.12}$$

for any $1 \leqslant i,j \leqslant n$.

COROLLARY. *Under these conditions, the norms* (A.6) *are equivalent to the corresponding norms* (A.1) *with the equivalence constant* R', *i.e.*

$$(R')^{-1}\cdot|\vec{f}| \leqslant \|\vec{f}\|_{\rightarrow} \leqslant R'\cdot|\vec{f}|, \qquad (R')^{-1}|\overleftarrow{g}| \leqslant \|\overleftarrow{g}\|_{\leftarrow} \leqslant R'\cdot|\overleftarrow{g}|. \tag{A.13}$$

STATEMENT A.5. If the positive matrix A has $R' \geqslant 1$ as the estimate of the ratio of rows, see (A.9), then

$$\sum_{1\leqslant j\leqslant n} e_j\cdot \inf_{1\leqslant i\leqslant n} \{a_{ij}(e_i)^{-1}\} \geqslant \lambda\cdot(R')^{-1}. \tag{A.14}$$

The left-hand side is called the *ergodicity coefficient*.

In $\mathfrak{F}$ and $\mathfrak{G}$, introduce subspaces $\mathfrak{F}_0$ and $\mathfrak{G}_0$ orthogonal to Perron vectors of the dual space:

$$\mathfrak{F}_0 = \{\vec{f}: (\overleftarrow{p},\vec{f}) = 0\}, \qquad \mathfrak{G}_0 = \{\overleftarrow{g}: (\overleftarrow{g},\vec{e}) = 0\}. \tag{A.15}$$

It is evident that there are decompositions

$$\vec{f} = (\overleftarrow{p},\vec{f})\cdot\vec{e} + [\vec{f} - (\overleftarrow{p},\vec{f})\cdot\vec{e}], \qquad \overleftarrow{g} = (\overleftarrow{g},\vec{e})\cdot\overleftarrow{p} + [\overleftarrow{g} - (\overleftarrow{g},\vec{e})\cdot\overleftarrow{p}]. \tag{A.16}$$

LEMMA A.6. *If* A *satisfies* (A.3) *and* (A.9) *then*

$$\|A\vec{f}\|_{\rightarrow} \leqslant \lambda\cdot(1 - (R')^{-2})\|\vec{f}\|_{\rightarrow} \tag{A.17}$$

for any column vector $\vec{f} \in \mathfrak{F}_0$.

REMARK. *If A is a stochastic matrix then a stronger inequality,* $\|A\vec{f}\|_{\rightarrow} \leqslant \lambda\cdot(1-(R')^{-1})\cdot\|\vec{f}\|_{\rightarrow}$, *holds for any* $\vec{f} \in \mathfrak{F}_0$.

STATEMENT A.7. If A satisfies (A.3) and (A.9) then for any row vector $\overleftarrow{g}$ from $\mathfrak{G}_0$ we have

$$\|\overleftarrow{g}A\|_{\leftarrow} \leqslant \lambda[1 - (R')^{-1}]\|\overleftarrow{g}\|_{\leftarrow}. \tag{A.18}$$

REMARK. *If* $\overleftarrow{g} \in \mathfrak{G}_0$, *then*

$$\sum_{i\in\Delta_-} |g_i|e_i = \sum_{i\in\Delta_+} |g_i|e_i = \tfrac{1}{2}\|\overleftarrow{g}\|_{\leftarrow}, \tag{A.19}$$

where

$$\Delta_- = \{i: g_i < 0\}, \qquad \Delta_+ = \{i: g_i > 0\}.$$

Let us move on to our main aim, i.e. the study of the action of matrices close to A. Let A and B be $n \times n$ matrices with positive entries. Following [15] we shall say that $R \geqslant 1$ is an *estimate of proximity* of these matrices if

$$(R)^{-1} \cdot a_{ij} \leqslant b_{ij} \leqslant R \cdot a_{ij} \text{ for all } 1 \leqslant i,j \leqslant n. \tag{A.20}$$

LEMMA A.8. *Let A be a matrix with positive entries and $R' \geqslant 1$ an estimate of the ratio of its rows, B be close to A with the constant of proximity $R \geqslant 1$ and $\overleftarrow{p}$ an eigen Perron row vector of A normed with respect to* (A.7), *λ the corresponding eigenvalue and $\overleftarrow{q}$ an arbitrary row vector with positive elements normed with respect to* (A.7). *Then*

$$(R)^{-1} \|\overleftarrow{p}A\|_{\leftarrow} \leqslant \|\overleftarrow{q} \cdot B\|_{\leftarrow} \leqslant R \cdot \|\overleftarrow{p}A\|_{\leftarrow}, \tag{A.21}$$

$$\|\overleftarrow{q}B - \overleftarrow{p}A\|_{\leftarrow} \leqslant \lambda \cdot (1 - (R')^{-1}) \|\overleftarrow{q} - \overleftarrow{q}\|_{\leftarrow} + \lambda \cdot (R - 1). \tag{A.22}$$

STATEMENT A.9. Under the conditions of Lemma A.8 we have

$$\left\| \frac{\overleftarrow{q}B}{\|\overleftarrow{q}B\|_{\leftarrow}} - \frac{\overleftarrow{p}A}{\|\overleftarrow{p}A\|_{\leftarrow}} \right\| \leqslant R \cdot (1 - (R')^{-1}) \|\overleftarrow{q} - \overleftarrow{p}\|_{\leftarrow} + 2 \cdot R \cdot (R - 1). \tag{A.23}$$

THEOREM A.10. *Let A be a matrix with positive entries, $R' \geqslant 1$ be an estimate of the ratio of rows, and $\overleftarrow{p}$ be an eigen Perron row vector normed by the formula* (A.7). *Let any matrix $B^{(k)}$, where $k = 1, 2, \ldots$, be close to A with the constant of proximity $R \geqslant 1$ and $\overleftarrow{q}$ be the row vector with positive coordinates normed by the formula* (A.7). *If*

$$\theta = 1 - (R')^{-1}, \qquad \theta \cdot R < 1 \tag{A.24}$$

then for the vector $\overleftarrow{q}^{(N)} = \overleftarrow{q} \cdot \prod_{1 \leqslant k \leqslant N} B^{(k)}$ in the norm (A.7) *for A we have*

$$\left\| \frac{\overleftarrow{q}^{(N)}}{\|\overleftarrow{q}^{(N)}\|_{\leftarrow}} - \overleftarrow{p} \right\|_{\leftarrow} \leqslant 2(\theta \cdot R)^N + 2 \cdot R \cdot (R - 1) \cdot (1 - \theta \cdot R)^{-1}. \tag{A.25}$$

COROLLARY. *Under the conditions of the theorem in the norm* (A.1) *we have*

$$\left| \frac{\overleftarrow{q}^{(N)}}{|\overleftarrow{q}^{(N)}|} - \overleftarrow{p} \right| \leqslant 4 \cdot R' \cdot (\theta \cdot R)^N + 4 \cdot R' \cdot R \cdot (R - 1) \cdot (1 - \theta \cdot R)^{-1}. \tag{A.26}$$

REMARK. *For the stochastic matrix A the right-hand side of* (A.26) *can be replaced by the right-hand side of* (A.25).

Acknowledgement

The author wishes to express her gratitude to Professor Ya. G. Sinai for posing the problem and for his constant encouragement.

References

1. Alekseev, V. M.: 'Quasirandom Dynamical Systems, I', *Math. USSR Sb.* **5** (1968), 73–128.
2. Alekseev, V. M.: 'Quasirandom Oscillations and Qualitative Problems in Celestial Mechanics', *9th Math. Summer School, Katziveli*, 1971, pp. 212–341 (in Russian).
3. Alekseev, V. M.: 'Symbolic Dynamics', *11th Math. Summer School, Kolomyya*, 1973, Ukrainian SSR Acad. Sci. Math. Inst., Kiev, 1976 (in Russian).
4. Čencova, N. N.: 'Investigation of a Model System of Quasistochastic Relaxation Oscillations', *Uspehi Mat. Nauk* **37** (1982), 205–206 (in Russian).
5. Smale, S.: 'Structural Stable Homeomorphism with Infinite Number of Periodic Points', *Proc. Int. Symp. Nonlinear Oscillations*, Vol. 2, Ukrainian SSR Acad. Sci. Math. Inst, Kiev, 1963, pp. 365–366.
6. Smale S.: 'Diffeomorphisms with Many Periodic Points, Differential and Combinatorial Topology', *in Symp. in Honor of Marston Morse*, Princeton University Press, Princeton, N.J., 1965, pp. 63–80.
7. Smale, S.: 'Differential Dynamical Systems', *Bull. Amer. Math. Soc.* **73** (1967), 747–817.
8. Devaney, R. and Nitecki, Z.: 'Shift Automorphism in the Henon Mapping', *Commun. Math. Phys.* **67** (1979), 137–146.
9. Moser, J.: *Stable and Random Motions in Dynamical Systems: Herman Weyl Lectures*, Princeton University Press, Princeton, N.J., 1973.
10. Nitecki, Z.: *Differential Dynamics*, MIT Press, Cambridge, Mass., 1971.
11. Shub, M.: 'Stabilité Globale des Systèmes Dynamiques', *Astérisque* **56**, (1978).
12. Sinai, Ya, G.: 'The Stochasticity of Dynamical Systems', *Select. Math. Sov.* **1** (1981), 100–119.
13. Sinai, Ya. G.: 'Markov Partitions and *U*-diffeomorphisms', *Funct. Anal. Appl.* **2** (1968), 64–89.
14. Sinai, Ya. G.: 'Some Rigorous Results on Decay of Correlations', Appendix to G. M. Zaslavskii, *Statistical Irreversibility in Nonlinear Systems*, Nauka, Moscow, 1970, pp. 114–139 (in Russian).
15. Sinai, Ya. G.: 'Gibbs Measures in Ergodic Theory', *Russian Math. Surveys* **27** (4) (1972), 21–69.
16. Anosov, D. V.: Flows on Closed Riemannian Manifolds with Negative Curvature', *Amer. Math. Soc. Transl.*, Providence, RI, 1969.
17. Anosov, D. V.: 'Tangential Fields of Transversal Foliations in *U*-systems', *Math. Notes* **2** (1967), 818–823.
18. Bowen, R. and Ruelle, D.: 'The Ergodic Theory of Axiom A Flows', *Inv. Math.* **29** (1975), 181–202.
19. Bowen, R.: *Equilibrium States and the Ergodic Theory of Anosov Diffeomorphisms*, Lecture Notes in Mathematics **470**, Springer-Verlag, Berlin, 1975.
20. Čencova, N. N.: 'Investigation of Quasi-stochastic Conditions of Autogenerator on a Tunnel Diode', In: *Methods of Qualitative Theory of Differential Equations*, Gorki, 1983, pp. 95–118.
21. Pikovskii, A. S. and Rabinovich, M. J.: 'A Simple Autogenerator with Stochastic Behaviour', *Dokl. Akad. Nauk SSSR* **239** (1978), 301–307 (in Russian).
22. Čencova, N. N.: 'A Natural Invariant Measure on Smale's Horseshoe', *Soviet Math. Dokl.* **23** (1981), 87–91.
23. Afraimovich, V. S., Bykov, V. V. and Shilnikov, L. P.: 'On the Appearance and Structure of Lorentz Attractor', *Soviet Phys. Dokl.* **22**, (1977).
24. Bunimovich, L. A. and Sinai, Ya. G.: 'Stochastic Properties of the Lorentz Attractor', In: *Nonlinear Waves*, Nauka, Moscow, 1979, pp. 212–226 (in Russian).
25. Rohlin, V. A.: 'On the Fundamental Concepts of Measure Theory', *Amer. Math. Soc. Transl.* **1** (10) (1962).
26. Rohlin, V. A.: 'Selected Topics from the Metric Theory of Dynamical Systems', *Amer. Math. Soc. Trans.* **2** (49), (1965), 171–240.
27. Cramér, H.: *Mathematical Methods of Statistics*, Princeton University Press, Princeton, N. J., 1946.
28. Kornfel'd, J. P., Sinai, Ya. G. and Fomin, S. V.: *Ergodic Theory*, Nauka, Moscow, 1980 (in Russian).
29. Gantmacher, F.: *The Theory of Matrices*, Vol. 2. Chelsea, New York, 1959.
30. Pesin, Ya. B. and Sinai, Ya. G.: 'Hyperbolicity and Stochasticity of Dynamical Systems', in *Soviet Science Review*, Pergamon Press, Oxford 1981, pp. 53–115.

Author Index

Subject Index